Studies in Fuzziness and Soft Computing

Volume 393

Series Editor

Janusz Kacprzyk, Systems Research Institute, Polish Academy of Sciences, Warsaw, Poland

The series "Studies in Fuzziness and Soft Computing" contains publications on various topics in the area of soft computing, which include fuzzy sets, rough sets, neural networks, evolutionary computation, probabilistic and evidential reasoning, multi-valued logic, and related fields. The publications within "Studies in Fuzziness and Soft Computing" are primarily monographs and edited volumes. They cover significant recent developments in the field, both of a foundational and applicable character. An important feature of the series is its short publication time and world-wide distribution. This permits a rapid and broad dissemination of research results.

Indexed by ISI, DBLP and Ulrichs, SCOPUS, Zentralblatt Math, GeoRef, Current Mathematical Publications, IngentaConnect, MetaPress and Springerlink. The books of the series are submitted for indexing to Web of Science.

More information about this series at http://www.springer.com/series/2941

Shahnaz N. Shahbazova · Janusz Kacprzyk ·
Valentina Emilia Balas · Vladik Kreinovich
Editors

Recent Developments and the New Direction in Soft-Computing Foundations and Applications

Selected Papers from the 7th World Conference on Soft Computing, May 29–31, 2018, Baku, Azerbaijan

Springer

Editors
Shahnaz N. Shahbazova
Department of Information Technology
and Programming
Azerbaijan Technical University
Baku, Azerbaijan

Valentina Emilia Balas
Aurel Vlaicu University of Arad
Arad, Romania

Janusz Kacprzyk
Systems Research Institute
Polish Academy of Sciences
Warsaw, Poland

Vladik Kreinovich
Department of Computer Science
University of Texas at El Paso
El Paso, TX, USA

ISSN 1434-9922 ISSN 1860-0808 (electronic)
Studies in Fuzziness and Soft Computing
ISBN 978-3-030-47123-1 ISBN 978-3-030-47124-8 (eBook)
https://doi.org/10.1007/978-3-030-47124-8

This Springer imprint is published by the registered company Springer Nature Switzerland AG
The registered company address is: Gewerbestrasse 11, 6330 Cham, Switzerland

Contents

Neural Networks

Granular Computing and Relations

Fuzzy Control Applications

Big Date and Fuzzy Pattern Recognition

Big Data and Fuzzy Based Techniques in Analysis of Social Networks

Shahnaz N. Shahbazova and Sabina Shahbazzade

Abstract Social networks—a phenomenon of an early 21st century—is perceived as a source of data generated by users about users themselves and things and dealings related to them. Because of that social networks are treated as an object of many academic, corporate and industrial research activities leading to a better understanding of individuals' behavior, (dis)likes and needs, as well as events and issues important for them. A very active involvement of individuals in social network means that every day millions of new pieces of information is generated. Analysis of this vast amount of available data requires methods and approaches taken from the domain of big data. The theory of fuzzy sets and systems, introduced in 1965, provides the researchers with techniques that are able to cope with imprecise information expressed linguistically. This theory constitutes a basis for designing and developing methodologies of processing data that are able to identify and understand views and judgments expressed in a unique, human way – the core of information generated by the users of social networks. The paper tries to recognize a few important example of extracting value from social network data that can benefit from application of fuzzy set and systems methodology.

Keywords Fuzzy sets · Social network · Data-driven manufacturing · Statistical approaches · Graph based approaches · Summarization · Sentiment analysis · Event detection

1 Introduction

More and more often social networks, or shall we say data generated by its users, is an object of research activities being conducted by variety of organization and

S. N. Shahbazova (✉)
Azerbaijan Technical University, Baku AZ1073, Azerbaijan
e-mail: shahbazova@gmail.com; shahbazova@berkeley.edu

S. Shahbazzade
The George Washington University, Washington, DC 20052, USA
e-mail: shahbazzade@gwu.edu

S. N. Shahbazova et al. (eds.), *Recent Developments and the New Direction in Soft-Computing Foundations and Applications*, Studies in Fuzziness and Soft Computing 393, https://doi.org/10.1007/978-3-030-47124-8_1

corporations in order to extract information and knowledge about multiple aspects characterizing activities, behavior, as well as interests and likes of users and group of users. We can find multiple examples of corporations and agencies putting enormous effort to analyze and understand data that is generated by actions, interactions, and conversations involving users, as well as by the users' views and opinions on almost everything what happens in their lives and surroundings.

Varieties of techniques are used to process data generated by users in social networks: statistical approaches, graph based approaches, and many others [8, 13, 20, 21, 22, 26]. However, a human nature is present in the social networks. This means that the networks are human-like—full of imprecise relations and connections between individuals, vague terms, groups and individuals with indefinite descriptions and characteristics of interests. It seems that many aspects of social networks resemble the ones of their users.

In the light of these statements, we would like to state that techniques of processing social networks of users and groups should reflect such human facets. These techniques should be based on a human-like methodology, and the theory of fuzzy sets and systems [29] is suitable for such a purpose. Its ability to deal with ambiguous data and facts, its ability to describe things in a human manner, and its ability to handle imprecision and ambiguity make fuzzy sets [11] one of the best tools for analyzing social network [17].

2 Analysis of Social Networks: Targeted Area

The nature of data generated by users of social networks allows us to identify multiple areas of human life, as well as industrial and corporate activities that could benefit from analyzing social network data. In the following subsections we describe just a few of them in the wide category of issues contributing to improvements in: quality of human life, manufacturing, and corporate visions and goals.

2.1 Quality of Life

The user's ability to observe and quickly react to different events—would they be positive or not—mean that analysis of data can be an important element of sophisticated and intelligent **Disaster Management** systems. An early detection of disasters, for example earthquakes or floods, would enable quick interventions that increase changes of minimizing or even mitigating the effects of disasters. Tools that allow visualization of social media data would allow for better understanding of dynamics of investigated calamities. They would allow for close, real-time monitoring of disasters and their progress, and also estimating effectiveness of preventing actions and provided help. This is closely related to another very important area that could benefit form data analysis: **Health Care**. In this case, application of big data techniques would have an enormous impact on coping with diseases—better understanding of their origin, better knowledge about their spreading, as well as better understanding of their impact on human behavior and actions. Systems for detecting diseases and tracking them could be of great importance.

Education is yet another domain where preforming big data analysis on social network data could play an important role. For example, it is already known that analysis of posts generated by pupils and students can lead to detecting bullying and even preventing tragedies caused by such a behavior. Development of systems monitoring and analyzing variety of issues related to ways of study as well as problems and issues encountered by students would mean creation of a better environment for education. Such an environment would increase effectiveness of education systems and contribute to better-educated societies.

Analysis of social network data could also lead to a more effective **Criminal Justice** system. More information extracted from social data that is related to actions violating laws and by-laws would definitely change the way law enforcement agencies work. Ability to detect undesired behaviors as well as presence of dangerous individuals would lead to safer communities.

2.2 *Manufacturing and Industrial Activities*

Facebook [32], Twitter [34], Pinterest [33] and other social sites continue to generate and provide an uninterrupted stream of data. Development of methods and approaches that analyze this data from a perspective of the users' opinions on different products and services will allow companies to identify the customers' preferences, as well as their needs and likes. This would lead to a so-called **data-driven manufacturing**—a scenario where existing and potential customers influence what is being manufactured. Analysis of the users' data would also help identifying weaknesses and strengths of manufactured goods. Social media sentiment investigations could determine if users intent to purchase specific products or if they dislike these products. All this would provide insight that can be explored and acted upon. Such analyses could also assess consumers' interest in a product before it is launched.

Introduction of tools and systems analyzing social network data would affect variety of **service and utility** companies. Such systems could change the way companies provide services to the users, and how these systems could react to changes in the demand and needs. For example, General Electric (GE) is about to release a system that uses analysis of social media to support estimation of potential problems in an electrical grid—it is called Grid IQ [35]. Many aspects of individuals' lives are being discussed on the forum of social networks therefore any company that provides services would benefit a lot from social network data. Another interesting aspect is related to a so-called geo-tagging where many aspects of collected data is location sensitive—and this alone will bring valuable information to be analyzed.

2.3 Corporate Visions and Goals

Understanding of the users' needs and their attitude to multiple services and products offered by companies are key elements of building corporate strategies and plans for future. Also here, vast amounts of data collected during the users' activities could provide corporations with valuable information. In such areas like **advertising and marketing** any indications regarding the users' moods, attitudes and opinions expressed directly or indirectly would be able to change policies and strategies of companies. The analysis would provide companies with indicators of their social presence, their ranking, and popularity among users.

One of the most interesting and intriguing aspects of analysis of social network data could be related to strategies and operations of **insurance and financial** companies. Information—in a form of opinions, facts or evaluations—that describe the users' behaviors, patterns and rules of actions would be of great interest to insurance companies. They could adopt and customize their policies and offer insurance packages to variety of customers trying to fit their specific needs. Financial companies could use social media data to improve returns on investment—analysis of the users' discussions and posts from the point of few of sentiments and attitudes towards investment and associated with it expectations can lead to identification of trends, patterns and motivations of the users. All this would translate into better strategies and policies of investment polices of companies.

3 Discussions and Conclusion

The above section provides a just few examples of areas that can benefit from processing, analyzing and modeling social network data. At this stage we would like to foresee how synthesis of big data methods with fuzzy technologies could contribute to analysis of social networks.

Most of the work dedicated to analysis of network data is targeting structural information of networks. The interconnection and relations between actors have been the main source of information [1, 2, 3, 5, 9, 15]. Some works [4, 6, 7, 14, 19, 23, 27, 28] show an attempt to use additional information that brings different aspects of analysis. It is well known that each of the nodes/actors as well as connections/relation is associated with supplementary information. It seems very reasonable that including that information in analysis procedure and methods would increase the scope of analysis and allow looking and discovering new insights regarding the users and their behavior.

A very important aspect of analysis of network data is to look into posts, comments, tweets, or any form of text generated by the users. They contain enormous amounts of information about users: what is being currently discussed, what are moods among people, what they like, what types of things invoke positive responses, what type of things are perceived as negative. It seems critical to be armed with the ability to find answers to these questions—this would result in detecting trends, popular and/or important topics, determine users' requirements and expectations regarding variety of items and events. In such a context, we would like to state that application of fuzzy-based approaches will enhance the capabilities of data analytical methods, will enable deeper and more semantic oriented analysis, and what is also essential will make the obtained outcomes more human-like. Fuzzy methods targeting **summarization** of texts [16], **sentiment analysis** [24, 25], **trust** inference and propagation [10, 12], and **event detection** [18, 30, 31], are just a few examples of research topics that constitute important issues that can be addressed with, and will benefit from application of fuzzy-based techniques and methods.

Acknowledgements This scientific work was supported by Grant a Science Development Foundation under the President of the Republic of Azerbaijan, Ministry of Communications and High Technologies of the Republic of Azerbaijan in order to support projects aimed at the development of ICT for Grants for 2013 for the 2nd joint ICT Contest (SDF-MCI-MQM-2/ICT-2-2013-7 (13)).

References

1. A.M. Abbasov, Information boom: new trends and expectations, in *Soft Computing: State of the Art Theory and Novel Applications (Studies in Fuzziness and Soft Computing)*, ed. by R.R. Yager, A.M. Abbasov, M.Z. Reformat, S.N. Shahbazova (Springer, Berlin, 2012), pp. 1–12
2. S.P. Borgatti, Centrality and network flow. Soc. Netw. **27**(1), 55–71 (2005)
3. J. Boyd, M. Everett, Relations, residuals, regular interiors, and relative regular equivalence. Soc. Netw. **21**(2), 147–165 (1999)
4. M. Brunelli, M. Fedrizzi, M. Fedrizzi, Fuzzy m-ary adjacency relations in social network analysis: Optimization and consensus evaluation. Inform. Fusion **17**, 36–45 (2014)
5. Y. Cao, J. Cao, M. Li, Distributed data distribution mechanism in social network based on fuzzy clustering, in *Foundations and Applications of Intelligent Systems, Advances in Intelligent Systems and Computing*, vol. 213, ed. by F. Sun, et al. (Springer, Berlin Heidelberg, 2014), pp. 603–620
6. T. Casasús-Estellés, R.R. Yager, Fuzzy concepts in small worlds and the identification of leaders in social networks, in *IPMU 2014, Part II, CCIS 443*, ed. by A. Laurent, et al. (Springer International Publishing, Switzerland, 2014), pp. 37–45
7. S. Elkosantini, D. Gien, A dynamic model for the behavior of an operator in a company, in *Proceedings of the 12th IFAC Symposium on Information Control Problems in Manufacturing, France* vol. 2, (2006), pp. 187–192
8. S. Elkosantini, D. Gien, Human behavior and social network simulation: fuzzy sets/logic and agents-based approach, in *Proceedings of the 2007 Spring Simulation Multi-conference SpringSim '07*, vol. 1 (2007), pp. 102–109
9. L. Freeman, *The Development of Social Network Analysis* (Empirical Press, Vancouver, 2006)
10. F. Hao, G. Min, M. Lin, C. Luo, L.T. Yang, IEEEMobiFuzzyTrust: an efficient fuzzy trust inference mechanism in mobile social networks. IEEE Trans. Parallel and Distrib. Syst. **25**(11), 2944–2955 (2014)
11. R. Hannemanand, M. Riddle, *Introduction to Social Network Methods* (University of California, Riverside, 2005)
12. S. Kim, S. Han, The method of inferring trust in web-based social network using fuzzy logic, in *Proceedings of the International Workshop on Machine Intelligence Research* (2009) pp. 140–144

13. G.J. Klir, B. Yuan, *Fuzzy Sets and Fuzzy Logic: Theory and Applications* (Pretience Hall, 1995)
14. P.N. Krivitsky, M.S. Handcock, A.E. Raftery, P.D. Hoff, Representing degree distributions, clustering, and homophily in social networks with latent cluster random effects models. Soc. Netw. **31**, 204–213 (2009)
15. M.J. Lanham, G.P. Morgan, K.M. Carley, Social network modeling and agent-based simulation in support of crisis de-escalation. IEEE Trans. Syst. Man Cybern. **44**(1), 103–110 (2014)
16. J. Liu, Fuzzy modularity and fuzzy community structure in networks. Eur. Phys. J. B **77**(4), 547–557 (2010)
17. X.H. Liu, Y.T. Li, F.R. Wei, M. Zhou, Graph-based multi-tweet summarization using social signals, in *Proceedings of COLING 2012* (2012), pp. 1699–1714
18. T. Matuszka, Z. Vincellér, S. Laki, On a keyword-lifecycle model for real-time event detection in social network data, in *Proceedings of 4th IEEE International Conference on Cognitive Infocommunications* (2013), pp. 453–458
19. W. Pedrycz, S.-M. Chen (eds.), *Social Networks: A Framework of Computational Intelligence, Studies in Computational Intelligence*, vol. 526 (Springer International Publishing Switzerland, 2014)
20. M. Z. Reformat, R. R. Yager, Using tagging in social networks to find groups of compatible users, in *Proceedings of Join IFSA World Congress and NAFIPS Annual Meeting (IFSA/NAFIPS), Edmonton, Canada* (2013), pp. 697–702
21. S.N. Shahbazova, Application of fuzzy sets for control of student knowledge//applied and computational mathematics. Int. J. **10**(1), 195–208 (2011). Special Issue on Fuzzy Set Theory and Applications
22. J. Scott, *Social Network Analysis. A Handbook* (Sage, London, 2000)
23. T.A.B. Snijders, C. Baerveldt, A multilevel network study of the effects of delinquent behavior on friendship evolution. J. Math. Sociol. **27**, 123–151 (2003)
24. T.A.B. Snijders, Statistical models for social networks. Annu. Rev. Soc. (2011)
25. G. Stakias, M. Psoras, and M. Glykas, Fuzzy cognitive maps in social and business network analysis, in *Business Process Management, SCI 444* ed. by M. Glykas (Springer, Berlin Heidelberg, 2013), pp. 241–279
26. D.N. Trung, J.J. Jung, L.A. Vu, A. Kiss, Towards modeling fuzzy propagation for sentiment analysis in online social networks: a case study on TweetScope, in *Proceedings of 4th IEEE International Conference on Cognitive Infocommunication* (2013), pp. 331–337
27. D.N. Trung, J.J. Jung, Sentiment analysis based on fuzzy propagation in online social networks: a case study on TweetScop. Comput. Sci. Inf. Syst. **11**(1), 215–228 (2014)
28. F. Vega-Redondo, *Complex Social Networks* (Cambridge University Press, 2007)
29. R.R. Yager, Intelligent social network analysis using granular computing. Int. J. Intell. Syst. **23**, 1196–1219 (2008)
30. R.R. Yager, M.Z. Reformat, Looking for like-minded individuals in social networks using tagging and fuzzy sets. IEEE Trans. Fuzzy Sets **21**(4), 672–687 (2013)
31. L.A. Zadeh, Fuzzy sets. Inf. Control **8**, 338–353 (1965)
32. https://www.facebook.com. Accessed 7 Nov 2015
33. https://www.pinterest.com. Accessed 7 Nov 2015
34. https://twitter.com. Accessed 7 Nov 2015
35. http://www.gegridsolutions.com/demandopt/Catalog/GridIQ.htm. Accessed 7 Nov 2015

Application of Fuzzy Pattern Recognition in the Recruitment of IT Specialists

Masuma Mammadova and Zarifa Jabrayilova

Abstract The article the task of the management of the IT specialists market reduced to decision-making on selecting the policy for the coordination of supply and demand. The fuzzy situational models for supply and demand are proposed. Decision support methods for the recruitment of IT specialists based on fuzzy pattern recognition taking into account the multivariate nature of the coordination of supply and demand is developed.

Keywords IT specialist · Fuzzy situational model · Decision-making methods · Fuzzy pattern recognition · Fuzzy similarity

1 Introduction

The emergence of information as a global resource has resulted in a sharply increased demand for IT specialists in the labor market, which is still significantly exceeding the supply in many countries [1–4].

The new demands on IT specialists emphasize a strong ability to adapt to change and innovations, flexibility, multifaceted skills, mobility, and ability to switch between different areas of activity, which correspondingly require continuous updating of professional knowledge and skills by means of both formal and informal education [5].

However, academic publications have failed to properly focus on the system analysis of the IT labor maker and the development of adequate technology and support techniques for managerial decision making in reconciling the supply and demand to reflect the specific features of this sector of the economy and the preferences of its main factors.

M. Mammadova · Z. Jabrayilova (✉)
Institute of Information Technology, National Academy of Science of Azerbaijan, Baku, Azerbaijan
e-mail: djabrailova_z@mail.ru

M. Mammadova
e-mail: mmg51@mail.ru

S. N. Shahbazova et al. (eds.), *Recent Developments and the New Direction in Soft-Computing Foundations and Applications*, Studies in Fuzziness and Soft Computing 393, https://doi.org/10.1007/978-3-030-47124-8_2

The specific features of the IT labor market, the uncertainty of information flows about its state, the much-varied profile of data on IT specialists and the difficulties of measuring it, and the ambiguity of indicators characterizing it predetermine the variety of possible fuzzy states of supply and demand for IT specialists and the multi-option pattern of reconciling them. Comparisons and evaluations of these conditions and the choice of the reconciliation policy can be efficiently accomplished with the use of intellectual methods and technologies [6–8]. The latter help to integrate the versatile information on the supply and demand conditions and facilitate the development of a set of alternatives for managerial decisions and the selection of the most efficient one among them. Management in such systems is considered as a process of identification of the demand and supply condition at a given point in time and decision making suitable in the current situation.

The listed features identify the problem of management of the labor market of IT specialists as semi-structured and difficultly formalized tasks [9–11]. Obviously, for the solution of such problems, the use of intelligent methods and technologies is the most effective. In this interpretation, solution of the management problem of the IT specialists market assumes:

- development of situational models of supply and demand for IT specialists;
- comparison and evaluation of fuzzy situation of supply and demand in the current situation;
- deciding on the choice of a policy for their harmonization [4].

For the most flexible expression of continuously changing management situations and the correlation of supply and demand, we use the model of fuzzy pattern recognition.

The study develops decision support methods for the recruitment of IT specialist on the multivariate coordination of supply and demand based on fuzzy pattern recognition.

2 Formulation of the Problem

Assume are two sets of fuzzy situations describing the conditions of demand $\tilde{V}_k$ and supply $\tilde{S}_q$ in the IT specialist labor market [4]:

$$\tilde{V}_i = \left\{ < \mu_{l_j}(V_i) >, < \mu_{c_r}(V_i) >, < \mu_{u_z}(V_i) > \right\} = \left\{ \mu_{V_i}(y) / y \right\}$$

$$\tilde{S}_g = \left\{ < \mu_{l_j}(S_g) >, < \mu_{c_r}(S_g) >, < \mu_{u_z}(S_g) > \right\} = \left\{ \mu_{S_g}(y) / y \right\}$$

Here:

- $\mu_{l_j}(V_i) : V \times L \rightarrow [0, 1]$—the degree of conformity of the vacancy V_i by the indicators L and $L = \{l_1, l_2, \ldots, l_n\}$—is the set of personal features (characteristics) required in a candidate to a particular position (job, workplace);

- $\mu_{c_f}(V_i) : V \times C \to [0, 1]$—the degree of conformity of the vacancy V_i by the indicators C and $C = \{c_1, c_2, \ldots, c_m\}$—is the set of competencies sought to fill the IT vacancy;
- $\mu_{u_{i\gamma}}(V_i) : V \times U \to [0, 1]$—the degree of conformity of the vacancy V_i by the indicators U and $U = \{u_1, u_2, \ldots, u_p\}$—is the set of conditions offered to the applicants to vacant IT jobs;
- $\mu_{l_j}(S_g) : S \times L \to [0, 1]$—the degree of a competency by a certain IT specialist S_g by the indicators L and $L = \{l_i\}, j = \overline{1, n}$ is the set of actual competencies characterizing the IT specialists;
- $\mu_{c_{gf}}(S_g) : S \times C \to [0, 1]$ —the degree of a competency by a certain IT specialist S_g by the indicators C and $C = \{c_f\}, f = \overline{1, m}$ is the set of actual competencies in each individual applicant to a vacancy;
- $\mu_{u_{g\gamma}}(S_g) : S \times U \to [0, 1]$—the degree of a competency by a certain IT specialist S_g by the indicators U and $U = \{u_\gamma\}, \gamma = \overline{1, p}$ is the set of preferences of the ITl specialist expressed as his or her requirements for the vacancy.

The set $\tilde{V}_k = \{\mu_{V_k}(y)/y\}, i = \overline{1, k}$ accounts for fuzzy reference situation (fuzzy reference image), and set $\tilde{S}_q = \{\mu_{S_q}(y)/y\}$ $g = \overline{1, q}$ accounts for fuzzy real situation (fuzzy real image).

The problem is to find the most acceptable pair of "fuzzy reference image—fuzzy real image" in terms of real search patterns sets of IT specialists (supply) and reference search patterns of the query (demand).

3 Problem Solution

For the intelligent management of supply and demand in the IT specialists labor market, the problem of the supply and demand conformity decision can be posed as a task of fuzzy pattern recognition. The problem of pattern recognition is based on a fuzzy situational analysis and determination of the degree of similarity of fuzzy situations.

The procedure of pattern recognition involves the following steps:

- real situations are determined (search patterns of IT specialists) in accordance with the indicator's values that characterize each applicant for the announced vacancy;
- reference situations are determined (search patterns of requests) in accordance with the indicators' values that characterize the employer's requirements to the applicant for the vacancy;
- calculate the degree of fuzzy similarity of the reference situation to each real situation in accordance with the chosen measure of assessment of similarity between two fuzzy situations;
- real situation is revealed, which has the greatest proximity rate with the reference situation. In other words, a decision is made to hire a IT specialist (supply)

for the announced vacancy (demand), which most closely corresponds to the requirements of the employer.

The works [10, 12] review various measures to determine proximity rate between two fuzzy situations, involving single-stage or multistage measurement procedures. The case study uses the degree fuzzy inclusion of situation $\tilde{S}_g$ in fuzzy situation $\tilde{V}_i$, and the degree of fuzzy equality of $\tilde{S}_g$ and $\tilde{V}_i$ as the measure of identifying the proximity rate of fuzzy real and reference situations [12].

4 Multi-scenario Approach to Decision-Making on the Coordination of Demand and Supply to IT Specialists

In the work [4], the authors identified possible scenarios for coordinating supply and demand for IT specialists during the recruitment. Hence, after the completion of the process of recognizing the most acceptable pair of "fuzzy reference image—fuzzy real image" in terms of proximity rate, there may be several possible scenarios among the sets of real search patterns of IT specialists (supply) and reference search patterns of the query (demand):

Scenario 1
One vacancy (employer's request)—one applicant (IT specialist), i.e. "one fuzzy reference image—one fuzzy real image".

Method for decision making. In this case, if the degree of fuzzy similarity of the two situations (i.e. $\tilde{S}_g$ and $\tilde{V}_i$) is not less than the threshold accepted by the employer (for example, $\psi \in [0,6; 1]$), then a hiring decision is made.

The degree of fuzzy inclusion of a fuzzy situation $\tilde{S}_g$ in a situation $\tilde{V}_i$ is determined based on the following formula:

$$\mu(\tilde{S}_g, \tilde{V}_i) = \&\mu(\mu_{S_g}(y), \mu_{V_i}(y)) = \underset{y \in Y}{\&} (\max(1 - \mu_{S_g}(y), \mu_{V_i}(y))) = \min(\max(1 - \mu_{S_g}(y), \mu_{V_i}(y))) \quad (1)$$

The situation $\tilde{S}_g$ is considered fuzzy inclusive in the situation $\tilde{V}_i$ $(..\underset{\sim}{\subset}\tilde{V}_i)$, if the degree of inclusion of $\tilde{S}_g$ in a situation $\tilde{V}_i$ is not less than the inclusion threshold ($\psi \in$ [0,6; 1] determined by the management conditions), i.e. $\mu(\tilde{S}_g, \tilde{V}_i) \geq \psi$.

The degree of fuzzy equality $\tilde{V}_i$ and $\tilde{S}_g$ is defined by the following formula:

$$\mu(\tilde{S}_g, \tilde{V}_i) = \vee(\tilde{S}_g, \tilde{V}_i) \& \vee (\tilde{V}_i, \tilde{S}_g) = \& \mu(\mu_{S_g}(y), \mu_{V_i}(y) = \min_{y \in Y}\left[\min(\max(1 - \mu_{S_g}(y), \mu_{V_i}(y)), \max(1 - \mu_{V_i}(y), \mu_{S_g}(y)))\right] \quad (2)$$

Situations $\tilde{S}_g$ and $\tilde{V}_i$ are considered to be fuzzy equal $\tilde{S}_g \approx \tilde{V}_i$, if $\mu(\tilde{S}_g, \tilde{V}_i) \geq \psi, \ \psi \in [0,7; \ 1]$, where ψ is a certain threshold of fuzzy equality of situations.

Scenario 2

A few applicants (IT specialists) are eligible to the employer in accordance with the accepted measure of the similarity of the two fuzzy situations, i.e. "one fuzzy reference image—more fuzzy real image". The applicants form subset of fuzzy situations (alternatives), the one of which must be chosen that corresponds to the most suitable candidate.

Therefore, there is a set $\left\{\mu(\tilde{S}_g, \ \tilde{V}_i), \ g = \overline{1, \ \eta}, \quad \eta \leq q)\right\}$, the elements of which satisfy the following condition: $(\mu(\tilde{S}_g, \ \tilde{V}_i), \ g = \overline{1, \ \eta}\ ) \geq \psi$.

In this case, the employer, acting as an expert (DM), can be offered various methods for decision making:

Scenario 2.1. The task of decision making is reduced to the comparison of similarity of reference and real situations by the degree of possessing the criteria characterizing the applicants for the vacancy. The best alternative (applicant) is that, who has the greatest proximity rate by coincidence of criteria and the degree of their possession.

Method for decision making. In this case, the sought is the real situation, which provides the maximum value in terms of the degree of reference similarity. I.e.:

$$\mu(\tilde{S}_d, \ \tilde{V}_i)^* = \max(\mu(\tilde{S}_g, \ \tilde{V}_i), \ g = \overline{1, \ \eta}\,, \quad \tilde{S}_d \in \left\{\tilde{S}_g, \ g = \overline{1, \ \eta}\right\}.$$

Scenario 2.2. The task of decision making is reduced to a multicriterion choice of the best solution (alternative), taking into account the relative importance of the criteria characterizing the IT specialists [7, 13–16].

Method for decision making. In this case, the task of decision-making is implemented in accordance with the following stages:

Step 1. The coefficients of relative importance of the indicators shall be defined [17, 18].

Assume that:

- $\omega_j, \ j = \overline{1, \ n}$—coefficients of relative importance of indicators characterizing the criterion L;
- $\omega_f, \ f = \overline{1, \ m}$—coefficients of relative importance of indicators characterizing the criterion C;
- $\omega_\gamma, \ \gamma = \overline{1, \ p}$—coefficients of relative importance of indicators characterizing the criterion U.

Step 2. Based on the aggregation of the degrees of possession of individual indicators (i.e)

$$\mu_{l_{gj}}\left(\tilde{S}_g, \ \tilde{V}_i\right), \ j = \overline{1, \ n}, \quad \mu_{c_{gf}}\left(\tilde{S}_g, \ \tilde{V}_i\right), \ f = \overline{1, m}, \quad \mu_{u_{g\gamma}}\left(\tilde{S}_g, \ \tilde{V}_i\right), \ \gamma = \overline{1, \ p}$$

the specific IT specialists determine the degree of fuzzy similarity of fuzzy real situations with the reference situation by the following step [19]:

(a) Based on the "convolution" $\mu_{l_{gj}}\left(\tilde{S}_g,\ \tilde{V}_i\right),\ j = \overline{1,\ n},$ the degree of fuzzy similarity between real and reference situations is determined by personal characteristics (L):

$$\mu_L\left(\tilde{S}_g,\ \tilde{V}_i\right) = \sum_{j=1}^{n} w_j\ \mu_{l_j}\left(\tilde{S}_g,\ \tilde{V}_i\right).$$

(b) Based on the "convolution" $\mu_{c_{gf}}\left(\tilde{S}_g,\ \tilde{V}_i\right),\ f = \overline{1, m}$—the degree of fuzzy similarity between real and reference situations is determined in terms of competences (C):

$$\mu_C\left(\tilde{S}_g,\ \tilde{V}_i\right) = \sum_{f=1}^{m} w_f \mu_{c_f}\left(\tilde{S}_g,\ \tilde{V}_i\right).$$

(c) Based on the "convolution" $\mu_{u_{g\gamma}}\left(\tilde{S}_g,\ \tilde{V}_i\right),\ \gamma = \overline{1,\ p}$—the degree of fuzzy similarity between real and reference situations is determined through the prism of the requirements for the vacancy U:

$$\mu_U\left(\tilde{S}_g,\ \tilde{V}_i\right) = \sum_{\gamma=1}^{p} w_\gamma \mu_{u_\gamma}\left(\tilde{S}_g,\ \tilde{V}_i\right).$$

(d) Based on the obtained results and coefficients of relative importance L, C and U—w_L, w_C, w_U, the similarity rates of the real situation with the reference ones are determined:

$$\mu_w\left(\tilde{S}_g,\ \tilde{V}_i\right) = \omega_L \cdot \mu_L\left(\tilde{S}_g,\ \tilde{V}_i\right) + \omega_C \cdot \mu_C\left(\tilde{S}_g,\ \tilde{V}_i\right) + \omega_U \cdot \mu_U\left(\tilde{S}_g,\ \tilde{V}_i\right).$$

Step 3. Fuzzy real situation with the maximum value is chosen:

$$\phi\left(\tilde{S}_g,\ \tilde{V}_i\right)^* = \max\left\{\phi\left(\tilde{S}_g,\ \tilde{V}_i\right),\ g = \overline{1, \eta}\right\}$$

The selected fuzzy real situation corresponds to the search pattern of the applicant, who has the greatest degree of fuzzy similarity with the reference pattern of the vacancy, and is accepted as the best solution.

Scenario 2.3. The list of evaluation criteria is extended, and the input situations are re-defined (re-examined) and the recognition procedures are repeated.

Scenario 3

Several employers are identified, who are interested in hiring one IT specialist, i.e. "more fuzzy reference image—one fuzzy real image".

In this case, there is an inverse task: a subset of fuzzy reference situations (alternatives) is presented, which is represented by the vacancies with the corresponding hiring conditions, among which a IT specialist has to make a choice according to his/her preferences. According to the preferences of the latter, there can be the variants of scenario 3 listed below.

Scenario 3.1. The proximity rate of the claims of the IT specialist with the criteria characterizing the conditions of employment shall be compared, and decision shall be made on the greatest coincidence of the degree of possession of the criteria.

Method for decision making. The task of decision making is reduced to the comparison of the similarities between real and standard situations in terms of the similarity rate of the conditions offered by employers and the claims of the applicant. The best vacancy has the greatest degree of similarity by the coincidence of the criteria that characterize the conditions offered by employers, and the applicant's claims. Thus, assume that:

$$\mu(\tilde{S}_d,\ \tilde{V}_z) = \max\left\{\mu(\tilde{S}_g,\ \tilde{V}_z),\ z = \overline{1,\ f},\ g = \overline{1,\ q}\right\}$$
$$V_z \in \left\{V_i,\ i = \overline{1, k}\right\},\quad S_d \in \left\{S_g,\ g = \overline{1,\ q}\right\},\ 2 \le f < k.$$

In this case, the pair with the maximum value of the degree of fuzzy similarity situations is defined by the following formula:

$$\mu(\tilde{S}_d,\ \tilde{V}_\beta)^* = \max\left\{\mu_U(\tilde{S}_d,\ \tilde{V}_z),\ z = \overline{1,\ f}\right\}.\quad V_\beta \in \left\{\tilde{V}_z,\ z = \overline{1,\ f}\right\}.$$

The fuzzy reference situation is accepted as the best solution, which is corresponding to the search pattern of the vacancy that has the greatest degree of fuzzy similarity with the applicant's real pattern.

Scenario 3.2. The task of decisions making shall be reduced to the multi-objective task of choosing the best solution, taking into account the relative importance of the criteria that characterize workplaces.

Method for decision making. The task of decisions making shall be reduced to the multi-objective task of choosing the best solution, taking into account the relative importance of the indicators characterizing the preferences of the IT specialist (U), expressed in terms of his/her requirements for a IT profile vacancy.

If $\omega_\gamma,\ \gamma = \overline{1,\ p}$—the coefficients of the relative importance of the indicators characterizing the criterion U, then a fuzzy reference situation that has the greatest degree of fuzzy similarity with the applicant's real image is determined based on the following formula:

$$\mu(\tilde{S}_d, \tilde{V}_z)^* = \max\left\{\sum_{\gamma=1}^{p} w_\gamma \cdot \mu_{u_\gamma}(\tilde{S}_d, \tilde{V}_z),\ z = \overline{1, f}\right\}.$$

The selected pair is taken as the best solution.

Scenario 3.3. The list of criteria for the workplace assessment shall be expanded, furthermore, the input situations shall be re-defined (re-examined) and the recognition procedure shall be repeated using formulas (1) or (2).

5 Implementation of the Coordination of Supply and Demand for IT Professionals Based on the Calculation of the Fuzzy Equality Degree of Situations

A phased implementation of the fuzzy equality method of two fuzzy situations is shown below in the example of solving the problem of real recruitment and hiring IT professionals for the specific vacancies.

Assume that there are two vacancies of "programmer" position $V = \{V_1, V_2\}$ for which 4 candidates $S = \{S_1, S_2, S_3, S_4\}$ are applying V_1—programmer; V_2—system administrator with advanced knowledge in networks. The implementation stages of the task of recruiting the candidates for these three vacancies out of four applicants using their fuzzy equality degree to determine the proximity of the two situations.

Stage 1: Determination of the reference situation model of the vacancy, which is reduced to the selecting of the indicators characterizing the employers' requirements for vacancies. To create a system of indicators that characterize the demand (employer's requirements), the IT vacancies in the Internet-recruiting environment are analyzed [20–22].

The indicators system for the vacancy of programmer V1:

- *personal qualities* (L): communication skills (l_1); responsibility (l_2); self-development (l_3); analytical thinking (l_4).
- *competences*, knowledge and skills (C): practical experience of using programming languages PL/SQL and JAVA (c_1); knowledge of intrabank processes (c_2); work experience and knowledge on development/projects (c_3).
- *requirements and conditions for applicants for a vacancy* (U): higher technical education (u_1); ability, if necessary, to work under busy schedule (u_2); certificate in programming (u_3).
- *Indicators system for vacancy V_2*: *personal qualities* (L): teamwork ability (l_1); energetic, purposeful (l_2); desire to grow and develop (l_3); stress-resistant (l_4).

- *competences, knowledge and skills* (C): administrating the Microsoft Windows servers (c_1); administrating the Linux and Unix servers (c_2); knowledge of virtualization systems (VMware, MS Hyper-V) (c_3); knowledge of administration and configuration of network equipment (switches, routers and Wi-Fi access points) (c_4);
- *Job requirements and conditions* (U): experience in storage systems (u_1); experience in backup systems (u_2); experience in the field of IP-Telephony (u_3).

Stage 2: To determine the degree of matching the separate indicators, which are describing the vacancies, by the applicants, the indicators must be formed. As follows from the above requirements for the vacancies, most of the indicators are specified at the verbal level (semantic concepts). This imposes the use of linguistic variables and their values in the form of verbal assessment scales for the formalization of such indicators [23, 24].

The latter reflect the increasing intensity of the representation of the indicators (excellent, good, medium, satisfactory, poor). The formal task of both the candidates for the vacancies and the employers' requirements for them makes it possible to model the interrelation between supply and demand and their matching for the employment task.

Table 1 demonstrates the 5-level values of the linguistic variable "Practical experience of using programming languages PL/SQL and JAVA" and the corresponding ranges of the variation of fuzzy degrees of matching this indicator in the interval [0, 1].

Stage 3: The linguistic assessments of the applicants for the vacancy V_1 are illustrated in Table 2.

Fuzzy degrees of matching the indicators, which are characterizing the vacancy V_1:

$\mu_{L_j}(S_i) \to [0, 1], \ j = \overline{1, 4}, \ \mu_{C_f}(S_i) \to [0, 1], \ f = \overline{1, 3}, \ \mu_{U_\gamma}(S_i) \to [0, 1], \ \gamma = \overline{1, 3}$ are given in Table 3.

According to Table 3, fuzzy real situations are formed, i.e. fuzzy descriptions of the job applicants V_1:

Table 1 Fuzzification of the indicator "Practical experience of using programming languages PL/SQL and JAVA"

Gradation of the indicator "Practical experience in programming languages PL/SQL and JAVA"	Linguistic assessment	The range of variation of fuzzy degrees in the interval [0; 1]
Wide-ranging experience	Excellent	[0.95–1]
Extensive experience	Good	[0.8–0.94]
Sufficient experience	Medium	[0.5–0.79]
Little experience	Satisfactory	[0.26–0.49]
No experience	Poor	[0–0.25]

Table 2 Linguistic assessments of the Applicants for the vacancy V_1

Indicators (V_1)	S_1	S_2	S_3	S_4
l_1	Excellent	Satisfactory	Satisfactory	Excellent
l_2	Good	Good	Good	Good
l_3	Good	Good	Satisfactory	Good
l_4	Good	Satisfactory	Good	Good
c_1	Excellent	Excellent	Good	Good
c_2	Satisfactory	Good	Satisfactory	Satisfactory
c_3	Satisfactory	Good	Satisfactory	Good
u_1	Excellent	Good	Excellent	Excellent
u_2	Excellent	Satisfactory	Satisfactory	Excellent
u_3	Excellent	Poor	Excellent	Good

Table 3 Fuzzy degrees of matching the indicators, which are characterizing the vacancy V_1, by the applicants

L, C, U	S_1	S_2	S_3	S_4
l_1	0.97	0.75	0.78	0.97
l_2	0.9	0.84	0.82	0.90
l_3	0.92	0.86	0.72	0.88
l_4	0.9	0.77	0.9	0.85
c_1	0.94	0.98	0.92	0.93
c_2	0.79	0.93	0.78	0.7
c_3	0.77	0.93	0.65	0.8
u_1	0.98	0.90	0.95	0.95
u_2	0.97	0.65	0.7	0.95
u_3	0.95	0.49	0.97	0.9

$$\tilde{S}_1 = \left\{ \begin{array}{l} 0.97/l_1; 0.9/l_2; 0.92/l_3; 0.9/l_4; 0.94/c_1; 0.79/c_2; \\ 0.77/c_3; 0.98/u_1; 0.97/u_2; 0.95/u_3 \end{array} \right\}$$

$$\tilde{S}_2 = \left\{ \begin{array}{l} 0.75/l_1; 0.84/l_2; 0.86/l_3; 0.77/l_4; 0.98/c_1; \\ 0.93/c_2; 0.93/c_3; 0.9/u_1; 0.65/u_2; 0.69/u_3 \end{array} \right\}$$

$$\tilde{S}_3 = \left\{ \begin{array}{l} 0.78/l_1; 0.82/l_2; 0.72/l_3; 0.9/l_4; 0.92/c_1; 0.78/c_2; \\ 0.65/c_3; 0.95/u_1; 0.7/u_2; 0.97/u_3 \end{array} \right\}$$

$$\tilde{S}_4 = \left\{ \begin{array}{l} 0.97/l_1; 0.9/l_2; 0.88/l_3; 0.85/l_4; 0.93/c_1; \\ 0.7/c_2; 0.8/c_3; 0.95/u_1; 0.95/u_2; 0.9/u_3 \end{array} \right\}$$

The reference fuzzy description of the vacancy V_1 is defined as follows:

$$\tilde{V}_1 = \left\{ \begin{array}{r} 0.97/l_1;\ 0.98/l_2;\ 0.99/l_3;\ 0.96/l_4; 0.98/c_1;\ 0.98/c_2;\ 0.96/c_3; \\ 0.98/u_1;\ 0.99/u_2;\ 0.96/u_3 \end{array} \right\}$$

Stage 4: Using the formula (2) fuzzy equality degrees of the reference $\tilde{V}_1$ and real situations $\tilde{S}_1,\ \tilde{S}_2,\ \tilde{S}_3,\ \tilde{S}_4$ are determined: (a) by personal indicators (L); (b) in terms of competences (C); and (c) from the prism of the requirements for the vacancy (U):

Determination of the fuzzy equality degree of the reference $\tilde{V}_1$ and real situations $\tilde{S}_1,\ \tilde{S}_2,\ \tilde{S}_3,\ \tilde{S}_4$, according to the personal indicators (L):

$$\begin{aligned} &\mu_L(\tilde{V}_1, \tilde{S}_1) = \min(\max(1-0.97, 0.97), \max(0.97, 1-0.97)) \\ &\&\min(\max(1-0.98, 0.9), \max(0.96, 1-0.9))\&\&\min(\max(1-0.99, 0.92), \\ &\max(0.99, 1-0.92))\&\min(\max(1-0.96, 0.9), \max(0.96, 1-0.9)) \\ &= \min(0.97, 0.97)\&\min(0.9, 0.98)\&\min(0.92, 0.99)\&\min(0.9, 0.96) \\ &= 0.97\&0.9\&0.92\&0.9 = 0.9 \end{aligned}$$

$$\begin{aligned} &\mu_L(\tilde{V}_1, \tilde{S}_1) = \min(\max(1-0.97,\ 0.97), \max(0.97, 1-0.97))\&\min(\max(1-0.98, 0.9), \\ &\max(0.96, 1-0.9))\&\&\min(\max(1-0.99, 0.92), \max(0.99, 1-0.92)) \\ &\&\min(\max(1-0.96, 0.9), \max(0.96, 1-0.9)) = \min(0.97, 0.97)\&\min(0.9, 0.98) \\ &\&\min(0.92, 0.99)\&\min(0.9, 0.96) = 0.97\&0.9\&0.92\&0.9 = 0.9 \end{aligned}$$

$$\mu_L(\tilde{V}_1,\ \tilde{S}_2)\ =\ 0.75\&0.84\&0.86\&0.77 = 0.75$$

$$\mu_L(\tilde{V}_1,\ \tilde{S}_3)\ =\ 0.78\&0.82\&0.72\&0.9 = 0.72$$

$$\mu_L(\tilde{V}_1,\ \tilde{S}_4)\ =\ 0.97\&0.9\&0.88\&0.85 = 0.85$$

Determination of the fuzzy equality (similarity) degree of the applicants with the vacancy requirements V_I in terms of their competencies (C):

$$\mu_C(\tilde{V}_1,\ \tilde{S}_1)\ =\ 0.94\&0.79\&0.77 = 0.77$$

$$\mu_C(\tilde{V}_1,\ \tilde{S}_2)\ =\ 0.98\&0.93\&0.93 = 0.93$$

$$\mu_C(\tilde{V}_1,\ \tilde{S}_3)\ =\ 0.92\&0.78\&0.65 = 0.65$$

$$\mu_C(\tilde{V}_1,\ \tilde{S}_4)\ =\ 0.93\&0.7\&0.8 = 0.7$$

Determination of the fuzzy equality degree of the applicants' parameters (the search image of the offer) with the parameters of the vacancy V_I (the search image of the request) by the indicators characterizing the requirements and conditions presented by the employer:

$$\mu_U(\tilde{V}_1, \tilde{S}_1) = 0.98\&0.97\&0.95 = 0.95$$

$$\mu_U(\tilde{V}_1, \tilde{S}_2) = 0.9\&0.65\&0.46 = 0.49$$

$$\mu_U(\tilde{V}_1, \tilde{S}_3) = 0.95\&0.7\&0.97 = 0.7$$

$$\mu_U(\tilde{V}_1, \tilde{S}_4) = 0.95\&0.95\&0.9 = 0.9$$

In accordance with the obtained results taking into account all indicators, fuzzy equality degree $\phi(\tilde{V}_1, \tilde{S}_g),\ g = \overline{1, 4}$ of the reference V_1 and search images of the real situations $\tilde{S}_1, \tilde{S}_2, \tilde{S}_3, \tilde{S}_4$ is determined:

$$\mu(\tilde{V}_1, \tilde{S}_1) = \mu_L(\tilde{V}_1, \tilde{S}_1)\&\mu_C(\tilde{V}_1, \tilde{S}_1)\&\mu_U(\tilde{V}_1, \tilde{S}_1) = 0.9\&0.77\&0.95 = 0.77$$
$$\mu(\tilde{V}_1, \tilde{S}_2) = \mu_L(\tilde{V}_1, \tilde{S}_2)\&\mu_C(\tilde{V}_1, \tilde{S}_2)\&\mu_U(\tilde{V}_1, \tilde{S}_2) = 0.75\&0.93\&0.49 = 0.49$$
$$\mu(\tilde{V}_1, \tilde{S}_3) = \mu_L(\tilde{V}_1, \tilde{S}_3)\&\mu_C(\tilde{V}_1, \tilde{S}_3)\&\mu_U(\tilde{V}_1, \tilde{S}_3) = 0.72\&0.65\&0.7 = 0.65$$
$$\mu(\tilde{V}_1, \tilde{S}_4) = \mu_L(\tilde{V}_1, \tilde{S}_4)\&\mu_C(\tilde{V}_1, \tilde{S}_4)\&\mu_U(\tilde{V}_1, \tilde{S}_4) = 0.85\&0.7\&0.9 = 0.7$$

Step 5: Correspondingly, fuzzy equality degree of the real situations $\tilde{S}_1, \tilde{S}_2, \tilde{S}_3, \tilde{S}_4$ (search images of the applicants) in the reference situation $\tilde{V}_2$ (the search image of the employer's request) is determined.

The results are summarized in Table 4.

As seen from Table 4, in terms of the fuzzy equality degree of the requirements of the vacancy V_1, the applicant S_1 is the most suitable ($\mu(\tilde{V}_1, \tilde{S}_1) = 0.77$), whereas the applicant S_4 is the most suitable for the parameters of the vacancy V_2 ($\mu(\tilde{V}_1, \tilde{S}_4) = 0.7$).

Table 4 Fuzzy equality degrees $\mu(\tilde{V}_1, \tilde{S}_g),\ g = \overline{1, 4}$ and $\mu(\tilde{V}_2, \tilde{S}_g),\ g = \overline{1, 4}$

	$\mu(\tilde{V}_1, \tilde{S}_g),\ g = \overline{1, 4}$	$\mu(\tilde{V}_2, \tilde{S}_g),\ g = \overline{1, 4}$
S_1	0.77	0.59
S_2	0.49	0.49
S_3	0.65	0.62
S_4	0.70	0.70

6 Conclusion

The method proposed in the article makes it possible to reduce subjectivity in the evaluation of heterogeneous information about the current state of demand and supply for IT specialists and provides the possibility of decisions making during the recruitment that are adequate to the current situation. This methodology is invariant for the management of human resources in various segments of the economy in terms of its adaptation to the characteristics of the explored segment, and it can be applied in formation of e-health as a main component of e-government [4].

References

1. European Commission, *EU Skills Panorama Analytical Highlight 'ICT Professionals'*, E-Skills UK (2012). http://euskillspanorama.ec.europa.eu
2. IT-Specialist Deficit Problem Has Global Character GosInformNauki Ukrainy, (2014) www.unian.net/society/871034-defitsit-it-spetsialistov-v-2015-godu-gosinformnauki.html
3. M.H. Mammadova, F.R. Mammadzada, Development of conceptual foundations of intellectual management of demand and supply of IT specialists labor market. Eastern Eur. J. Adv. Technol. **76**(4/3), 53–67 (2015). https://doi.org/10.15587/1729-4061.2015.47458
4. M.H. Mammadova, Z.Q. Jabrayilova, F.R. Mammadzada, Managing the IT labor market in conditions of fuzzy information. Autom. Control Comput. Sci. **49**(2), 88–93 (2015)
5. W. Bartlett, *European Training Foundation, 2011. Skills Anticipation and Matching systems in transition and developing countries: Conditions and challenges*. Working paper for the European Training Foundation. www.etf.europa.eu
6. M.H. Mammadova, Z.Q. Jabrayilova, F.R. Mammadzada, Fuzzy decision-making support methods for the selection of IT-professionals. Int. J. Eng. Innov. Technol. (IJEIT) **3**(7), 169–175 (2014)
7. M.H. Mammadova, Z.Q. Jabrayilova, F.R. Mammadzada, Fuzzy multicriterial methods for the selection of IT-professionals. Int. J. Intell. Syst. Appl. Eng. **3**(2), 40–45 (2015)
8. M.H. Mammadova, Z.G. Jabrayilova, F.R. Mammadzada, Application of fuzzy situational analysis for IT-professionals labor market management, in *Proceedings of the 2nd International Conference on Information Science and Control Engineering (ICISCE 2015)*, 24–26 April (2015), Shanghai, China, pp. 143–146
9. H.A. Simon, The structure of ill structured problems. Artif. Intell. **4**(3–4), 181–201 (1973)
10. M.H. Mammadova, *Decision-Making Based on Knowledge Bases with a Fuzzy Relational Structure* (ELM, Baku, 1997), p. 296
11. L.A. Zadeh, Fuzzy sets. Inf. Control **8**(3), 338–353 (1965)
12. A.N. Melikhov, L.S. Bernshtein, S.Y. Korovin, Situational advisory systems with fuzzy logic. Nauka 272 (1990)
13. M.H. Mammadova, Z.Q. Jabrayilova, F.R. Mammadzada, Fuzzy multi-scenario approach to decision-making support in human resource management. Stud. Fuzziness Soft Comput. **342**, 19–36 (2016)
14. M.H. Mammadova, Z.G. Jabrailova, S. Nobari, Application of TOPSIS method in support of decisions made in staff management issues, in *Proceedings of the IV IEEE International Conference "Problems of Cybernetics and Informatics" (PCI-2012)*, vol. IV, (2012), Baku. pp. 195–198
15. M.G. Mammadova, Z.G. Jabrayilova, Decision-making support for human resources management on the basis of multi-criteria optimization method, in *Proceedings of the 9th International*

Conference on Application of Information and Communication Technologies (Aict), Russia, Rostov-on-Don, 14–16 October (2015), pp. 579–582
16. M.H. Mammadova, Z.G. Djabrailova, Decision-making support in human resource management based on multi-objective optimization. TWMS J. Pure Appl. Math. **9**(1), 53–73 (2018)
17. T.L. Saaty, Decision-making. The method of analyzing hierarchies. Radio Commun. 320 (1993)
18. M.H. Mammadova, Z.G. Jabrayilova, S.M. Nobari, Use of information about the importance of the criteria in the solution of personnel management problems, in *International Conference on "Problems of Cybermetics and Informatics, Baku*, 6–8 Sept. (2010), pp. 83–86
19. D. Neyman, O. Morgenstern, Game theory and economic behavior. Nauka 708 (1970)
20. www.boss.az
21. www.jobsearch
22. www.rabota.az
23. O.I. Laricev, Verbal decision analysis. Nauka 181 (2006)
24. R. Bellman, L.A. Zadeh, Decision- making in fuzzy environment. Manage. Sci. **17**, 141–164 (1970)

Recognition of Sarcastic Sentences in Sentiment Analysis of a Text that Describes the Appearance of a Person

A. V. Dolbin, V. L. Rozaliev, Y. A. Orlova, and A. V. Zaboleeva-Zotova

Abstract This article is devoted to the method of recognition of sarcasm in the text in order to improve the accuracy of sentiment analysis. As a domain area for the experiment, the task of determination the sentiment level of a text that describes the appearance of a person was chosen. At first, references to the personality and elements that describes appearance from the text written in a natural language are detected using the method of latent semantic analysis. Frame model of knowledge representation was chosen as the information structure of the human appearance. The next step is to evaluate the attitude to the person in the text using the pre-labeled sentiment dictionary. At this stage, the method of recognizing sarcastic sentences that contains a description of the appearance is used. The sentiment level should be re-evaluated in the slots of the frame instance. The F-measure was used as an efficiency criterion for correct sarcasm recognition. At the end of the article, the results of experiment are provided.

This work was partially supported by RFBR (Grants 18-07-00220, 19-47-343001, 19-47-343002, 19-47-340009, 19-47-340013, 19-37-90060, 20-07-00502).

A. V. Dolbin · V. L. Rozaliev (✉)
CAD Department, Volgograd State Technical University, Volgograd, Volgograd Region, Russian Federation
e-mail: vladimir.rozaliev@gmail.com

A. V. Dolbin
e-mail: sizeof.void34@gmail.com

Y. A. Orlova · A. V. Zaboleeva-Zotova
SAS Department, Volgograd State Technical University, Volgograd, Volgograd Region, Russian Federation
e-mail: yulia.orlova@gmail.com

A. V. Zaboleeva-Zotova
e-mail: zabzot@gmail.com

S. N. Shahbazova et al. (eds.), *Recent Developments and the New Direction in Soft-Computing Foundations and Applications*, Studies in Fuzziness and Soft Computing 393, https://doi.org/10.1007/978-3-030-47124-8_3

1 Introduction

Sentiment analysis of the text belongs to the category of information retrieval tasks. The importance of an effective solution to this problem grows over time, since the amount of information that needs to be processed by semantic analysis systems is continuously increasing. At the moment there are quite effective methods for sentiment analysis of the text, but there are a number of directions, the solution of which will make it possible to achieve greater accuracy of correct recognition. One such direction is the recognition of sarcasm. Sarcasm can be classified as an implicit approach to the expression of opposing emotions. However, even a person cannot always determine reliably whether this phrase is a sarcasm.

The task of automating the definition of sarcasm itself is of little practical value. Typically, you need a limited application area to apply the sentiment analysis. And most often the development is carried out in the following areas:

(1) sentiment analysis of users reviews,
(2) analysis of comments posted on social media resources [1].

The problem of recognizing sarcastic sentences in the text in natural language was considered in the context of searching for elements of a person's appearance and determining the sentiment class. This named entity was chosen not by chance, since it is a quite complex task to recognize it with high accuracy due to a large number of approaches to co-referencing through pronouns in the third person.

The aim of this work was to examine modern methods for determining the author's relationship to the described person by performing the sentiment analysis. The most obvious area of application of the development, considered in this article, is the analysis of comments on photos in social networks. Using the methods of machine learning, it is possible to construct a model that is able to recognize a positive or negative attitude to the appearance of the person depicted in the photograph. The main contribution of the authors of the article is the adaptation of existing methods of assessing the sentiment in the field of recognition of a person's appearance in the text in natural language [2].

2 Basic Personality Search Model

First of all, it was required to develop an information model of a person's appearance. This model must meet the following requirements:

(1) extensibility,
(2) visibility,
(3) completeness of the description.

A frame presentation of knowledge is perfectly suitable for this description. Figure 1 shows the final model of a person's appearance using the frame repre-

```
(frame Human_Appearance
(Height       ( value (M) ) ( IF_ADDED(sentiment_analysis) ) )
(Body         ( value (M) ) ( IF_ADDED(sentiment_analysis) ) )
(Head         ( value (M) ) ( IF_ADDED(sentiment_analysis) ) )
(Hair         ( value (M) ) ( IF_ADDED(sentiment_analysis) ) )
(Face         ( value (M) ) ( IF_ADDED(sentiment_analysis) ) )
(Forehead     ( value (M) ) ( IF_ADDED(sentiment_analysis) ) )
(Eyebrows     ( value (M) ) ( IF_ADDED(sentiment_analysis) ) )
(Eyes         ( value (M) ) ( IF_ADDED(sentiment_analysis) ) )
(Eyelashes    ( value (M) ) ( IF_ADDED(sentiment_analysis) ) )
(Nose         ( value (M) ) ( IF_ADDED(sentiment_analysis) ) )
(Lips         ( value (M) ) ( IF_ADDED(sentiment_analysis) ) )
(Chin         ( value (M) ) ( IF_ADDED(sentiment_analysis) ) )
(Teeth        ( value (M) ) ( IF_ADDED(sentiment_analysis) ) )
(Neck         ( value (M) ) ( IF_ADDED(sentiment_analysis) ) )
(Shoulders    ( value (M) ) ( IF_ADDED(sentiment_analysis) ) )
(Chest        ( value (M) ) ( IF_ADDED(sentiment_analysis) ) )
(Back         ( value (M) ) ( IF_ADDED(sentiment_analysis) ) )
(Legs         ( value (M) ) ( IF_ADDED(sentiment_analysis) ) )
(Arms         ( value (M) ) ( IF_ADDED(sentiment_analysis) ) )
)
```

Fig. 1 Appearance model in FRL notation

sentation language notation. There are the main components on which it is possible to compose a complete description of a person's appearance in the frame slots. Slots for the model were compiled by the authors of the article. In Fig. 1, "M" is the set of valid values for the description elements of the appearance for each slot. The specialty of the FRL notation is that it is permissible to join special procedures-demons to it. The only procedure is the determination of sentiment level with the subsequent resolution of sarcasm. It is worth noting that each non-empty slot must correspond to the sentences from which the facts were extracted for the frame. This is required for further recognition of the presence of sarcasm in the text [3].

To fill the frame, a method of latent semantic analysis was used, or abbreviated LSA, as it has proved itself in the field of machine learning. Methods that do not use a pre-tagged training sample for the learning process show a smaller effectiveness in terms of recognition. The method of latent semantic analysis can be characterized as establishing the relationship between the vectors of the features of the analyzed documents to the words that serve as the keys. Thus, to use the method of semantic analysis of text in natural language, the slots of the frame should be used as search keys [3].

The latent-semantic algorithm is as follows:

(1) create a list of all keywords that will be searched in the text,
(2) create a frequency matrix A, in cells of which the count of how many times does this word occur in the text,

(3) apply TF-IDF method on a frequency matrix to ensure that results are relevant [4],
(4) apply a singular matrix decomposition: algorithm divides the transformed frequency matrix A into three composite matrices U, V^t and S according to the (1).

$$A = U \times S \times V^t. \tag{1}$$

(5) matrix U contains the coordinates of keywords and Vt—coordinates of documents.

Singular value decomposition of the matrix allows you to get rid of unnecessary noise, which significantly increases the efficiency of the method. The number of rows and columns that can be discarded before the subsequent analysis can be selected experimentally. It is now possible to obtain the nearest documents, which has the same semantic meaning as specified keyword, and then fill in the frame slots.

3 Sentiment Analysis Based on Vocabulary

All approaches to determination of sentiment class are divided into three main groups:

(1) compilation of a sentiment vocabulary.
(2) the use of various classifiers.
(3) the use of compiled contextual rules.

A rule-based approach shows the most accurate results, but it requires very high costs and colossal linguistic work for compilation. The main drawback of this approach is that it is extremely difficult to compose universal rules that are suitable for all domains. To achieve the most effective evaluation of the tonality, the rules are compiled for a specific application area.

In this experiment, an approach based on the valence dictionary was applied, since it shows a fairly high percentage of correct recognition. The task is greatly simplified if there is a source for compiling a dictionary of valences belonging to the domain under study. Such dictionary was compiled on the basis of the corpus of the Russian language OpenCorpora all the phrases that are marked as "Qual" were chosen from this dictionary. Further, only those word forms that can be used for describing a person's appearance have been filtered out. To simplify the task of sentiment analysis, it was decided that the valences would correspond to available sentiment class. Table 1 shows an example of a sentiment dictionary.

For the study, the basic five sentiment classes were compiled:

(1) negative.
(2) strongly negative.
(3) positive.

Table 1 A part of compiled sentiment dictionary

Keyword	Valence
Friendly	2
Unfriendly	−2
Shy	0
Impartial	1
Mean	−1

(4) strongly positive.
(5) neutral.

To determine the sentiment, the naive Bayes method was used. This method has proved itself in the field of machine learning. A naïve Bayesian algorithm is a classification algorithm based on the Bayes theorem with the assumption of independence of features. The classifier assumes that the presence of any feature in the class is not related to the presence of any other attribute. Let the $P(d|c)$ be the probability of finding a document in all the documents of a given class. The basis of the naive Bayesian classifier is the corresponding theorem (2). In (2) $P(c)$ is the probability of certain document can be found among all data set and $P(d)$—probability the document occurs throughout the whole corpus.

$$P(c|d) = \frac{P(d|c) * P(c)}{P(d)} \tag{2}$$

Thus, the naive Bayes method is based on the problem of finding the maximum probability of a document belonging to a certain sentiment class. Thus, the sentiment level for each key element of a person's appearance can be determined by (3) [5].

Classification using naïve Bayes is easy and fast and requires less training data. Also, it is better suited for classification based on categories (sentiment analysis with separate defined classes refers to such cases). However, if there is some value of a category characteristic in the data set that was not found in the training samples, then the model will assign a zero probability to this frame slot. Sentiment class for each key element of a person's appearance can be determined by (3), where $P(w|c)$ is probability of occurrence of a certain term in a document:

$$P\text{max} = \arg\max\left[P(c)\prod_{i=1}^{n} P(w_i|c)\right]. \tag{3}$$

4 Approach to Determination of Sarcastic Sentences

The issue of sarcasm recognition in a sentence requires the training of another classifier. To solve this issue, the method of k-nearest neighbors is used [6]. To classify each of the test sample objects, you must perform the following steps sequentially:

(1) calculate the distance to each of the training sample objects,
(2) choose k training sample objects, the distance to which is minimal,
(3) class of the object being classified is the class most often encountered among the k nearest neighbors.

The following set of parameters for the vector of singularities was compiled:

(1) the presence of word forms, which are specific for sarcasm (such expressions include common words from the Internet slang),
(2) the presence of quotes in the text (if there are quotes, it is most likely that the text contains a certain degree of irony),
(3) high frequency of punctuation,
(4) the presence in the text of words that are most often used in conjunction with sarcasm for a particular language, which are taken from training samples [7].

For this case, the weight is given as a function of the distance to the nearest neighbors. In (4) $d(x, x_{(i)})$ is a function which determines the distance between elements in a vector space. Equation (5) finally determines whether or not the text being analyzed contains sarcasm, where Z_i is a sum of weights for all of the available classes. If so, then the class of the slot must be changed to the opposite.

Empirically, it was revealed that the classifier gives the best efficiency in terms of accuracy if it analyzes the nearest-neighbor number K equal to the number of sentiment classes.

To obtain more plausible results, you should filter out the most frequent words in the model. This step removes unnecessary noise that could affect the final result of the study. In addition, before using the K nearest neighbors method, the volume of aggregated sentiment information should be considered. In this study the results for unigrams and trigrams are provided [8].

$$w(x(i)) = w(d(x, x_{(i)})) \tag{4}$$

$$C = \arg\max Z_i \tag{5}$$

Table 2 Experiment results

	Recall	Precision	F
Unigrams, without sarcasm	0.80	0.82	0.810
Trigrams, without sarcasm	0.85	0.84	0.844
Unigrams, with sarcasm	0.86	0.68	0.76
Trigrams, without sarcasm	0.87	0.77	0.82

5 Conducting the Experiment and Analysis of the Results

To decide whether a sentiment recognition effectiveness is better or worse using the method of sarcasm determination, a numerical metric is needed. For most modern algorithms based on machine learning, metrics of accuracy and completeness of search are used. The accuracy of the search determines the proportion of documents that really belong to a given sentiment class across all documents of this class. The completeness of the search determines the ratio of the found classifiers of documents belonging to this class to all documents in the sample. Since in real practice of machine learning the maximum accuracy and completeness of search are unattainable simultaneously, the analysis of results using the F-measure will be the most acceptable. The F-measure is calculated as follows (6):

$$F = 2 \times \frac{\text{Presision} \times \text{Recall}}{\text{Precision} + \text{Recall}}. \tag{6}$$

A training set consists of 500 samples was compiled: 150 of them were marked as “containing sarcasm” and 350 were marked as “not containing sarcasm”. This ratio between classes was not chosen randomly, since the likelihood of evaluating the sentiment class of the text as positive or negative is much higher than sarcastic. The experiment was conducted on a sample of 100 texts, which are supposed to be a description to the different photos with no more than 200 words in length and contains only the information about person’s appearance.

As can be seen from the obtained results (Table 2), the method of sarcasm recognition in the text slightly lowers accuracy due to a relatively large number of false positives. It can be concluded that lexical features and punctuation signs are not enough to train the classifier at a sufficient level. Most often, sentences have a complex structure, which cannot be treated as a “bag of words” and requires the use of contextual syntactic rules [9].

6 Conclusion

As a result of the experiment, it can be concluded that the resolution of the task of recognizing sarcasm in a text containing a description of the appearance of a person cannot be effectively resolved only using the methods of machine learning with supervision. As a further study, it requires the development of contextual rules based on the syntactic structure of the text. At this stage, the F-measure estimation showed that the method slightly reduces effectiveness due to a relatively large number of false positives.

References

1. A. Ritter, S. Clark, Mausam, O. Etzioni, Named entity recognition in tweets: an experimental study, in *Proceeding of the conference on empirical methods in natural language processing, Association for Computational Linguistics* (2011), pp. 1524–1534
2. A. Dmitriev, A. Zaboleeva-Zotova, Y. Orlova, V. Rozaliev, Automatic identification of time and space categories in the natural language text, in *Proceedings of the IADIS Internation Conference* (2013), pp. 23–25
3. A. Dolbin, V. Rozaliev, Y. Orlova, Recognition of a person named entity from the text written in a natural language. J. Phys. Conf. Ser. **803**, 5 (2017)
4. B. Gebre, M Zampieri, P. Wittenburg, T. Heskes, Improving native language identification with TF-IDF weighting, in *8th NAACL Workshop on Innovative Use of NLP for Building Educational Applications* (2013), pp. 216–223
5. H. Shimodaira, Text classification using naïve bayes. Learn. Data Note **7**, 1–9 (2014)
6. D. Davidov, O. Tsur, A. Rappoport, Semi-supervised recognition of sarcastic sentences in Twitter and Amazon, in *CoNLL'10 Proceedings of the Fourteenth Conference on Computational Natural Language Learning* (2010), pp. 107–116
7. P. Carvalho, L. Sarmento, M.J. Silva, E. Oliveira, Clues for detecting irony in user-generated contents: oh…!! it's so easy, in *Proceedings of the 1st International CIKM Workshop on Topic-Sentiment Analysis for Mass Opinion* (2009), pp. 53–56
8. A. Reyes, P. Rosso, Mining subjective knowledge from customer reviews: a specific case of irony detection, in *Proceedings of the 2nd Workshop on Computational Approaches to Subjectivity and Sentiment Analysis* (2011), pp. 118–124
9. R. Gonzalez-Ibanez, S. Muresan, N. Wacholder, Identifying sarcasm in Twitter: a closer look, in *Proceedings of the 49th Annual Meeting of the Association for Computational Linguistics*, vol. 2 (2011), pp. 581–586

Cognitive Knowledge Seeding Using Collective Uncertainty Shaping

Nasibeh Rady Raz and Mohammad R. Akbarzadeh-T

Abstract Advanced Therapy Medicinal Products (ATMPs) are a part of medicine that use carriers for targeted delivery of genes, tissues, cells, and drugs. In this regard, knowledge has a great effect in enhancing the delivery process. Knowledge, that can be transferred or gained, is an understanding of someone or something which is acquired through perceiving, discovering, inferencing, experiencing, and learning. In this paper, a swarm of nano drug carriers are guided towards cancer sites. At first, the drug carriers' (nanomachines') intention is to reach from the self-knowledge (*pure knowledge state*) by doing simple fuzzy inferencing to the more sophisticated collective knowledge (*flourishing knowledge state*) by joining the closing swarm to the site of cancer, and then further enhancing their knowledge through knowledge seeding. The Vascular Endothelial Growth Factor (VEGF) concentration is used here as cancer marker since it is high around cancer sites based on in vivo data. As soon as nanomachines reach the flourishing knowledge state, the swarm intention changes from cancer reaching to the cost reduction by doing rule reduction. This change of intention based on the current need is defined as a cognitive knowledge seeding based on uncertainty shaping. To do this, we do the rule reduction using a Singular Value Decomposition (SVD) in a collective manner. The proposed method outperforms the state of the art in having resilient rule reduction, less computation, and more energy saving, while having better therapeutic result.

Keywords Targeted cancer nano-therapy · Knowledge seeding · Uncertainty shaping · Cognitive science

N. Rady Raz · M. R. Akbarzadeh-T (✉)
Department of Computer Engineering, Center of Excellence on Soft Computing and Intelligent Information Processing, Ferdowsi University of Mashhad, Mashhad, Iran
e-mail: akbazar@um.ac.ir

N. Rady Raz
e-mail: radyraz@mail.um.ac.ir

S. N. Shahbazova et al. (eds.), *Recent Developments and the New Direction in Soft-Computing Foundations and Applications*, Studies in Fuzziness and Soft Computing 393, https://doi.org/10.1007/978-3-030-47124-8_4

1 Introduction

Positive uncertainty or the uncertainty for the sake of knowledge will put an end to many endeavors regarding pure uncertainty removal tasks. Great decision makers are those that consider a trade-off between certainty and uncertainty. A diverse bunch of uncertainty works as a kind of exploratory task for exploiting better decisions. This process happens by using cognitive operations such as decision making based on the current knowledge of the system. In precise applications i.e., targeted delivery of genes, tissues, cells, or drug in Advanced Therapy Medicinal Products (ATMPs) due to limited amount of resources, this process shows its importance.

There is a good literature on uncertainty handling techniques. In [1] an uncertainty handling method based on a distributed adaptive sliding mode is presented for controlling Platoons. Platoons or grouping vehicles control is used for automated highway system. Moreover, redundancy resolution task is done in [2] using parameter identification and neural control in a simultaneous manner with the help of online adaptive projection neural network. Furthermore, uncertainty handling in a granular computing framework is mentioned in [3] for video tracking using an adaptive rule generation based on rough rule base and granular flow graph. A new sampling technique for addressing the scalability of high-dimensional uncertain system computation is discussed in [4] using a multivariate probabilistic collocation method (M-PCM).

Uncertainty handling in sensors' decision-making using a stochastic optimal control law is designed in [5]. A radial basis function neural network is used for a class of pure-feedback nonlinear systems with full state constraints and dynamic uncertainties in [6]. The bounded input uncertainty handling in consensus tracking control of multi-agent systems is developed in [7] using sliding mode surface with bounded convergence. In order to overcome the parametric uncertainty, a tube-based robust Model Predictive Control (MPC) is described in [8] for constrained linear systems.

Moreover, in [9] a Type-2 Fuzzy Logic (T2FL) approach is presented for EEG pattern recognition in the field of automated analysis and interpretation of electrical brain activity and uncertainty handling for brain computer interface systems. A method for quantifying and treatment of uncertainty by using rank-based selection operator of evolutionary algorithms is implemented in [10] for uncertainty handling in the context of the Covariance Matrix Adaptation Evolution Strategy (CMA-ES).

Earlier than that, there was a try in [11] to provide a qualitative treatment of uncertainty for the manifestations of two fault presence cases including more or less certainly present, or more or less possibly present using an uncertainty representation based on Zadeh's possibility theory and fuzzy sets. Furthermore, an uncertainty handling method using rule reduction task for Takagi-Sugeno fuzzy systems is designed in [12]. Finally, uncertainty handling of the conflict concept for the sake of cooperation tuning in Mobile Ad hock Networks (MANETs) is introduced in [13].

We previously proposed a bio inspired nanomachines for doing cancer collective therapy in [14], then we enhanced them in order to show their emergent complex behavior in [15]. After that we cognitively coordinate their swarm in [16]. In this paper, we propose a collective uncertainty shaping using a cognitive knowledge seeding concept. To do this, we consider a group of nanomachines carrying drug to the place of cancer by sensing their environment. At first nanomachines' intention is to reach from the pure knowledge by doing the fuzzy inferencing to the flourishing state by joining the closing swarm to the site of cancer, and completing their knowledge through knowledge seeding.

Considering the closeness of swarms' lactations to cancer site, the initial pure knowledge will be updated and distributed among members. Cancer site is presented by the Vascular Endothelial Growth Factor (VEGF) concentration which is high around the cancer site based on in vivo data in [17]. As soon as they reach the flourishing state, the intention of the swarm changes from cancer reaching task to the cost reduction task by doing rule reduction. This change of intention based on the current need is defined as a cognitive knowledge seeding. The proposed knowledge seeding is based on uncertainty shaping by applying rule reduction strategy in a collective manner i.e., a Singular Value Decomposition (SVD) strategy. Less computation and more resiliency in swarm are two important results of the method.

The rest of the paper is organized into 5 sections. In Sect. 2 basic concepts are reviewed. The proposed technique is presented in Sect. 3. Experimental results are presented in Sect. 4, and finally Sect. 5 summarizes this paper.

2 Basic Concept

In this section we introduce the following relevant basic concepts: advanced therapy medicinal products, knowledge seeding, fuzzy approach and fuzzy rule reduction.

Advanced Therapy Medicinal Products

Advanced Therapy Medicinal Products (ATMPs) are a type of innovation therapies that include gene therapy [18], somatic cell therapy [18] , tissue-engineered products [19], and targeted drug delivery [20] as presented in Fig. 1a–d, respectively. ATMPs are used for various diseases in the case of limited or no therapeutic options [21, 22]. This cutting-edge therapy needs smart carriers for delivering its product to the targeted places of the body. In this paper aiming to do the targeted delivery, we present a framework for nano drug delivery based on the cognitive knowledge seeding (Fig. 1).

Knowledge Seeding

Knowledge is an understanding of someone or something which is acquired through several tasks i.e., perceiving, discovering, analyzing, experiencing, educating, inferencing and learning. The process to create, capture, handle and transfer knowledge is called Knowledge Management (KM) [27].

Keeping its originality, knowledge can transfer from agent to agent in a direct or an indirect one to one, one to many, and many to many components. Besides transferring, knowledge can be obtained through the concept of knowledge seeding

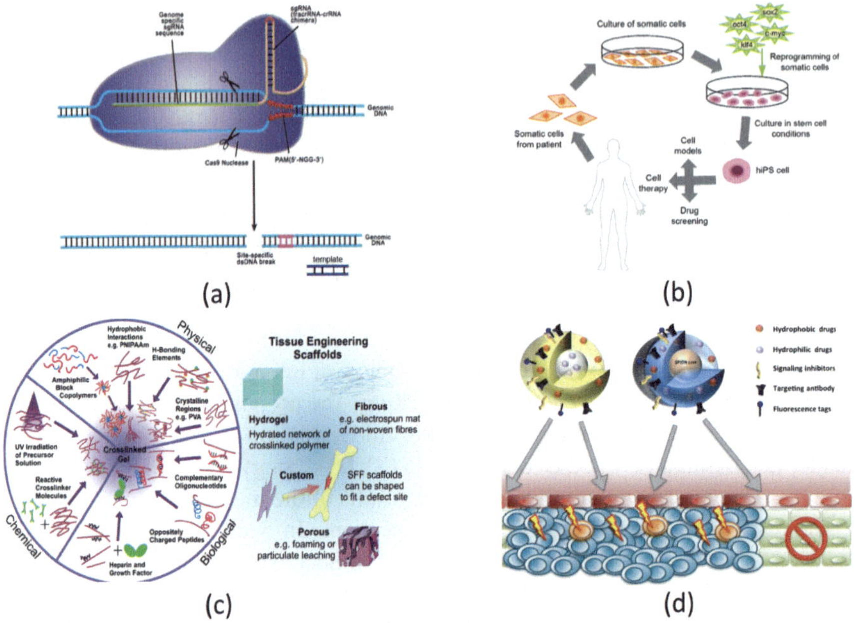

Fig. 1 Advanced therapy medicinal products. **a** Gene therapy [23], **b** somatic cell therapy [24], **c** tissue-engineered products [25], and **d** targeted drug delivery [26]

in which a relevant knowledge is obtained. Since we live in an age of competence, not only knowledge sharing and knowledge learning are two important ability among humans, but we also need these key abilities among artificial systems. This is what we are going to try in this paper.

Fuzzy Approach

One of the uncertainty handling tools is Fuzzy logic which is introduced by Lotfi Zadeh in [28]. FT1 is uses a set function on universe X into [0, 1]. The Membership Function (MF) [29] of A which is called $\mu_A(x)$ is presented in (1).

$$A = \{(x, \mu_A(x)) | x \in X\} \tag{1}$$

FT1 [18] has p inputs and one output, and the system can be presented using M rules. Then these IF-THEN rules are combined using fuzzy logic principles in the inference engine. Each rule is interpreted as a fuzzy implication.

Fuzzy Rule Reduction There are two recent fuzzy rule reduction methods including Singular Value Decomposition (SVD), and the Combs method for rapid inference [29]. SVD uses linear algebra and coordinate transformation for producing a reduced mapping in a different coordination system. Combs method uses a logical Boolean set-theoretic proof for doing the transforms. Using error analysis, it is possible in the

SVD to choose the amount of reduction, while in the Combs method the scalability, transparency of ruled, and simulation times is considered.

In fuzzy rule reduction process the main part is Membership Function (MF). In order to use SVD, MFs should have three conditions including having degree of overlap, mapping membership values on the unit interval, and having a prototypical value [29].

3 Proposed Techniques

Here, we propose a collective uncertainty shaping using cognitive knowledge seeding. To do this, we consider a group of nanomachines carrying drug to the cancer site by sensing their environment. Their initial pure knowledge of the environment which is made by the fuzzy inferencing will be updated based on the closeness of the swarms' location to the place of cancer. Place of cancer is presented by the Vascular Endothelial Growth Factor (VEGF) concentration which is high around the cancer site based on in vivo data. Depending on the diversity of the information each swarm is shaped. The aim of each nanomachine is to gain more knowledge about the cancer state by joining the closing swarm to the site of cancer, and completing their knowledge.

The swarm which is belong to this group is called flourishing swarm. In this state the swarm's internal behavior is to reach from the pure knowledge state to the flourishing knowledge state through use of knowledge seeding and the swarm's external behavior is to reach the cancer site.

To do this, the swarms can reshape their fuzzy rule bases by collectively doing rule reduction using Singular Value Decomposition (SVD) strategy. In this phase the intention of swarm changes from cancer reaching to the cost reduction by starting the rule reduction process. This changes of intention of the swarm based on their current need is defined as a cognitive knowledge seeding based on uncertainty shaping. This will result in less computation, more energy saving, and more swarms resiliency in the case of uncertainty facing. Figure 2 shows the proposed method. In the following parts pure knowledge and uncertainty shaping phases are presented.

Pure Knowledge Phase

Inferencing is a type of knowledge making. This type of knowledge is divided into explicit knowledge which is distinguishable by behavior and can be learned, and implicit knowledge which is hidden and before learning one should first have intention to gain that knowledge. In this paper in order to gain implicit knowledge, we use knowledge seeding concept. To do this, each nanomachine uses a FIS with two inputs and an output. Vascular Endothelial Growth Factor (VEGF) concentration $V(n)$, and oxygen concentration $O(n)$ are the inputs, and cancer state $CK(n)$ is the output. Figure 3. Shows their original membership functions. In vivo data in [17] are used for parameters' value, and the used equations for data gathering are presented in Table 1.

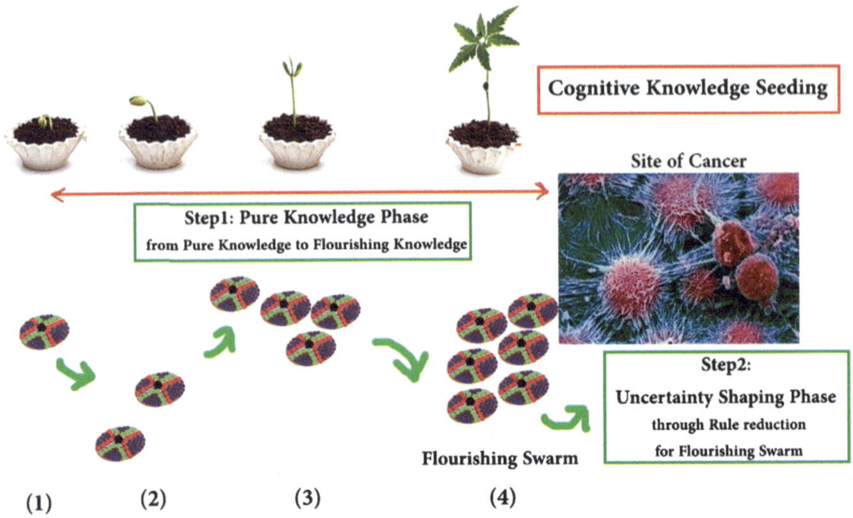

Fig. 2 The proposed method

Depending on the diversity of the information each swarm is shaped. The aim of each nanomachine is to gain more knowledge about the cancer state by join swarm with different information to complete their knowledge. Considering these states of knowledge, the ultimate state is defined as the flourishing state and the swarm which is belong to this group is called flourishing swarm.

In this state the swarm's internal behavior is to reach from the pure knowledge state to the flourishing knowledge state through use of knowledge seeding and the swarm's external behavior is to reach the cancer site.

Uncertainty Shaping by Rule Reduction

This phase starts as soon as the swarm reaches the flourishing knowledge state, and reaches the place close to the cancer site. In this phase the intention of swarm changes from cancer reaching to the cost reduction by starting the rule reduction process. This changes of intention of the swarm based on their current need is defined as a cognitive knowledge seeding based on uncertainty shaping. To do the uncertainty shaping, fuzzy rules are proper tools in this regard. Each precise rule is a key toward gaining more certainty. However, the growth in the number of rules accompanies by high computational cost, especially in the complex systems. Making a trade-off between *gaining certainty* and *uncertainty resiliency* can be a good solution toward this problem. To do this, we use a rule reduction technique which is called singular value decomposition (SVD) [30].

Considering the rule base in (4), the consequence matrix is decomposed to (5) using the SVD method,

$$Rule : If(A(x_1) and (B(x_2)) then Z \tag{4}$$

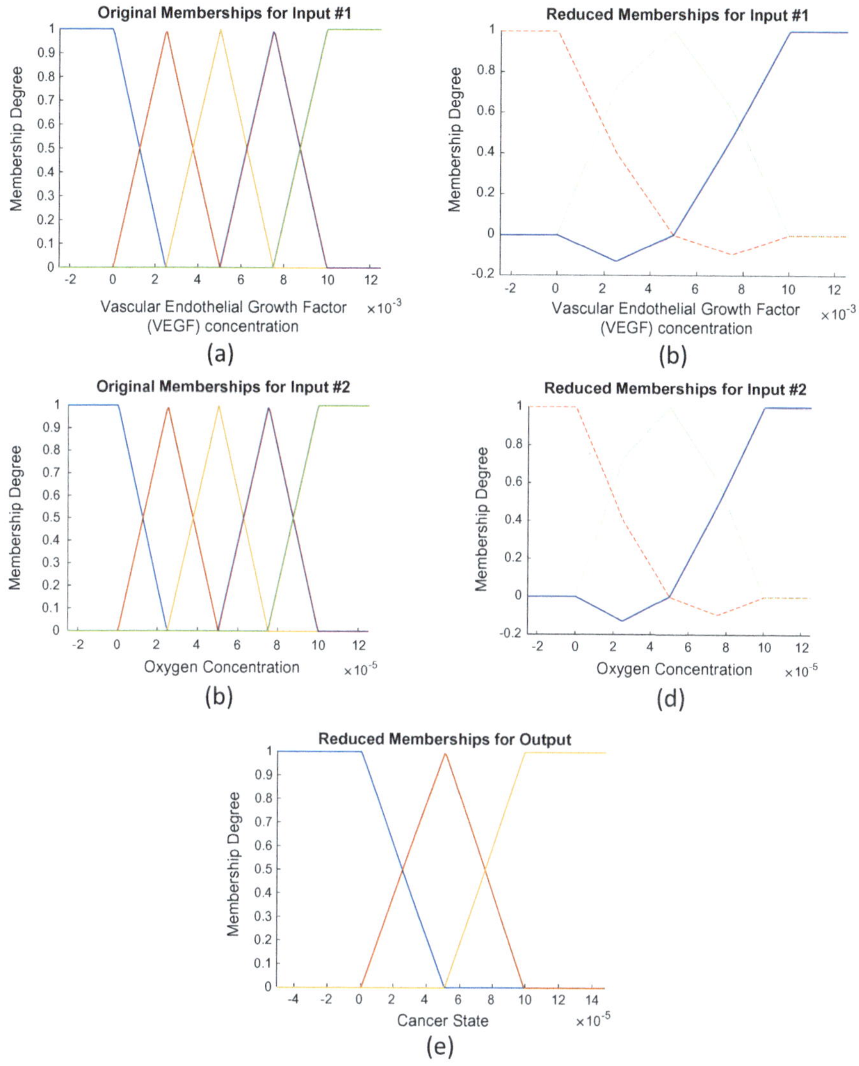

Fig. 3 Original membership functions, and reduced membership functions. **a** Original membership function of the first input, **b** reduced membership function of the first input, **c** original membership function of the second input, **d** reduced membership function of the second input, **e** reduced membership function of the output

where A, and B are fuzzy sets for input values x_1, x_2, and Z is the consequence matrix.

$$Z = U \Sigma V^T \tag{5}$$

Table 1 Data gathering's equations [14]

Equations name	Equations	#
Oxygen concentration	$\frac{\partial w}{\partial t}(x,t) = D_w \frac{\partial^2 w}{\partial x^2} + \alpha_w m(w_{max} - w) - \beta_w(n + h + m)w - \gamma_w w.$	(2)
VEGF concentration	$\frac{\partial g}{\partial t}(x,t) = D_g \frac{\partial^2 g}{\partial x^2} + \alpha_g h - \beta_g mg$	(3)

Then choosing the most important r singular values, Z is approximated as (6)

$$Z \approx \bar{z} = U_r \Sigma_r V_r^T \tag{6}$$

Two U_r and V_r matrices are conditioned to become A_r and B_r, and Σ_r is updated to become Z_r. This results in achieving the new approximation which is presented in (7).

$$Z \approx \bar{z} = A_r Z_r B_r^T \tag{7}$$

where the two columns of A_r, and B_r are the new fuzzy sets for input values, and Z_r is the reduced rule consequent matrix i.e. which is presented in (8).

$$NewRule: If (A_r(x_1) and (B_r(x_2)) then Z_r \tag{8}$$

4 Result Analyses

In this part, we simulate the cognitive knowledge seeding with collective uncertainty shaping method using 100 nanomachines. Figure 3a–e show the original membership functions and the reduced membership functions. As the results show, the number of rules is reduced from 25 rules to 9 rules using the proposed method.

Figure 4 shows the comparison of the hypoxic cell density and the endothelial cell density in both of the cognitive knowledge seeding using collective uncertainty shaping method and therapy model in [17]. The results show the merits of the proposed method in decreasing both of the hypoxic cell density, and endothelial cell density.

As the nanomachines' sources of energy are very limited, this reduction in the rules and its accompanied cost reduction can be useful and leads to a more reliable therapy. Moreover, the proposed method outperform the therapy model in [17] by performing as a type of smart ATMPs, and possessing all its benefits such as being self-coordinated and self-learning targeted therapy.

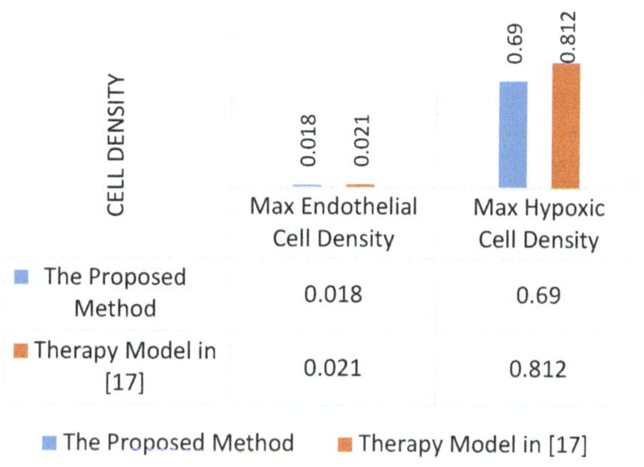

Fig. 4 Comparing results of the proposed method with therapy models in [17]

5 Conclusion

Advanced Therapy Medicinal Products (ATMPs) are part of medicines that use genes, tissues, cells, and drug as a living therapeutic agent. There is a great need to design proper carriers for doing targeted delivery in this field. In order to prepare such a device, knowledge and uncertainty handling are two important effective concepts that work hand in hand. Knowledge is an understanding of someone or something which is acquired through e.g., inferencing, and can be transferred or gained. Uncertainty handling includes different techniques in order to surf the trade-off between certainty and uncertainty.

Here, swarms of nano drug carriers are guided toward cancer site. Their first intention is to reach from the pure knowledge by doing simple fuzzy inferencing to the flourishing state by joining the closing swarm to the cancer site, and completing their knowledge through knowledge seeding process. Place of cancer is presented by the Vascular Endothelial Growth Factor (VEGF) concentration which is high around the cancer site based on the in vivo data. When the swarm is in the flourishing state, the intention of swarm changes from cancer site reaching task to the cost reduction task by doing fuzzy rule reduction. This changes of intention based on the current need is defined as a cognitive knowledge seeding based on uncertainty shaping. To do this, we do the rule reduction using a singular value decomposition (SVD) in a collective manner. Results show the remedy of the method in having resilience rule reduction, less computational cost, and more energy saving comparing the state of the art uncertainty handling techniques, and better therapeutic result comparing the mathematical therapy model.

Acknowledgements The authors would like to express their gratitude to Iran National Science Foundation (INSF) for supporting the research under the Grant number 96016045.

References

1. F. Gao, X. Hu, S.E. Li, K. Li, Q. Sun, Distributed adaptive sliding mode control of vehicular platoon with uncertain interaction topology. IEEE Trans. Industr. Electron. **65**(8), 6352–6361 (2018)
2. S. Zhang, S. Li. Chen, Z. Zhang, adaptive projection neural network for kinematic control of redundant manipulators with unknown physical parameters. IEEE Trans. Ind. Electron. **65**(6), 4909–4920 (2018)
3. S.K. Pal, D.B. Chakraborty, Granular flow graph, adaptive rule generation and tracking. IEEE Trans. Cybern. **47**(12), 4096–4107 (2017)
4. J. Xie, Y. Wan, K. Mills, J.J. Filliben, F.L. Lewis, A scalable sampling method to high-dimensional uncertainties for optimal and reinforcement learning-based controls. IEEE Control Syst. Lett. **1**(1), 98–103 (2017)
5. I. Yadav, H.G. Tanner, Controlled mobile radiation detection under stochastic uncertainty. IEEE Control Syst. Lett. **1**(1), 194–199 (2017)
6. T. Zhang, M. Xia, Y. Yi, Q. Shen, Adaptive neural dynamic surface control of pure-feedback nonlinear systems with full state constraints and dynamic uncertainties. IEEE Trans. Syst. Man Cybern. Syst. **47**(8), 2378–2387 (2017)
7. Y. Huang, Y. Jia, Fixed-time consensus tracking control for second-order multi-agent systems with bounded input uncertainties via NFFTSM. IET Control Theory Appl. **11**(16), 2900–2909 (2017)
8. S. Jafari Fesharaki, M. Kamali, F. Sheikholeslam, Adaptive tube-based model predictive control for linear systems with parametric uncertainty. IET Control Theory Appl. **11**(17), 2947–2953 (2017)
9. P.A. Herman, G. Prasad, T.M. McGinnity, Designing an interval type-2 fuzzy logic system for handling uncertainty effects in brain-computer interface classification of motor imagery induced EEG patterns. IEEE Trans. Fuzzy Syst. **25**(1), 29–42 (2017)
10. N. Hansen, A.S.P. Niederberger, L. Guzzella, P. Koumoutsakos, A method for handling uncertainty in evolutionary optimization with an application to feedback control of combustion. IEEE Trans. Evol. Comput. **13**(1), 180–197 (2009)
11. D. Cayrac, D. Dubois, H. Prade, Handling uncertainty with possibility theory and fuzzy sets in a satellite fault diagnosis application. IEEE Trans. Fuzzy Syst. **4**(3), 251–269 (1996)
12. T. Taniguchi, K. Tanaka, H. Ohtake, H.O. Wang, Model construction, rule reduction, and robust compensation for generalized form of Takagi-Sugeno fuzzy systems. IEEE Trans. Fuzzy Syst. **9**(4), 525–538 (2001)
13. N. Rady Raz, M. R. Akbarzadeh-T, Cooperation tuning in MANETs: a fuzzy approach Fuzzy behaviors of node in the presence of conflict, in *2013 13th Iranian Conference on Fuzzy Systems (IFSC)* (2013), Qazvin, pp. 1–6
14. N.R. Raz, M.R. Akbarzadeh-T, M. Tafaghodi, Bioinspired nanonetworks for targeted cancer drug delivery. IEEE Trans. Nano Biosci. **14**(8), 894–906 (2015)
15. N.R. Raz, M.R. Akbarzadeh-T, Fuzzy-CA model for an in-silico cancer cell line: A journey from simple cellular pattern to an emergent complex behavior, in *2016 IEEE International Conference on Fuzzy Systems* (*FUZZ-IEEE*), (Vancouver, BC, 2016), pp. 730–737
16. N.R. Raz, M.R. Akbarzadeh-T, Swarm fuzzy-reinforcement coordination using bloom's taxonomy of the cognitive domain, in *2017 Joint 17th World Congress of International Fuzzy Systems Association and 9th International Conference on Soft Computing and Intelligent Systems (IFSA-SCIS)*, Otsu (2017), pp. 1–5
17. P. Hinow, P. Gerlee, L. McCawley, M. Ciabanu, S. Wang, J. Graham, A spatial model of tumor-host interaction: application of chemotherapy. Math. Biosci. Eng. **6**(3), 521–542 (2009)
18. P. Sampara, R.R. Banala, S. Kumar Vemuri, G. Reddy, Understanding the molecular biology of intervertebral disc degeneration and potential gene therapy strategies for regeneration: a review. Gene Therapy (2018)
19. A. Khademhosseini, R. Langer, A decade of progress in tissue engineering. Nat. Protoc. **11**, 1775–1781 (2016)

20. A. Rodzinski, R. Guduru, P. Liang, A. Hadjikhani, S. Khizroev, Targeted and controlled anticancer drug delivery and release with magnetoelectric nanoparticles. Sci. Rep. **6**(20867) (2016)
21. C.K. Schneider, P. Celis, Challenges with advanced therapy medicinal products and how to meet them. Nat. Rev. Drug Discov. **9**, 195–201 (2010)
22. S. de Wilde, D G.M. Coppens, J. Hoekman, M.L. de Bruin, H.G.M. Leufkens, H. Guchelaar, P. Meij, EU decision-making for marketing authorization of advanced therapy medicinal products: a case study. Drug Discov. Today (2018)
23. https://www.genecopoeia.com/product/crispr-cas9/
24. https://www.tuni.fi/en/research/ips-cells
25. E. Caló, V.V. Khutoryanskiy, Biomedical applications of hydrogels: A review of patents and commercial products. Eur. Polymer J. **65**, 252–267 (2015)
26. http://www.xarxbio.com/pro/proc-383.html
27. S. Petrov, Dynamics properties of knowledge acquisition. Cogn. Syst. Res. **47**, 12–15 (2018)
28. L.A. Zadeh, Fuzzy sets. Inf. Control **8**(3), 338–353 (1965)
29. T. Ross, *Fuzzy Logic with Engineering Applications* (Wiley, 2004)
30. Y. Yam, P. Baranyi, C. Yang, Reduction of fuzzy rule base via singular value decomposition. IEEE Trans. Fuzzy Syst. **7**(2), 120–132 (1999)

Data Processing, Uncertainty and Type-2

Interval Type-2 Fuzzy Dynamic Parameter Adaptation in Bee Colony Optimization for Autonomous Mobile Robot Navigation

Oscar Castillo

Abstract This paper describes a comparative study of Interval Type-2 Membership Functions applied for dynamic parameter adaptation in Bee Colony Optimization Algorithm for the controlling of the trajectory in an autonomous mobile robot. The main goal of the work is comparing the impact of the results when different Interval Type-2 Membership Functions are used in this study case to adapt some parameters of bee colony optimization (BCO) algorithm. Two trajectories are optimized to test the methods. Perturbation is added in the model, and a comparative study is performed.

Keywords Type-2 membership functions · Fuzzy controller · Bee colony optimization · Interval type-2 fuzzy logic system

1 Introduction

In main goal in this work consists in develop the impact that to present the used of different types of Interval Type-2 membership functions used for dynamic parameter adaptation in BCO, the used of adjustment dynamic of parameters is presented in [2]; this work was a generalization from a system used in [1]. So the proposed of this work is the analysis with experiments the difference when different types of Interval Type-2 membership functions are used.

Zadeh was the first to mention the initial idea in the fuzzy in 1965 [28] allows to model problems using linguistic information that is related with the problem, and each linguistic label is represented by MFs. In problems with major complexity, the interval type-2 fuzzy logic systems (IT2FLS) are better to evaluate, because; this fuzzy sets present more precise when levels of noise or disturbances are aggregated in non-linear plants [18]. In this work on based in the comparing the different type-2 membership functions to find α and β parameters which represents the exploitation and exploration, respectively in the BCO algorithm. Four interval type-2 MFs are presented which are: Generalized Bell, Gaussian, Triangular, and Trapezoidal.

O. Castillo (✉)
Division of Graduate Studies, Tijuana Institute of Technology, Tijuana, Mexico
e-mail: ocastillo@tectijuana.mx

S. N. Shahbazova et al. (eds.), *Recent Developments and the New Direction in Soft-Computing Foundations and Applications*, Studies in Fuzziness and Soft Computing 393, https://doi.org/10.1007/978-3-030-47124-8_5

The BCO algorithm is a metaheuristic based on the smart behavior of the bees to locate food sources [1, 7, 27]. Various works have been presented that demonstrate the approach of the BCO which is a methodology for solving problems of optimization, just to mention some in [2, 3, 7, 25, 27].

This paper is based on a comparative study to dynamically adapt parameters in the BCO algorithm using four type-2 of MFs applied the optimization technique of intelligent and collaborative behavior that the bees have in the problem of controlling the trajectory in an autonomous mobile robot.

The rest of this paper is divided as follows. Section 2 shows an overview of Related Works important for this work. Section 3 shows a general description of the interval type-2 fuzzy logic systems. Section 4 presents the problem statement. Section 5 describes the bio-inspired algorithm used. Various simulation results are presented in Sect. 6. Finally, Sect. 7 outlines the conclusions and future works.

2 Related Works

Be sure to choose the type of MF that shows the best results in the problem solving can only be to found through trial and error, this is why various researchers have realized comparative studies with the different MFs.

Works where the comparative of the different types of the MFs is an issue of interest in the community of the fuzzy logic. In this section, a list of the several works with the previously mentioned objective is presented in [1, 9, 11, 21, 22].

In the area of stabilization in fuzzy control particularly in the trajectory of the autonomous mobile robot there exists works like that of Castillo [5], in which an optimization of MFs for type-1 and type 2 fuzzy controllers of an autonomous mobile robot using PSO is designed; Hagras [12], in which A hierarchical type-2 fuzzy logic control architecture for autonomous mobile robots is presented; Kanayama et al. [13], in which a stable tracking control method for an autonomous mobile robot. In Robotics and Automation is performed; Larin [14] a Stabilization of a wheeled robotic vehicle is presented; Park and Hashimoto [23] an Autonomous mobile robot navigation using passive RFID in indoor environment is designed; and Pervozvanski and Freidovich [24] a Robust stabilization of robotic manipulators by PID controllers is performed. The main goal of this paper is to help us to determine the type of membership function to show a difference in the results in terms of performance. In this work, four interval type-2 MFs are used and compared, which are; Triangular, Generalized Bell, Gaussian, and Trapezoidal.

3 General Description of the Interval Type-2 Fuzzy Logic Systems

An interval type-2 fuzzy set is characterized by a fuzzy MF, i.e., the membership grade for each element of this set is a fuzzy set in [0, 1], unlike a type-1 set where the membership grade is a crisp number in [0, 1] [28, 29]. The applications of an interval type-2 fuzzy sets can be in situations where there is uncertainty, i.e. when in fuzzy controller is added various levels of perturbation [19, 20]. Figure 1 illustrates the general description of IT2FLS, and the Eq. (1) represents the expression mathematical [20, 26].

$$\tilde{A} = \left\{\left((x, u), \mu_{\tilde{A}}(x, u)\right) | \forall x \in X, \forall u \in J_x \subseteq J_x[0, 1]\right\} \quad (1)$$

Various types of MFs for Interval type-2 FLS exist, and for this work four are used, Triangular, Generalized Bell, Trapezoidal and Gaussian. Figure 2 illustrates the four FMs.

3.1 *Fuzzy Logic Controllers*

Mamdani was the first that used the theory of Zadeh in 1973 to represent the fuzzy controllers [15, 17]. The representation of a generic fuzzy controller is illustrates in Fig. 3.

Many researchers have shown an interest in the study and stabilization of fuzzy controllers, just to mention a few in [1–5, 12–15, 17, 23].

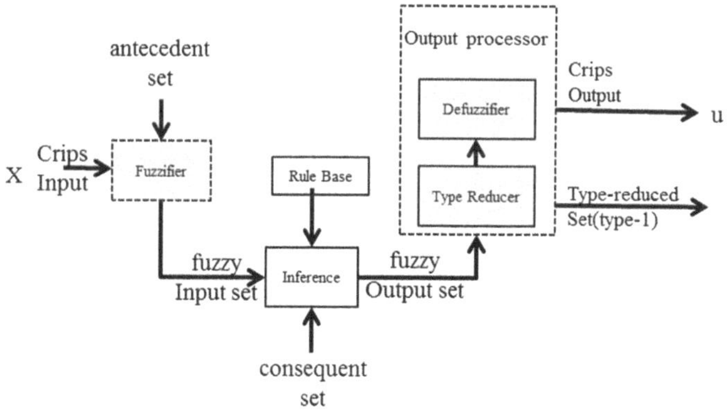

Fig. 1 Architecture of interval type-2 fuzzy sets

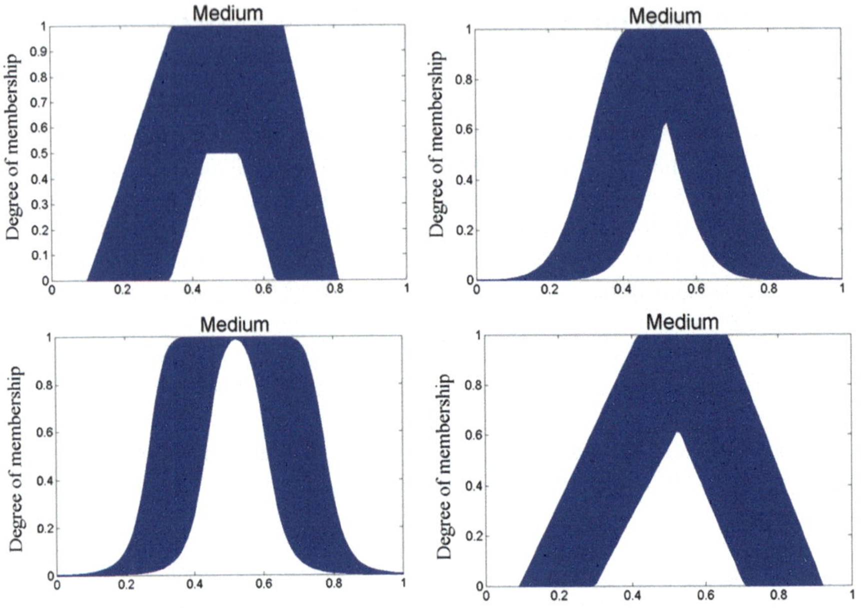

Fig. 2 Different types of type-2 MFs

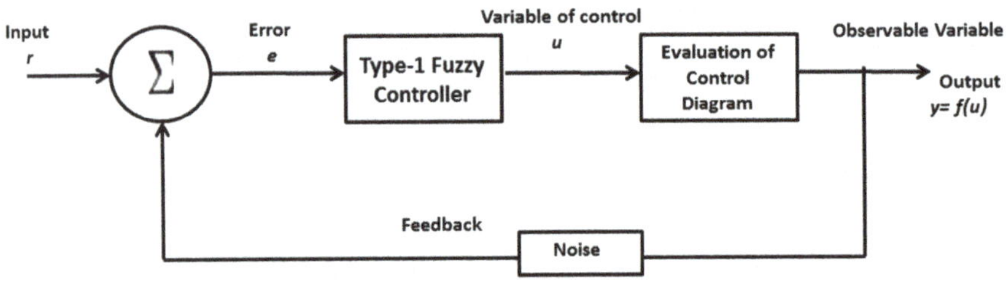

Fig. 3 Graphical representation of fuzzy controller

4 Problem Statement

In the robotics area, one of the problems particularly studied is the control and manipulation of the trajectories of autonomous mobile robots. This paper focuses in a comparative study of Interval Type-2 MFs applied to find the values that affect the performance of the BCO algorithm which are: α and β values is shown for the problem statement the controlling the trajectories in an autonomous mobile robot. The model that is used in simulation is presented in following subsections.

4.1 Definition of the Problem

The model used is composed by an unicycle mobile robot [1, 5, 6], which contains two driving wheels located on the same axis and a front free wheel Fig. 4 illustrates the graphic idea of the robot model.

The robot model assumes that the motion of the free wheel can be ignored in its dynamics, as shown in Eqs. (2) and (3).

$$M(q)\dot{v} + C(q, \dot{q})v + Dv = \tau + P(t) \tag{2}$$

where,

$q = (x, y, \theta)^T$ is the vector of the configuration coordinates,

$\upsilon = (v, w)^T$ is the vector of velocities,

$\tau = (\tau_1, \tau_2)$ is the vector of torques applied to the wheels of the robot where τ_1 and τ_2 denote the torques of the right and left wheel,

$P \in R^2$ is the uniformly bounded disturbance vector,

$M(q) \in R^{2\times 2}$ is the positive-definite inertia matrix,

$C(q, \dot{q})\vartheta$ is the vector of centripetal and Coriolis forces, and

$D \in R^{2\times 2}$ is a diagonal positive-definite damping matrix.

The kinematic system is represented by Eq. (3).

$$\dot{q} = \underbrace{\begin{bmatrix} \cos\theta & 0 \\ \sin\theta & 0 \\ 0 & 1 \end{bmatrix}}_{J(q)} \underbrace{\begin{bmatrix} v \\ w \end{bmatrix}}_{\upsilon} \tag{3}$$

where,

(x, y) is the position in the $\mathrm{X} - \mathrm{Y}$ (world) reference frame,

θ is the angle between the heading direction and the x-axis,

v and w are the linear and angular velocities.

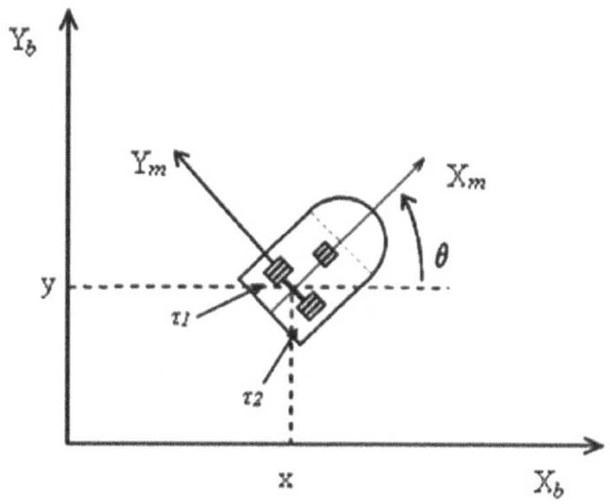

Fig. 4 Mobile robot model

Also, Eq. (4) indicates the non-holonomic constraint which this system has, which corresponds to a no-slip wheel condition preventing the robot from moving sideways.

$$\dot{y}\cos\theta - \dot{x}\sin\theta = 0 \tag{4}$$

4.2 Characteristic of the Fuzzy Logic Controller

The controlling of the two wheels in the mobile robot is the main problem, being thus, it will be possible to have a better control in the trajectory of the autonomous mobile robot. The design of the fuzzy logic system is illustrated in Fig. 5.

The Type-1 FLS is designed with 2 inputs, the first is called **ev** (angular velocity) with three MFs with trapezoidal in the extremes and triangular MF in the middle, the same of equal distribution the second input is called **ew** (linear velocity). For the outputs; the first is called **T1** (Torque 1) which represents the left wheel and the second is called **T2** (Torque 2) for the right wheel, both outputs are triangular MFs with linguistic value of N, Z, and P, respectively.

The initial rules for controlling the moves in each wheel are presented in the following Table 1, and the model that represents the simulation of the trajectories in autonomous mobile robots is shown in Fig. 6.

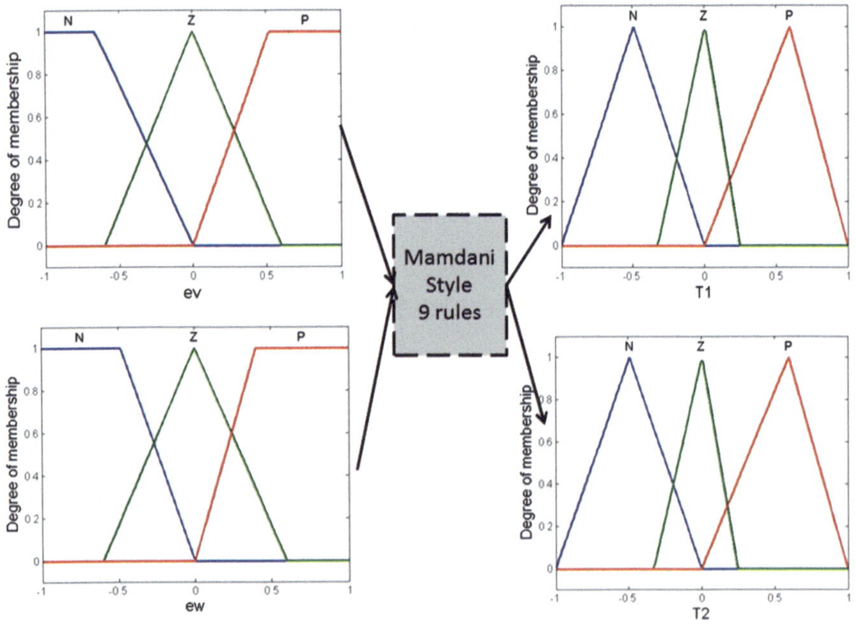

Fig. 5 Design of the type-1 fuzzy logic system

Table 1 Fuzzy rules for the study case

# Rules	Inputs		Outputs	
	ev	ew	T1	T2
1	N	N	N	N
2	N	Z	N	Z
3	N	P	N	P
4	Z	N	Z	N
5	Z	Z	Z	Z
6	Z	P	Z	P
7	P	N	P	N
8	P	Z	P	Z
9	P	P	P	P

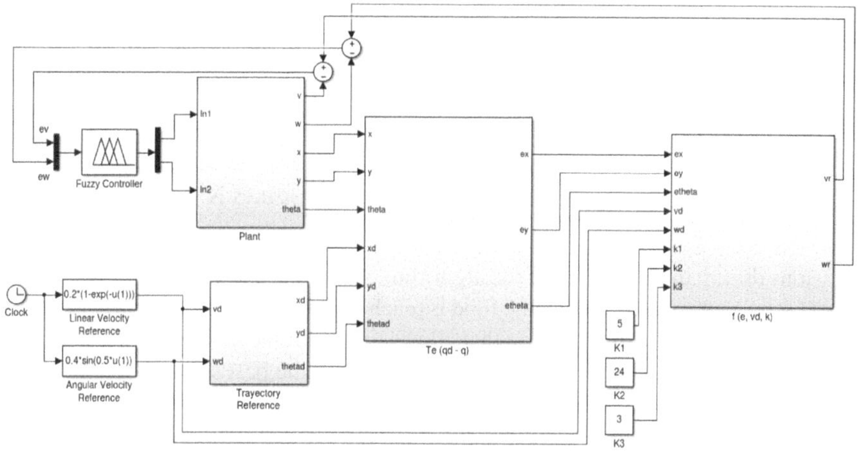

Fig. 6 Representation in simulation of the model that controls the trajectories

Two trajectories were optimized for each interval type-2 MFs. Figure 7a shows the first trajectory (normal) and the Fig. 7b (inverted trajectory) the second trajectory.

The green line indicates the trajectory that the autonomous mobile robot should follow and the red line indicates the trajectory that the proposed methods have found in the simulations.

5 Fuzzy Bee Colony Optimization Algorithm

This algorithm is based on the collaborative behavior that bees have in nature. In this algorithm a bee generates one solution to the problem. Each bee is making its own

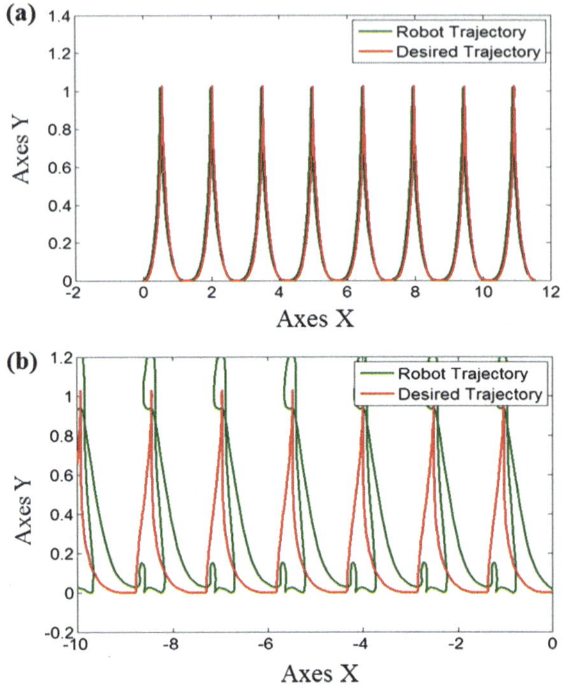

Fig. 7 Tracking trajectories by the autonomous mobile robot

decisions during the execution of the algorithm, such as; the main function consist to identify the nectar source, when the food is reached, the bee realized a dance (waggle dance), which; is a quality function that allows to communicate to the other bees that a good solution is found, this process is repeated until the best solution found by the best bee is reached [16, 27]. When the BCO algorithm start, all the bees are in the hive. Figure 8 shows the general idea for changing the important parameters for the BCO algorithm using an interval Type-2 Fuzzy System that influence the quality of the solution.

Equations (5)–(8) represent the dynamics of the BCO algorithm; Eq. (5) shows the probability that a bee has found the best solution [1, 2, 7, 10, 27]:

$$Pij, n = \frac{[\rho ij, n]^{\alpha}.[\frac{1}{dij}]^{\beta}}{\sum\limits_{j \in Ai,n} [\rho ij, n]^{\alpha}.[\frac{1}{dij}]^{\beta}} \tag{5}$$

$$D_i = K.\frac{Pf_i}{Pf_{colony}} \tag{6}$$

$$Pf_i = \frac{1}{L_I}, \; L_i = \text{Tour Length} \tag{7}$$

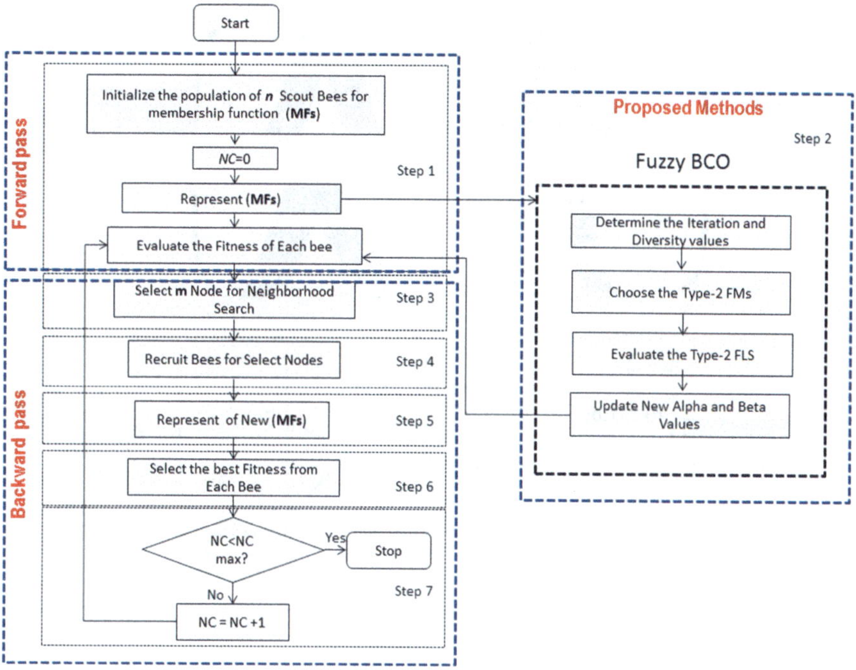

Fig. 8 Flowchart of the general idea in the evaluation for each type-2 MFs in the BCO algorithm

$$Pf_{colony} = \frac{1}{N_{Bee}} \sum_{i=1}^{N_{Bee}} Pf_i \tag{8}$$

The state transition probability, $P_{ij,n}$, gives the likelihood to move from node i to node j after n transitions. Equation (7) represents the probability that a bee to find a good selection, and β is the probability to visit the following position in the space search (β value is a parameter that affects the exploration process in the algorithm). Note that the ρ_{ij} is inversely proportional to the distance. d_{ij} representing the distance of node i until node j. Finally ∝ is a binary variable that is used to find better solutions in the algorithm. The ∝ value affects the process of exploitation in the algorithm.

All moves that the bees realized in BCO algorithms is called the waggle dances, this moves is represented by the intensity with which a bee finds a good possible solution [4]. Bee whit intensity higher indicates that the best solution in this time is found [8]. Equation (8) represents the fact that a waggle dance will last for a certain duration, where K denotes the waggle dance scaling factor, Pf_i denotes the profitability scores of bee i as defined in Eq. (9) and Pf_{colony} denotes the bee colony's average profitability as in Eq. (10), and is updated after each bee completes its tour. Thus, a higher quantity of nectar will be collected if a bee travels along a shorter route. Therefore, Pf_i is defined to be inversely proportional to the tour length.

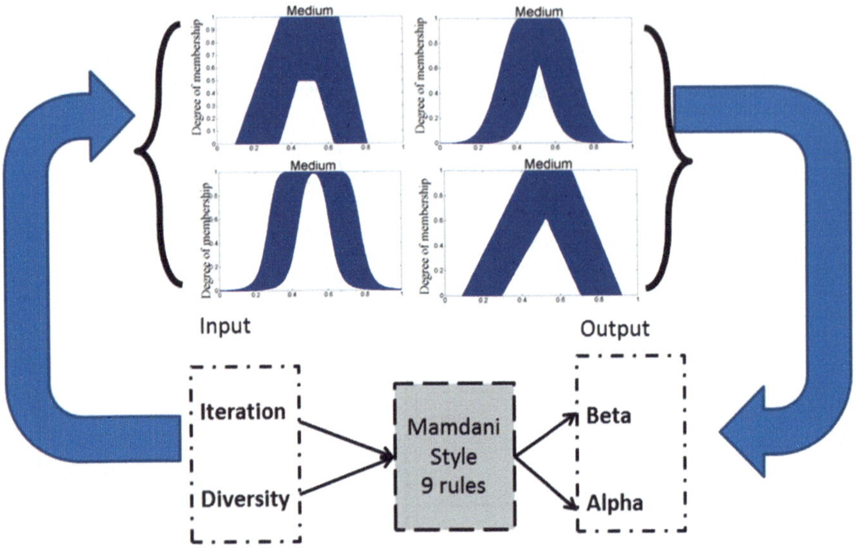

Fig. 9 General idea for representation of four IT2FLS in dynamic adaptation of parameters in the BCO algorithm

The methodology for the adaptation dynamic of parameters in BCO algorithm is shown in [2], in this paper; four type-2 membership functions are used and compared. Figure 9 illustrates a general idea of the fuzzy BCO algorithm.

Inputs: Iteration has a range from [0, 1], and is expressed by Eq. (11). This variable has three MFs with linguist value of Low, Medium and High.

$$\text{Iteration} = \frac{\text{Current Iteration}}{\text{Maximum of Iterations}} \tag{9}$$

Diversity represents the dispersion of the bees in the space of search, with the range of [0, 1] (normalized), and is expressed by Eq. (12), this variable has three MFs with linguistic values of Low, Medium and High.

$$Diversity(S(t)) = \frac{1}{n_s}\sum_{i=1}^{n_x}\sqrt{X_{ij}(t) - \bar{X}_j(t)^2} \tag{10}$$

Outputs: Beta indicates the parameter β (see Eq. 7) of BCO algorithm, with a range of [2, 5], in previous works has been demonstrated that those range indicates the optimal possible values for β according [1, 2], this variable affect in the exploration process, and this variable is granulated into five MFs with linguist values of Low, Medium Low, Medium, Medium High and High.

Alpha; this variable indicates the parameter α (see Eq. 7) of BCO algorithm, is a weight that affects in the exploitation process in the algorithm, this a range of [0,

Table 2 Rules of the fuzzy BCO algorithm

	Input		Output	
# Rules	Iteration	Diversity	Beta	Alpha
1	Low	Low	High	Low
2	Low	Medium	Medium high	Medium
3	Low	High	Medium high	Medium low
4	Medium	Low	Medium high	Medium low
5	Medium	Medium	Medium	Medium
6	Medium	High	Medium low	Medium high
7	High	Low	Medium	High
8	High	Medium	Medium low	Medium high
9	High	High	Low	High

1], with variable has five MFs with linguist values of Low, Medium Low, Medium, Medium High and High.

The design of fuzzy rule is shown in Table 2. Taking the initial idea that when α is higher that β, this improves the exploitation of the best area the space of search found so far, and when the α is lower that β, this improves the exploration of space search [2], and the fitness function is represented by Eq. (12).

$$RMSE = \sqrt{\frac{1}{N}\sum\nolimits_{i=1}^{N}\left(\overline{Y}_i - Y_i\right)^2} \tag{11}$$

6 Simulation Results

All the design of IT2FLS developed are integrated in a BCO algorithm to dynamically adapt α and β parameters, with 30 experiments for each type membership functions the averages are shown, and each system is tested with model which controls the study case implemented.The initial parameter of the BCO algorithm is shown in Table 3.

Table 3 Settings parameters for the BCO algorithm

Parameter	Values
Population	50
# of bee employed	30
Iterations	25
Alpha	Dynamic
Beta	Dynamic

Tables 4 and 5 show the results to found by BCO algorithm with the trajectory illustrated in Fig. 7a (normal), and (reverse), respectively.

Table 4 shows that when trapezoidal MFs is used with the normal trajectory the best results are found, and for the reverse trajectory the best type of MFs is the generalized Bell with the average RMSE of **1.594** and the best of **0.077**.

The convergence for the best results shown in Table 5 is shown in Fig. 10, the behavior of the trajectory using type-2 generalized bell MFs is shown in Figs. 11 and

Table 4 Simulation results with the first trajectory

Performance index	Output (Trajectory)			
	Triangular MFs	Gaussian MFs	Generalized Bell MFs	Trapezoidal MFs
Average	1.026	0.794	0.794	**0.695**
Standard deviation (σ)	0.312	0.145	0.145	0.198
Best	0.081	0.152	0.152	**0.078**
Worst	2.928	3.028	3.028	1.905

Table 5 Simulation results with the second trajectory

Performance index	Output (Trajectory)			
	Triangular MFs	Gaussian MFs	Generalized Bell MFs	Trapezoidal MFs
Average	1.735	1.998	**1.594**	2.001
Standard deviation (σ)	0.247	0.412	0.388	0.409
Best	0.160	0.185	**0.077**	0.197
Worst	2.830	2.129	2.134	2.129

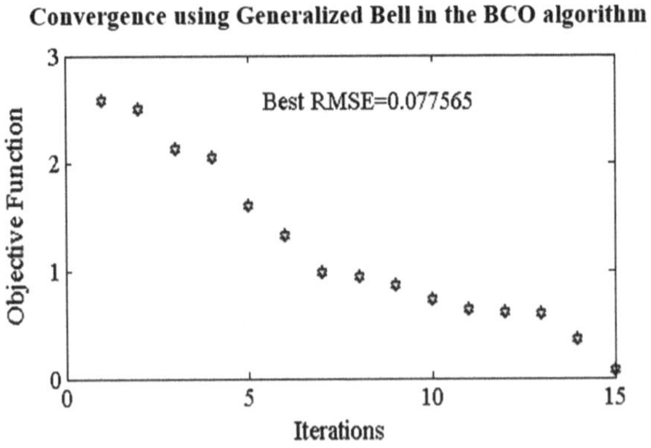

Fig. 10 Best convergence of the proposed methods

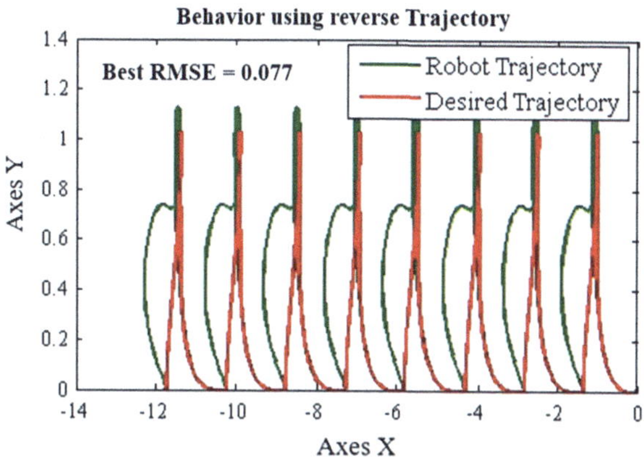

Fig. 11 Best reverse trajectory using generalized bell MFs

12 illustrate the best experiment using trapezoidal MFs without perturbation in the model.

In the experimentation the two trajectories was added level of noise; a pulse generator noise with phase delay, amplitude, pulse width (%), and period are set

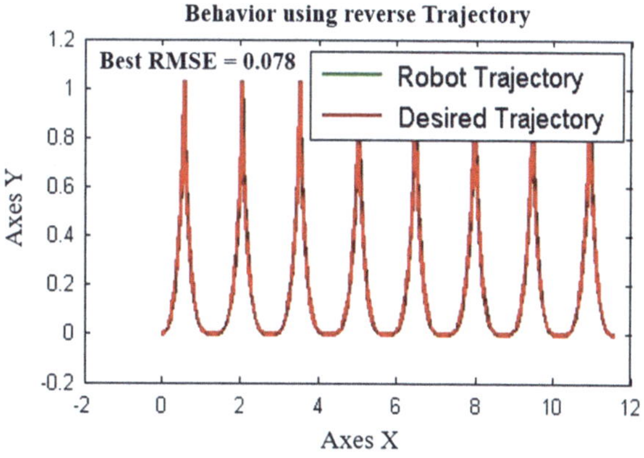

Fig. 12 Best normal trajectory using trapezoidal bell MFs

Table 6 Simulation results with the first trajectory applying perturbation

Performance index	Output (Trajectory)			
	Triangular MFs	Gaussian MFs	Generalized Bell MFs	Trapezoidal MFs
Average	1.134	**0.813**	1.257	0.894
Standard deviation (σ)	0.187	0.128	0.551	**0.096**
Best	0.467	0.252	0.456	0.329
Worst	1.513	1.709	2.632	1.278

Table 7 Simulation results with the second trajectory applying perturbation

Performance index	Output (Trajectory)			
	Triangular MFs	Gaussian MFs	Generalized Bell MFs	Trapezoidal MFs
Average	1.511	**1.475**	1.557	1.528
Standard deviation (σ)	0.241	0.206	0.332	0.234
Best	0.448	0.514	0.408	0.502
Worst	2.294	3.273	2.355	2.137

to 0, 1, 1.5 and 10 (seconds). Tables 6 and 7 illustrate the simulation error when perturbation is added in the model for each Type-2 MFs.

The perturbation allow that Type-2 FLS to found lowers errors, this is because it handles more precise levels of uncertainty; i.e.; using the normal trajectory in Table 4 the standard deviation (σ) is the **0.198** compared to Table 6 using perturbation the standard deviation (σ) is of **0.096**, this metric indicates that the all errors to found for the method are similar. In average when perturbation is added the errors are lower.

The α and β values to found by the best evaluation in Table 6 with Trapezoidal MFs is shown in Fig. 13.

Figure 13 shows that the range of β values representing by blue line is about of [3.2, 4.8], and the range of α values representing by pink line is about of [0.1, 0.8].

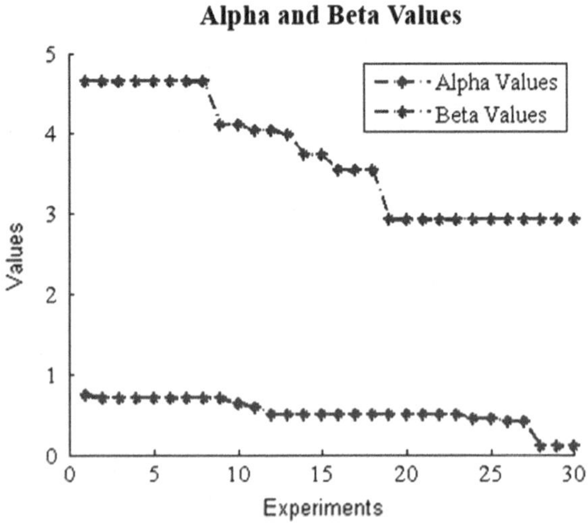

Fig. 13 α and β values with trapezoidal MFs in the dynamic adjustment

7 Comparative Results

All results mentioned in the previous section indicate that trapezoidal and generalized Bell membership functions without perturbation and Gaussian MFs is better when perturbation is added in the model, thus; allow keep better the stability in the model with the study case analyzed in this work.

A total of 30 experiments are realized. Figure 14 illustrates a comparison with the average using the normal trajectory without perturbation in the model, i.e., Table 4 the average RMSE was of **0.695** using Type-2 Trapezoidal MFs using the normal trajectory compared to Type-2 Generalized Bell MFs was of **1.594** using the reverse trajectory.

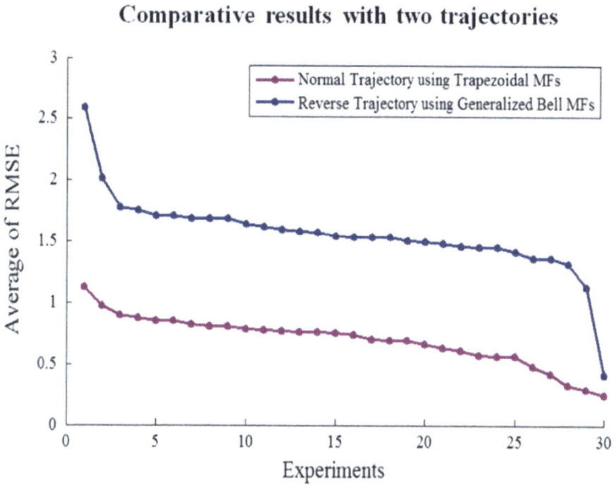

Fig. 14 Comparative results with two trajectories

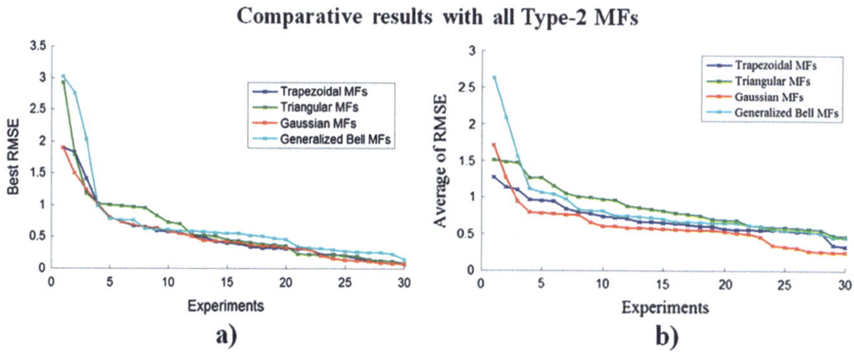

Fig. 15 Comparative results with all type-2 MFs applying perturbation in the model

Table 8 Comparative with standard deviations

Type of trajectory	Output (Trajectory)				
	Triangular MFs	Gaussian MFs	Generalized Bell MFs	Trapezoidal MFs	Perturbation
Normal	0.187	0.128	0.551	**0.096**	Yes
Reverse	0.241	0.206	0.332	0.234	Yes
Normal	0.312	0.145	0.145	**0.198**	No
Reverse	0.247	0.411	0.388	0.408	No

To observe the behavior by each Type-2 MFs Fig. 15a illustrates the best RMSE for each experiment without perturbation and Fig. 15b illustrates with perturbation in the model using the normal trajectory.

Other hand to analyze the effects that the Type of MFs has is with the metric of standard deviation, Table 8 show the comparative of each MFs with perturbation in the model for both trajectories.

Table 8 can to observe that when the perturbation is added in the model better standard deviation are found, i.e., when trapezoidal MFs is used without perturbation the average of the σ is of 0.198 compared to when the perturbation is added in the model the average is of 0.096. This analysis allows to demonstrate the effect that the Trapezoidal MFs has to determine the dynamic values for solve the problem of to control the trajectory in an autonomous mobile robot.

8 Conclusions

With the experiments analyzed, the main conclusion is that the proposed method is an excellent technique of optimization applied in the fuzzy controllers and when several type-2 MFs are used in the proposed method for the study case the stabilization on the model is found. In this work, when perturbation is added in the model it was observed that Gaussian MFs presents lower errors in the model, and on the other hand, when is no used perturbation in the model trapezoidal and generalized bell MFs presents the better results. Another important conclusion is found the optimal range for α and β parameters by each Type-2 MFs that allow minimizing the error in the study case analyzed. Some future works is to realize more comparisons with several generalized type-2 membership functions to find the optimal values in the important parameters that affect in the performance of the bio-inspired algorithm with the objective of can handle better the uncertainty that is a characteristic that Generalized Type-2 Fuzzy Logic System presents, others future work is the simulation of the study case with dynamic trajectories, this will allow the α and β values obtained to be more generalized for any trajectory to be analyzed.

References

1. L. Amador-Angulo, O. Castillo, Comparative analysis of designing differents types of membership functions using bee colony optimization in the stabilization of fuzzy controllers, in *Nature-Inspired Design of Hybrid Intelligent Systems* (Springer International Publishing, 2017), pp. 551–571
2. L. Amador-Angulo, O. Castillo, A new fuzzy bee colony optimization with dynamic adaptation of parameters using interval type-2 fuzzy logic for tuning fuzzy controllers. Soft Comput. 1–24
3. U.S. Banu, G. Uma, Tuning of PID controller using internal model control with the filter constant optimized using bee colony optimization technique, in *International Conference on Swarm, Evolutionary, and Memetic Computing* (Springer, Berlin, Heidelberg, 2010), pp. 648–655
4. J.C. Biesmeijer, T.D. Seeley, The use of waggle dance information by honey bees throughout their foranging careers. Behav. Eco. Sociobiol. **59**(1), 133–142 (2005)
5. O. Castillo, Optimization of membership functions for type-1 and type 2 fuzzy controllers of an autonomous mobile robot using PSO, in *Recent Advances on Hybrid Intelligent Systems* (Springer, Berlin, Heidelberg, 2013), pp. 97–104
6. O. Castillo, R. Martinez-Marroquin, P. Melin, P. Valdez, J. Soria, Comparative study of bio-inspired algorithms applied to the optimization of type-1 and type-2 fuzzy controllers for an autonomous mobile robot. Inf. Sci. **192**(1), 19–38 (2010)
7. B. Crawford, R. Soto, R. Cuesta, F. Paredes, Using the bee colony optimization method to solve the weighted set covering problem, in *International Conference on Human-Computer Interaction* (Springer International Publishing, 2014)
8. F.C. Dyler, The biology of the dance language. Ann. Rev. Entomol. **47**, 917–949 (2002)
9. F. Gaxiola, P. Melin, F. Valdez, Comparison of neural networks with different membership functions in the type-2 fuzzy weights, in *Intelligent Systems' 2014* (Springer International Publishing, 2015), pp. 707–713
10. A.S. Girsang, C.W. Tsai, C.S. Yang, A hybrid ant-bee colony optimization for solving traveling salesman problem with competitive agents, in *Mobile, Ubiquitous, and Intelligent Computing* (Springer, Berlin, Heidelberg, 2014), pp. 643–648
11. D. Hidalgo, P. Melin, O. Castillo, G. Licea, Comparative study of type-2 fuzzy inference system optimization based on the uncertainty of membership functions, in *Soft Computing for Recognition Based on Biometrics* (Springer, Berlin, Heidelberg, 2010), pp. 103–120
12. H.A. Hagras, A hierarchical type-2 fuzzy logic control architecture for autonomous mobile robots. IEEE Trans. Fuzzy Syst. **12**(4), 524–539 (2004)
13. Y. Kanayama, Y. Kimura, F. Miyazaki, T. Noguchi, A stable tracking control method for an autonomous mobile robot, in *Proceedings of 1990 IEEE International Conference on Robotics and Automation, 1990* (IEEE), pp. 384–389
14. V.B. Larin, Stabilization of a wheeled robotic vehicle. International Applied Mechanics **43**(7), 800–808 (2007)
15. C.C. Lee, Fuzzy logic in control systems: fuzzy logic controller. I. IEEE Trans. Syst. Man Cybern. **20**(2), 404–418 (1990)
16. J. Ning, B. Zhang, T. Liu, C. Zhang, An archive-based artificial bee colony optimization algorithm for multi-objective continuous optimization problem. Neural Comput. Appl. 1–11
17. E.H. Mamdani, Application of fuzzy algorithms for control of simple dynamic plant, in *Proceedings of the Institution of Electrical Engineers*, vol. 121, No. 12 (IET, 1974), pp. 1585–1588
18. P. Melin, A comparative study of type-2 fuzzy system optimization based on parameter uncertainty of membership functions, in *Modular Neural Networks and Type-2 Fuzzy Systems for Pattern Recognition* (Springer, Berlin, Heidelberg, 2012), pp. 145–161
19. J.M. Mendel, Advances in type-2 fuzzy sets and systems. Inf. Sci. **177**(1), 84–110 (2007)
20. J.M. Mendel, R.I. John, F. Liu, Interval type-2 fuzzy logic systems made simple. IEEE Trans. Fuzzy Syst. **14**(6), 808–821 (2006)

21. F. Olivas, F. Valdez, O. Castillo, A comparative study of membership functions for an interval type-2 fuzzy system used for dynamic parameter adaptation in particle swarm optimization, in *Recent Developments and New Direction in Soft-Computing Foundations and Applications* (Springer International Publishing, 2016), pp. 373–385
22. T. Ozen, J.M. Garibaldi, Effect of type-2 fuzzy membership function shape on modelling variation in human decision making, in *Proceedings of 2004 IEEE International Conference on Fuzzy Systems, 2004*, vol. 2 (IEEE), pp. 971–976
23. S. Park, S. Hashimoto, Autonomous mobile robot navigation using passive RFID in indoor environment. IEEE Trans. Industr. Electron. **56**(7), 2366–2373 (2009)
24. A.A. Pervozvanski, L.B. Freidovich, Robust stabilization of robotic manipulators by PID controllers. Dyn. Control **9**(3), 203–222 (1999)
25. A.P. Pertiwi, Globally evolved dynamic bee colony optimization, in *International Conference on Knowledge-Based and Intelligent Information and Engineering Systems* (Springer Berlin Heidelberg, 2011), pp. 52–61
26. F.C.H. Rhee, B.I. Choi, Interval type-2 fuzzy membership function generation methods for representing sample data, in *Advances in Type-2 Fuzzy Sets and Systems* (Springer New York, 2013), pp. 165–184
27. D. Teodorovic, M. Dell'Orco, Bee colony optimization—A cooperative learning approach to complex transportation problems. Adv. OR AI Methods Transp. 51–60
28. L. Zadeh, Fuzzy sets. Inf. Control **8**(338) (1965)
29. L.A. Zadeh, Fuzzy sets as a basis for a theory of possibility. Fuzzy Sets Syst. **1**(1), 3–28 (1978)

Type-2 Fuzzy Analysis Explains Ubiquity of Triangular and Trapezoid Membership Functions

Olga Kosheleva, Vladik Kreinovich, and Shahnaz Shahbazova

Abstract In principle, we can have many different membership functions. Interestingly, however, in many practical applications, triangular and trapezoidal membership functions are the most efficient ones. In this paper, we use fuzzy approach to explain this empirical phenomenon.

1 Introduction

Triangular and trapezoid membership functions are ubiquitous: why? One of the main ideas behind fuzzy logic is to represent an imprecise ("fuzzy") natural-language property P like "small" by its *membership function*, i.e., a function $\mu(x)$ that assigns, to each possible value x of the corresponding property, the degree $\mu(x) \in [0, 1]$ this value satisfies the property P (e.g., to what extent x is small); see, e.g., [1–3, 5, 6, 9].

According to this definition, we can have many different membership functions. However, in many applications of fuzzy techniques, the simplest piece-wise linear membership functions—e.g., triangular and trapezoid ones—works very well.

Why?

What we do in this paper. In this paper, we use fuzzy techniques to analyze this question. As a result of this analysis, we show that indeed triangular and trapezoid membership functions are the most reasonable ones. Thus, our analysis explains the ubiquity of triangular and trapezoid membership functions.

O. Kosheleva · V. Kreinovich (✉)
University of Texas at El Paso, 500 W. University, El Paso, TX 79968, USA
e-mail: vladik@utep.edu

O. Kosheleva
e-mail: olgak@utep.edu

S. Shahbazova
Azerbaijan Technical University, Baku, Azerbaijan
e-mail: shahbazova@gmail.com

S. N. Shahbazova et al. (eds.), *Recent Developments and the New Direction in Soft-Computing Foundations and Applications*, Studies in Fuzziness and Soft Computing 393, https://doi.org/10.1007/978-3-030-47124-8_6

2 Analysis of the Problem

How can we analyze the problem: need for a type-2 approach. Traditionally—e.g., in control applications—fuzzy logic is used to select a value of the corresponding quantity, e.g., the value of control u. To come up with such a value:

- first, we use the experts' rules to come up, for each possible control value u, with a degree $d(u)$ to which this control value is reasonable;
- then, we select one of the control values u—e.g., the one for which the degree of reasonableness is the largest:

$$d(u) \to \max_u .$$

In our problem, instead of selecting a single *value* u, we select the whole membership *function* $\mu(x)$. To use fuzzy techniques for selecting μ, we thus need to do the following:

- first, we need to use experts' rules to assign, to each possible membership function $\mu(x)$, a degree $d(\mu)$ to which this membership function is reasonable, and
- then, out of all possible members functions, we select the one which is the most reasonable, i.e., for which the degree of reasonableness $d(\mu)$ is the largest:

$$d(\mu) \to \max_\mu .$$

Let us follow this path.

Comment. Traditionally:

- situations in which we use fuzzy to reason about real values is known as *type-1* fuzzy; while
- situations in which we use fuzzy to reason about fuzzy is known as *type-2 fuzzy* approach; see, e.g., [3, 4].

From this viewpoint, what we plan to use is an example of the type-2 fuzzy approach.

Expert rules. First, we need to select expert rules. We consider the problem in its utmost generality, we want rules that will be applicable to all possible fuzzy properties. In this case, the only appropriate rule that comes to mind is the following natural natural-language rule:

- if x and x' are close,
- then $\mu(x)$ and $\mu(x')$ should be close.

This rule exemplifies the whole idea of fuzziness: instead of abruptly changing the degree of confidence from 0 to 1 as would happen if we consider crisp properties (like $x \geq 0$), we have a smooth transition from 0 to 1.

How can we formalize this expert rule? Since there are infinitely many possible values of x and x', the above rule consists of infinitely many implications—one implication for each pair (x, x'). Dealing with *infinitely many* rules is difficult. It is therefore desirable to try to limit ourselves to *finite* number of rules.

Such a limitation is indeed possible. Indeed, theoretically, we can consider all infinitely many possible values x. However, in practice, the values of any physical quantity are bounded: e.g., locations on the Earth are bounded by the Earth's diameter, speeds are limited by the speed of light, etc. Thus, it is reasonable to assume that all possible values x are within some interval $[\underline{x}, \overline{x}]$.

Second, we only know x and x' with a certain accuracy $\varepsilon > 0$. From this viewpoint, there is no need to consider all infinitely many values, it is sufficient to consider only values on the grid of width ε, i.e., values

$$x_0 = \underline{x}, x_1 = \underline{x} + \varepsilon, x_2 = \underline{x} + 2\varepsilon, \ldots, x_n = \underline{x} + n \cdot \varepsilon = \overline{x},$$

where we denoted $n \stackrel{\text{def}}{=} \dfrac{\overline{x} - \underline{x}}{\varepsilon}$. In view of this, it is sufficient to describe the values $\mu_i \stackrel{\text{def}}{=} \mu(x_i)$ of the desired membership function at the points $x_0, x_1, \ldots, x_n$. We will call these values *discrete (d-)membership function*.

For these values, it is sufficient to formulate the above "closeness" rule only for neighboring values μ_i and μ_{i+1}. To be more precise, this rule now takes the following form:

For all i, μ_i is close to μ_{i+1},

i.e., in other words,

$$(\mu_1 \text{ is close to } \mu_2) \text{ and } \ldots \text{and } (\mu_{n-1} \text{ is close to } \mu_n). \tag{1}$$

Formula (1) can be formalized according to the usual fuzzy methodology. Intuitively, closeness of two numbers x and x' is equivalent to the requirement that the absolute value $d = |x - x'|$ of their difference is small. Thus, to express closeness, we need to select a membership function $s(d)$ describing "small". The larger the difference, the less small it is, so it is reasonable to require that the membership function $s(d)$ be strictly decreasing—at least until it reaches value 0 for the differences d which are clearly *not* small.

Since n is usually large, and thus, $1/n$ is small, without losing generality, we can safely assume that the distance $1/n$ is small, i.e., that $s(1/n) > 0$.

In terms of the selected membership function $s(d)$, for each i, the degree to which μ_i is close to μ_{i+1} is equal to $s(|\mu_i - \mu_{i+1}|)$. To find the degree $d(\mu)$ to which a given d-membership function $\mu = (\mu_1, \ldots, \mu_n)$ is reasonable (i.e., satisfies the above experts' rule), we need to apply some "and"-operation (t-norm) $f_\&(a, b)$ to these degrees, and get

$$d(\mu) = f_\&(s(|\mu_0 - \mu_1|), \ldots, s(|\mu_{n-1} - \mu_n|)).$$

It is reasonable to consider the simplest "and"-operation $f_{\&}(a,b)=\min(a,b)$, then we get

$$d(\mu)=\min(s(|\mu_0-\mu_1|),\ldots,s(|\mu_{n-1}-\mu_n|)). \tag{2}$$

Now, we are ready to formulate the problem in precise terms.

3 Definitions and the Main Results

Definition 1 Let n be a positive integer, and let $s(d)$ be a function from non-negative numbers to $[0,1]$ which is strictly increasing until it reaches 0 and for which $s(1/n)>0$.

- By a *discrete (d-) membership function*, we mean a tuple $\mu=(\mu_0,\ldots,\mu_n)$.
- By a *degree of reasonableness* $d(\mu)$ of a d-membership function μ, we mean the value

$$d(\mu)=\min(s(|\mu_0-\mu_1|),\ldots,s(|\mu_{n-1}-\mu_n|)). \tag{2}$$

- Let M be a class of d-membership functions. We say that a d-membership function $\mu_{\rm opt}\in M$ is *the most reasonable* d-membership function from the class M if

$$d(\mu_{\rm opt})=\sup_{\mu\in M} d(\mu).$$

- Let $\mathcal{M}$ be a class of membership functions defined on an interval $[\underline{x},\overline{x}]$. We say that a membership function $\mu(x)\in\mathcal{M}$ is *the most reasonable* membership function from the class $\mathcal{M}$ if for a sequence $n_k\to\infty$, the corresponding d-membership functions are the most reasonable.

Proposition 1 *Among all d-membership functions for which* $\mu_0=0$ *and* $\mu_n=1$, *the most reasonable d-membership function is* $\mu_i=\dfrac{i}{n}$.

Comments.

- Notice that our result does not depend on the selection of the membership function $s(d)$.
- For reader's convenience, all the proofs are presented in the special proofs section.

Corollary 1 *Among all membership functions on the interval* $[\underline{x},\overline{x}]$ *for which* $\mu(\underline{x})=0$ *and* $\mu(\overline{x})=1$, *the most reasonable membership function is* $\mu(x)=\dfrac{x-\underline{x}}{\overline{x}-\underline{x}}$.

Comment. Thus, the most reasonable membership function is linear.

Proposition 2 *Among all d-membership functions for which* $\mu_0=1$ *and* $\mu_n=0$, *the most reasonable d-membership function is* $\mu_i=\dfrac{n-i}{n}$.

Corollary 2 *Among all membership functions on the interval* $[\underline{x}, \overline{x}]$ *for which* $\mu(\underline{x}) = 1$ *and* $\mu(\overline{x}) = 0$, *the most reasonable membership function is* $\mu(x) = \dfrac{\overline{x} - x}{\overline{x} - \underline{x}}$.

Comment. Thus, here also, the most reasonable membership function is linear.

This explains ubiquity of trapezoid membership functions. Let us consider a property P like "medium", for which:

- the property P is absolutely true for all values x from some interval $[\underline{t}, \overline{t}]$, and
- the property P is absolutely false for all x outside a wider interval $[\underline{T}, \overline{T}]$

Such properties are common.

In terms of membership degrees, the above condition means that:

- $\mu(x) = 0$ for $x \leq \underline{T}$,
- $\mu(x) = 1$ for $\underline{t} \leq x \leq \overline{t}$, and
- $\mu(x) = 0$ for $x \geq \overline{T}$.

On the intervals $[\underline{T}, \underline{t}]$ and $[\overline{t}, \overline{T}]$, we do not know the values of the membership function. On both these subintervals, it is reasonable to select the most reasonable membership function.

Definition 2 We say that a d-membership function $\mu = (\mu_0, \ldots, \mu_n)$ is *normalized* if $\mu_i = 1$ for some i.

Proposition 3 *Among all normalized d-membership functions for which* $\mu_0 = \mu_{2k} = 0$, *the most reasonable d-membership function is the following one:*

- $\mu_i = \dfrac{i}{k}$ *when* $i \leq k$, *and*
- $\mu_i = \dfrac{2k - i}{k}$ *when* $i \geq k$.

Corollary 3 *Among all normalized membership functions on the interval* $[\underline{x}, \overline{x}]$ *with midpoint* $\widetilde{x}$, *for which* $\mu(\underline{x}) = \mu(\overline{x}) = 0$, *the most reasonable membership function is:*

$$\mu(x) = \frac{x - \widetilde{x}}{\widetilde{x} - \underline{x}} \text{ for } x \leq \widetilde{x} \text{ and } \mu(x) = \frac{\overline{x} - x}{\overline{x} - \widetilde{x}} \text{ for } x \geq \widetilde{x}.$$

Comment. Thus, here also, the most reasonable membership function is a triangular one.

Discussion. How robust are these results? To answer this question, let us show that under two somewhat different approaches, trapezoid and linear membership functions are still the most reasonable ones.

4 First Set of Auxiliary Results: What if We Use a Different "And"-Operation, E.g., Product?

Discussion. In the previous section, we used the min "and"-operation. What if we use a different "and"-operation—e.g., the algebraic product $f_\&(a, b) = a \cdot b$, an operation also proposed by L. Zadeh in his original paper?

In this case, the result depends, in general, on the selection of the membership function $s(d)$ for "small". All we know about "small" is that 0 is definitely absolutely small, and that there exists some value D which is definitely not small. This is one of the cases discussed in the previous section, so let us use the results of the previous section to select the most reasonable membership function for small:

$$s_0(d) = 1 - \frac{d}{D}$$

for $d \le D$ and $s_0(d) = 0$ for $d \ge D$. For this selection, we get the following results.

Definition 3 Let n be a positive integer, and let $D > 0$ be a positive real number. Let $s_0(d) = 1 - \frac{d}{D}$ for $d \le D$ and $s_0(d) = 0$ for $d \ge D$.

- By a *product-based degree of reasonableness* $d_0(\mu)$ of a d-membership function μ, we mean the value

$$d_0(\mu) = s_0(|\mu_0 - \mu_1|) \cdot \ldots \cdot s_0(|\mu_{n-1} - \mu_n|).$$

- Let M be a class of d-membership functions. We say that a membership function $\mu_{\text{opt}} \in M$ is *the most product-based reasonable* membership function from the class M if

$$d_0(\mu_{\text{opt}}) = \sup_{\mu \in M} d_0(\mu).$$

- Let $\mathcal{M}$ be a class of membership functions defined on an interval $[\underline{x}, \overline{x}]$. We say that a membership function $\mu(x) \in \mathcal{M}$ is *the most product-based reasonable* membership function from the class $\mathcal{M}$ if for a sequence $n_k \to \infty$, the corresponding d-membership function are the most product-based reasonable.

Proposition 4 *Among all d-membership functions for which* $\mu_0 = 0$ *and* $\mu_n = 1$, *the most product-based reasonable d-membership function is* $\mu_i = \frac{i}{n}$.

Corollary 4 *Among all membership functions on the interval* $[\underline{x}, \overline{x}]$ *for which* $\mu(\underline{x}) = 0$ *and* $\mu(\overline{x}) = 1$, *the most product-based reasonable membership function is*

$$\mu(x) = \frac{x - \underline{x}}{\overline{x} - \underline{x}}.$$

Comment. Thus, the most reasonable membership function is linear.

Proposition 5 *Among all d-membership functions for which* $\mu_0 = 1$ *and* $\mu_n = 0$, *the most product-based reasonable d-membership function is* $\mu_i = \frac{n-i}{n}$.

Corollary 5 *Among all membership functions on the interval* $[\underline{x}, \overline{x}]$ *for which* $\mu(\underline{x}) = 1$ *and* $\mu(\overline{x}) = 0$, *the most product-based reasonable membership function is*

$$\mu(x) = \frac{\overline{x} - x}{\overline{x} - \underline{x}}.$$

Comment. Thus, here also, the most reasonable membership function is linear. Similarly to the previous section, this explains the ubiquity of trapezoid membership functions.

Proposition 6 *Among all normalized membership functions for which* $\mu_0 = \mu_{2k} = 0$, *the most product-based reasonable d-membership function is the following one:*

- $\mu_i = \frac{i}{k}$ *when* $i \le k$, *and*
- $\mu_i = \frac{2k - i}{k}$ *when* $i \ge k$.

Corollary 6 *Among all normalized membership functions on the interval* $[\underline{x}, \overline{x}]$ *with midpoint* $\widetilde{x}$, *for which* $\mu(\underline{x}) = \mu(\overline{x}) = 0$, *the most product-based reasonable membership function is:*

$$\mu(x) = \frac{x - \widetilde{x}}{\widetilde{x} - \underline{x}} \text{ for } x \le \widetilde{x} \text{ and } \mu(x) = \frac{\overline{x} - x}{\overline{x} - \widetilde{x}} \text{ for } x \ge \widetilde{x}.$$

Comment. Thus, here also, the most reasonable membership function is a triangular one.

5 Second Set of Auxiliary Results: What if We Use Statistics-Motivated Least Squares Approach to Select the Most Reasonable Membership Function

Discussion. In the above sections, we used fuzzy techniques to determine the degree to which a d-membership function is reasonable, i.e., a degree to which $\mu_1 - \mu_0$ is small, $\mu_2 - \mu_1$ is small, etc. Intuitively, small means close to 0, i.e., being approximately equal to 0. In other words, we determine a degree to which the following system of approximate equalities hold:

$$\mu_1 - \mu_0 \approx 0, \ldots, \mu_n - \mu_{n-1} \approx 0.$$

It is worth noticing that such systems of approximate equation are well known in traditional statistical data analysis, where the usual way of dealing with such system is to use the Least Squares approach (see, e.g., [7, 8]), i.e., to look for the solutions for which the sum of the squares of the approximation errors is the smallest possible:

$$(\mu_1 - \mu_0)^2 + \cdots + (\mu_n - \mu_{n-1})^2 \to \min.$$

Thus, we arrive at the following definitions.

Definition 4 Let n be a positive integer.

- By the *least-squares degree of reasonableness* $d_1(\mu)$ of a d-membership function μ, we mean the value

$$d_1(\mu) = (\mu_0 - \mu_1)^2 + \cdots + (\mu_{n-1} - \mu_n)^2.$$

- Let M be a class of d-membership functions. We say that a membership function $\mu_{\text{opt}} \in M$ is *the most least-squares reasonable* membership function from the class M if

$$d_1(\mu_{\text{opt}}) = \inf_{\mu \in M} d_1(\mu).$$

- Let $\mathcal{M}$ be a class of membership functions defined on an interval $[\underline{x}, \overline{x}]$. We say that a membership function $\mu(x) \in \mathcal{M}$ is *the most least-squares reasonable reasonable* membership function from the class $\mathcal{M}$ if for a sequence $n_k \to \infty$, the corresponding d-membership functions are the most least-squares reasonable.

Proposition 7 *Among all d-membership functions for which* $\mu_0 = 0$ *and* $\mu_n = 1$, *the most least-squares reasonable d-membership function is* $\mu_i = \dfrac{i}{n}$.

Corollary 7 *Among all membership functions on the interval* $[\underline{x}, \overline{x}]$ *for which* $\mu(\underline{x}) = 0$ *and* $\mu(\overline{x}) = 1$, *the most least-squares reasonable membership function is*

$$\mu(x) = \frac{x - \underline{x}}{\overline{x} - \underline{x}}.$$

Comment. Thus, the most reasonable membership function is linear.

Proposition 8 *Among all d-membership functions for which* $\mu_0 = 1$ *and* $\mu_n = 0$, *the most least-squares reasonable d-membership function is* $\mu_i = \dfrac{n - i}{n}$.

Corollary 8 *Among all membership functions on the interval* $[\underline{x}, \overline{x}]$ *for which* $\mu(\underline{x}) = 1$ *and* $\mu(\overline{x}) = 0$, *the most least-squares reasonable membership function is*

$$\mu(x) = \frac{\overline{x} - x}{\overline{x} - \underline{x}}.$$

Comment. Thus, here also, the most reasonable membership function is linear. Similarly to the previous section, this explains the ubiquity of trapezoid membership functions.

Proposition 9 *Among all normalized membership functions for which* $\mu_0 = \mu_{2k} = 0$, *the most least-squares reasonable d-membership function is the following one:*

- $\mu_i = \frac{i}{k}$ *when* $i \le k$, *and*
- $\mu_i = \frac{2k - i}{k}$ *when* $i \ge k$.

Corollary 9 *Among all normalized membership functions on the interval* $[\underline{x}, \overline{x}]$ *with midpoint* $\widetilde{x}$, *for which* $\mu(\underline{x}) = \mu(\overline{x}) = 0$, *the most least-squares reasonable membership function is:*

$$\mu(x) = \frac{x - \widetilde{x}}{\widetilde{x} - \underline{x}} \text{ for } x \le \widetilde{x} \text{ and } \mu(x) = \frac{\overline{x} - x}{\overline{x} - \widetilde{x}} \text{ for } x \ge \widetilde{x}.$$

Comment. Thus, here also, the most reasonable membership function is a triangular one.

6 Proofs

Proof of Proposition 1.
1. Let us first prove, by contradiction, that for every d-membership function μ from the class M, we have $d(\mu) \le s\left(\frac{1}{n}\right)$.

Indeed, if we had $d(\mu) > s\left(\frac{1}{n}\right)$, then, by definition of the degree $d(\mu)$, this would means that $s(|\mu_i - \mu_{i+1}|) > s\left(\frac{1}{n}\right)$ for all i. Since the function $s(d)$ is strictly decreasing, this implies that $|\mu_i - \mu_{i+1}| < \frac{1}{n}$ for all i. However, we always have

$$|a + \cdots + b| \le |a| + \cdots + |b|.$$

Here,

$$\mu_0 - \mu_n = (\mu_0 - \mu_1) + \cdots + (\mu_{n-1} - \mu_n),$$

hence

$$|\mu_0 - \mu_n| \le |\mu_0 - \mu_1| + \cdots + |\mu_{n-1} - \mu_n|. \quad (3)$$

However, the left-hand side is equal to $|0 - 1| = 1$, while the right hand side is the sum of n terms each of which is smaller than $\frac{1}{n}$, hence the sum is smaller than 1. This contradiction shows that the case $d(\mu) > s\left(\frac{1}{n}\right)$ is indeed impossible.

2. One can easily check that for $x_i = \frac{i}{n}$, we have $|\mu_i - \mu_{i+1}| = \frac{1}{n}$, hence

$$s(|x_i - x_{i+1}|) = s\left(\frac{1}{n}\right)$$

for all i, and $d(\mu) = s\left(\frac{1}{n}\right)$.

3. Let us prove that, vice versa, if $d(\mu) = s\left(\frac{1}{n}\right)$, then $x_i = \frac{i}{n}$ for all i.

Indeed, if $d(\mu) = s\left(\frac{1}{n}\right)$, then for each i, we have $s(|\mu_i - \mu_{i+1}|) \geq s\left(\frac{1}{n}\right)$, hence, due to strict monotonicity of the function $s(d)$, we have $|\mu_i - \mu_{i+1}| \leq \frac{1}{n}$. If one of the values $|\mu_i - \mu_{i+1}|$ was smaller than $\frac{1}{n}$, then the sum

$$|\mu_0 - \mu_1| + \cdots + |\mu_{n-1} - \mu_n|$$

would be smaller than 1, which contradicts to the inequality (3). Similarly, a difference $\mu_{i+1} - \mu_i$ cannot be negative, since then the sum of all the values $\mu_{i+1} - \mu_i$, which is equal to $\mu_n - \mu_0 = 1$, would be smaller than 1.

Thus, $\mu_{i+1} - \mu_n = \frac{1}{n}$ for all i, hence, for each i, we have:

$$\mu_i = \mu_0 + (\mu_1 - \mu_0) + \cdots + (\mu_i - \mu_{i-1}) = \frac{1}{n} + \cdots + \frac{1}{n}(i\text{times}) = \frac{i}{n}.$$

The proposition is proven.

Proof of Corollary 1. If we take into account that $\mu_i = \mu(x_i)$ and $x_i = \underline{x} + i \cdot \varepsilon$, we conclude that $i = \frac{x_i - \underline{x}}{\varepsilon}$. Substituting this expression for i into the formula $\mu(x_i) = \frac{i}{n}$, we conclude that $\mu(x_i) = \frac{x_i - \underline{x}}{n \cdot \varepsilon}$. Here, by definition of n, we have $n \cdot \varepsilon = \overline{x} - \underline{x}$, hence $\mu(x_i) = \frac{x_i - \underline{x}}{\overline{x} - \underline{x}}$. The corollary is proven.

Proofs of Proposition 2 and Corollary 2 are similar to the proofs of Proposition 1 and Corollary 1.

Proof of Proposition 3. Let i_0 denote the value for which $\mu_{i_0} = 1$. Then, we have $d(\mu) = \min(d^-(\mu), d^+(\mu))$, where we denoted

$$d^-(\mu) = \min(s(|\mu_0 - \mu_1|), \ldots, s(|\mu_{i_0-1} - \mu_{i_0}|))$$

and

$$d^+(\mu) = \min(s(|\mu_{i_0} - \mu_{i_0+1}|), \ldots, s(|\mu_{n-1} - \mu_n|))$$

Similarly to the proof of Proposition 1, we can conclude that $d^-(\mu) \le s\left(\frac{1}{i_0}\right)$ and

$$d^+(\mu) \le s\left(\frac{1}{n - i_0}\right).$$

If $i_0 < \frac{n}{2}$, then $n - i_0 > \frac{n}{2}$, hence

$$d(\mu) \le d^+(\mu) \le s\left(\frac{1}{n - i_0}\right) < s\left(\frac{2}{n}\right).$$

Similarly, if $i_0 > \frac{n}{2}$, then

$$d(\mu) \le d^-(\mu) \le s\left(\frac{1}{i_0}\right) < s\left(\frac{2}{n}\right).$$

On the other hand, when $i_0 = \frac{n}{2}$ and values μ_i are equidistant, we get $s(\mu) = s\left(\frac{2}{n}\right)$. Thus, for the optimal d-membership function, we must have $i_0 = \frac{n}{2}$. In this case, similarly to the proofs of Propositions 1 and 2, we can conclude that the only way to get $d(\mu) = s\left(\frac{2}{n}\right)$ is to have values μ_i uniformly changing on each of the intervals $[0, i_0]$ and $[i_0, n]$, i.e., to have $\mu_i = \frac{i}{k}$ for $i \le k = \frac{n}{2}$ and $\mu_i = \frac{n - i}{k}$ for $i \ge k$. The proposition is proven.

Proof of Corollary 3 is similar to the proofs of Corollaries 1 and 2: we plug in the expression for i in terms of x_i into the formula for $\mu_i = \mu(x_i)$.

Proof of Proposition 4. For $\mu_i = \frac{i}{n}$, we have $|\mu_i - \mu_{i+1}| = \frac{1}{n}$, thus $s_0(|\mu_i - \mu_{i+1}|) = 1 - \frac{1}{n \cdot D}$ and

$$d_1(\mu) = \left(1 - \frac{1}{n \cdot D}\right)^n.$$

Let us prove that for all other d-membership functions, we have smaller values of $d_0(\mu)$. Indeed, it is known that the geometric mean is always smaller than or equal than the arithmetic mean, and the only time when they are equal is when all the values are the same. For the values $s_0(|\mu_i - \mu_{i+1}|) = 1 - \frac{1}{n \cdot D}$, the geometric mean is equal to $(d_0(\mu))^{1/n}$, while the arithmetic mean a is equal to

$$a = 1 - \frac{|\mu_0 - \mu_1| + \cdots + |\mu_{n-1} - \mu_n|}{n \cdot D}.$$

As we have proven earlier, the sum $|\mu_0 - \mu_1| + \cdots + |\mu_{n-1} - \mu_n|$ is always greater than or equal to 1, and it is equal to 1 only if all the differences $\mu_{i+1} - \mu_i$ are positive. Thus, the arithmetic mean is always smaller than or equal to $1 - \frac{1}{n \cdot D}$. So, the fact that the geometric mean is smaller than or equal to the arithmetic mean implies that

$$(s_0(\mu))^{1/n} \le a \le 1 - \frac{1}{n \cdot D}.$$

Thus,

$$s_0(\mu) \le a^n \le \left(1 - \frac{1}{n \cdot D}\right)^n.$$

So, we have the desired strict inequality except for the case when all the differences $\mu_{i+1} - \mu_i$ are equal and positive. This is exactly the case of $\mu_i = \frac{i}{n}$. The proposition is proven.

Proofs of Propositions 5 and 6 and Corollaries 4–6 are similar.

Proof of Proposition 7. To find the values μ_i minimizing the sum of the squares, we can differentiate this sum with respect to μ_i and equate the derivative to 0. We then conclude that $(\mu_i - \mu_{i-1}) + (\mu_i - \mu_{i+1}) = 0$, i.e., that $\mu_i - \mu_{i-1} = \mu_{i+1} - \mu_i$. So, all the differences $\mu_{i+1} - \mu_i$ are equal to each other. Since their sum is equal to 1, each difference is equal to $\frac{1}{n}$, hence $\mu_i = \frac{i}{n}$. The proposition is proven.

Proofs of Propositions 8 and 9 and Corollaries 7–9 are similar.

Acknowledgements This work was supported in part by the US National Science Foundation grant HRD-1242122.

References

1. R. Belohlavek, J.W. Dauben, G.J. Klir, *Fuzzy Logic and Mathematics: A Historical Perspective* (Oxford University Press, New York, 2017)
2. G. Klir, B. Yuan, *Fuzzy Sets and Fuzzy Logic* (Prentice Hall, Upper Saddle River, New Jersey, 1995)
3. J.M. Mendel, *Uncertain Rule-Based Fuzzy Systems: Introduction and New Directions* (Springer, Cham, Switzerland, 2017)
4. J.M. Mendel, D. Wu, *Perceptual Computing: Aiding People in Making Subjective Judgments* (IEEE Press and Wiley, New York, 2010)
5. H.T. Nguyen, E.A. Walker, *A First Course in Fuzzy Logic* (Chapman and Hall/CRC, Boca Raton, Florida, 2006)

6. V. Novák, I. Perfilieva, J. Močkoř, *Mathematical Principles of Fuzzy Logic* (Kluwer, Boston, Dordrecht, 1999)
7. S.G. Rabinovich, *Measurement Errors and Uncertainty: Theory and Practice* (Springer, Berlin, 2005)
8. D.J. Sheskin, *Handbook of Parametric and Nonparametric Statistical Procedures* (Chapman and Hall/CRC, Boca Raton, Florida, 2011)
9. L.A. Zadeh, Fuzzy sets. Inf. Control **8**, 338–353 (1965)

SCR-Apriori for Mining 'Sets of Contrasting Rules'

Marharyta Aleksandrova and Oleg Chertov

Abstract In this paper, we propose an efficient algorithm for mining novel 'Set of Contrasting Rules'-pattern (SCR-pattern), which consists of several association rules. This pattern is of high interest due to the guaranteed quality of the rules forming it and its ability to discover useful knowledge. However, SCR-pattern has no efficient mining algorithm. We propose SCR-Apriori algorithm, which results in the same set of SCR-patterns as the state-of-the-art approach, but is less computationally expensive. We also show experimentally that by incorporating the knowledge about the pattern structure into Apriori algorithm, SCR-Apriori can significantly prune the search space of frequent itemsets to be analysed.

1 Introduction

Association rules learning is a popular technique in data mining [1]. However, it is known that finding rules of high quality is not always an easy task [2]. This issue is even more significant in domains where the reliability of the obtained knowledge is required to be high (for example, in medicine). Also, association rules mining techniques usually generate a huge number of rules that have to be analysed by a human in order to choose meaningful and useful ones [3].

Recently a novel pattern called 'Sets of Contrasting Rules' was proposed [4]. This pattern was proven to produce rules of high-quality [5]. It was also shown that its structure allows identifying meaningful knowledge, in particular, *local differences* between subgroups of the dataset and *trigger factors*. However, no efficient algorithm for mining SCR-patterns was proposed so far. The state-of-the-art procedure of constructing SCR-patterns performs filtering of association rules obtained by other

M. Aleksandrova · O. Chertov (✉)
National Technical University of Ukraine 'Igor Sikorsky Kyiv Polytechnic Institute', Kyiv, Ukraine
e-mail: chertov@i.ua

M. Aleksandrova
e-mail: marharyta.aleksandrova@gmail.com

S. N. Shahbazova et al. (eds.), *Recent Developments and the New Direction in Soft-Computing Foundations and Applications*, Studies in Fuzziness and Soft Computing 393, https://doi.org/10.1007/978-3-030-47124-8_7

techniques and chooses those that can form this pattern [6]. Although this approach allows discovering all SCR-patterns in the dataset, it has high level of redundancy, as a vast majority of mined rules is filtered out.

In this paper, we address a question of efficient mining of SCR-patterns. We propose a novel SCR-Apriori algorithm that prunes the search space via incorporating knowledge about the structure of the pattern. The proposed algorithm is a modified version of existing Apriori [7] and CAR-Apriori [8] algorithms and incorporates advantages of both of them.

2 Association Patterns

2.1 Association Rules

The core notion of association patterns mining is *item*. Item is defined as a value of a particular attribute in a given record. A set of items X is referred to as an *itemset* and an *association rule* is defined as an induction rule of the form $X \rightarrow Y$, where X and Y are itemsets and $X \cap Y = \emptyset$ [7]. The left-hand side of the association rule X is called *antecedent*, and its right-hand side Y is called *consequent*.

The *support number* of an itemset X on the dataset D is defined as a number of records in D that contain X. Similarly, the *support number* of an association rule $X \rightarrow Y$ is defined as a support number of the union $X \cup Y$. Often an alternative measure called *support* is used. It is defined as the ratio of the support number to the total number of records in the dataset. The ratio of the support of the rule to the support of its antecedent defines the *confidence* of the rule on a particular dataset, that is $conf_D(X \rightarrow Y) = supp_D(X \cup Y)/supp_D(X)$. Support and confidence are two basic quality measures of association rules [2]. Most of association rules mining algorithms search for *frequent* and *confident* rules, that is for rules whose support and confidence are above or equal to a user-defined thresholds ($supp_D(X \rightarrow Y) \geq minSupp$ and $conf_D(X \rightarrow Y) \geq minConf$).

2.2 Classification Rules and Contrast Patterns

In many applications, the elements of the dataset are distributed among multiple classes. We restrict ourselves to the case of only two classes Cl_1 and Cl_2 in this work for the sake of simplicity. We also assume that the belonging of an element to a class is specified by the value of a particular attribute, which we refer to as *class attribute* $Att_{Cl} = Cl_k$. If the consequent of an association rule is restricted to be composed of only the class attribute, then such rules are called *classification rules* [9]. These rules can be used to address a very important task in class-labeled data: the task of identification of patterns that explain differences between classes of the dataset, or *classification*.

A group of patterns called *contrast patterns* [10] was proposed as an alternative to classification rules. These patterns were also designed with the aim to understand differences between classes, however they are defined in a different way. An itemset is said to be a contrast pattern if it changes significantly its support from one class to another [11]. This change can be measured in terms of, for example, *growth rate*: $GrowthRate(X, Cl_1, Cl_2) = supp_{Cl_1}(X)/supp_{Cl_2}(X)$, where $supp_{Cl_k}(X)$ stands for support of itemset X in class Cl_k. Given ρ (with $\rho > 1$) a growth rate threshold, if $GrowthRate(X, Cl_1, Cl_2) \geq \rho$, then the itemset X is said to be a ρ-emerging pattern from class Cl_2 to class Cl_1 [12].

Some works suggested that contrast patterns differ meaningfully from itemsets and thus association and classification rules [10]. However, it was shown in [13] that a straightforward application of an existing association rules discovery algorithm can be successfully used to mine contrast patterns. It was also proven analytically in [5] that classification rules and contrast patterns can be considered as equivalent. To be more precise, there exists a functional dependence between the value of the growth rate and the value of the confidence. For every given confidence threshold α, it is possible to fix a growth rate threshold ρ in such a way that all ρ-emerging patterns will also form confident classification rules with $minConf = \alpha$.

2.3 'Sets of Contrasting Rules'-Pattern

Contrary to many state-of-the-art patterns, 'Sets of Contrasting Rules'-pattern is composed of multiple rules and is based on notions of invariant and variable attributes [4, 14]. An attribute is said to be *varying* if its value can be changed externally to the system within the specified application task, and *invariant* otherwise. For example, assume that we need to analyse medical data with an attribute *drug* that specifies medicines taken by a patient, and another attribute *age* corresponding to his age. If it is possible to change the medication then the attribute *drug* is varying. At the same time, the value of the parameter *age* cannot be changed. That is it belongs to the group of invariant attributes. The SCR-pattern is defined as follows.

Definition 1 For a parameter α, a pair of classification rules $R1$ and $R2$ is called a pair of α-contrasting rules if:

(1) $conf(R1) \geq \alpha$ & $conf(R2) \geq \alpha$;
(2) $R1$ and $R2$ correspond to different classes;
(3) antecedents of the rules are made up of the same attributes;
(4) the number of attributes is at least two and at least one of the attributes is varying;
(5) values of all invariant attributes are the same for both rules;
(6) if there are no invariant attributes, then the values of at least one varying attribute is the same for both rules;
(7) at least one varying attribute has different values in the pair.

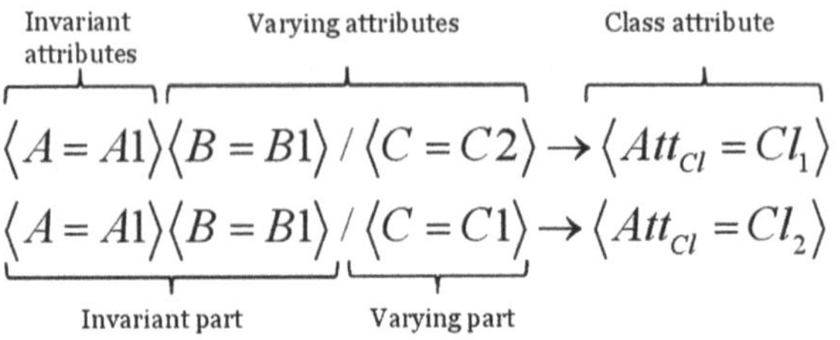

Fig. 1 Example of an SCR-pattern

We say that the rule $R2$ is a *contrast pair* for the rule $R1$ and vice versa. An example of a pair of contrasting rules forming an SCR-pattern is given in Fig. 1. The antecedents of these rules contain one invariant attribute (A) and two varying attributes (B and C). Sign '/' in the figure is used to visually separate invariant part of the pattern (attributes with same values) and its varying part (attributes with different values).

It was shown that, depending upon application, the varying part of the rule can be interpreted as *local differences* between classes of the dataset [5] or *trigger factors* [4]. The term *trigger factors* is used to refer to the factors that can stimulate transfer of data elements from one class to another. It was also analytically proven that rules forming SCR-pattern are guaranteed to be of high-quality [5]. This characteristic can be crucial for certain domains that require only highly confident knowledge (for example, in medicine).

For discovering SCR-patterns authors proposed to perform post-processing of association rules mined by classical algorithms [4]. However, only a very small fraction of discovered rules have contrast pair and thus can be used. Thereby, despite its advantages 'Sets of Contrasting Rules'-pattern has no efficient mining algorithm. In this paper we aim to solve this problem and to propose an efficient solution.

3 Mining Association Patterns

Agrawal et al. in [7] introduced the problem of association rules mining as *the problem of mining all association rules that satisfy user-specified minimum support* $minSupp$ *and minimum confidence* $minConf$ *conditions*. This task is usually divided into two steps [1]:

(1) mining all frequent itemsets,
(2) forming confident rules of these itemsets.

Forming confident rules of frequent itemsets is straightforward and most of the algorithms concentrate on the first task. In contrast, the task of constructing all frequent itemsets is considered to be computationally expensive [15, 16]. It is also known that its complexity grows exponentially with the number of attributes in the dataset. However, the search space of all possible itemsets can be effectively pruned basing on the downward-closure property: *all subsets of a frequent itemset are themselves frequent itemsets* [7]. It means that every non-frequent itemset cannot be a part of any larger frequent itemset and it is possible to find a border that separates frequent and non-frequent itemsets. Thereby, the first task simply comes to searching for this border [16, 17].

The search through the set of items can be done in depth-first or in width-first manners [18]. Both approaches result in the same set of frequent itemsets [19]. The width-first strategy forms the basis of multiple algorithms. Among them, there is Apriori algorithm proposed by Agrawal and Srikant [9]. The latter one is very popular and proved its efficiency in a great variety of applications.

3.1 Apriori: Mining Frequent Itemsets

Let us examine the procedure of constructing frequent itemsets used in Apriori algorithm (see Fig. 2). An itemset containing exactly p attributes is referred to as *p-itemset* and set L_p stands for a set of all p-itemsets. The procedure starts with the initialisation of L_1 with the set of all possible 1-itemsets (all values of each attribute), see **step 1**. After that, all non-frequent itemsets are excluded from L_1. This is done via comparing their support with a defined $minSupp$ value (corresponds to **step 2**). If L_1 contains frequent itemsets (**step 3**), then these itemsets are added to the list of all frequent itemsets L_{all} (**step 4**). Finally, the set of all 2-itemsets L_2 is constructed on the **step 5**. From the downward-closure property, it follows that a 2-itemset is not frequent if at least one its subsets does not belong to the set of frequent 1-itemsets (L_1). Thereby, the set L_2 can be constructed from the set L_1 through the operation of self-joining (see [9] for more details). This allows reducing the number of 2-itemset to be considered by eliminating all those, that are known to be non-frequent. Afterwards, the procedure restarts from the step 2 processing the set L_2 in a similar way.

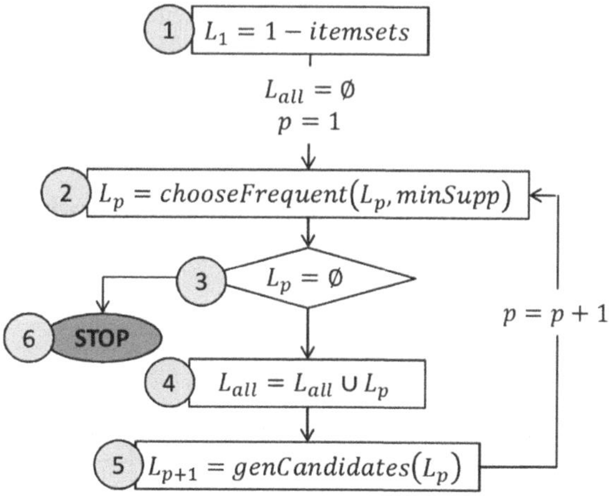

Fig. 2 Apriori: mining frequent itemsets

The procedure finishes when a newly generated on step 5 and filtered on step 2 set L_p is empty. The resulting set L_{all} consists of all frequent itemsets of different sizes.

Finally, all possible confident association rules are constructed by exploiting the set of generated frequent itemsets. Suppose we have a frequent itemset c and its non-empty subset β. Then an association rule is constructed as follows: $(c - \beta) \rightarrow \beta$. If the resulting rule meets the $minConf$ condition, then it is added to the list of confident rules.

3.2 CAR-Apriori: Mining Frequent Ruleitems

Apriori algorithm can be used for the identification of classification rules through post-processing of the obtained association rules: we choose only those rules, whose consequents are composed of only class attribute. However, computational costs can be reduced if the search of classification rules is incorporated into Apriori algorithm. Such an algorithm, CAR-Apriori, was proposed in [8]. This algorithm is based on a new concept: *ruleitem*. A ruleitem is a construction of the form $condset < Att_{Cl} = Cl_k >$. A $condset$ is essentially an itemset, and expression $< Att_{Cl} = Cl_k >$ specifies the value of the class attribute. The difference between a condset and an itemset is that the latter can contain class attributes Att_{Cl} while the former one no.

The support of a ruleitem is defined as a ratio of the number of elements containing the specified condset and belonging to the class Cl_k to the number of all elements in the dataset. Each ruleitem essentially represents a rule of the form $condset \rightarrow Att_{Cl} = Cl_k$ with the support equal to the support of the ruleitem and the confidence calculated as a fraction of the support of the ruleitem to the support of the corresponding condset.

The algorithm for mining classification rules is the same as Apriori with the difference that instead of frequent itemsets frequent rule items are mined. Also, classification rules are formed directly from ruleitems. Both Apriori and CAR-Apriori result in the same set of classification rules. However, the incorporation of class-specific information that is done in CAR-Apriori allows to speed-up the process.

4 SCR-Apriori: Mining Frequent and Contrast SCR-Ruleitems

4.1 Intuition

Similar to using Apriori for mining classification rules, it is possible to use CAR-Apriori for mining 'Sets of Contrasting Rules'-pattern via filtering out all classification rules that have no contrast pair. However, this solution is not efficient.

We propose to further extend the idea of CAR-Apriori algorithm for direct mining of ruleitems that can potentially form SCR-pattern. Apart from excluding all non-frequent itemsets (as it is done in Apriori) and all itemsets that cannot form classification rules (as it is done in CAR-Apriori), we propose to further exclude those itemsets that cannot be used for constructing SCR-patterns, even if they can form frequent and confident classification rules on their own.

Recall that an SCR-pattern consists of two contrasting rules. Thereby, if for a certain frequent ruleitem there is no other frequent ruleitem that can potentially form a pair of contrast classification rules with the first ruleitem, then it can be excluded from the search space regardless of its frequency. In order to fulfil this task, we propose a new algorithm SCR-Apriori. This algorithm aims at mining all frequent and contrast *SCR-ruleitems* and forms SCR-patterns of them.

4.2 SCR-Apriori Algorithm

We define a *SCR-ruleitem* as follows: $condset{<}supp_{Cl_1}, supp_{Cl_2}{>}$, where $supp_{Cl_k}$ stands for support of the condset with respect to the class Cl_k. Essentially an SCR-ruleitem is a condset accompanied with its support values in all classes defined on the dataset D. We also define a *contrast pair* for an SCR-ruleitem in a similar way as we defined a contrast pair for a rule (see Definition 1) with the following differences:

- SCR-ruleitems in the pair can be composed of only one attribute that can be either invariant or varying (compare with condition 4 in Definition 1);
- all varying attributes can have different values even if there are no invariant attributes in the condsets of the considered SCR-ruleitems (compare with condition 6).

The procedure of mining frequent and contrast SCR-ruleitems is very similar to the procedure of mining frequent itemsets and frequent ruleitems presented in Fig. 2. The only difference is in **step 2**. On this step SCR-ruleitems are filtered out from the set L_p not only depending on the value of their support on different classes but also on the fact if a given SCR-ruleitem has a *contrast pair* frequent in the opposite class. Considering this, the function $chooseFrequent$ in the algorithm in Fig. 2 should be replaced with the function $chooseFrequentAndContrast$. This function decides whether to exclude or not an SCR-ruleitem from L_p according to the algorithm given in Fig. 3.

We analyse the structure and visualise different branches of the proposed algorithm using two examples. We utilize 2 sets of records consisting of 16 and 14 elements respectively (see Fig. 4). All records are defined on 3 attributes A, B, and C. Each of these attributes has 2 possible values which we distinguish by numbers, e.g. the possible values of the first attribute are $A1$ and $A2$. Among these 3 attributes the first one (A) is invariant and the rest of the attributes (B and C) are varying. There are 2 classes Cl_1 and Cl_2 defined on each set of records. Records belonging to the first class are in **bold** while records belonging to the second class are in wine-red color.

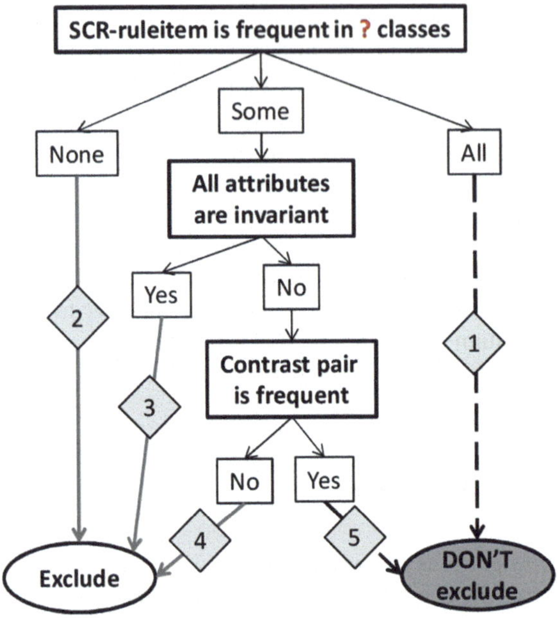

Fig. 3 Function for choosing frequent and contrast SCR-ruleitems

Example 1							
1	A1,B1,C1	5	A1,B1,C2	9	A1,B2,C1	13	A2,B1,C2
2	A1,B1,C1	6	A1,B2,C1	10	A1,B2,C1	14	A2,B1,C2
3	A1,B1,C1	7	A1,B2,C1	11	A1,B2,C2	15	A2,B2,C1
4	A1,B1,C2	8	A1,B2,C1	12	A2,B1,C2	16	A2,B2,C1
Example 2							
1	A1,B1,C1	5	A1,B2,C1	9	A2,B1,C1	13	A2,B2,C2
2	A1,B1,C1	6	A1,B2,C1	10	A2,B2,C2	14	A2,B2,C2
3	A1,B1,C1	7	A1,B2,C1	11	A2,B2,C2		
4	A1,B2,C1	8	A2,B1,C1	9	A2,B2,C2		

Fig. 4 List of records for Example 1 and Example 2

Figures 5a and b show all SCR-ruleitems for the first (a) and the second (b) examples respectively. Each square represents a particular SCR-ruleitem. For the sake of simplicity we present not the support of a condset on each class, but its support number. For both examples, we fix the minimum support number threshold equal to 2.

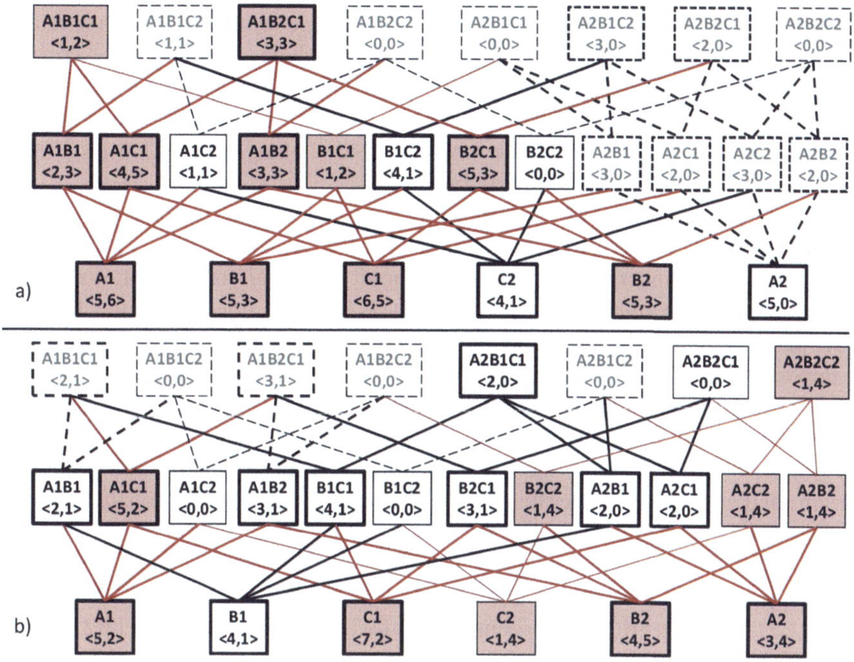

Fig. 5 Contrast ruleitems for Example 1 (**a**) and Example 2 (**b**)

The fact that a particular condset is frequent in Cl_1 or Cl_2 is visualised with a think border or a wine-red background color respectively. A particular SCR-ruleitem having both a thick border and a wine-red background is frequent in both classes. A dashed border means that a particular SCR-ruleitem was not constructed by our algorithm due to exclusion of one or more of its subsets.

The shape of links between SCR-ruleitems also reflects their frequency. In particular, a link up-coming from an SCR-ruleitem is thick or/and of wine-red color, if this SCR-ruleitem is frequent on the first or/and on the second class. Additionally, an up-coming link is dashed if the SCR-ruleitem was marked by our algorithm as the one that cannot form an SCR-pattern.

Let us now analyse the algorithm given in Fig. 3. First, this algorithm examines the values of support of an SCR-ruleitem in all classes defined on the dataset.

Branch 1. If it is frequent on all classes then it can form an SCR-pattern even if it has no frequent contrast pair. Indeed, let us have a look on SCR-ruleitem $A1C1<4,5>$ in Fig. 5a. There is only one SCR-ruleitem that can form a contrast pair for it: $A1C2<1,1>$ and it is not frequent on any of 2 classes. However, we can form SCR-ruleitems $A1B1C1<1,2>$ and $A1B2C1<3,3>$ which will result in SCR-pattern $\{A1C1/B1 \rightarrow Att_{Cl} = Cl_2, conf = 0.67 \ : \ A1C1/B2 \rightarrow Att_{Cl} = Cl_1, conf = 0.5\}$. Thereby, an SCR-ruleitem that is frequent on all classes should not be excluded.

Branch 2 of the algorithm corresponds to the case when the SCR-ruleitem is not frequent on any of the classes and suggests to exclude it. Indeed, if the condset of the SCR-ruleitem is not frequent, then no frequent rules can be formed neither from this condset itself, nor from its super-sets. And, thereby, no SCR-patterns can be produced (see for example SCR-ruleitems $A1C2$<1, 1> and $B2C2$<0, 0> in Fig. 5a or $A1C2$<0, 0> and $B1C2$<0, 0> in Fig. 5b).

If the SCR-ruleitem is frequent on only a fraction of classes, then the nature of its attributes should be considered.

Branch 3. Let us consider the case when all attributes forming SCR-ruleitem are invariant, for example SCR-ruleitem $A2$<5, 0> in Fig. 5a. Both rules from an SCR-pattern should have the same values of all invariant attributes. It means that it is impossible to construct an SCR-pattern in this case as some components of the rules forming this SCR-pattern will be frequent on only one class. In our example $A2$ is frequent only in the first class.

Finally, if the condset of the SCR-ruleitem consists of either varying or both varying and invariant attributes then the frequency of its contrast pairs on the opposite class is checked.

Branch 4 considers the case when SCR-ruleitem is frequent in one class but has no contrast pair frequent in another class. Let us consider as an example SCR-ruleitem $A1B1$<2, 1> in Fig. 5b. The only contrast pair for this SCR-ruleitem is $A1B2$<3, 1>. No SCR-patterns can be produced of this pair as no frequent classification rule can be formed for the second class. From the downward-closure property it also follows that all supersets of $A1B1$<2, 1> will not be frequent on the second class either, for example, $A1B1C1$<2, 1>. Thereby, supersets of the considered SCR-ruleitem can potentially form an SCR-pattern only with those SCR-ruleitems that have different values of at least one invariant attribute from the considered SCR-ruleitem. They have to be also frequent on the other class. However, each of such SCR-ruleitems has a subset that forms a contrast pair with the considered SCR-ruleitem. But we know that all of them are non-frequent on the other class. This means that no SCR-patterns can be formed from the supersets either. In the considered example, SCR-ruleitem $A1B1$<2, 1> has only one invariant attribute B with the value $B1$. Thereby, $A1B1C1$<2, 1> as a superset of $A1B1$<2, 1> can potentially form SCR-pattern with either $A1B2C1$<3, 1> or $A1B2C2$<0, 0>. However, as we know that $A1B2$<3, 1> is not frequent on the second class, we can conclude the same about its supersets without calculating their support explicitly. The same reasoning is valid for another superset of $A1B1$<2, 1>, namely for $A1B1C2$<0, 0>.

Finally, **branch 5** of our algorithm considers the case of partially frequent SCR-ruleitems that have contrast pairs. We can show on an example that in this case it is possible to construct SCR-patterns and thus such SCR-ruleitems should not be excluded. Let us consider $B1C1$<4, 1> in Fig. 5b which is frequent on the first class. Its contrast pair $B2C2$<1, 4> is frequent on the second class. Using the supersets of these SCR-ruleitems we can construct SCR-pattern $\{A2/B1C1 \rightarrow Att_{Cl} = Cl_1, conf = 1 : A2/B2C2 \rightarrow Att_{Cl} = Cl_2, conf = 0.8\}$. Note, that SCR-ruleitem $C2$<4, 1> in Fig. 5a is considered in a similar way. It has a contrast pair $C1$<6, 5> that is frequent on the second class. And it is also possible to construct SCR-pattern of their supersets $B1C1$<1, 2> and $B1C2$<4, 1>.

Using the algorithm given in Fig. 3 it is possible to perform similar analysis for all SCR-ruleitems in Fig. 5 and choose those that satisfy conditions of frequency and contrast. Note that, for example, SCR-ruleitem $A2B1C2$<3, 0> in Fig. 5a is not considered by our algorithm even though it can form a frequent and confident association rule. No SCR-patterns can be formed of this SCR-ruleitem and it is excluded from the consideration because its subset $A2$ does not satisfy the conditions of our algorithm (branch 3).

When all frequent and contrast SCR-ruleitems are discovered, we can construct SCR-patterns directly using SCR-ruleitems and their contrast pairs. Note, however, that not all constructed SCR-ruleitems will form SCR-patterns. For example, if the parameter α is set to $\alpha = 0.6$, no SCR-patterns can be produced from a pair of SCR-ruleitems $A1B1C1$<1, 2> and $A1B2C1$<3, 3> in Fig. 5a.

4.3 *Performance Evaluation*

In order to quantify the gain in performance of SCR-Apriori as compared to CAR-Apriori we used the same dataset as in state-of-the-art works [4, 6, 14]: 5-percent sample of the California census dataset for the year 2000.[1] We also performed the same preprocessing steps as in the cited papers. After that, we generated SCR-patterns with SCR-Apriori and via post-filtering of classification rules obtained with CAR-Apriori, as suggested in the state-of-the-art.

We mined SCR-patterns with $minConf = 0.5$ and $minSupp = 0.07$. The value of support is chosen to be so small in accordance with previous works [6, 14] and with the statement that *contrast patterns with small support can be of particular interest* [14].

We obtained the same set of SCR-patterns with both approaches. However, the number of mined SCR-ruleitems corresponds to 58% of the number of frequent ruleitems identified with CAR-Apriori. Also, the number of rules generated by SCR-Apriori corresponds to only 6.4% of the number of classification rules obtained with the state-of-the-art procedure. Thereby, the proposed algorithm allows obtaining the same results but with the substantial decrease of computation costs.

5 Conclusions

In this paper, we proposed a novel SCR-Appriori algorithm for direct mining of 'Sets of Contrasting Rules'-pattern. Through experimental results on real dataset, we showed that SCR-Apriori allows decreasing significantly computational costs, as compared to the state-of-the-art approach.

---

[1] https://www.census.gov/prod/cen2000/doc/pums.pdf.

The proposed algorithm incorporates the knowledge about the structure of SCR-pattern into the mining process and is essentially a modified version of known Apriori and CAR-Apriori algorithms.

References

1. S. Kotsiantis, D. Kanellopoulos, Association rules mining: a recent overview. GESTS Int. Trans. Comput. Sci. Eng. **32**(1), 71–82 (2006)
2. P. Lenca, P. Meyer, B. Vaillant, S. Lallich, On selecting interestingness measures for association rules: user oriented description and multiple criteria decision aid. Eur. J. Oper. Res. **184**(2), 610–626 (2008)
3. K. Techapichetvanich, A. Datta, Visual mining of market basket association rules, in *Computational Science and Its Applications-ICCSA, 2004* (Springer, Berlin, 2004), pp. 479–488
4. M. Aleksandrova, A. Brun, O. Chertov, A. Boyer, Sets of contrasting rules to identify trigger factors, in *ECAI 2016: 22nd European Conference on Artificial Intelligence* (IOS Press, 2016)
5. M. Aleksandrova, O. Chertov, A. Brun, A. Boyer, Contrast classification rules for mining local differences in medical data, in *2017 9th IEEE International Conference on Intelligent Data Acquisition and Advanced Computing Systems: Technology and Applications (IDAACS)*, vol. 2, (IEEE, 2017), pp. 880–883
6. M. Aleksandrova, A. Brun, O. Chertov, A. Boyer, Sets of contrasting rules: a supervised descriptive rule induction pattern for identification of trigger factors, in *2016 IEEE 28th International Conference on Tools with Artificial Intelligence (ICTAI)* (IEEE, 2016) pp. 431–435
7. R. Agrawal, T. Imieliński, A. Swami, Mining association rules between sets of items in large databases. ACM SIGMOD Rec. **22**(2), 207–216 (1993)
8. B. Liu, W. Hsu, M. Yiming, Integrating classification and association rule mining, in *Proceedings of the fourth international conference on knowledge discovery and data mining* (1998), pp. 80–86
9. R. Agrawal, R. Srikant et al., Fast algorithms for mining association rules, in *Proceedings 20th International Conference very large data bases, VLDB*, vol. 1215 (1994), pp. 487–499
10. K. Ramamohanarao, J. Bailey, H. Fan, Efficient mining of contrast patterns and their applications to classification, in *2005 3rd International Conference on Intelligent Sensing and Information Processing* (IEEE, 2005), pp. 39–47
11. P.K. Novak, N. Lavrač, G.I. Webb, Supervised descriptive rule discovery: a unifying survey of contrast set, emerging pattern and subgroup mining. J. Mach. Learn. Res. **10**, 377–403 (2009)
12. G. Dong, J. Li, Efficient mining of emerging patterns: discovering trends and differences, in *Proceedings of the fifth ACM SIGKDD international conference on Knowledge discovery and data mining*. (ACM, 1999), pp. 43–52
13. G.I. Webb, S. Butler, D. Newlands, On detecting differences between groups, in *Proceedings of the Ninth ACM SIGKDD International Conference on Knowledge Discovery and Data Mining* (ACM, 2003), pp. 256–265
14. O. Chertov, M. Aleksandrova, Fuzzy clustering with prototype extraction for census data analysis, in *Soft Computing: State of the Art Theory and Novel Applications* (Springer, Berlin, 2013), pp. 289–313
15. M. Girotra, K. Nagpal, S. Minocha, N. Sharma, Comparative survey on association rule mining algorithms. Int. J. Comput. Appl. **84**(10), (2013)
16. J. Hipp, U. Güntzer, G. Nakhaeizadeh, Algorithms for association rule mining a general survey and comparison. ACM Sigkdd Explor. Newsl. **2**(1), 58–64 (2000)

17. C.C. Aggarwal, M.A. Bhuiyan, M. Al Hasan, Frequent pattern mining algorithms: a survey, in *Frequent pattern mining* (Springer, Berlin, 2014), pp. 19–64
18. M.J. Zaki, S. Parthasarathy, M. Ogihara, W. Li et al., New algorithms for fast discovery of association rules, in *KDD*, vol. 97 (1997), pp. 283–286
19. M.J. Zaki, Scalable algorithms for association mining. IEEE Trans. Knowl. Data Eng. **12**(3), 372–390 (2000)

When Is Data Processing Under Interval and Fuzzy Uncertainty Feasible: What if Few Inputs Interact? Does Feasibility Depend on How We Describe Interaction?

Milan Hladík, Michal Černý, and Vladik Kreinovich

Abstract It is known that, in general, data processing under interval and fuzzy uncertainty is NP-hard—which means that, unless P = NP, no feasible algorithm is possible for computing the accuracy of the result of data processing. It is also known that the corresponding problem becomes feasible if the inputs do not interact with each other, i.e., if the data processing algorithm computes the sum of n functions, each depending on only one of the n inputs. In general, inputs x_i and x_j interact. If we take into account all possible interactions, and we use bilinear functions $x_i \cdot x_j$ to describe this interaction, we get an NP-hard problem. This raises two natural questions: what if only a few inputs interact? What if the interaction is described by some other functions? In this paper, we show that the problem remains NP-hard if we use different formulas to describe the inputs' interaction, and it becomes feasible if we only have $O(\log(n))$ interacting inputs—but remains NP-hard if the number of inputs is $O(n^{\varepsilon})$ for any $\varepsilon > 0$.

1 Introduction

Need for data processing. In many practical situations, we are interested in the value of a quantity y which is difficult or even impossible to measure directly. For example, we may be interested:

M. Hladík
Faculty of Mathematics and Physics, Department of Applied Mathematics, Charles University, Malostranské Nám. 25, 11800 Prague, Czech Republic
e-mail: milan.hladik@matfyz.cz

M. Černý
Department of Econometrics, University of Economics Prague, W. Churchill's sq. 4, 130 67 Prague, Czech Republic
e-mail: cerym@vse.cz

V. Kreinovich (✉)
Department of Computer Science, University of Texas at El Paso, 500 W. University, El Paso, TX 79968, USA
e-mail: vladik@utep.edu

S. N. Shahbazova et al. (eds.), *Recent Developments and the New Direction in Soft-Computing Foundations and Applications*, Studies in Fuzziness and Soft Computing 393, https://doi.org/10.1007/978-3-030-47124-8_8

- in a distance to a faraway star, or
- in tomorrow's temperature.

Since we cannot measure y directly, we measure it indirectly: namely, we find some easier-to-estimate quantities $x_1, \ldots, x_n$ which are related to y by a known dependence $y = f(x_1, \ldots, x_n)$, and then we use the results $\widetilde{x}_i$ of measuring or estimating the quantities x_i to estimate y as $\widetilde{y} = f(\widetilde{x}_1, \ldots, \widetilde{x}_n)$.

Need to take uncertainty into account. Measurements are never absolutely accurate. As a result, the measurement results $\widetilde{x}_i$ are, in general, different from the actual (unknown) values x_i of the corresponding quantities. Thus, even if the relation $y = f(x_1, \ldots, x_n)$ is precise, the result $\widetilde{y}$ of applying the algorithm f to the measurement results is, in general, different from the actual value y. It is desirable to find out how accurate is the estimate $\widetilde{y}$, i.e., what can we conclude about the measurement errors $\Delta y \stackrel{\text{def}}{=} \widetilde{y} - y$.

This is definitely important. For example, if we predict tomorrow's temperature to be $\widetilde{y} = -2°$ C, and the accuracy of this prediction is $\pm 1°$, then we know that tomorrow will be freezing, with the possibility of ice on the road, so we need to send a warning to the public, put sand (or salt) on the roads, and prepare the corresponding equipment. On the other hand, if the accuracy is ± 10 degrees, we may still alert the public, but it is better to wait until we get closer—and thus, get more accurate information—before we start placing sand (or salt) on the roads.

This is even more important for a spaceship sent to Mars: we want to make sure that, with all the uncertainty taken into account, the spaceship will land in the desired Martian region.

Case of interval uncertainty. In many practical situations, the only information that we have about the measurement errors $\Delta x_i \stackrel{\text{def}}{=} \widetilde{x}_i - x_i$ is the upper bound Δ_i on its absolute value: $|\Delta x_i| \le \Delta_i$; see e.g., [1]. In this case, once we know the measurement result $\widetilde{x}_i$, the only information that we have about the actual (unknown) values x_i is that x_i belongs to the interval $[\underline{x}_i, \overline{x}_i] \stackrel{\text{def}}{=} [\widetilde{x}_i - \Delta_i, \widetilde{x}_i + \Delta_i]$.

Different values x_i from the corresponding intervals lead, in general, to different values $y = f(x_1, \ldots, x_n)$. In this case, we would like to find the range of all possible values of y:

$$[\underline{y}, \overline{y}] = f([\underline{x}_1, \overline{x}_1], \ldots, [\underline{x}_n, \overline{x}_n]) \stackrel{\text{def}}{=} \{f(x_1, \ldots, x_n) : x_1 \in [\underline{x}_1, \overline{x}_1], \ldots, x_n \in [\underline{x}_n, \overline{x}_n]\}.$$

The problem of computing this range is known as the problem of *interval computations*; see, e.g., [2–4].

Already for interval uncertainty, the corresponding problem is often computationally intractable (NP-hard). When the function $f(x_1, \ldots, x_n)$ is linear

$$f(x_1, \ldots, x_n) = a_0 + \sum_{i=1}^{n} a_i \cdot x_i,$$

then we have explicit formulas for the corresponding range: $\underline{y} = \widetilde{y} - \Delta$ and $\overline{y} = \widetilde{y} + \Delta$, where

$$\Delta = \sum_{i=1}^{n} |a_i| \cdot \Delta_i.$$

However, already for quadratic functions $f(x_1, \ldots, x_n)$, the problem of computing the range $[\underline{y}, \overline{y}]$ is, in general, NP-hard—which means, if P$\neq$NP (as most computer scientists believe) that no feasible algorithms is possible that would solve all particular cases of this problem; see, e.g., [5].

Case of fuzzy uncertainty. In many practical situations, in addition to the upper bounds Δ_i on the measurement error, experts also tell us which values from the corresponding interval $[-\Delta_i, \Delta_i]$ are more probable and which are less probable. This information is usually not given in terms of probabilities, but rather in terms of imprecise ("fuzzy") words form natural language, such as "somewhat probable", "very probable", etc. Such imprecise knowledge is ubiquitous.

To describe such knowledge in precise computer-understandable terms, Zadeh invented the technique of *fuzzy logic*; see, e.g., [6–11]. In this technique, to describe each imprecise property like "somewhat probable", we ask the expert to mark, on a scale from 0 to 10, to what extent the corresponding value is possible. If an expert marks 7, we take 7/10 as the degree to which the corresponding value is possible.

As a result, in addition to the interval $[-\Delta_i, \Delta_i]$, we also have, for each value Δx_i from this interval, a degree $\mu_i(\Delta x_i)$ to which this value is possible. The function that assigns, to each value Δx_i, the corresponding degree, is known as the *membership function*.

Data processing under fuzzy uncertainty. A value y is possible if $y = f(x_1, \ldots, x_n)$ for some tuples for which:

- x_1 is a possible value of the first input and
- x_2 is a possible value of the second inputs, etc.

We know the degrees $\mu_i(x_i)$ to which each x_i is a possible value of the ith input. To estimate the degree to which x_1 is possible *and* x_2 is possible, etc., it is reasonable to use a corresponding "and"-operation $_{\&}(a, b)$ (t-norm) of fuzzy logic, resulting in

$$f_{\&}(\mu_1(x_1), \ldots, \mu_n(x_n)).$$

The simplest such operation is $f_{\&}(a, b) = \min(a, b)$, in which case the corresponding inputs has the form

$$\min(\mu_1(x_1), \ldots, \mu_n(x_n)).$$

To find the degree $\mu(y)$ corresponding to the possibility of having either one tuple *or* another, we can similarly apply an "or"-operation (t-conorm) $f_\vee(a, b)$, the simplest of which is $f_\vee(a, b) = \max(a, b)$. Then, we get

$$\mu(y) = \max\{\min(\mu_1(x_1), \ldots, \mu_n(x_n)) : f(x_, \ldots, x_n) = y\}.$$

This formula was originally proposed by Zadeh and is thus known as *Zadeh's extension principle*.

Data processing under fuzzy uncertainty: computational aspects. From the computational viewpoint, this formula can be described in terms of *α-cuts*

$$\mathbf{x}_i(\alpha) \stackrel{\text{def}}{=} \{x_i : \mu_i(x_i) \geq \alpha\} :$$

for every α, we have $\mathbf{y}(\alpha) = f(\mathbf{x}_1(\alpha), \ldots, \mathbf{x}_n(\alpha))$. Thus, from the computational viewpoint, propagation of fuzzy uncertainty can be reduced to several interval computation problems corresponding, e.g., to $\alpha = 0, 0.1, \ldots, 0.9, 1.0$.

Because of this reduction, in the following text, we will consider only the case of interval uncertainty.

How to describe the dependence? In some cases, we know the dependence $f(x_1, \ldots, x_n)$ from physics. In many other cases, however, we need to determine this dependence experimentally. For this, we need to first select a reasonable finite-parametric family of functions, and then find the parameters from the experiments.

When we analyze the dependence of the desired quantity y on the auxiliary quantities $x_1, \ldots, x_n$, the first thing we usually do is analyze how y changes if we change only of these inputs. For each input x_i, we thus get a dependence $y = f_i(x_i)$. This dependence may be linear, may be quadratic, etc.

In some cases, inputs are independent—in the sense that the changes in y caused by each inputs x_i do not depend on the values of all the other inputs x_j with $j \neq i$. In this case, the resulting dependence has the form $f(x_1, \ldots, x_n) = \sum\limits_{i=1}^{n} f_i(x_i)$. One can easily check that in this case, the desired range $[\underline{y}, \overline{y}]$ is equal to the sum of the ranges corresponding to each of the inputs:

$$\underline{y} = \underline{y}_1 + \cdots + \underline{y}_n \text{ and } \overline{y} = \overline{y}_1 + \cdots + \overline{y}_n,$$

where $[\underline{y}_i, \overline{y}_i] \stackrel{\text{def}}{=} \{f_i(x_i) : x_i \in [\underline{x}_i, \overline{x}_i]\}$. For simple functions $f_i(x_i)$ like linear or quadratic, the range is easy to compute; thus, the corresponding interval computations problem is feasible.

In practice, inputs often interact. A natural idea is to use bilinear terms $x_i \cdot x_j$ to describe such an interaction. In this case, we get a general quadratic formula, for which the corresponding problems are NP-hard.

2 Formulation of the Problems

First question. NP-hardness comes from considering the case when all inputs interact with each other. What if only a few inputs interact?

This is the first question for which we provide an answer in this paper.

Second question. In the NP-hardness result, we assume that the interaction is described by bilinear terms. However, other expressions are also possible. For example, in chemical kinetics, for small concentrations of the corresponding chemicals, the interactions are described by a bilinear formula $x_i \cdot x_j$, but for very strong concentrations, the interaction is described by a different formula $\min(x_i, x_j)$ (see, e.g., [12–14]), and for intermediate concentration, we get a more complex formula; see, e.g., [15, 16]. Will the general result remain NP-hard if we consider such interaction?

This is the second question for which we provide the answer.

3 Answer to the First Question

Main result. Our first result is that if we have a quadratic form with only $O(\log(n))$ pairs of interacting inputs, then we have a feasible algorithm for estimating the range $[\underline{y}, \overline{y}]$.

Comment. Detailed proofs of all our results—as well as several related results—are presented in an accompanying mathematical paper [17]. In the present paper, we describe the main ideas behind these proofs.

Main idea of the proof. Since only $v = O(\log(n))$ many inputs $x_{i_1}, \ldots, x_{i_v}$ are involved in the interaction, we can describe the desired quadratic function as the sum

$$f(x_1, \ldots, x_n) = \sum_{i \neq i_k} f_i(x_i) + r(x_{i_1}, \ldots, x_{i_v}),$$

where $f_i(x_i)$ is a quadratic function of one variable, and $r(x_{i_1}, \ldots, x_{i_v})$ is a quadratic function of v variables.

Since each of the terms in the above sum depends on each own inputs, we conclude that

$$\underline{y} = \sum_{i \neq i_k} \underline{y}_i + \underline{r} \text{ and } \overline{y} = \sum_{i \neq i_k} \overline{y}_i + \overline{r},$$

where $\underline{y}_i \stackrel{\text{def}}{=} \min\{f(x_i) : x_i \in [\underline{x}_i, \overline{x}_i]\}$, $\overline{y}_i \stackrel{\text{def}}{=} \max\{f(x_i) : x_i \in [\underline{x}_i, \overline{x}_i]\}$,

$$\underline{r} = \min\{r(x_{i_1}, \ldots, x_{i_v}) : z_{i_1} \in [\underline{x}_{i_1}, \overline{x}_{i_1}], \ldots, x_{i_v} \in [\underline{x}_{i_v}, \overline{x}_{i_v}]\}, \text{ and}$$

$$\overline{r} = \max\{r(x_{i_1}, \ldots, x_{i_v}) : x_{i_1} \in [\underline{x}_{i_1}, \overline{x}_{i_1}], \ldots, x_{i_v} \in [\underline{x}_{i_v}, \overline{x}_{i_v}]\}.$$

Minima and maxima $\underline{y}_i$ and $\overline{y}_i$ of a quadratic function $f_i(x_i)$ over an interval are easy to compute. Thus, to show that the computation of the range $[\underline{y}, \overline{y}]$ is feasible, we need to show how to feasibly compute the minimum and the maximum of the function $r(x_{i_1}, \ldots, x_{i_v})$ over the box $[\underline{x}_{i_1}, \overline{x}_{i_1}] \times \cdots \times [\underline{x}_{i_v}, \overline{x}_{i_v}]$.

According to calculus, a maximum or a minimum of a function $F(z)$ on an interval $[\underline{z}, \overline{z}]$ is attained:

- either at a point which is inside the interval $(\underline{z}, \overline{z})$, in which case $\frac{dF}{dz} = 0$;
- or at the left endpoint $z = \underline{z}$ of the give interval,
- or at the right endpoint $z = \overline{z}$ of this interval.

Similarly, the maximum or minimum of a function $F(z_1, \ldots, z_v)$ on a box

$$[\underline{z}_1, \overline{z}_1] \times \cdots \times [\underline{z}_v, \overline{z}_v]$$

is attained when for each of the v variables z_i, one of the following three situations happens:

- either the corresponding value z_i is inside the interval $(\underline{z}_i, \overline{z}_i)$, in which case
$$\frac{\partial F}{\partial z_i} = 0;$$
- or the optimizing value is at the left end of the corresponding interval $z_i = \underline{z}_i$,
- or the optimizing value is at the right end of the corresponding interval $z_i = \overline{z}_i$.

For each variable, we have 3 options. Thus:

- for two variables, we have $3 \cdot 3 = 9$ possible options, and
- for v variables, we have 3^v possible options.

In each of these 3^v options, for each variables z_i, we have either $z_i = \underline{z}_i$, or $z_i = \overline{z}_i$, or $\frac{\partial F}{\partial z_i} = 0$. The first two equations are clearly linear in z_i. In our case, when $z_k = x_{i_k}$ and the function $F(z_1, \ldots, z_v) = r(z_1, \ldots, z_v)$ is quadratic, each derivative of a quadratic expression is linear, thus, the equation $\frac{\partial F}{\partial z_i} = 0$ is also linear in $z_1, \ldots, z_v$.

So, in each of the 3^v cases, we have a system of linear equations to find the corresponding values $z_1, \ldots, z_v$—a system which can be feasible solved. Out of all cases for which each component z_i of the solution is within the corresponding interval, we choose:

- the smallest as $\underline{r}$ and
- the largest as $\overline{r}$.

When $v = O(\log(n))$, i.e., $v \le C \cdot \log(n)$ for some constant C, we have

$$3^v \le 3^{C \cdot \log(n)} = n^{\log(3) \cdot C}$$

linear systems. Thus, the number of linear system is polynomial in n. Hence, the overall time for solving all these systems is also bounded by a polynomial in n—and is, thus, feasible. This proves our main result.

Analysis of the problem. What if we have more interacting inputs? It is known that $\log(n)$ can be viewed as a limit of power functions n^ε when $\varepsilon \to 0$. So, a natural next question is: what if we have n^ε interacting inputs, for some small ε?

Auxiliary result. If we allow n^ε interacting inputs, then, in general, the problem of computing the range $[\underline{y}, \overline{y}]$ remains NP-hard.

Idea of the proof. Indeed, formally, NP-hard means that any problem from a class NP can be reduced to this problem; see, e.g., [5]. Thus, if we can reduce a known NP-hard problem to a new problem, this means, by transitivity of reduction, that every problem from the class NP can be reduced to the new problem as well, and thus, that the new problem is also NP-hard.

We know that the problem of estimating the range of a general quadratic function over a given box is NP-hard. Let us reduce this known NP-hard problem to our new problem—of estimating the range of a quadratic function in which at most n^ε inputs interact.

For this, let us start with any original quadratic form $Q(x_1, \ldots, x_m)$. Then, we add $M = n^{1/\varepsilon}$ new variables $v_1, \ldots, v_M$, and consider a new quadratic function

$$f(x_1, \ldots, x_m, v_1, \ldots, v_M) = Q(x_1, \ldots, x_m) + \sum_{j=1}^{M} v_j.$$

For this function, only inputs $x_1, \ldots, x_m$ interact, so out of $n = m + M$ variables, only $O(n^\varepsilon)$ interact with each other.

On the other hand, since the new function f is the sum of expressions each of which depends only on its own variables, we conclude that its range $[\underline{y}, \overline{y}]$ has the form

$$\underline{y} = \underline{q} + \sum_{j=1}^{M} \underline{v}_j \text{ and } \overline{y} = \overline{q} + \sum_{j=1}^{M} \overline{v}_j,$$

where $\underline{q}$ and $\overline{q}$ are the minimum and maximum of the original quadratic expression $Q(x_1, \ldots, x_m)$ on the corresponding box.

So, if we know the bounds for f, we can easily find the bounds for Q, and vice versa. Thus, computing the range of f is indeed feasibly equivalent to computing the range of Q—so we have the desired reduction, and thus, the problem is indeed NP-hard.

4 Answer to the Second Question

Analysis of the problem. Instead of the usual interaction terms $x_i \cdot x_j$, we allow more general terms $f_{ij}(x_i, x_j)$. If one of the inputs is absent, i.e., if $x_i = 0$, then there is usually no interaction, so we can safely assume that $f_{ij}(0, x_j) = f_{ij}(x_i, 0) = 0$ for all x_i and x_j.

To make the comparison with the product term (for which $f_{ij}(1, 1) = 1$) easier, we can divide and multiply the expression $f_{ij}(x_i, x_j)$ by $a_{ij} \stackrel{\text{def}}{=} f_{ij}(1, 1)$, then the corresponding terms takes the form $f_{ij}(x_i, x_j) = a_{ij} \cdot T_{ij}(x_i, x_j)$, where

$$T_{ij}(x_i, x_j) \stackrel{\text{def}}{=} \frac{f_{ij}(x_i, x_j)}{a_{ij}}.$$

It is reasonable to require that small changes in x_i and x_j should lead to small changes in T_{ij}, i.e., that

$$|T_{ij}(x_i, x_j) - T_{ij}(x_i', x_j')| \le L \cdot (|x_i - x_i'| + |x_j - x_j'|),$$

for some Lipschitz constant L.

In this case, we consider expressions of the type

$$f(x_, \ldots, x_n) = \sum_{i=1}^{n} f_i(x_i) + \sum_{i \ne j} a_{ij} \cdot T_{ij}(x_i, x_j),$$

where $f_i(x_i)$ are quadratic functions of one variable.

Main result. The problem of computing the range $[\underline{y}, \overline{y}]$ of a function of the above type over a given box is NP-hard.

Idea of the proof. It is known that the problem of computing the range of a quadratic function over a given box is NP-hard already for the case when the corresponding quadratic form is positive definite (i.e., when the corresponding function is convex) and the range of each variable is $[\underline{x}_i, \overline{x}_i] = [0, 1]$; see, e.g., [5]. (Reduction to $[0, 1]$ can be easily achieved by a linear transformation of each variable.)

To be more precise, for convex functions, computing the minimum $\underline{y}$ is feasible, but computing the maximum $\overline{y}$ is NP-hard.

Let us reduce the NP-hard problem of computing this maximum to the new problem. Let us start with a general convex quadratic expression

$$f(x_1, \ldots, x_n) = a_0 + \sum_{i=1}^{n} a_i \cdot x_i + \sum_{i,j} a_{ij} \cdot x_i \cdot x_j.$$

By separating quadratic terms corresponding to $i = j$ and $i \ne j$, we get

$$f(x_1, \ldots, x_n) = a_0 + \sum_{i=1}^{n} a_i \cdot x_i + \sum_{i=1}^{n} a_{ii} \cdot x_i^2 + \sum_{i \ne j} a_{ij} \cdot x_i \cdot x_j.$$

Let us consider a new function

$$F(x_1, \ldots, x_n) = a_0 + \sum_{i=1}^{n} a_i \cdot x_i + \sum_{i=1}^{n} a_{ii} \cdot x_i^2 + \sum_{i \neq j} a_{ij} \cdot T_{ij}(x_i, x_j) + \beta \cdot \sum_{i=1}^{n} (2x_i - 1)^2,$$

for some $\beta > 0$.

Due to the Lipschitz condition, for sufficiently large β, the function $F(x_1, \ldots, x_n)$ is convex. For a convex function, the maximum $\overline{Y}$ on a convex set $[0, 1]^n$ is attained at one of the vertices, i.e., when each of the values x_i is equal to 0 or 1:

$$\overline{Y} = \max_{x_i \in \{0,1\}} F(x_1, \ldots, x_n).$$

On each vertex, however, $T_{ij}(x_i, x_j) = x_i \cdot x_j$ and $(2x_i - 1)^2 = 1$. So, for vertices $(x_1, \ldots, x_n)$, we have $F(x_1, \ldots, x_n) = f(x_1, \ldots, x_n) + \beta \cdot n$. The maximum $\overline{y}$ of the original convex quadratic function $f(x_1, \ldots, x_n)$ is also attained at one of the vertices:

$$\overline{y} = \max_{x_i \in \{0,1\}} f(x_1, \ldots, x_n).$$

Thus,

$$\overline{Y} = \max_{x_i \in \{0,1\}} F(x_1, \ldots, x_n) = \max_{x_i \in \{0,1\}} (f(x_1, \ldots, x_n) + \beta \cdot n) =$$

$$\max_{x_i \in \{0,1\}} f(x_1, \ldots, x_n) + \beta \cdot n = \overline{y} = \beta \cdot n.$$

So, we get $\overline{Y} = \overline{y} + \beta \cdot n$. Thus, the computation of $\overline{y}$ is indeed feasibly reduced to computing $\overline{Y}$. This reduction shows that our problem is also NP-hard.

Acknowledgements M. Hladík was supported by the Czech Science Foundation Grant P403-18-04735S. The work of M. Cerny was supported by the Czech Science Foundation under Grant P402/12/G097. V. Kreinovich was supported in part by the National Science Foundation grant HRD-1242122 (Cyber-ShARE Center).

References

1. S.G. Rabinovich, *Measurement Errors and Uncertainty: Theory and Practice* (Springer, Berlin, 2005)
2. L. Jaulin, M. Kiefer, O. Didrit, E. Walter, *Applied Interval Analysis, with Examples in Parameter and State Estimation, Robust Control, and Robotics* (Springer, London, 2001)
3. G. Mayer, *Interval Analysis and Automatic Result Verification* (de Gruyter, Berlin, 2017)
4. R.E. Moore, R.B. Kearfott, M.J. Cloud, *Introduction to Interval Analysis* (SIAM, Philadelphia, 2009)
5. V. Kreinovich, A. Lakeyev, J. Rohn, P. Kahl, *Computational Complexity and Feasibility of Data Processing and Interval Computations* (Kluwer, Dordrecht, 1998)
6. R. Belohlavek, J.W. Dauben, G.J. Klir, *Fuzzy Logic and Mathematics: A Historical Perspective* (Oxford University Press, New York, 2017)
7. G. Klir, B. Yuan, *Fuzzy Sets and Fuzzy Logic* (Prentice Hall, Upper Saddle River, New Jersey, 1995)

8. J.M. Mendel, *Uncertain Rule-Based Fuzzy Systems: Introduction and New Directions* (Springer, Cham, Switzerland, 2017)
9. H.T. Nguyen, E.A. Walker, *A First Course in Fuzzy Logic* (Chapman and Hall/CRC, Boca Raton, Florida, 2006)
10. V. Novák, I. Perfilieva, J. Močkoř, *Mathematical Principles of Fuzzy Logic* (Kluwer, Boston, Dordrecht, 1999)
11. L.A. Zadeh, Fuzzy sets. Inf. Control **8**, 338–353 (1965)
12. V. Kreinovich, S. Maslov's iterative method: 15 years later (freedom of choice, neural networks, numerical optimization, uncertainty reasoning, and chemical computing), in *Problems of Reducing the Exhaustive Search*, ed. by V. Kreinovich, G. Mints (American Mathematical Society, Providence, RI, 1997), pp. 175–189
13. V. Kreinovich, O. Fuentes, Simulation of chemical kinetics—a promising approach to inference engines, in vol. 3, ed. by J. Liebowitz, *Proceedings of the World Congress on Expert Systems* (Orlando, Florida, Pergamon Press, N.Y., 1991), pp. 1510–1517
14. V. Kreinovich, O. Fuentes, High-concentration chemical computing techniques for solving hard-to-solve problems, and their relation to numerical optimization, neural computing, reasoning under uncertainty, and freedom of choice, in *Molecular and Supramolecular Information Processing: From Molecular Switches to Logical Systems*, ed. by E. Katz (Wiley-VCH, Wienheim, Germany, 2012), pp. 209–235
15. L.C. de Barros, R.C. Bassanezi, W.A. Lodwick, *A First Course in Fuzzy Logic, Fuzzy Dynamical Systems, and Biomathematics: Theory and Applications* (Springer, Berlin, Heidelberg, 2018)
16. O. Kosheleva, V. Kreinovich, L.C.+ Barros, Chemical kinetics in situations intermediate between usual and high concentrations: fuzzy-motivated derivation of the formulas, in *Proceedings of the 2016 World Conference on Soft Computing*, (Berkeley, California, May 22–25, 2016), pp. 332–335
17. M. Hladík, M. Černý, V. Kreinovich, Optimization of quadratic forms and t-norm forms on interval domain and computational complexity, *These Proceedings*

Optimization of Quadratic Forms and t-norm Forms on Interval Domain and Computational Complexity

Milan Hladík, Michal Černý, and Vladik Kreinovich

Abstract We consider the problem of maximization of a quadratic form over a box. We identify the NP-hardness boundary for sparse quadratic forms: the problem is polynomially solvable for $O(\log n)$ nonzero entries, but it is NP-hard if the number of nonzero entries is of the order n^{ε} for an arbitrarily small $\varepsilon > 0$. Then we inspect further polynomially solvable cases. We define a sunflower graph over the quadratic form and study efficiently solvable cases according to the shape of this graph (e.g. the case with small sunflower leaves or the case with a restricted number of negative entries). Finally, we define a generalized quadratic form, called t-norm form, where the quadratic terms are replaced by t-norms. We prove that the optimization problem remains NP-hard with an *arbitrary* Lipschitz continuous t-norm.

1 Introduction

In this paper we elaborate on the problems outlined in [4] in more details. In that work we studied processing imprecise data from multiple sources which interact together. The interaction among inputs $x_1, \ldots, x_n$ is formalized by a function $f(x_1, \ldots, x_n)$ which cannot be written in the separable form as $\sum_{i=1}^{n} f_i(x_i)$ (for some functions f_i). An example is a quadratic form $x^T Ax$ with nonzero off-diagonal entries, which is

M. Hladík
Faculty of Mathematics and Physics, Department of Applied Mathematics,
Charles University, Malostranské Nám. 25, 11800 Prague, Czech Republic
e-mail: milan.hladik@matfyz.cz

M. Černý
Department of Econometrics, University of Economics Prague,
W. Churchill's sq., 4 130 67 Prague, Czech Republic
e-mail: cernym@vse.cz

V. Kreinovich (✉)
Department of Computer Science, University of Texas at El Paso,
500 W. University, El Paso, TX 79968, USA
e-mail: vladik@utep.edu

S. N. Shahbazova et al. (eds.), *Recent Developments and the New Direction in Soft-Computing Foundations and Applications*, Studies in Fuzziness and Soft Computing 393, https://doi.org/10.1007/978-3-030-47124-8_9

studied in this paper. Then we consider a more general form of pairwise interactions: formally, we replace the bilinear terms $x_i x_j$ ($i \neq j$) from $x^T Ax$ by so-called t-norms (which can be regarded as generalizations of the "AND" logical connective).

The general question is: when the inputs $x_1, \ldots, x_n$ are imprecise but are known to be in given compact intervals

$$\boldsymbol{x}_1 = [\underline{x}_1, \overline{x}_1], \ \ldots, \ \boldsymbol{x}_n = [\underline{x}_n, \overline{x}_n],$$

and we are given a function $f : \mathbb{R}^n \to \mathbb{R}$, can we find tight bounds for $f(x_1, \ldots, x_n)$? Formally, denoting

$$\boldsymbol{x} = \boldsymbol{x}_1 \times \cdots \times \boldsymbol{x}_n, \tag{1}$$

the problem reduces to the computation of

$$\sup_{x \in \mathbb{R}^n} \{f(x) \mid x \in \boldsymbol{x}\} \text{ and } \inf_{x \in \mathbb{R}^n} \{f(x) \mid x \in \boldsymbol{x}\}.$$

Here, the expression "to find the bounds" refers to computational complexity: We are to determine under which conditions the bounds can be evaluated in polynomial time and when the computation is NP-hard.

Recall that in general, finding tight bounds for a general function f need not be recursive. This is why various classes of functions of interest in data processing need to be studied separately.

In this text, bold symbols—such as $\boldsymbol{x}$—refer to n-dimensional intervals of the form (1). The real n-vectors of lower and upper bounds are denoted by $\underline{x}$ and $\overline{x}$, respectively, and we write $\boldsymbol{x} = [\underline{x}, \overline{x}]$ for short.

Basics in computational complexity and interval computing can be found, e.g., in [5].

2 Quadratic Forms on Interval Domain

Consider a quadratic form $f \colon \mathbb{R}^n \to \mathbb{R}$

$$f(x) = b^T x + x^T Ax = \sum_{i=1}^{n} b_i x_i + \sum_{i,j=1}^{n} a_{ij} x_i x_j$$

restricted to a given interval domain $x = [\underline{x}, \overline{x}]$. It is known that computing the range of f on x, i.e.,

$$\underline{f} := \min \ f(x) \ \text{ subject to } \ x \in \mathbf{x},$$
$$\overline{f} := \max \ f(x) \ \text{ subject to } \ x \in \mathbf{x}$$

is an NP-hard problem. This is true even for A positive definite, in which case computing $\underline{f}$ is polynomial whereas computing $\overline{f}$ is NP-hard.

Assumption. For simplicity of exposition, we focus only on computation of $\overline{f}$ in the following text. We will also assume for the remainder of the paper that $f(x)$ is convex (i.e., that A is positive semidefinite).

3 Sparse Quadratic Forms

Suppose that the matrix A is sparse, that is, most of the off-diagonal entries are zero. Then, does the problem problem of computing the range become tractable?

In this section it is sufficient to fix

$$x = [0, 1]^n.$$

Proposition 1 *The problem of computing $\overline{f}$ remains NP-hard even when the number of off-diagonal non-zeros in A is bounded by $\mathcal{O}(n^{1/k})$.*

Proof Let $f(x) = b^T x + x^T Ax$ be a quadratic function on $\mathbb{R}^n$. Consider the quadratic form

$$g(x, y) := f(x) + \sum_{i=1}^{m} (2y_i - 1)^2.$$

Then the maximum of $g(x, y)$ on $[0, 1]^{n+m}$ is the same as the maximum of $f(x)$, shifted by the amount of m. That is,

$$\overline{g} = \overline{f} + m. \tag{2}$$

Putting $m := n^{2k}$ we get that the quadratic form $g(x, y)$ of dimension $d = n + m$ has $\mathcal{O}(d^{1/k})$, non-zero off-diagonal entries in the corresponding matrix. Since $f(x)$ was an arbitrary quadratic form, computing the range of $g(x, y)$ is NP-hard, too. ■

Corollary (to the proof). Under the assumption of Proposition 1, *it is NP-hard to approximate $\overline{f}$ with a given (arbitrarily large) absolute error*. This follows from the fact that the maximum of a quadratic form is known to be NP-hard to approximate with an absolute error [1], and (2) does not change the absolute error.

On the other hand, approximating $\overline{f}$ with a relative error can be done efficiently via semidefinite relaxation even for a nonconvex $f(x)$, see [6].

Proposition 2 *The problem becomes polynomial if the number of off-diagonal non-zeros in A is bounded by $\mathcal{O}(\log n)$.*

Proof Denote

$$I := \{i = 1, \ldots, n \mid \exists j \neq i : a_{ij} \neq 0\}.$$

Now, $f(x)$ can be expressed as

$$f(x) = \sum_{i \notin I} (b_i x_i + a_{ii} x_i^2) + \sum_{i,j \in I} a_{ij} x_i x_j,$$

and its maximum as

$$\overline{f} = \max_{x \in \boldsymbol{x}} \left(\sum_{i \notin I} (b_i x_i + a_{ii} x_i^2) \right) + \max_{x \in \boldsymbol{x}} \left(\sum_{i,j \in I} a_{ij} x_i x_j \right). \tag{3}$$

The first term in (3) is computed easily as

$$\max_{x \in \boldsymbol{x}} \left(\sum_{i \notin I} (b_i x_i + a_{ii} x_i^2) \right) = \sum_{i \notin I} \max_{x_i \in \boldsymbol{x}_i} (b_i x_i + a_{ii} x_i^2),$$

and maximizing a univariate quadratic function is a trivial task.

The second term in (3) requires maximizing a quadratic function on an interval domain in dimension $\mathcal{O}(\log n)$. Hence, by brute force, we find the maximum [2] in exponential time w.r.t. $\mathcal{O}(\log n)$, which is polynomial w.r.t. n. ∎

4 Polynomial Cases Based on Sunflower Graphs

Without loss of generality assume that A is upper triangular and that $\boldsymbol{x} = [0, 1]^n$. Consider the graph $G = (V, E)$, where $V = \{x_1, \ldots, x_n\}$ and $\{x_i, x_j\}$ is an edge of G if and only if $a_{ij} \neq 0$. So we are in fact maximizing a quadratic form $f(x)$ on the graph G (see Chap. 10 of [3]).

Let $D \subseteq V$ be a vertex cut such that the graph $G' = (V \setminus D, E')$ after removing the cut D consists of connected components of vertex size $\mathcal{O}(\log n)$. Suppose further that the size of the cut is $|D| = \mathcal{O}(\log n)$. (A graph with such cut is sometimes called *sunflower graph*, see Fig. 1.) Then the cut is associated with $|D|$ variables. Hence we can process all 0/1-assignments of these variables. There are at most $2^{|D|}$ such assignments. For every such assignments, we resolve the problem by brute force in each of the components. Therefore, the overall time complexity is

$$2^{|D|}(T_1 + T_2 + \cdots + T_k) \leq 2^{\mathcal{O}(\log n)}(2^{\mathcal{O}(\log n)} + 2^{\mathcal{O}(\log n)} + \cdots + 2^{\mathcal{O}(\log n)}) \leq \text{poly}(n),$$

where T_i is time complexity of maximization over ith component, $k \leq n$ is the number of components, and poly(n) is a polynomial in n.

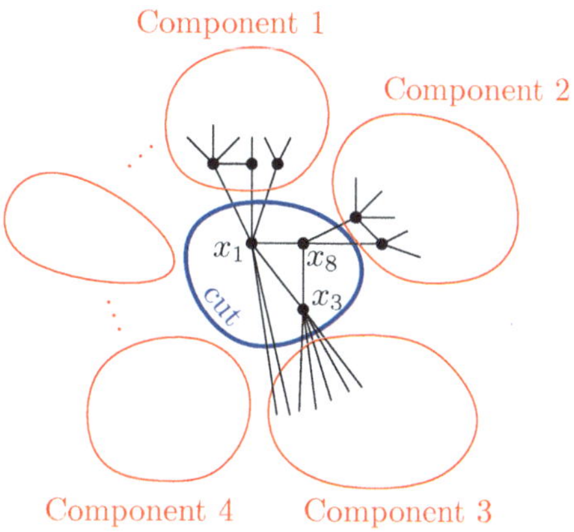

Fig. 1 A sunflower graph with a cut of size $\mathcal{O}(\log n)$ and components of size $\mathcal{O}(\log n)$

A problem. How to find a suitable cut? This is an open challenging question. Notice that minimum cut splitting graph G into two components can be found efficiently by means of linear programming. Nevertheless, incorporating restrictions on size of the components seems a hard problem.

Special graphs

The above reasoning can be extended even to the components larger than $\mathcal{O}(\log n)$, but having a special structure. So, we will now discuss a few of special graphs possessing a suitable structure. For the sake of simplicity of exposition, we will illustrate it on the graph $G = (V, E)$.

Few negative coefficients: Provided that all coefficients are nonnegative, that is, $b_i \geq 0$ and $a_{ij} \geq 0$ for all $i, j \in \{1, \ldots, n\}$, then the optimal solution is simply $x = (1, \ldots, 1)^T$. If it is not the case, we can still effectively compute an optimal solution as long as the number of negative coefficients b_i and a_{ij} is small. Define a cut D to contain all variables incident with negative coefficients:

$$D := \{x_i;\ b_i < 0 \text{ or } a_{ij} < 0 \text{ for some } j\}.$$

If $|D| = \mathcal{O}(\log n)$, then we are done since applying the cut we obtain a subproblem with nonnegative coefficients and the 0/1-variables in D can be tested brute-force in time $2^{|D|} = poly(n)$.

Other special graphs

Assume now for simplicity in the remainder of this section that the domain of variables is $x = [-1, 1]^n$. Further assume that $b_i = 0$ for every i.

Trees: If G is a tree, then maximizing the quadratic function on G is easy: Take an arbitrary vertex in $x_i \in G$ as a root, and distinguish two assignments $x_i = \pm 1$. For each assignment, the remaining variables associated with G have determined values. Sorting the vertices according to some tree search algorithm, we put $x_j := \operatorname{sgn}(a_{ij}x_i)$ when x_i precedes x_j.

Planar graphs: The above class can be extended to planar graphs with $\mathcal{O}(\log n)$ faces because by removing $\mathcal{O}(\log n)$ vertices we obtain a tree.

Bipartite graphs: Complete bipartite graphs $K_{m,n}$ and their subgraphs are also efficiently processed provided $a_{ij} \leq 0$ for $i \neq j$. The variables associated with the first set of vertices will be set as $x_i := 1$, and the others $x_i := -1$.

If the assumption $a_{ij} \leq 0$ for $i \neq j$ is not satisfied, then the bipartite graph is still efficiently processed as long as $m = \mathcal{O}(\log n)$, in which case the vertex cut D is the smaller of those two subsets.

Remark. For related results see [7].

5 t-norm Forms

Recall that a t-norm is a function $T\colon [0, 1]^2 \to [0, 1]$ satisfying:

- commutativity: $T(a, b) = T(b, a)$,
- monotonicity: $a \leq c,\ b \leq d \ \Rightarrow\ T(a, b) \leq T(c, d)$,
- associativity: $T(a, (T(b, c))) = T(T(a, b), c)$,
- 1 is identity element: $T(a, 1) = a$.

From the definition, we immediately have

$$T(0, 0) = T(0, 1) = T(1, 0) = 0, \quad T(1, 1) = 1. \tag{4}$$

Given t-norms T_{ij}, the question is how easy is evaluation of the t-norm form

$$f_T(x) = \sum_{i=1}^{n}(b_i x_i + a_{ii}x_i^2) + \sum_{i\neq j} a_{ij} T_{ij}(x_i, x_j)$$

on a given interval domain x.

Proposition 3 *Maximizing a t-norm form on $x = [0, 1]^n$ is NP-hard even if we choose and fix for every T_{ij} a Lipschitz continuous t-norm, that is,*

$$|T_{ij}(x) - T_{ij}(x')| \leq \alpha \cdot \|x - x'\|,$$

where α is a Lipschitz constant and $\| \cdot \|$ is any vector norm.

Proof Let $f(x) = b^T x + x^T Ax$ be a convex quadratic function on $\mathbb{R}^n$. Consider the t-norm form

$$f_T(x) := b^T x + \sum_{i=1}^{n} a_{ii} x_i^2 + \beta \sum_{i=1}^{n} (2x_i - 1)^2 + \sum_{i \neq j} a_{ij} T_{ij}(x_i, x_j).$$

By the Lipschitz continuity assumption, for sufficiently large β the function $f_T(x)$ is convex. Thus the maximum of $f_T(x)$ is attained in a vertex of x. However, on a set of vertices $x \in \{0, 1\}^n$,

$$f_T(x) = f(x) + \beta n$$

since $T_{ij}(x_i, x_j) = x_i x_j$ and $(2x_i - 1)^2 = 1$. This means that the maximum of $f_T(x)$ is the same as the maximum of $f(x)$, shifted by the amount of βn. Since maximizing $f(x)$ on x is NP-hard, maximizing t-norm forms on x is NP-hard, too. ∎

Remark 1 It is interesting that the proof does not require all the axioms of a *t*-norm. Basically, we used (4) only. Thus the statement holds true for any Lipschitz continuous functions T_{ij} satisfying (4).

Notice that the commonly used t-norms satisfy the assumption of the proposition:

- product t-norm $T(x, y) = xy$ (in this case, the t-norm form is a quadratic form),
- minimum t-norm $T(x, y) = \min\{x, y\}$,
- Łukasiewicz t-norm

$$T(x, y) = \max\{0, x + y - 1\},$$

- nilpotent minimum t-norm

$$T(x, y) = \begin{cases} \min\{x, y\} & \text{if } x + y > 1, \\ 0 & \text{otherwise,} \end{cases}$$

- Hamacher product t-norm

$$T(x, y) = \begin{cases} 0 & \text{if } x = y = 0, \\ \frac{xy}{x+y-xy} & \text{otherwise.} \end{cases}$$

On the other hand, the drastic t-norm defined as

$$T(x, y) = \begin{cases} \min\{x, y\} & \text{if } \max\{x, y\} = 1, \\ 0 & \text{otherwise} \end{cases}$$

does not satisfy the assumption.

Acknowledgements M. Hladík was supported by the Czech Science Foundation Grant P403-18-04735S. M. Černý was supported by the Czech Science Foundation Grant P402/12/G097. V. Kreinovich was supported in part by the National Science Foundation Grant HRD-1242122 (Cyber-ShARE Center of Excellence).

References

1. M. Černý, M. Hladík, The complexity of computation and approximation of the t-ratio over one-dimensional interval data. Comput. Stat. Data Anal. **80**, 26–43 (2014)
2. J.-A. Ferrez, K. Fukuda, T. Liebling, Solving the fixed rank convex quadratic maximization in binary variables by a parallel zonotope construction algorithm. Eur. J. Oper. Res. **166**, 35–50 (2005)
3. B. Gärtner, J. Matoušek, *Approximation Algorithms and Semidefinite Programming* (Springer, 2012)
4. M. Hladík, M. Černý, V. Kreinovich, When is data processing under interval and fuzzy uncertainty feasible: what if few inputs interact? Does feasibility depend on how we describe interaction? *This Volume*
5. V. Kreinovich, A. Lakeyev, J. Rohn, P. Kahl, *Computational Complexity and Feasibility of Data Processing and Interval Computations* (Kluwer, Dordrecht, 1998)
6. Yu. Nesterov, Semidefinite relaxation and nonconvex quadratic optimization. Optim. Methods Softw. **9**(1–3), 141–160 (1998)
7. G.-D. Yu, Quadratic forms on graphs with applications to minimizing the least eigenvalue of signless Laplacian over bicyclic graphs. Electron. J. Linear Algebr. **27**(13) (2014)

Fuzzy Knowledge and Fuzzy Estemation

Asymptotically Optimal Learning in Fuzzy Environments

V. L. Stefanuk

Abstract The paper continues our approach for measuring fuzzy values with a game of learning finite automata that are able to make an optimal choice among several alternatives. In principle, with such a game it is possible to measure any fuzzy values with certain precision. For the mentioned game we used previously a linear tactics automaton as the construction with asymptotically optimal behavior. However, our study showed that its asymptotical optimality is achieved under certain restriction concerning the fuzzy environment. In practice the mentioned automata allow to measure fuzzy value only when it is greater than 0.5. Presently, architecture has been chosen, referred to as trusting automata. The architecture of such an automaton has been proposed many years ago by Prof. V. I. Krinsky. In this paper we proved mathematically that the trusting automata do have the asymptotically optimal property for arbitrary values of memberships in the fuzzy environment. Consequently, the trusting automata may be used to measure arbitrary membership values in the game procedure described above without any restriction.

Keywords Fuzzy measurement · Finite automata · Asymptotic optimality · Game approach

1 Introduction

For measuring fuzzy membership functions [1] in an experimental situation it was proposed in [2] to apply learning automata that are able to make an optimal choice among several alternatives due to use of certain architecture. In the beginning for this goal a finite state automaton, referred to as an automaton with linear tactics, have been considered, that was proposed by Tsetlin [3] in his attempts to model mathematically the real learning used in biological systems.

V. L. Stefanuk (✉)
Institute for Information Transmission Problems, Bolshoy Karetny Per. 19, 127051 Moscow, Russia
e-mail: stefanuk@iitp.ru

Peoples' Friendship University of Russia, Miklucho-Maklaya Str. 6, 117198 Moscow, Russia

S. N. Shahbazova et al. (eds.), *Recent Developments and the New Direction in Soft-Computing Foundations and Applications*, Studies in Fuzziness and Soft Computing 393, https://doi.org/10.1007/978-3-030-47124-8_10

1.1 The Linear Tactics Automaton in a Fuzzy Environment

The linear tactics automaton has 2 actions and two sets of n states. Each set corresponds to performing one of its actions. The architecture of the automaton is demonstrated in Fig. 1, where two graphs are shown.

The arrows on the upper graph correspond to state transitions in case the automaton is obtaining a penalty for its actions. The arrows in the bottom graph corresponds to transition of its states in case when the automaton is rewarded.

Originally this architecture was intended to be used in a probability environment. In our publications [2] it was proposed to be placed into a fuzzy environment, when the penalties are certain fuzzy values $\lambda^{(1)}$ and $\lambda^{(2)}$. The rewards are correspondingly $(1 - \lambda^{(1)})$ and $(1 - \lambda^{(2)})$. The goal was to show that the automaton by its activity in time is able to learn whether $\lambda^{(1)} \leq \lambda^{(2)}$, or that $\lambda^{(1)} > \lambda^{(2)}$. It was proposed in [4] to use the result of learning in a fuzzy measuring system, that is able to measure the fuzzy value some new value μ with a certain precision.

However, the provided in [2] mathematical analyses demonstrated that the linear tactics automaton in fuzzy environment has the property of asymptotic optimality only when a certain supplementary condition is valid, which makes it difficult its application for measuring small membership values. Namely, it was shown that the following condition

$$\frac{1 - \lambda^{(i)}}{\lambda^{(i)}} \leq 1, i = 1, 2, \tag{1}$$

i.e. at least either $\lambda^{(1)} < \frac{1}{2}$ or $\lambda^{(2)} < \frac{1}{2}$ must be valid.

The condition (1) makes it impossible to measure values of new λ in case when its exceeds 0.5, when both penalties are greater than 0.5.

For these reason in present paper a behavior of a modified architecture proposed by V. I. Krinsky is under study shown in Fig. 2.

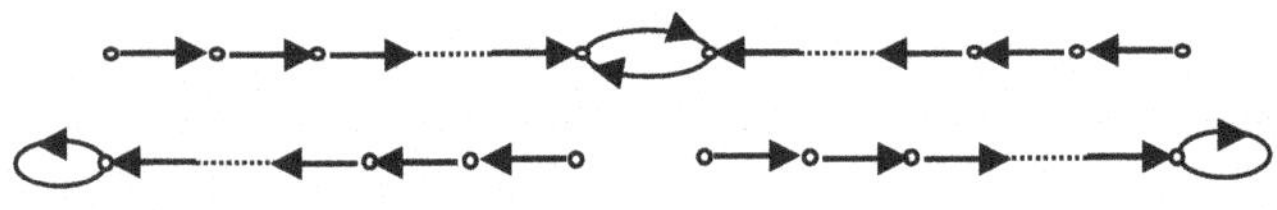

Fig. 1 Linear tactics automaton

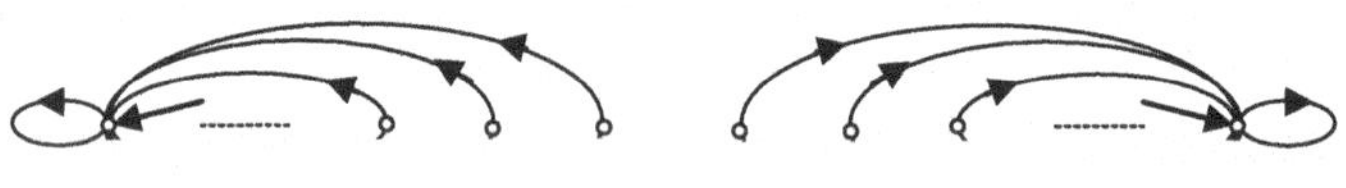

Fig. 2 The trusting finite automaton

1.2 *The Trusting Automaton in a Fuzzy Environment*

In Fig. 2 it is shown only the part of the graph describing its behavior when the automata is rewarded. The graph for a penalty is the same as in Fig. 1 (upper graph in Fig. 1). The diagram of Fig. 2 explains why this automaton was called in Russian literature the trusting automaton: when the automaton in Fig. 2 obtains a reward it jumps to the deepest its state corresponding to the action that was rewarded.

From mathematical point of view it is more complicated that is demonstrated in the next chapter.

2 Mathematical Study of the Trusting Automaton

In our previous papers it was proved that the automaton of Fig. 1 has the property to be asymptotically optimal only when condition (1) is true. In our measuring system described in [4], when the system applied to unknown μ it means that the measuring membership functions below 0.5 is not guaranteed, even if one takes the automata with large memory depth n.

The property of asymptotic optimality was defined in [3] for a probability environment. It means that when $n \to \infty$ then the automaton eventually performs only the action with minimal value for punishment. It will be shown below that the trusted automata has the property of asymptotic optimality in an arbitrary fuzzy environment.

Unfortunately the analysis used in [2] is not directly applicable for the trusting automaton for some technical reasons. Another way is proposed below, where first some simple examples will be used, and the final results will be achieved by induction over value of n.

The main role is played by the Markov-Stefanuk chain [6] that is generalization of Simple Markov chain that allows to go out of probability theory typical for the Markov chain and all its generalizations. Another fact that will be used below is the problem if summarization of fuzzy evidences, introduced in [7], where it was demonstrated that the difference of our axiomatically obtained formula with *max-min* techniques by Zadeh [1] is not essential and maybe neglected.

The advantage of our axiomatics is the simplicity of mathematical operations with evidences. Using Markov-Stefanuk chain one may built a one step matrix of transitions from one states to another in correspondence with Fig. 2.

For example, from the left state of number n the automaton goes to the state n when it obtained the reward with the fuzzy value $1 - \lambda^{(1)}$ or goes to the state $n - 1$, when it obtained a penalty with the fuzzy value $\lambda^{(1)}$ all other elements of the line equals to zero. It may be seen that sums of elements all the in each line of the matrix is equal to one. Nevertheless this is not exactly stochastic matrix as the elements of the matrix are not probabilities. Actually, this transition matrix corresponds to Markov-Stefanuk.

Yet this matrix has obviously the ergodic property and hence does produce in the long run the final values for each of its states in accordance with [6] that will be denoted as

$$\mu_1^{(1)}, \ldots, \mu_n^{(1)}; \mu_1^{(2)}, \ldots, \mu_n^{(2)} \tag{3}$$

As punishment $\lambda^{(2)}$ and rewarding $1-\lambda^{(2)}$ values are considered to be membership functions the list of (3) might be also membership values.

First, let us consider some simple cases.

Case 1 (*i* = *1*, *j* = **1**). Consider the flows of fuzzy value from state one for the action **1** to the state 1 for the action **2** and back. They must be equal.

By applying a relation in accordance with Markov-Stefanuk theory [7] one obtains the following important relation:

$$\boxed{(1-\mu_1^{(1)}) = (1-\mu_1^{(2)})\frac{(1-\lambda^{(2)})}{(1-\lambda^{(1)})}} \tag{4}$$

Hence, if the penalty $\lambda^{(1)}$ for the action **1** is bigger than that for the action **2**, i.e. $\frac{(1-\lambda^{(2)})}{(1-\lambda^{(1)})} >> 1$, then fuzzy membership for the state 1 for the action **2** would be bigger than that for the action **1**, i.e.

$$(1-\mu_1^{(1)}) >> (1-\mu_1^{(2)})$$
$$or \quad \mu_1^{(1)} << \mu_1^{(2)} \tag{5}$$

Case 2 (*i* = *2*, *j* = **1**). Considering flows between states 1 and 2 for the action **1** (see Fig. 2) and using an analog of formula (4) we obtain the following (for simplicity we assume that the fuzzy penalty and reward are supplementary values, i.e. if the fuzzy penalty is $\lambda^{(i)}$, then the fuzzy reward is $(1-\lambda^{(i)})$.

One obtained

$$\boxed{(1-\mu_1^{(1)}) = (1-\mu_2^{(1)})\frac{(1-\lambda^{(1)})}{\lambda^{(1)}}} \tag{6}$$

Hence, if the penalty $\lambda^{(1)}$ for the action **1** exceeds 0.5 and $\frac{(1-\lambda^{(1)})}{\lambda^{(1)}} < 1$ is true, then one has

$$(1-\mu_1^{(1)}) > (1-\mu_2^{(1)})$$
$$or \quad \mu_1^{(1)} < \mu_2^{(1)} \tag{7}$$

In general case one might expect the relation for the states corresponding to action **1**.

$$\boxed{(1-\mu_k^{(1)}) = \left(\frac{\lambda^{(1)}}{(1-\lambda^{(1)})}\right)^{k-1} (1-\mu_1^{(1)})^{k-1}} \tag{8}$$

The same calculation for the states belonging to the action **2** we have

$$\boxed{(1-\mu_k^{(2)}) = \left(\frac{\lambda^{(2)}}{(1-\lambda^{(2)})}\right)^{k-1} (1-\mu_1^{(2)})^{k-1}} \tag{9}$$

Let $M^{(1)}$ and $M^{(2)}$ be membership values for actions **1** and **2**. Thus $M^{(1)}$—is a fuzzy sum of membership functions for all the states corresponding to the action **1**.

$$M^{(1)} = [1-(1-\mu_1^{(1)})\left\{\frac{\lambda^{(1)}}{1-\lambda^{(1)}}\left(\frac{\lambda^{(1)}}{1-\lambda^{(1)}}\right)^2 \cdots \left(\frac{\lambda^{(1)}}{1-\lambda^{(1)}}\right)^{n-1}\right\} \tag{10}$$

Similarly one has

$$M^{(2)} = [1-(1-\mu_1^{(2)})\left\{\frac{\lambda^{(2)}}{1-\lambda^{(2)}}\left(\frac{\lambda^{(2)}}{1-\lambda^{(2)}}\right)^2 \cdots \left(\frac{\lambda^{(2)}}{1-\lambda^{(2)}}\right)^{n-1}\right\}] \tag{11}$$

Assuming $\lambda^{(1)} > \lambda^{(2)}$ one obtains finally

$$\left(\frac{\frac{\lambda^{(2)}}{1-\lambda^{(2)}}}{\frac{\lambda^{(1)}}{1-\lambda^{(1)}}}\right)^{\frac{n(n-1)}{2}} \leq \frac{(1-\lambda^{(1)})}{(1-\lambda^{(2)})} \tag{12}$$

$$\left(\frac{\lambda^{(2)}}{1-\lambda^{(2)}} < \frac{\lambda^{(1)}}{1-\lambda^{(1)}}\ in\ case\ of\ \lambda^{(2)} < \lambda^{(1)}\right) \tag{13}$$

The value in the left in (12) is less then 1 due to (13). Hence we have not only $M^{(1)} < M^{(2)}$ but also we proved the asymptotic optimality of automata shown in Fig. 2, i.e. when $n \to \infty$ one has

$$M^{(1)} \to 0\ and\ M^{(2)} \to 1 \tag{14}$$

The last expression proves the asymptotically optimal property of trusting automata to be used in the game described in [5] to measure the fuzzy values without any restrictions.

3 Conclusion

Thus, it was proved the asymptotically optimality property of trusting automata, that should be used in the game described into measure the fuzzy values without any restrictions.

Actually created game procedure is somehow reminiscent of collecting statistics in Probability Theory [8].

References

1. L.A. Zadeh, Fuzzy sets. Inf. Control (8), 338–348, USA (1965)
2. V.L. Stefanuk, Behavior of Tsetlin's learning automata in a fuzzy environment, in *Second World Conference on Soft Computing (WConSC)* (Letterpress, Azerbaijan, Baku, 2012), pp. 511–513
3. M.L. Tsetlin, Some problems of finite automata behaviour. Doklady USSR Acad. Sci. **139**(4), Moscow (Некоторые проблемы поведения конечных автоматов.) (1961)
4. V.L. Stefanuk, How to measure qualitative data, in *Proceedings of American Fuzzy Information Processing Society NAFIPS'2015 and 5th World Conference on Soft Computing*, Redmond, USA (2015), pp. 37–40
5. V.L. Stefanuk, Game approach to fuzzy measurement, in *World Conference on Soft Computing*, Berkeley, May 22–25 (2016), pp. 149–154
6. V.L. Stefanuk, Deterministic Markovian chains. Inf. Process. **11**(4), 702–709 Moscow (Детерминированные цепи Маркова) (2011)
7. V.L. Stefanuk, Should one trust evidences? in *Proceedings of the all-country AI conference* **1**, pp. 406–410, Moscow (Доверять ли свидетельствам) (1988)
8. A.N. Kolmogorov, Zur Theorie der Markoffschen Ketten. Math. Ann. **101**, 126–136 (1929)

Multi-source Fusion Using Neural Networks and Genetic Algorithms Towards Ego-Lane Estimation

Jan-Ole Perschewski, Tran Tuan Nguyen, Jens Spehr, Jonas Krüsemann, Sebastian Zug, and Rudolf Kruse

Abstract An important task of automated driving is to keep the vehicle in the host lane. For a robust lane estimation, the information of multiple data sources needs to be combined to overcome the drawbacks of each individual sensor [1]. For that reason, this paper proposes a new mediated perception approach towards lane keeping function. Additionally to artificial neural networks (ANNs) used by Nguyen et al. [2], genetic programming (GP) is applied to estimate the parameters of an approximation of a clothoid, which is used to represent the ego-lane. Therefor, a set of lane marking detections, information about the ego-vehicle and the leading vehicle and information about the current location are used as input for the estimators. Compared to the reference, the resulting ANN and GP estimators mostly achieve an angle deviation of smaller than 2° at a distance of 30 m. Hereby, both approaches achieve an overall availability of around 90%. Surprisingly, GP surpasses all compared approaches based on lane markings or a deep learning approach, which directly estimates the ego-lane from camera images, with an overall availability of 0.91. For that reason, ANNs and GPs can be applied to solve the lane keeping task so that future research can lead to further increases of the availability.

1 Introduction

Although huge advances in technology occurred since the beginnings of the automotive industry, the development of safe vehicles is far from over. Even in 2016 around 3000 people died in traffic accidents in Germany [3]. Around half of these people died in car accidents. Moreover, around half of these accidents could have been prevented by driver assistant systems, since the driver could have reacted appropriately considering the additional information provided by these systems. For instance, 30%

J.-O. Perschewski · S. Zug · R. Kruse (✉)
Otto-Von-Guericke University, Universitätsplatz 2, 39106 Magdeburg, Germany
e-mail: rudolf.kruse@ovgu.de

T. T. Nguyen · J. Spehr · J. Krüsemann
Volkswagen Group, Berliner Ring 2, 38440 Wolfsburg, Germany

S. N. Shahbazova et al. (eds.), *Recent Developments and the New Direction in Soft-Computing Foundations and Applications*, Studies in Fuzziness and Soft Computing 393, https://doi.org/10.1007/978-3-030-47124-8_11

of the avoidable accidents were caused maneuvering the car, e.g., turning, driving backwards, keeping to the speed limit or holding distance. Most of these aspects are already supported by systems like cruise control or blind spot monitoring. The biggest weak spot is the human driver, not being able to perceive all important information towards safe driving. This is where autonomous driving comes into play relieving the driver of this increasingly complex task and reducing traffic related accidents. Other advantages of autonomous driving include allowing the driver to use their time in the vehicle differently and the possibility for more efficient road usage, since shorter reaction times entail shorter safety distances.

In the context of automated driving one important issue is road estimation. To reach an availability greater than simple lane marking following we employ AI techniques to learn the estimation of a clothoid road model. Here, availability refers to the lane hypothesis having an angle deviation of less than 2° compared to recorded drives.

In the next section, related works are presented, followed by the overall concept in Sect. 3. Then, artificial neural networks and genetic programming towards lane estimation are introduced in Sect. 4 and Sect. 5 respectively. Finally, the two approaches are evaluated in Sect. 6, followed by the conclusion in Sect. 7.

2 Related Work

In the field of autonomous driving, there exist various approaches to control vehicles based on the perceived environment. They can be divided into three categories visualized in Fig. 1 [4]. In the following, we illustrate these categories by presenting several relevant approaches.

(a) Behavior Reflex Approaches: The simplest way to achieve autonomous driving regarding the design process is to apply an AI-technique that directly derives operations for the actuators. An early approach using a neural network was proposed by Pomerleau [5]. For that, an ANN consisting of three layers is trained, which gets its input from subregions of a camera image. Each neuron in the output layer represents a steering angle, which is used to determine the final steering. A more sophisticated approach to achieve autonomous driving using ANNs was proposed by Bojarski et al.

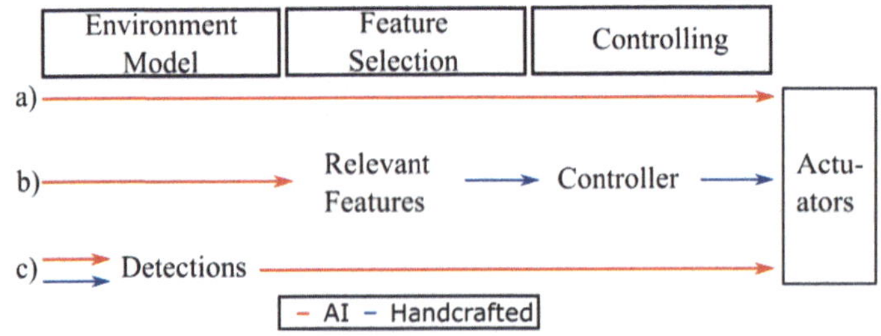

Fig. 1 Perception models. **a** Behavior reflex. **b** Direct perception and **c** Mediated perception

[6], which improves the work of Pomerleau [5] by using recent advances in ANNs. A multi-layer convolution network is trained on images from three cameras to imitate the steering angle of a human driver. In their evaluation, they reach an autonomy level of 98%, which is defined as $(1 - \frac{interventions \cdot 6}{elapsed}) \cdot 100$.

A general problem using behavior reflex approaches is that the frame by frame evaluation can result in small errors amplifying in each frame [7]. Moreover, the imitation of driver behavior leads to copying human faults into the estimation process [7]. Finally, it is very hard to install a fail-safe for this approach. This can result in accidents in unknown environments and endanger other traffic participants.

(b) Direct Perception Approaches: For *direct perception approaches* AI-techniques determine a set of predefined features, which are used by simple controllers to decide which action the vehicle should take. The following approaches are presented with focus on the overall process on learning the controllers.

Overall Process: The authors in [4] introduce a direct perception approach for autonomous driving in highway scenarios, which minimizes the distance to the center line of the lane and a safe distance to other traffic participants. For this purpose, they choose a set of thirteen features determined by a CNN to represent the current environment to construct a controller. Al-Qizwini et al. [8] provide an approach reducing the parameters by Chen et al. [4] to five parameters and using GoogLeNet. Even though these approaches seem promising, they suffer from the lack of complexity in comparison to general real-world scenarios.

Controller: Since there are many ways to design even simple controllers regarding a fixed set of features, the development of an optimal controller is an important and challenging task. Yakami et al. [9] therefore evolve genetic programs for two specific scenarios, where a controlled faster vehicle advances to a slower one. The fitness for their genetic programs minimizes the lateral acceleration and distance to the lane center line. Obviously, the discussed scenario represents but a tiny fraction of possible real world scenarios. Therefore, this paper proposes to generalize this procedure and increase the expressiveness.

(c) Mediated Perception Approaches: *Mediated perception approaches* distinguish themselves by creating an environment model, which is used by an AI-based approach to determine the action.

Creating Environment Models: Creating an environment model often depends on multiple types of sensors to determine information about the environment. Hence, Valente et al. propose a real-time method to segment a general road model from a sequence of camera images using texture orientation and vanishing points [10]. Since deep learning experiences rising popularity in computer vision tasks, Badrinarayanan et al. [11] propose a network for image segmentation which determines the label per pixel of a road scene. Furthermore, Prabhakar et al. [12] propose a detection and classification method creating a model in using faster R-CNN to determine bounding boxes and labels. To detect moving obstacles Zhou et al. [13] present a technique to detect moving objects, from a moving camera. A more specific environment model is presented by the authors in [14], who propose a technique to estimate lane markings with convolutional neural networks.

AI Controlling: In [15] Li et al. propose a framework for fusion of LIDAR and camera data to achieve robust lane estimation on structured and unstructured roads. Nguyen et al. [1, 16] suggest a reliability-aware fusion framework for lane estimation. Therein, they use an environment model of lane markings, leading vehicle trajectories, ego information and contextual data fetched from a map using GPS. At the beginning, multiple hypotheses of the course of the road are created. Afterwards, different classifiers estimate the reliabilities of these hypotheses depending on the current environment model for a reliability-aware fusion.

In this work, we build upon this framework, merging reliability estimation and fusion into a single task performed by neural networks and genetic programming respectively.

3 Overall Concept

In this paper, we extend the alternative system for ego-lane estimation [1, 2] displayed in Fig. 2. For this purpose, we apply genetic programming to estimate the lane and compare it to the performance of the artificial neural networks.

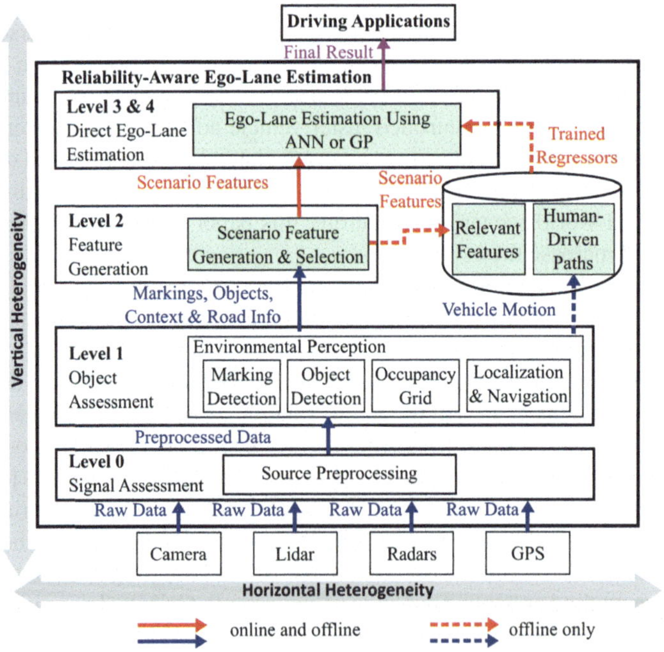

Fig. 2 Direct ego-lane estimation using artificial neural networks (ANN) and genetic programming (GP) [2]

(a) Data: For training and evaluating the ANNs and GPs, we use real-world data since this is the only way to achieve roadworthy performance. Specifically, we use data recorded by prototype vehicles manually driven throughout Wolfsburg and surrounding areas. Before the recorded data is used as ground truth in training and evaluation, it is cleansed of any intersections and situations without detections since for these purposes driver input is required. We use the common vehicle coordinate system with the origin in the center of the rear axle of the car facing towards positive infinity on the x-axis.

(b) Models of Lane Markings and Trajectories: Lane markings and trajectories are modeled uniformly to enable easy comparison. Additionally, the representation needs to be able to model straight lanes as well as curves for displaying the road environment in all scenarios sufficiently. Therefore, we chose the clothoid model, which is a commonly used curve in lane estimation tasks [17, 18]. A clothoid is derived from vehicle dynamics and can be sufficiently approximated by Eq. 1 as long as the yaw angle is below 15° [17].

$$y(x) \approx y_0 + \phi_0 \cdot x + \frac{C_0}{2}x^2 + \frac{C_1}{6}x^3 \tag{1}$$

(c) Sources: Next, the available information sources for the ego-lane estimation are introduced. For lane marking detections, the information of three separate camera systems is used. Two systems consisting of one monocular camera respectively supply redundant information and a surround view system combines the information of four fisheye cameras. Each camera system detects both right and left lane markings and provides us with clothoid approximations of lane and markings. Additionally, for each marking the distance from vehicle to starting point, the total length of the marking, the distance to the next obstacle along the clothoid as well as width, marking type and an existence value are given. For the ego-vehicle we track position, yaw angle, yaw rate, velocity, traveled lateral and longitudinal distance and velocity, rotation and the driven trajectory. Similar information is obtained about the leading vehicle, its trajectory given by a clothoid approximation (Eq. 1). Moreover, it is assigned an existence value and offsets to the left and right lane markings. Lastly, lane type and class as well as speed limits are determined (Fig. 3).

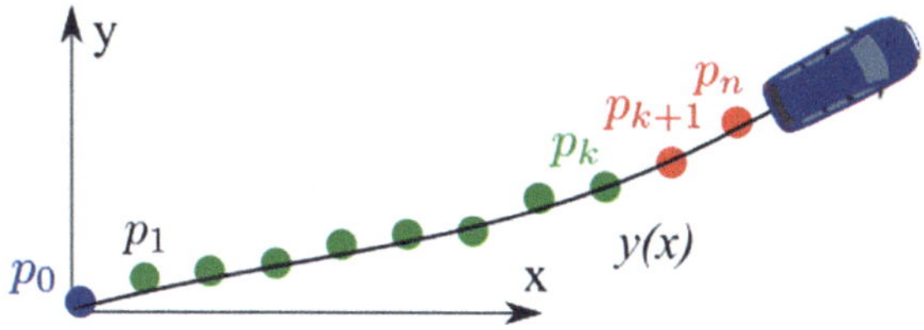

Fig. 3 The ground truth at p_0 is acquired by using linear polynomial regression of $p_0, \ldots, p_k$. The points $p_{k+1}, \ldots, p_n$ are available but they are left out due to exceeding the maximal distance or angle [2]

(d) Ground truth: As the ground truth we use the same as in [2]. Hereby, the formula $y(x) \approx \phi_0 \cdot x + \frac{C_0}{2}x^2 + \frac{C_1}{6}x^3$ is used since an estimation always has to start in the vehicle origin, such that the lateral offset can be omitted. This ground truth allows the comparison to the neural networks in [2]. Furthermore, we can use the same procedure of linear regression to create the ground truth values. Additionally, the curvature C_0 can be used to detect a curve if $|C_0| > 0.01$. Hence, the vehicle is in a left curve if $C_0 < -0.01$ and in a right curve if $C_0 > 0.01$. Otherwise, the ego-lane is approximately a straight line.

4 Data-Driven Lane Estimation with ANN

ANNs are successfully applied in a variety of perception tasks regarding autonomous driving, such as segmenting images [11] or extracting lane-marking information [14]. Therefore, we shortly introduce the approach of Nguyen et al. [2] for estimating the ego-lane since they directly compare to GP.

(a) Normalization: In the paper by Nguyen et al. [2] the input data for the ANNs is normalized to reduce the influence of artificial semantics and varying ranges. For that reason, they apply one-hot-encoding to the categorical input data transforming a categorical feature with n categories into an n-dimensional vector, where each entry is set to one if the index corresponds to the category and to zero otherwise. The varying ranges in the data are resolved by applying a min-max-scaling to the data set, so that all values are in the interval $[-1, 1]$. In addition to the normalization, a subset of the data is used consisting in equal parts of highway, rural, urban and connection scenarios.

(b) Structure: To determine the parameters of the estimation Nguyen et al. [2] use one network for each parameter. Each ANN consists of seven layers, where the first layer contains 80, the second 60, the third 40, the fourth 20, the fifth 10, the sixth 5 and the last one neuron. All layers except the last layer have the ReLU as activation function. The last layer has the identity function as activation function to enable an output of arbitrary real numbers. This structure should be able to deal with non-linear separable data and generalize the scenario features in small steps. Therefore, the ANNs are expected to achieve sufficient results.

(c) Training: For training, Nguyen et al. [2] use the stochastic gradient descent for 100,000 iterations on a training data set that is resampled regarding the road type to minimize the empirical risk. They use a batch size of $|B| = 25$ to prevent overfitting the data. A learning rate of $\eta = 0.0001$ is chosen that decreases to 0.9 times itself every 10,000 iterations. Additionally, a momentum weight of $\gamma = 0.0000001$ that is multiplied by 0.01 after the same amount of iterations is applied. Lastly, they scale the learning targets by multiplying them by 10,000 to lessen the effects of floating point errors.

5 Data-Driven Lane Estimation with GP

In the context of ego-lane estimation, a major problem with EAs is that they are only hardly applicable in a real-time environment. Hence, we generate a controller offline that is later used for estimating the ego-lane parameters online. We employ a standard evolutionary algorithm by Kruse et al. [19] and define the encoding, the fitness measure and the operators as follows:

(a) Encoding: All solutions in the search space need to be executable and able to deal with programs of arbitrary sizes. For that reason, we use expression trees by Koza [20] for representing the controller. A resulting requirement is that an expression tree for each parameter in the approximative clothoid is needed. We use the terminals in Table 1, which are an extension of the functionals used by Yakami et al. [9], where operators / and ln are the protected version of the operation that return one if the operation is undefined. Thus, our search space consists of all valid expression trees that can be created by using these functionals.

In addition to the protected operations, this encoding satisfies most of the important properties necessary for successful evolution. First, expression trees only differing in small subtrees are likely to have the same output, though the protected operators are somewhat of an exception. Second, the expression trees do not leave the search space unless the arity is violated.

(b) Fitness: Since the fitness of a tree has to measure how good it performs in real-world situations, we choose to minimize the error of each expression tree regarding the respective ground truth parameter. Therefor, we evaluate the tree for each trainings example and sum up the absolute differences between the result of the tree $\omega(k)$ and the ground truth value $groundtruth(k)$.

Table 1 Functionals

Operation	Arity	Functionality
$\begin{cases} Y & X \geq 0 \\ Z & \text{else} \end{cases}$	3	Case differentiation
max(X, Y)	2	Maximum
min(X, Y)	2	Minimum
X + Y	2	Adding two elements
X − Y	2	Subtracting two elements
X * Y	2	Multiplying two elements
X/Y	2	Protected division
exp(X)	1	e function
ln(X)	1	Protected natural logarithm
$\mathbb{R}$	0	Constant value in [−1, 1]
AT	0	Access to a certain feature

$$f_{abs}(\omega) = \sum_{k \in Trainingdata} |\omega(k) - groundtruth(k)| \tag{2}$$

Here, *Training data* is a subsample, consisting of left curves, right curves and straight lanes all in equal parts so that the expression trees learn to represent all curve scenarios. One problem that arises using f_{abs} as fitness is that this measure prefers solutions returning values near zero since the parameters are in a small interval around zero. However, a precision of at least two decimals is desired, so that we choose to increase the selective pressure by multiplying the ground truth values with 100. Hence, the less significant digits become more significant, such that possible solutions with more exact results are preferred. This leads to the following fitness measure:

$$f_{error}(\omega) = \sum_{k \in Trainingdata} |\omega(k) - 100 \cdot groundtruth(k)|$$

This, of course, makes it necessary for the tree to be normalized by replacing root node of the genetic program by a /-node, where the first child is the solution tree and the second child is 100. Another problem is that the trees become very large during the algorithm, which leads to a longer runtime.

$$f_{depth}(\omega) = \begin{cases} penalty^{depth(\omega)-depth} & , depth(\omega) > depth \\ 1 & , else \end{cases}$$
$$fitness(\omega) = f_{depth}(\omega) \cdot f_{error}(\omega) \tag{3}$$

For that reason, we introduce a penalty for trees, that exceed a certain threshold $depth$. The penalty f_{depth} grows depending on how big the difference between $depth$ and the depth of the tree $depth(\omega)$ is. Therefore, depending on the $penalty$ a small excess can be reasonable. In this paper, we use a $penalty$ of 1.01 and a $depth$ of seven.

(c) Algorithm: To complete the need to be defined. We use the typical crossover operation of swapping subtrees of two solution trees by Koza [20]. The mutation operator is implemented such that it traverses the expression tree randomly and replaces the chosen node by a small randomly grown tree. For the overall algorithm we take the basic evolutionary algorithm and adapt it to genetic programs. At the beginning, we randomly initialize trees, then mutation, crossover and Tournament Selection are applied iteratively for 200 iterations. We use a population containing 200 trees. During each iteration in the evolutionary process, we create 180 trees by mutation and 20 by crossover. The Tournament Selections always takes the best one out of five randomly selected trees.

6 Experimental Results

In this section we present the experimental results of our concept.[1] The used database is built with many driving hours in and around Wolfsburg with the following distribution: 50% urban roads, 25% highways, 15% rural and 10% on- and exit-ramps (connections). We examine if the proposed application of GP and ANNs results in an angle deviation of smaller than 2°.

(a) Evaluation Concept for the Ego-Lane Estimation: For evaluation, the ego-lane estimation E has to be compared to the ground truth GT. Therefor, we compare control points $cp_0, \ldots, cp_n$ on GT to the corresponding control points $c_0, \ldots, c_n$ on E after run lengths of 13, 16, ..., 37 and 40. Afterwards, the angle deviation metric introduced by Nguyen et al. [18] is applied to each pair of control points (cp_i, c_i) (Fig. 4).

The estimation E starts in $(x_{E,1}, y_{E,1})$ and the control point is $(x_{E,2}, y_{E,2})$. Analogously, the ground truth GT starts in $(x_{GT,1}, y_{GT,1})$ with the corresponding control point $(x_{GT,2}, y_{GT,2})$. Next, the deviation can be calculated as in Eq. 4.

$$\Delta\alpha = \left|\arctan\left(\frac{y_{E,2} - y_{E,1}}{x_{E,2} - x_{E,1}}\right) - \arctan\left(\frac{y_{GT}}{x_{GT}}\right)\right| \tag{4}$$

Additionally to the angle deviation the availability is observed as the proportion of estimations that achieve an angle deviation below 2°. We observe the availability in different scenarios since they require a different subset of data. For example, in a highway scenario, visible lane markings can usually be expected, whereas in an urban environment, there often are no lane markings so that the curbstones or the leading vehicle can prove more reliable. For that reason, we chose not only to evaluate the data set as a whole but to analyze the ego-lane estimation availability by partitioning the data into urban, highway, connection and rural scenarios.

(b) Results: In the following, we present the results of the estimations of both, GP (GEN) and ANN. These two approaches are compared to ego-lane estimations using the left lane markings and right lane markings of the first camera (FLH, FRH), of the second camera (SLH, SRH) and the third camera (TLH, TRH), to an estimation using the trail of the leading vehicle (VH) and finally to an existing deep learning

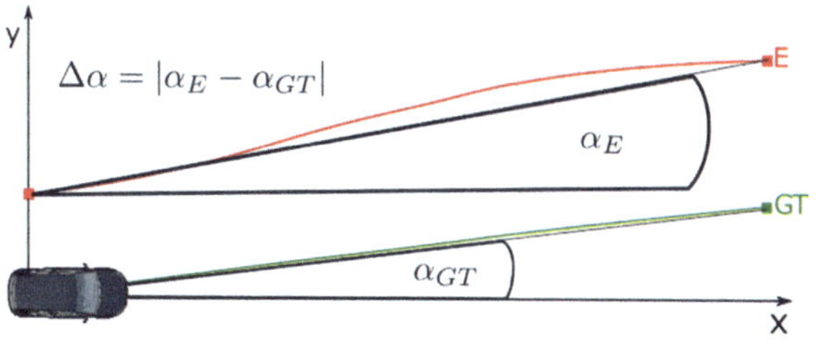

Fig. 4 Metric by Nguyen et al. [21, 22]

[1]The results of this work are not related to the series cars of Volkswagen.

approach (DL). The evaluation is performed on the data that is resampled regarding the curvature analogous to the GP training but on a distinct testing data set. By that, rarer (more difficult) curves appear more frequently and have a heavier impact on the evaluation.

Interestingly, an angle deviation smaller than 2° is nearly achieved Fig. 5. More specifically, the ANNs have a median slightly over 1° after 13 m whereas GEN's median is well below 1°. However, *most* of the estimations do not cross this threshold with VH performing best with a deviation that is firmly below 1°. Also, the performances of GP and ANN suffer greatly with growing distance, clearly visible in Fig. 5f, since small errors in the encoding of the estimation have a huge impact on the overall result. This implies that these techniques benefit from starting the estimation in the vehicle origin. Additionally, it can be seen that the performance of GEN exceeds ANN by 0.2°, showing that a simple model can be sufficient due to the generalization of the GP. Overall, ANN and GEN outperform all estimations in the beginning but are surpassed by DL at 28 m, which suffers from the later beginning of the estimation.

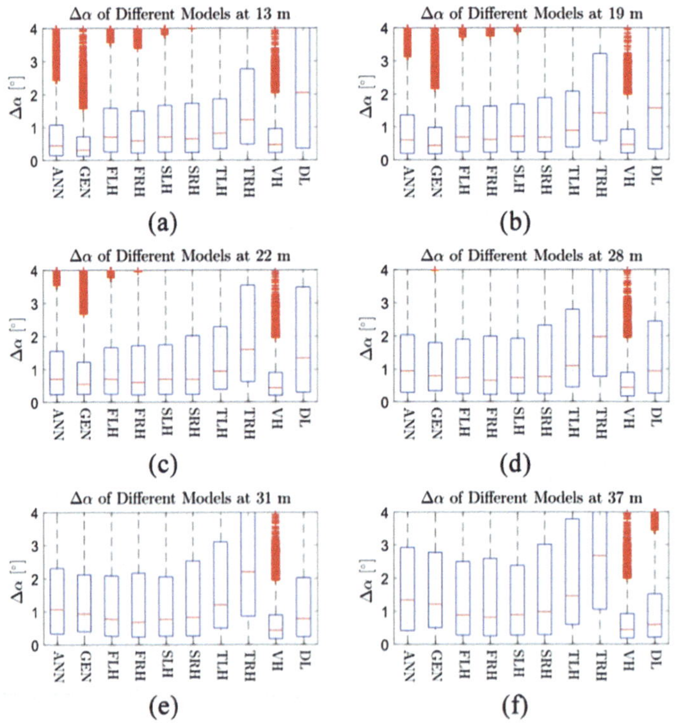

Fig. 5 Angle deviation of distinct ego-lane estimations at multiple distances. ANN and GEN denote the proposed techniques. LH and RH are estimations only considering the right and the left lane markings accordingly, whereas DL is a deep learning approach

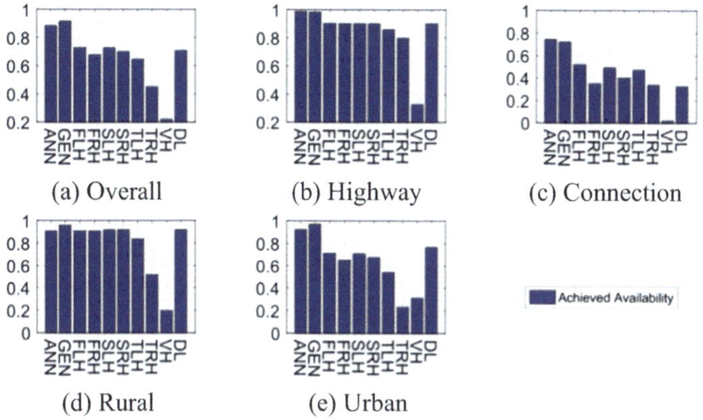

Fig. 6 Availability in multiple scenarios. An estimation is judged as available if in $\Delta\alpha$ Eq. 4 is smaller than 2° at a distance of 30 m

Next, we analyze the availability of the ego-lane estimations. Starting with the overall performance on a data set, where all scenarios are equally represented in Fig. 6a, GEN outperforms the other estimators by achieving an availability of 0.91, followed by ANN with an availability of 0.89. Availabilities of the lane-marking-dependent estimations and DL are below 0.75. Moreover, despite its performance in angle deviation the leading vehicle hypothesis has the worst availability by a large margin due to the rareness of a leading vehicle. Breaking these results down into different scenarios, it can be seen that both ANN and GEN perform especially well in highway scenarios with an availability of above 0.99 and 0.98 respectively. The lane marking hypotheses of the first and second camera and DL profit similarly from the high presence of easily detectable lane markings and achieve availabilities of 0.91. The performance of the surround view system (TLH, TRH), on the other hand, could be explained by the shorter vision range or by errors of reversing the effect of the fisheye cameras. In contrast to the highway scenarios, in connections (Fig. 6c) the general availability is significant lower with at most 0.74 for ANN, slightly higher than GEN with 0.72. These results of lane marking hypotheses probably are due to the insufficient detections of lane markings in tight road courses with the camera systems. In contrast, ANN and GEN can incorporate the information of all cameras and the current status of the vehicle. In rural scenarios in Fig. 6d most estimators achieve an availability of around 0.9 except for TRH and VH. While the low performance of VH is expected, the performance of TRH is not and could be due a wrong calibration or a disadvantageous subset of the data. The low performance of TRH continues in Fig. 6e, but, this time, is expected since the right markings or curbstones are often occluded by parking vehicles so that the fisheye is not able to capture them. In these scenarios, ANN and GEN exceed the availability of all other estimators with availabilities over 0.9, followed by DL with 0.77. The other estimators achieve an availability of around 0.7.

(c) Discussion: In the following, we take a closer look at weaknesses of the general concept. Starting with the ground truth acquisition, it is observable that the usage of the polynomial is not as sufficient as expected since the error in the regression and evaluation is amplified. For instance, if the errors in the estimations of the according parameters are ϵ_{ϕ_0}, ϵ_{C_0} and ϵ_{C_1}, then the absolute error can be calculated as $||\epsilon_{\phi_0} \cdot x + \frac{\epsilon_{C_0}}{2}x^2 + \frac{\epsilon_{C_1}}{6}x^3)||$. Hence, it becomes clear that an error in C_1 has a devastating effect on the performance. For example, an estimation error of error of $\epsilon_{C_1} = 0.001$ can lead to a lateral offset of 4.5 m, which can lead to leaving the lane. For that reason, a different representation for the ego-lane estimation could lead to improvements. Another problem is the evaluation of the ego-lane estimation using the real-world recordings of a prototype vehicle. Using this approach, exceptional situations are rarely recorded because data of dangerous situation and the recovery cannot be obtained while considering the safety of the driver during the recording. Moreover, the driven trajectory is not perfect due to the imperfect human driver which leads to the decision to remove such cases from the data set.

7 Conclusion

In this paper, it is shown that ANNs and GP are able to achieve the specified goal of achieving an angle deviation of at most 2°. The results also show an increased availability of the ego-lane estimations of around 90% compared to simple lane following or deep learning approaches on images of a single camera. Therefore, research considering both approaches with an improved representation of the ego-lane could lead towards further advances in autonomous driving.

In addition to the results of this work, there are multiple ways to improve the presented ego-lane estimation with AI-techniques. First, the use of an approximation of a clothoid by a polynomial is not recommendable due to the error amplification. The reference can be improved by choosing a different representation that reacts less to errors in the estimated values. Second, recurrent artificial neural networks can be applied so that the temporal information and the last estimation can be incorporated in the estimation process. Third, GP can be improved by increasing the allowed depth to enable a higher complexity of the expression trees and therefore a higher estimation quality. A further possibility to increase the performance is to apply a feature selection to remove unnecessary features and decrease the search space. Fourth, the estimation of the GP can be implemented as a hypothesis in the roadfusion framework in [1, 16].

References

1. T.T. Nguyen, J. Spehr, D. Vock, M. Baum, S. Zug, R. Kruse, A general reliability-aware fusion concept using DST and supervised learning with its applications in multi-source road estimation, in *2018 IEEE Intelligent Vehicles Symposium (IV)* (2018)
2. T.T. Nguyen, J.-O. Perschewski, F. Engel, J. Kruesemann, J. Sitzmann, J. Spehr, S. Zug, R. Kruse, Reliability-aware and robust multi-sensor fusion towards ego-lane estimation using artificial neural networks, in *Information quality in Information Fusion and decision making*, ed. by G. Rogova, E. Bosse (Springer, 2018)
3. *Statistisches Jahrbuch Deutschland 2017*, 1st edn. Statistisches Bundesamt, Wiesbaden, 2017)
4. C. Chen, A. Seff, A. Kornhauser, J. Xiao, DeepDriving: learning affordance for direct perception in autonomous driving, in *IEEE Inernational Conference on Computer Vision*, 2015, pp. 2722–2730
5. D.A. Pomerleau, Efficient training of artificial neural networks for autonomous navigation. Neural Comput. **3**(1), 88–97 (1991)
6. M. Bojarski, D.D. Testa, D. Dworakowski, B. Firner, B. Flepp, P. Goyal, L.D. Jackel, M. Monfort, U. Muller, J. Zhang, X. Zhang, J. Zhao, K. Zieba, End to end learning for self-driving cars. CoRR, vol. abs/1604.07316 (2016)
7. Z. Chen, X. Huang, End-to-end learning for lane keeping of selfdriving cars, in *2017 IEEE Intelligent Vehicles Symposium (IV)*, 2017, pp. 1856–1860
8. M. Al-Qizwini, I. Barjasteh, H. Al-Qassab, H. Radha, Deep learning algorithm for autonomous driving using googlenet, in *2017 IEEE Intelligent Vehicles Symposium (IV)*, 2017, pp. 89–96
9. G. Yakami, I. Tanev, K. Shimohara, S. Katagiri, M. Ohsaki, Automobile driving support system evolved by genetic programming, in *TENCON 2016—IEEE Region 10 Conference*, 2016, pp. 255–258
10. M. Valente, B. Stanciulescu, Real-time method for general road segmentation, in *IEEE Intelligent Vehicles Symposium (IV)*, 2017, pp. 443–447
11. V. Badrinarayanan, A. Kendall, R. Cipolla, Segnet: a deep convolutional encoder-decoder architecture for image segmentation. IEEE Trans. Pattern Anal. Mach. Intell. (2017)
12. G. Prabhakar, B. Kailath, S. Natarajan, R. Kumar, Obstacle detection and classification using deep learning for tracking in high-speed autonomous driving, in *2017 IEEE Region 10 Symposium (TENSYMP)*, 2017, pp. 1–6
13. D. Zhou, V. Frémont, B. Quost, Y. Dai, H. Li, Moving object detection and segmentation in urban environments from a moving platform. Image Vis. Comput. **68**, 76–87 (2017)
14. J. Kim, C. Park, End-to-end ego lane estimation based on sequential transfer learning for self-driving cars, in *2017 IEEE Conference on Computer Vision and Pattern Recognition Workshops (CVPRW)*, 2017, pp. 1194–1202
15. Q. Li, L. Chen, M. Li, S.L. Shaw, A. Nüchter, A sensor-fusion drivable-region and lane-detection system for autonomous vehicle navigation in challenging road scenarios. IEEE Trans. Veh. Technol. **63**(2), 540–555 (2014)
16. T.T. Nguyen, J. Spehr, J. Xiong, M. Baum, S. Zug, R. Kruse, Online reliability assessment and reliability-aware fusion for ego-lane detection using influence diagram and bayes filter, in *IEEE Conference on Multisensor Fusion and Integration for Intelligent Systems*, 2017, pp. 7–14
17. E.D. Dickmanns, B.D. Mysliwetz, Recursive 3-d road and relative ego-state recognition. IEEE Trans. Pattern Anal. Mach. Intell. **14**(2), 199–213 (1992)
18. T.T. Nguyen, J. Spehr, J. Xiong, M. Baum, S. Zug, R. Kruse, A survey of performance measures to evaluate ego-lane estimation and a novel sensor-independent measure along with its applications, in *IEEE Conference on Multisensor Fusion and Integration for Intelligent Systems*, 2017, pp. 239–246
19. R. Kruse, C. Borgelt, C. Braune, S. Mostaghim, M. Steinbrecher, *Computational Intelligence: A Methodological Introduction*, 2nd edn., ser. Texts in Computer Science (Springer London and Imprint: Springer, London, 2016)

20. J.R. Koza, Hierarchical genetic algorithms operating on populations of computer programs, in *Proceedings of the 11th International. Joint Conference on Artificial Intelligence*, vol. 1, ser. IJCAI'89 (Morgan Kaufmann Publishers Inc, San Francisco, CA, USA, 1989), pp. 768–774
21. T.T. Nguyen, J. Spehr, J. Sitzmann, M. Baum, S. Zug, R. Kruse, Improving ego-lane detection by incorporating source reliability, in *Multisensor Fusion and Integration in the Wake of Big Data, Deep Learning and Cyber Physical System*, ser. Lecture notes in electrical engineering, vol. 501, ed. by S. Lee, H. Ko, U.D. Hanebeck, F. Sun, S. Oh (Springer Intl. Publishing, 2018)
22. T.T. Nguyen, J. Spehr, M. Uhlemann, S. Zug, R. Kruse, Learning of lane information reliability for intelligent vehicles, in *IEEE International Conference on Multisensor Fusion and Integration for Intelligent Systems*, 2016, pp. 142–147

Two Phase Approach to Detection of Software Projects with Similar Architecture Based on Clustering and Ontological Methods

Nadezhda Yarushkina, Pavel Dudarin, and Gleb Guskov

Abstract Everyday in open repositories like GitHub, GitLab and others lots of software projects are created and updated. A method to determine similar source code or even better reusing project architecture could reduce an amount of wasted resources. In this paper two phase approach to detection of software projects with similar architecture based on NLP, clustering and ontological methods is proposed. This approach starts with natural language processing techniques and clustering to narrow the scope of processed software projects. And then goes stage of a novel described method for comparing software projects based on ontological representation of their architecture. This method allows to obtain project's architecture automatically from the projects source code. In order to estimate results several similarity metrics of similarity between projects are proposed.

1 Introduction

Well known that human resources in modern software development are the most valuable. Nevertheless often happens that the same tasks are solved by independent engineers multiple times, this leads to an ineffective software development process organization. Existed approaches allow to reuse certain functions and classes only and do not allow reveal the projects similarity on the basis of their subject area.

Knowledges about architecture extracted from already implemented projects in the same subject area could allow borrowing and reusing much larger parts of projects and avoiding conceptually incorrect solutions in the future.

N. Yarushkina · P. Dudarin · G. Guskov (✉)
Information Systems Department, Ulyanovsk State Technical University, Ulyanovksk 432027, Russia
e-mail: g.guskov@ulstu.ru

N. Yarushkina
e-mail: jng@ulstu.ru

P. Dudarin
e-mail: p.dudarin@ulstu.ru

S. N. Shahbazova et al. (eds.), *Recent Developments and the New Direction in Soft-Computing Foundations and Applications*, Studies in Fuzziness and Soft Computing 393, https://doi.org/10.1007/978-3-030-47124-8_12

Search on open repositories is carried out on the basis of keywords. This search could result in returning several thousand projects, which can not be handled by hand. The choice of projects set based on project subject area and inner architectural solutions is a promising approach. A tool capable of determining similarities between projects could store, extract and compare projects structure. Project subject area could be obtained by analyzing project description which is usually done in "readme.txt" file in versions control system repository and by scanning issue forums of the project. This task requires implementation of natural language processing (NLP) which is widely used in artificial intelligence and data science world. An interesting approach to forum analysis could be found in [1]. The state of the art technique in this area is word2vec models of natural language [2, 3]. Nowadays there are many available pre-trained word2vec models [4, 5], so there only thing to do is additionally tune word2vec model for specific task.

Further, to implement projects comparison based on their structure a tool for architectural concept extraction is needed. The UML language were developed to describe the project architecture with the required abstraction level. Based on the results of our previous research [6], could be concluded that uml-diagrams could been used as source for architecture extraction. Ontology in OWL 2 format could well act as a knowledge storage system for the project analysis tool. Attempts to integrate ontologies into software development were carried out at different levels: technical documents [7–10], maintenance and testing of the source code [11], UML diagrams [12, 13].

The minimal structural UML diagram elements, such as classes, interfaces, objects themselves, weakly convey the semantics and architectural solutions of the project. But combination of such an elements is much better describes the architecture. Stable combinations of structural elements are known as design patterns, this term exists in information technology for a long time but it is still relevant. Design patterns are actively used by the developer community, thus representing a reliable benchmark in the software project analysis. In addition, it makes sense to create local design patterns that solve specific task in a given subject area. There is a complete approach to development based on a domain known as development based on the subject area [14–16].

2 Formulation of the Problem

The current research is a continuation of previous authors' research described in [6]. The system described in that research made it possible to extract information from conceptual models and save it as an ontology of a certain format. But the life cycle and practices of development in IT companies show that UML diagrams are created mostly once at the beginning of the project and in rare cases updated at the beginning of each stage of the project.

Comparison of information obtained from conceptual models of a new project at the design stage and information obtained from the source code of projects that have already been implemented could allow to determine projects structural similarity.

In order to be able to analyze and measure the projects structural similarity, it is necessary to transform information about projects from different sources to a single format. The most convenient way of presentation of the extracted information is a form of ontology using OWL format. OWL ontology format allows to preserve semantics of complex architectural solutions and to perform logical operations on statements.

Before the OWL transformation phase there is a preliminary filtering phase. This phase filters out non appropriate software projects based on their subject area. As long as there is no any common tag system or any widely used classifier of software projects the only possible way is to analyze projects' descriptions. This could be done by means of natural language processing techniques and clustering algorithm. In this paper a combination of approaches suggested in papers [17, 18] for short text clustering based on semantic similarity measure obtained from wod2vec pre-trained model is adopted and developed.

3 Software Projects Preliminary Filtering

3.1 *Fuzzy Hierarchical Classifier Constructing*

To achieve the goal of construction fuzzy hierarchical classifier an approach from authors' previous paper [18] was adopted. This algorithm designed for a set of sentences (short text fragments). For example, project descriptions, forum discussions of the project, set of comments from source code, etc. could be taken as sentences. For the experiment there were selected projects with a set of keywords ("api", "java", "mobile", "sdk") from source code repositories like GitHub, GitLab and others. Totally there were 473 projects in the input dataset, and their general project descriptions have been taken as sentences. As long as there is not possibility to show all the input data, it Table 1 were placed the most demonstrative samples translated into English. Some of them are obviously appropriate, and after the clustering procedure will be searched for design patterns. While the others are obviously are not appropriate and will be moved to different clusters and thus filtered out.

Procedure of tokenization and lemmatization were performed for all the sentences. Fuzzy graph based on semantic similarity measure between obtained terms based on pre-trained word2vec model could be constructed. The last step is to perform clustering by ϵ-clustering algorithm which will result in obtaining hierarchical fuzzy graph. This graph could be treated as a fuzzy hierarchical classifier, see Fig. 1. On this figure only two brunches of hierarchy are presented for the reason of space and clear visibility. These two sub-hierarchy shows two semantically related groups of words, the first one for the programming and API and the second one for the music. In

Table 1 A set of sentences for software projects preliminary filtering

Project name	Sentence
VKCOM/vk-java-sdk	Java library for VK API interaction, includes OAuth 2.0 authorization and API methods. Full VK API features documentation can be found here
dewarder/HoldingButton	Button which is visible while user holds it. Main use case is controlling audio recording state (like in Telegram, Viber, VK)
korobitsyn/VKOpenRobot	VK Open Bot is a library for bot creation for VK social network. Main features: mass friends collection, mass group searching and aggregation, user detailed information, user status detection
gleb-kosteiko/vkb	Script allows you to automate the searching and participation in random reposts competitions in vk.com. Check friends walls for competition posts and repost these posts
PhoenixDevTeam/Phoenix-for-VK	First open-sourced VK client for Android inspired by Material Design
petersamokhin/vk-bot-java-sdk	Comfortable and simple library for creating bots for VK
strelnikovkirill/VKPhotoApp	Android OS + VK Api. VK application for a surfing in user news feed, but this news feed build only on posted photo
vladgolubev/nowplayingVk	This app broadcasts currently playing song from last.fm account to vk.com status
shavkunov/vk-analyzer	Application used to analyze wall of VK user or community and save results to internal database
asaskevich/VK-Small-API	Small Java API used for work with VK. Example of using is in VK Example.java. Authorization Counters of new messages, friends, answers and groups Total count of friends
akveo/cordova-vk	You can use this plugin to authenticate user via VK application rather than via webview. It makes use of official VkSDKs for iOS and Android
MLSDev/DroidFM	This application shows how you can integrate the RxJava, Realm, LastFM API, VK API for information on popular artists, their songs and albums
Try4W/VKontakteAPI	Simple, light weighted binding VK API for Java based on official Android SDK

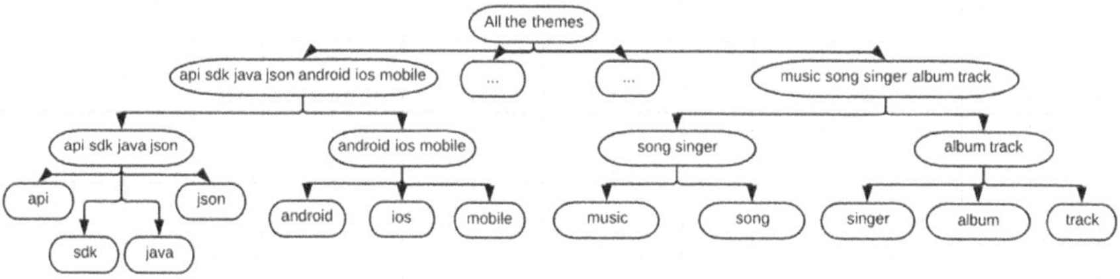

Fig. 1 Extract from hierarchical classifier of software project terms

the clustering result part below will be shown that these two groups lead to forming clusters dedicated to VK APIs and VK Players respectively.

3.2 Feature Construction

Obtained classifier could be treated as a data source to construct feature vectors. One possible form of features is a set of vectors where each vector coordinate equals to 0 or 1 depends on whether or not the current sentence have at least on of the words in common with a set of words assosiated with the current leaf node. So the dimension of these vectors would be equal to the count of leaves in the hierarchy. But this transformation loses any information about hierarchy, and treats all the words as equidistant.

To preserve information about words relations the transformation discussed in [?] could be used. In this transformation parent nodes are included into the feature vectors as additional coordinates and a special weight function is used to smooth out the effect from the root and other nodes that are close to the root one. But this transformation also has some disadvantages. The first one is a necessity to calibrate weight function to each type of hierarchy. The second one is that all the siblings of each node are treated equally. For example, if words ‘song’ and ‘singer’ are children of the node ‘song, singer’, and words ‘phone’ and ‘tv’ are children of the node ‘phone, tv, cable’, they will have the same pattern in their vector form. Their vectors will be look like $[\ldots, 1, 1, f(2), \ldots, 0, 0, 0, \ldots]$ and $[\ldots, 0, 0, 0, \ldots, 1, 1, f(2), \ldots]$. This happens because of the function f only depend on amount of the children and all the leaves are treated as orthogonal vectors to each other. The third disadvantage is that the vectors' dimension is twice higher because of non-leaf nodes.

In this paper another transformation is proposed. Let assume that there are two sibling nodes in hierarchy with the same parent node, then instead of assigning them vectors $[0, \ldots 0, 1, 0, 0, \ldots, 0]$ and $[0, \ldots 0, 0, 1, 0, \ldots, 0]$, as it was in the first approach, the vectors $[0, \ldots 0, x_1, x_2, 0, \ldots, 0]$ and $[0, \ldots 0, x_2, x_1, 0, \ldots, 0]$ could be used. Thus instead of distance $\sqrt{(1^2 - 0^2) + (0^2 - 1^2)} = \sqrt{2}$?, more appropriate distance could be used $\sqrt{(x_1^2 - x_2^2) + (x_2^2 - x_1^2)} = d(x_1, x_2)$, where $d(x_1, x_2) \in$

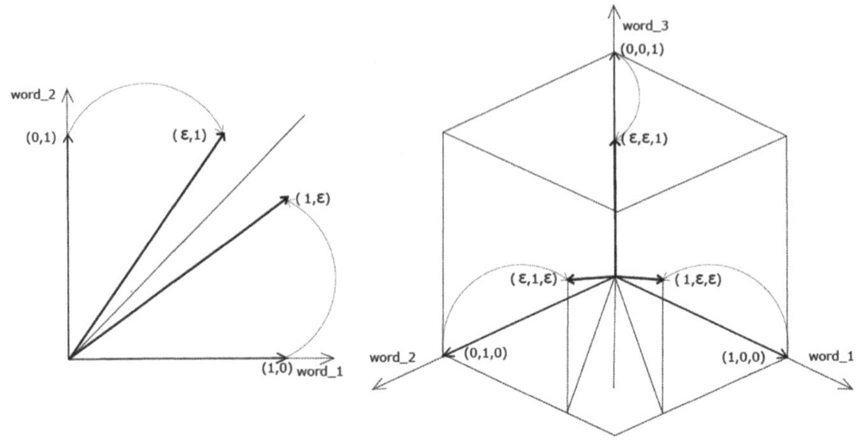

Fig. 2 Features constraction for 2- and 3- children nodes cases

[0, 1] and reflects the semantic distance between the words associated with the current leaf nodes. This transformation is illustrated in Fig. 2 where initially orthogonal vectors are moved closer to each other.

The algorithm of the transformation is following:

(1) Assume that every node except leaf ones has calculated minimum of semantic distances between its children nodes(words) denoted as ϵ_i. Actually, this value calculated during hierarchy construction part as a basis for $\epsilon - clustering$ algorithm.
(2) Start with the root node. Let it has k_1 children nodes, so for its children nodes define vectors $\big([1, \epsilon_1, \ldots, \epsilon_1], [\epsilon_1, 1, \epsilon_1, \ldots, \epsilon_1], \ldots, [\epsilon_1, \ldots, \epsilon_1, 1]\big)$, where each vector has k_1 coordinates.
(3) Get the first children node with vector $[1, \epsilon_1, \ldots, \epsilon_1]$. Let it has k_2 children nodes and minimum of semantic distances between its children is ϵ_2. Thus its vector transformed to the set of vectors :

$$\big([\underbrace{1, \epsilon_2, \ldots, \epsilon_2,}_{k_2\, coordinates}\ \underbrace{\epsilon_1, \ldots, \epsilon_1}_{k_1 - 1\, coordinates}], \tag{1}$$

$$[\epsilon_2, 1, \epsilon_2, \ldots, \epsilon_2, \epsilon_1, \ldots, \epsilon_1], \ldots, [\epsilon_2, \ldots, \epsilon_2, 1, \epsilon_1, \ldots, \epsilon_1]\big)$$

Vectors for the other $k_1 - 1$ nodes are transformed one to one to the set:

$$([\underbrace{\epsilon_1, \ldots, \epsilon_1}_{k_2\, coordinates}, \underbrace{1, \epsilon_1, \ldots, \epsilon_1}_{k_1 - 1\, coordinates}],$$

$$[\underbrace{\epsilon_1, \ldots, \epsilon_1}_{k_2\, coordinates}, \epsilon_1, 1, \epsilon_1, \ldots, \epsilon_1], \tag{2}$$

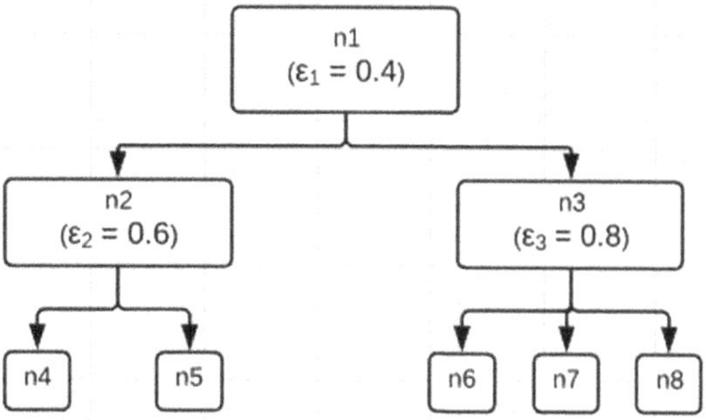

Fig. 3 Sample hierarchy to be vectorized

Table 2 Vectors for sample hierarch

Vectors/Nodes	n4	n5	n6	n7	n8
Vector n4	1	$\epsilon 2$	$\epsilon 1$	$\epsilon 1$	$\epsilon 1$
Vector n5	$\epsilon 2$	1	$\epsilon 1$	$\epsilon 1$	$\epsilon 1$
Vector n6	$\epsilon 1$	$\epsilon 1$	1	$\epsilon 3$	$\epsilon 3$
Vector n7	$\epsilon 1$	$\epsilon 1$	$\epsilon 3$	1	$\epsilon 3$
Vector n8	$\epsilon 1$	$\epsilon 1$	$\epsilon 3$	$\epsilon 3$	1

Table 3 Vectors for sample hierarchy 2

Vectors/Nodes	n4	n5	n6	n7	n8
Vector n4	1	0,6	0,4	0,4	0,4
Vector n5	0,6	1	0,4	0,4	0,4
Vector n6	0,4	0,4	1	0,8	0,8
Vector n7	0,4	0,4	0,8	1	0,8
Vector n8	0,4	0,4	0,8	0,8	1

$$\ldots, [\underbrace{\epsilon_1, \ldots, \epsilon_1}_{k_2\, coordinates}, \epsilon_1, \ldots, \epsilon_1, 1]$$

(4) Repeat step 3 for other nodes of the level.
(5) Recursively repeat steps 3 and 4 for all the levels in the hierarchy.

To give the intuition of this process short example could be discussed. Let us have a small sample hierarchy that will be vectorized. This hierarchy is shown in the Fig. 3. Its vectorized form could be found in Tables 2, 3, where the pre-calculated and calculated vectors are shown for the better understanding. In Tables 4, 5 euclidean and cosine distances for these nodes are calculated. As it could be seen these distances are quite adequate and preserve hierarchical information. The next step is to find the most appropriate groups of similar software projects.

Table 4 Euclidean and cosine distances for nodes of sample hierarchy

Nodes	n4	n5	n6	n7	n8
n4	0,00	0,57	1,04	1,04	1,04
n5	0,57	0,00	1,04	1,04	1,04
n6	1,04	1,04	0,00	0,28	0,28
n7	1,04	1,04	0,28	0,00	0,28
n8	1,04	1,04	0,28	0,28	0,00

Table 5 Euclidean and cosine distances for nodes of sample hierarchy 2

Nodes	n4	n5	n6	n7	n8
n4	0,00	0,09	0,23	0,23	0,23
n5	0,09	0,00	0,23	0,23	0,23
n6	0,23	0,23	0,00	0,02	0,02
n7	0,23	0,23	0,02	0,00	0,02
n8	0,23	0,23	0,02	0,02	0,00

3.3 Projects Clustering

As a clustering algorithm an HDBScan [19] has been chosen for the experiment. This clustering algorithm combined with features constructed from hierarchical classifier has been chosen for its ability to return quite $accurate$ and $pure$ clusters, mainly at the expense of $precision$. Metrics for precision, accuracy and purity are defined for each class as follows:

$$Precision_i = \frac{\max\limits_j CM_{i,j}}{\sum\limits_j CM_{i,j}} \tag{3}$$

$$Accuracy_i = \frac{\max\limits_j CM_{i,j}}{\sum\limits_k CM_{k,j_{\max}}} \tag{4}$$

$$Purity_i = \frac{\max\limits_j CM_{j,i}}{\sum\limits_j CM_{j,i}}, \tag{5}$$

where CM = [$cm_{i,j} = (\omega_i \cap c_j)$, ω_i - class with number i, c_j - cluster with number j].

Total $precision$, $accuracy$ and $purity$ are calculated as average values.

Fig. 4 Clustering results

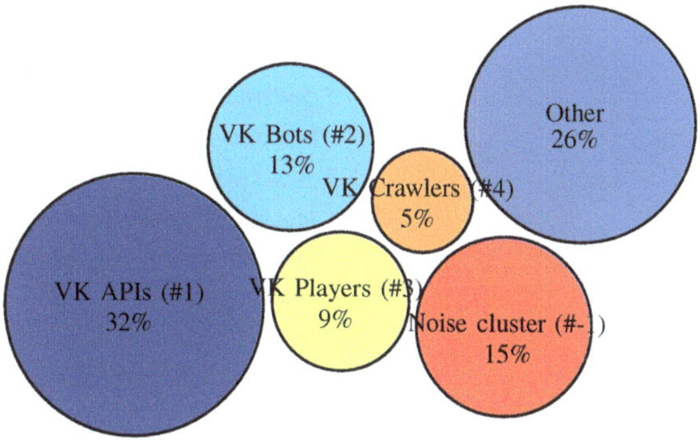

This method has shown quite a good performance results for a similar task which is discussed in paper [17].

During the clustering process 22 clusters we determined, main clusters with general description are presented on the Fig. 4. HDBScan algorithm was performed with parameter $min\_cluster\_size = 3$. Algorithm HDBScan as a true clustering algorithm always forms 'noise cluster' with number -1, where all the samples that could not be grouped are moved to.

In our example the 'noise cluster' is quite big, the main reason for this are too poor projects' description and high dimensional clustering features. More detailed discussion for this could be found in [17]. The 'other' cluster also have significant amount of projects because of many outliers in the input dataset.

Clustering labels for previously shown sentences could be found in Table 6. As it could be seen projects are grouped quite accurate. Calculated quality metrics based on expert evaluation are following: $precision = 0.689$, $accuracy = 0.985$, $purity = 0.972$. High level of accuracy and purity quite important in cases when expert evaluation follows the clustering process. In these cases homogeneous clusters are preferred to $precise$ ones. In our case clustering has been made to facilitate the process of software projects filtering, but the final decision is on expert.

For the next phase a software projects selection strategy could be different. One possible strategy is to choose one example from each cluster in order to have a good variety of possible architectures. The opposite strategy is to chose a few entire clusters in case of getting some clusters that generally satisfy business purposes of a new software project. In the experiment part of this paper the second strategy has been chosen. For the further processing sentences from clusters with numbers 1, 2 and 3 have been chosen.

Table 6 Sample of clustering results

Project name	Cluster
VKCOM/vk-java-sdk	1
dewarder/HoldingButton	−1
korobitsyn/VKOpenRobot	2
gleb-kosteiko/vkb	2
PhoenixDevTeam/Phoenix-for-VK	−1
petersamokhin/vk-bot-java-sdk	2
strelnikovkirill/VKPhotoApp	1
vladgolubev/nowplayingVk	3
shavkunov/vk-analyzer	4
asaskevich/VK-Small-API	1
akveo/cordova-vk	1
MLSDev/DroidFM	3
Try4W/VKontakteAPI	1

4 Software Design Ontology

If class diagram for software was built during design stage the structure analysis could be done. To complete this task ontology design approach was used and described below.

4.1 *UML Meta-Model Based Ontology*

As a target for storing knowlege from UML class diagrams has been chosen an OWL ontology format. The class diagram elements should be translated into ontology as concepts with their semantics consideration. Semantics of the whole diagram is being formed from the semantics of the diagram elements and the semantics of their relations. That is why the ontology was built on the basis of the UML meta-scheme, and not as a formal set of translated elements.

Ontology contains concepts that describe the most basic elements of the class diagram, but it could be expanded if necessary. During translation of the UML meta-scheme the following notations were applied.

Formally, the ontology of project diagrams is represented as a set:

$$O^{prj} = \langle C^{prj}, R^{prj}, F^{prj} \rangle, \tag{6}$$

where : $C^{prj} = \{c_1^{prj}, \ldots .c_i^{prj}\}$ – is a set of concepts that define main UML diagram elements such as : "Class", "Object", "Interface", "Relationship" and others;

R^{prj} – the set of connections between ontology concepts. These relationships allow to describe correctly rules of UML notation.

F^{prj} – is the set of interpretation functions defined on the relationships R^{prj}.

4.2 Design Patterns as Structural Parts of Software Projects

Design patterns are inserted into ontology as a set of individuals based on the ontology concepts described above. Semantic constraints and properties of design patterns are specified by the ObjectProperties and DatatypeProperties of OWL ontology. One of the most commonly used design patterns is the Builder.

Ontological representation of the design pattern:

$$O_{tmp_i}^{prj} = \{inst(C_1^{prj}), \ldots inst(r_1^{prj}), \ldots, r_{sameAs}\}, \tag{7}$$

In fact, the ontological representation of a single design pattern is a set of individuals of concepts and relations from the ontology of project diagrams.

To calculate the structural similarity of projects based on developed ontology, the following evaluation functions were proposed. The first metric gives priority to the maximum single expressed design pattern in both diagrams:

$$\mu_{dc_\gamma, dc_\delta} = \bigvee_{tmp \in (dc_\gamma \cap dc_\delta)} \mu_{dc_\gamma \cap dc_\delta}(tmp), \tag{8}$$

where dc_γ and dc_δ is projects class diagrams presented as UML metamodel ontology Abox expressions, $\mu_{dc_\gamma, dc_\delta(tmp)}$ - measure of expression the design pattern in project diagram.

The second metric considers the coincidence of all design patterns in equal proportions and does not considers design patterns with a measure of expression less than 0.3:

$$\mu_{dc_\gamma, dc_\delta} = (\sum_{tmp \in (dc_\gamma \cap dc_\delta) \geq 0.3} \mu_{dc_\gamma \cap dc_\delta})/N, \tag{9}$$

where N - count of design patterns with a measure of expression greater than 0.3 for both of projects.

The third metric works in the same way as the second one, but the contribution to the evaluation by design patterns depends on the number of elements in the design pattern (the design pattern with 20 elements means more than a design pattern with 5 elements):

$$\mu_{dc_\gamma, dc_\delta} = \Big(\sum_{tmp \in (dc_\gamma \cap dc_\delta) \geq 0.3} \widetilde{\mu}_{dc_\gamma \cap dc_\delta} \Big) / N, \tag{10}$$

where $\widetilde{\mu}_{dc_\gamma \cap dc_\delta}$ - weighted measure of expression.

5 The Results of Searching Structurally Similar Software Projects

5.1 Searching Design Patterns in Projects

To determine the measure of similarity between two projects, it is necessary to calculate an expression degree for each design pattern in each project. The expression measure of the design pattern in the project can be calculated by mapping a project ontology Abox on a design pattern ontology Abox. The Table 7 contains expression degree for each design pattern in each project.

5.2 Results of Searching Structurally Similar Software Projects by Different Metrics

This estimations are normalized from 0 to 1. For the first metric estimations are always equal to 1. This could be easily explained because first metric chooses the most expressed design pattern in both projects. Among the design patterns participating in the test there design patterns with a small number of elements, for example: Abstract superclass, interface and delegator. Such design patterns ware have measure of experssion equal 1 by first metric at both compared project. The estimations of

Table 7 Expression of design patterns in projects

Project name/Design pattern name	Delegator	Adapter	Builder	Abstract superclass	Interface
Android-MVP	1.0	0.875	0.83	1.0	1.0
cordova-social-vk	1.0	0.875	0.83	1.0	0.8
cvk	1.0	0.875	0.83	1.0	0.8
DroidFM	1.0	0.875	0.92	1.0	1.0
VK-Small-API	1	0.625	0.42	0.33	0.6
VKontakteAPI	1.0	0.875	0.83	1.0	0.8
VK_TEST	1	0.75	0.58	0.66	0.6

Table 8 Similarity between projects by second metric

Project 1/Project 2	android	cordova-vk	cvk	droidfm	vk-smallapi	vkontakteapi	vk_test
android-mvp	–	0.96	0.96	0.98	0.78	0.96	0.78
cordova-vk	0.96	–	1	0.94	0.85	1	0.83
cvk	0.96	1	–	0.94	0.85	1	0.83
droidfm	0.98	0.94	0.94	–	0.78	0.94	0.78
vk-small-api	0.78	0.85	0.85	0.78	–	0.85	0.96
vkontakte api	0.96	1	1	0.94	0.85	–	0.83
hline vk_test	0.79	0.83	0.83	0.78	0.95	0.83	–

Table 9 Similarity between projects by third metric

Project 1/Project 2	android-mvp	cordova-vk	cvk	droidfm	vk-small-api	vkontakte api	vk_test
android-mvp	–	0.96	0.96	0.96	0.64	0.96	0.77
cordova-vk	0.96	–	0.99	0.93	0.67	0.99	0.80
cvk	0.97	0.99	–	0.93	0.67	0.99	0.80
droidfm	0.97	0.93	0.93	–	0.61	0.93	0.74
vk-small-api	0.64	0.67	0.68	0.61	–	0.67	0.87
vkontakte api	0.97	0.99	0.99	0.93	0.67	–	0.80
vk_test	0.77	0.80	0.80	0.74	0.87	0.80	–

similarity calculation between projects by second and third metrics are presented in the Tables 8 and 9, respectively.

The results for the second and third metrics are also quite high. Design patterns with a expression degree less than 0.3 were excluded from consideration. All projects that participated in the comparison are downloaded from the open repository Github and realize interaction with the public API of the russian well-known social network vkontakte or with it's music or mobile API.

High estimations of similarity by second and third metrics are explained by applying the results of research first part an NLP procedures. Selection of projects was carried out by NLP procedures based on information obtained from: readme files, source code comments, version control system comments and issues excludes projects from other subject areas.

6 Conclusions

In this paper two phase approach of artificial intelligence to detection of software projects with similar architecture based on clustering and ontological methods is presented. Natural language processing, clustering analysis and ontology construction allow to find and explore projects with similar purposes and architecture. In the experimental part there is an application of proposed method to different projects. In order to measure obtained result several similarity metrics are proposed. Moreover, proposed approach could be seen as a framework to expand the ontology with new design patterns and compare more complex architectures. Also, expanding the system by using ontologies of subject areas can significantly increase the search result relevance.

Acknowledgements This work is supported by the RFBR and the Ulyanovsk region in the scientific projects 19-47-730006 and 18-47-730019.

References

1. X. Han, J. Ma, Y. Wu, C. Cui, A novel machine learning approach to rank web forum posts, Soft Comput. Springer:Berlin/Heidelberg **18**(5), 941–959 (2008)
2. T. Mikolov, I. Sutskever, K. Chen, G. Corrado, J. Dean, Distributed representations of words and phrases and their compositionality, in *Proceedings of the 26th International Conference on Neural Information Processing Systems* (Lake Tahoe, Nevada, 05–10 December 2013), pp. 3111–3119
3. Q. Le, T. Mikolov, Distributed representations of sentences and documents, in *Proceedings of the 31st International Conference on Machine Learning*, PMLR vol. 32 Issue 2, pp. 1188–1196 (2014)
4. A. Kutuzov, E. Kuzmenko, WebVectors: a toolkit for building web interfaces for vector semantic models, in *Proceedings International Conference on Analysis of Images, Social Networks and Texts (AIST 2016)* vol. 661(Springer:Cham, Moskow, Russian Federation, 27-29 July 2017), pp. 155–161
5. M. Pelevina, N. Arefyev, C. Biemann, A. Panchenko, Making Sense of Word Embeddings, in *Proceedings of the 1st Workshop on Representation Learning for NLP co-located with the ACL conference* (Berlin, Germany, 10 August 2017). arXiv:1708.03390
6. G. Guskov, A. Namestnikov, N. Yarushkina, Approach to the search for similar software projects based on the UML ontology, in *Proceedings of the Second International Scientific Conference "Intelligent Information Technologies for Industry" (IITI 2017)*, vol. 680 (Springer:Cham, Varna, Bulgaria, 14–16 September 2017), pp. 3–10
7. A. Namestnikov, A. Filippov, V. Avvakumova, An ontology based model of technical documentation fuzzy structuring, in *Proceedings CEUR Workshop, SCAKD 2016*, vol. 1687 (Moscow, Russian Federation, 18–22 July 2016), pp. 63–74
8. A. Koukias, D. Kiritsis, Rule-based mechanism to optimize asset management using a technical documentation ontology, in *15th IFAC Symposium on Information Control Problems in Manufacturing 2015, IFAC-PapersOnLine* vol. 48, Issue 3, pp. 1001–1006
9. U. Zagorulko, I. Ahmadeeva, D. Mouromtsev, Ontology-based information extraction for populating the intelligent scientific internet resources, in *International Conference on Knowledge Engineering and the Semantic Web* vol. 649, (Springer: Cham, Prague, Czech Republic, 21–23 September 2016), pp. 119–128

10. N. Yarushkina, A. Filippov, V. Moshkin, Development of the unified technological platform for constructing the domain knowledge base through the context analysis, in *Proceedings of the Conference on Creativity in Intelligent Technologies and Data Science (CIT&DS 2017)*, vol. 754 (Springer:Cham, Volgograd, Russian Federation, 12–14 September 2017), pp. 62–72
11. S. Hossein, K. Sartipi, Dynamic analysis of software systems using execution pattern mining, in *Proceedings 14th IEEE International Conference on Program Comprehension* (ICPC, IEEE Computer Society, Athens, Greece, 14–16 June 2006), pp. 84–88
12. J. Zedlitz, J. Jorke, N. Luttenberger, From UML to OWL 2, in *Proceedings of Third Knowledge Technology Week (KTW 2011)*, vol. 295 (Springer:Berlin/Heidelberg, Kajang, Malaysia 18–22 July 2011), pp. 154–163
13. F. Bobillo, U. Straccia, Representing fuzzy ontologies in OWL 2, in *Proceedings International Conference on Fuzzy Systems* (Barcelona, Spain, 18–19 July 2010), pp. 2695–2700
14. P. Wongthongtham, U. Pakdeetrakulwong, S. Marzooq, Ontology annotation for software engineering project management in multisite distributed software development environments, in *Software Project Management for Distributed Computing* (Zaigham Mahmood; Springer:Cham, 2017), pp. 315–343, 978-3-319-54325-3
15. A. Emdad, Use of ontologies in software engineering, in *Proceedings of 17th International Conference on Software Engineering and Data Engineering* (Los Angeles, California, USA, 30 June–2 July 2008), pp. 145–150
16. T. Dillon, E. Chang, P. Wongthongtham, Ontology-based software engineering-software engineering 2.0, in *Proceedings of Australian Software Engineering Conference, IEEE Computer Society* (26–28 March 2008), pp. 13–23
17. P. Dudarin, N. Yarushkina, Features construction from hierarchical classifier for short text fragments clustering. Fuzzy Syst. Soft Comput. **12**, 87–96 (2017)
18. P. Dudarin, N. Yarushkina, An approach to fuzzy hierarchical clustering of short text fragments based on fuzzy graph clustering, in *Proceedings of the Second International Scientific Conference "Intelligent Information Technologies for Industry" (IITI 2017)*, vol. 679 (Springer:Cham, Varna, Bulgaria, 14–16 September 2017), pp 295–304
19. L. McInnes, J. Healy, S. Astels, Hdbscan: hierarchical density based clustering. J. Open Source Softw., Open J. **2**(11) (2017)

Development of a System Based on a Fuzzy Knowledge Base Filled from Various Formats Documents

Nadezhda Yarushkina, A. A. Filippov, V. S. Moshkin, and Gleb Guskov

Abstract The article describes the fuzzy knowledge base developing process. The fuzzy knowledge base content represents a knowledge extracted from set of various documents by subject area. Set of documents contains of the wiki-resources, uml-diagrams, text documents and projects source code. A fuzzy knowledge base was implemented on the Neo4j database. An attempt to implement the mechanism of logical inference based on concepts, relationships and rules of a graph database was made. This mechanism is used for generate the screen forms of the user interface dynamically. The contexts allow to represents the content of the fuzzy knowledge base in space and time. Each unique context was assigned as a linguistic label, for example, low, middle, high. This label determines the competence of the expert in the subject area. Time contexts allows storing the history of knowledge base content changes. It allows holding and obtaining a specific state of the knowledge base contents.

This work is supported by the RFBR and the Ulyanovsk region in the scientific projects 19-47-730006 and 18-47-730019.

N. Yarushkina · A. A. Filippov · V. S. Moshkin · G. Guskov (✉)
Department of Information Systems, Ulyanovsk State Technical University, Ulyanovsk 432027, Russia
e-mail: g.guskov@ulstu.ru

N. Yarushkina
e-mail: jng@ulstu.ru

A. A. Filippov
e-mail: al.filippov@ulstu.ru

V. S. Moshkin
e-mail: v.moshkin@ulstu.ru

S. N. Shahbazova et al. (eds.), *Recent Developments and the New Direction in Soft-Computing Foundations and Applications*, Studies in Fuzziness and Soft Computing 393, https://doi.org/10.1007/978-3-030-47124-8_13

1 Introduction

People from all over the world operates with huge volumes of information both in everyday and professional activities. A large amount of information causes difficulties in decisions making within strict time constraints.

To solve this problem, many software tools are using to automate human activity. However, it is necessary to adapt them to the specifics of a particular subject area and its contexts for the effective operation of such tools [1–7].

Thus, customized software for automation solves such tasks more efficiently, but customizing procedure require considerable human and temporary resources.

In this article, an attempt to construct a fuzzy knowledge base as a core of the intelligent system Athene platform [8] was made. The content of the fuzzy knowledge base represent as an applied ontology. The basic requirements for fuzzy knowledge base are:

- Adaptation to the specifics of subject area based on contexts;
- Reliability and speed of ontology storage;
- The presence of a mechanism of logical inference;
- Availability of mechanisms for importing data from external information resources [8].

The contexts allow to represents the content of the fuzzy knowledge base in space and time [9, 10].

Space contexts allow to takes into account the competence of the subject area expert. Each space context was assigned a linguistic label, for example, low, middle, high. This label determines the competence of the expert in the given subject area. The membership function has defined for each linguistic label. The core ontology based on linguistic labels values. Time contexts allow to stores the history of knowledge base content changes. It allow holding and obtaining to a specific state of the knowledge base contents.

2 The Organization of the Ontology Store of Fuzzy Knowledge Base

Ontology of the subject area represented in the form of a.

Ontology storage of knowledge base based on graph-oriented database management system (Graph DBMS) Neo4j. Neo4j is currently one of the most popular graph databases and it has the following advantages:

1. Free community version.
2. Native format for data storage.
3. One copy of Neo4j can work with graphs containing billions of nodes and relationships.
4. The presence of a graph-oriented query language Cypher.

5. Availability of transaction support [11].

Figure 1 shows the system architecture that consists from the following subsystems:

Ontology storage:

- Neo4j [12];
- Content management module;
- ontology import/export module (RDF, OWL).

Inference subsystem;

A subsystem for importing knowledge from relational database managements systems;

A subsystem for importing knowledge from software source code and uml-diagram:

- Module for importing knowledge from software source code;
- Module for importing knowledge from uml-diagram in .xmi format.

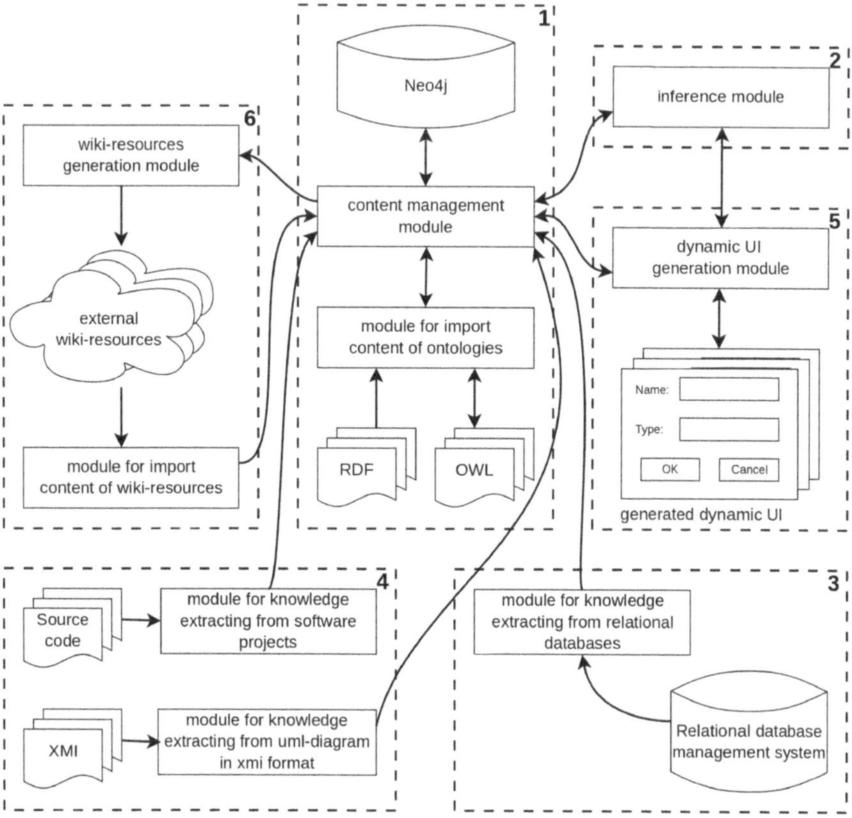

Fig. 1 Architecture of a fuzzy knowledge base of the Athene platform

A subsystem for interaction with users consists from screen forms generation module.

A subsystem for importing data from external information resources:

- Module for importing data from external wiki-resources;
- Module for filling ontology from external wiki-resources.

Neo4j ware chosen to store the subject area description in the applied ontology form. It's possible because ontology can be presented as a graph. In this case, it is only necessary to limit the set of nodes and graph relations of ontologies in RDF or OWL format for successful transformation.

The context of an ontology is some state of ontology, obtained during building an ontology using different "points of view" or from many versions of ontology.

Formally, the ontology can been represented by the following equation:

$$O = \left\langle T, C^{T_i}, I^{T_i}, P^{T_i}, S^{T_i}, F^{T_i}, R^{T_i} \right\rangle, \ i = \overline{1, t}, \tag{1}$$

where t is a number of the ontology contexts,

$T = \{T_1, T_2, \ldots, T_n\}$ is a set of ontology contexts,

$C^{T_i} = \left\{C_1^{T_i}, C_2^{T_i}, \ldots, C_n^{T_i}\right\}$ is a set of ontology classes within the ith context,

$I^{T_i} = \left\{I_1^{T_i}, I_2^{T_i}, \ldots, I_n^{T_i}\right\}$ is a set of ontology objects within the ith context,

$P^{T_i} = \left\{P_1^{T_i}, P_2^{T_i}, \ldots, P_n^{T_i}\right\}$ is a set of ontology classes properties within the ith context,

$S^{T_i} = \left\{S_1^{T_i}, S_2^{T_i}, \ldots, S_n^{T_i}\right\}$ is a set of ontology objects states within the ith context,

$F^{T_i} = \left\{F_1^{T_i}, F_2^{T_i}, \ldots, F_n^{T_i}\right\}$ is a set of the logical rules fixed in the ontology within the ith context,

R^{T_i} is a set of the logical rules fixed in the ontology within the ith context,

$$R^{T_i} = \left\{ R_C^{T_i}, R_I^{T_i}, R_P^{T_i}, R_S^{T_i}, R_F^{T_i} \right\},$$

where $R_C^{T_i}$ is a set of relations defining hierarchy of ontology classes within the ith context,

$R_I^{T_i}$ is a set of relations defining the 'class-object' ontology tie within the ith context,

$R_P^{T_i}$ is a set of relations defining the 'class-class property' ontology tie within the ith context,

$R_S^{T_i}$ is a set of relations defining the 'object-object state' ontology tie within the ith context,

$R_F^{T_i}$ is a set of relations generated on the basis of logical ontology rules in the context of ith context.

Fuzzy knowledge base main principles are similar with object-oriented programming paradigm:

- Ontology classes are concepts of the subject area;
- Classes can have properties; the child-class inherits properties of the parent class;
- Objects of ontology describe instances of the concepts of the subject ontology;
- Specific values for the properties of objects inherited from the parent class are determined by the states;
- Logical rules are used to implement the functions of inference by the content of fuzzy knowledge base.

3 The Inference on the Contents of Fuzzy Knowledge Base

The inference is the process of reasoning from the premises to the conclusion. Reasoners are used to implement the function of inference. Reasoners form logical consequences on the basis of many statements, facts and axioms. The most popular at the moment reasoners are:

- Pellet [13];
- FaCT++;
- Hermit;
- Racer, etc. [14].

These reasoners are actively used in the development of intelligent software. However, Neo4j does not assume the possibility of using similar default reasoners. Thus, there is a need to develop a mechanism for inference based on the content of a fuzzy KB.

Currently the Semantic Web Rule Language (SWRL) is use for recording logical rules.

These SWRL rules describe the conditions under which object a has "nephew-uncle" relation with object c. Formally, the logical rule of the ontology of the fuzzy knowledge base is:

$$F^{T_i} = \langle A^{Tree}, A^{SWRL}, A^{Cypher} \rangle,$$

where T_i ith context of the ontology of the fuzzy knowledge base;
A^{Tree} a tree-like representation of a logical rule F^{T_i};
A^{SWRL} SWRL representation of the logical rule F^{T_i};
A^{Cypher} Cypher representation of the logical rule F^{T_i}.

The tree-view A^{Tree} of a logical rule F^{T_i} is:

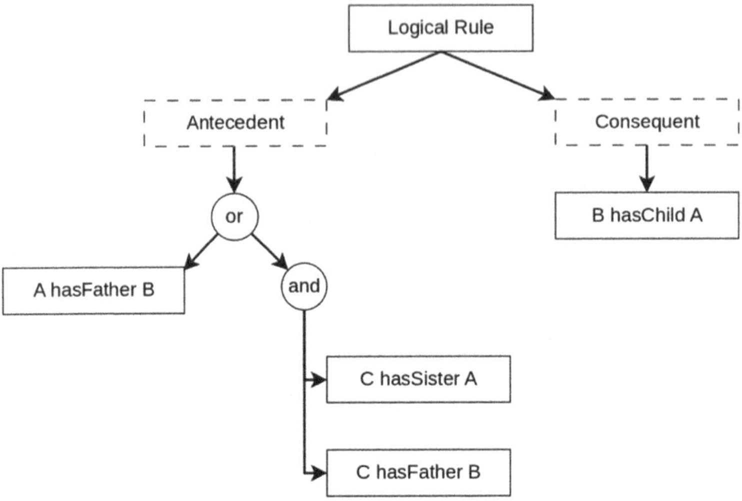

Fig. 2 Example of a tree-like representation of a logical rule

$$A^{Tree} = \langle Ant,\ Cons \rangle,$$

where

$Ant = Ant_1 \Theta Ant_2 \Theta \ldots Ant_n$	is the antecedent (condition) of the logical rule F^{T_i};
$\Theta \in \{AND,\ OR\}$	is a set of permissible logical operations between antecedent atoms;
$Cons$	consequent (consequence) of a logical rule F^{T_i}.

Figure 2 shows an example of a tree-like representation of two logical rules for the ontology of family relations. That rules describes the father-child relationships.

The tree-like logical rule is translated into the following SWRL (Fig. 2):

hasFather(?a,?b) => hasChild(?b,?a)
hasSister(?c,?a) & hasFather(?c,?b) => hasChild(?b,?a)
and the following Cypher view:
MATCH (s1:Statement{name: "hasChild", lr: true})
MATCH (r1a)<-[:Domain]-(:Statement{name: "hasFather"})-[:Range]->(r1b)
MERGE (r1b)-[:Domain]->(s1)
MERGE (r1a)-[:Range]->(s1)
MATCH (s1:Statement{name: "hasChild", lr: true})
MATCH (r2c)<-[:Domain]-(:Statement{name: "hasSister"})-[:Range]->(r2a)
MATCH (r2c)<-[:Domain]-(:Statement{name: "hasFather"})-[:Range]->(r2b)
MERGE (r2b)-[:Domain]->(s1)
MERGE (r2a)-[:Range]->(s1)

Thus, the rules are translated into their tree-view when imported into the knowledge base of logical rules on SWRL language.

The presence of a tree-like representation of a logical rule allows to form both a SWRL-representation of a logical rule and a Cypher-representation based on it.

Relations of a special type formed by using Cypher to represent the logical rule between entities of the ontology of the fuzzy knowledge base. These relations correspond to the antecedent atoms of the logical rule. Formed relationships provide the inference from the contents of the fuzzy knowledge base.

4 Extracting Knowledge from Relational Databases

Relational databases are widely used for data storing and contains subject area description in the form of interconnected tables. Nowadays, researchers of various scientific groups are involved in solving the problem of extracting knowledge from relational databases [15, 16].

The relational data model can be represented as the following expression:

$$RDM = (E, R),$$

where $E = \{E_1, E_2, \ldots, E_t\}$—set of database tables (entities), $R = \{R_1, R_2, \ldots, R_r\}$—set of relationships between database tables

$$R_i = E_j \frac{F(x)}{G(x)} E_k,$$

where E_j, E_k database entities,
$F(x)$ relationship between object E_j and object E_k, $G(x)$—relationship between object E_k and object E_j

Scope of functions $F(x)$ and $G(x)$ is U—single relationship and N—multiply relationship.

For mapping of relational database structure with knowledge base ontology structure special function was used:

$$F(RDM, O) : \{E^{RDM}, R^{RDM}\} \rightarrow \{C^O, P^O, R_P^O\},$$

where $\{E^{RDM}, R^{RDM}\}$ set of entities and relationships between them,
$\{C^O, P^O, R_P^O\}$ set of knowledge base ontology components.

In the process of mapping the structure of a relational database to the knowledge base ontology structure, it is necessary:

1. Create ith context of knowledge base ontology—T_i.
2. Create a set of ontology classes on the basis of C^{T_i} in the context of T_i. The number of classes should be equal to the number of tables in the relational database. The name of the class will be the name of the table.

Table 1 The correspondence between the relational database components and the ontology of the fuzzy knowledge base

The element of relation database	The ontology element of knowledge base
Table	Class
Table row	Object
Relations between tables, table column	Class property
External key	Object states

3. Create a set of class properties P^{T_i} in the T_i context based on existing relational database links and columns.
4. Create a set of relations $R_P^{T_i}$ linking the previously created classes and properties of the knowledge base ontology.

Importing data from a relational database to the knowledge base ontology were finish after mapping the structure of the relational database to the ontology structures of the knowledge base ends. Set of objects, states and relations of ontology: I^{T_i}, S^{T_i}, $R_I^{T_i}$ and $R_S^{T_i}$ are created during the import of data basis from the relational database (Row set) to the T_i context. Table 1 contains a comparison of knowledge base ontology components with relational database entities.

Thus, it becomes possible to extract knowledge from the contents of relational databases and present the extracted knowledge as a fuzzy ontology.

5 UML Meta-Model Based Ontology

Subsystem of extracting knowledge from UML-diagrams and source code based on specific ontology. T-box of this ontology corresponds to UML meta-model.

UML class diagram contains all knowledge from project. If class diagram for the project is not exists, representation for ontology could been obtained from project source code. Considered subsystem works with projects written on Java language [17].

As a structure for storing UML class diagrams was chosen an OWL ontology, because this format is the most expressive for representation of knowledge from complex subject areas [18]. The class diagram elements should be translated into ontology as concepts with considering to their semantics. Semantics of the whole diagram is being formed from the semantics of diagram elements and the semantics of their interaction. That is why the ontology was built on the basis of the UML meta-scheme, not a formal set of translated elements [19, 20].

To solve the problem of intellectual analysis of project diagrams, included in the project documentation, it is necessary to have knowledge in the area of construction of formalized diagrams [21, 22].

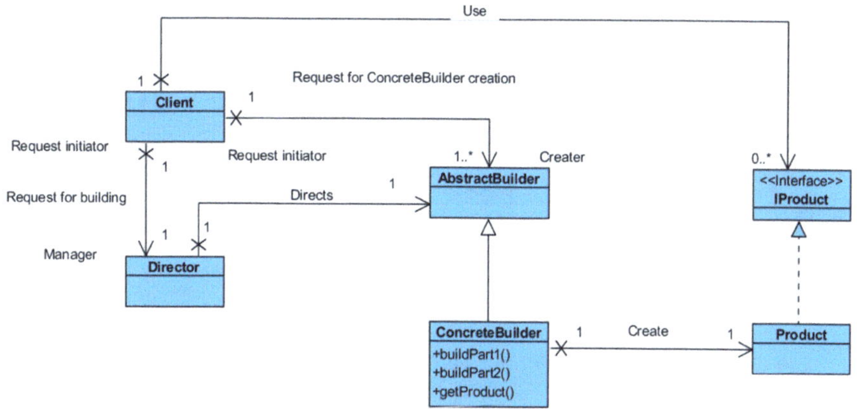

Fig. 3 Class diagram for builder design pattern

Ontology contains concepts that describe the most basic elements of the class diagram, but it can be expanded if necessary. When translating the meta-scheme of UML, the following notations were applied.

Formally, the ontology of project diagrams is represented as a set:

$$O^{PRJ} = \left(C^{PRJ}, R^{PRJ}, F^{PRJ}\right)$$

where $C^{PRJ} = \left\{c_1^{PRJ}, \ldots, c_i^{PRJ}\right\}$ is a set of concepts—is a set of concepts that define main UML diagram elements such as: "Class", "Object", "Interface", "Relationship" and others;

- R^{PRJ} the set of connections between ontology concepts. These relationships allow us to correctly describe the rules of UML notation.
- F^{PRJ} is the set of interpretation functions defined on the relationships R^{PRJ}.

On Fig. 3 represented UML class diagram for project that realize design pattern builder.

Figure 4 shows the A-Box ontology for representing a project that contains a Builder design pattern.

6 Building a Graphical User Interface Based on the Contents of a Fuzzy Knowledge Base

The dynamic graphical user interface (GUI) mechanism is used to simplify the work with knowledge base of untrained users and control of user input [23–25].

You need to map the fuzzy knowledge base ontology entities to the GUI elements to build a GUI based on the contents of the fuzzy knowledge base. Formally, the GUI

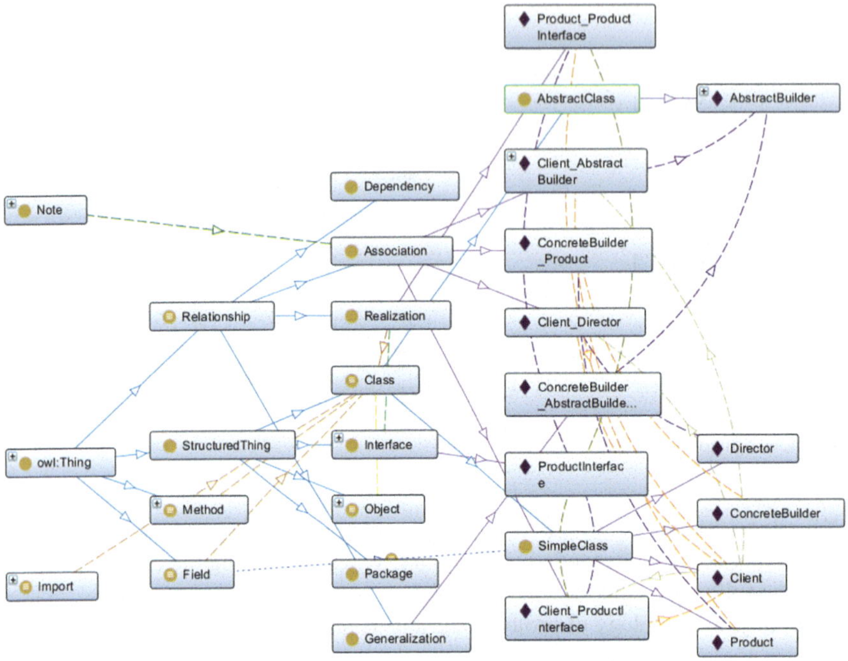

Fig. 4 "Builder" design pattern ontology presentation in Protege editor

model can be represented as follows:

$$UI = \langle L, C, I, P, S \rangle, \tag{2}$$

where $L = \{L_1, L_2, \ldots, L_n\}$ is a set of graphical GUI components (for example, ListBox, TextBox, ComboBox, etc.);

$C = \{C_1, C_2, \ldots, C_n\}$—is a set of ontology classes;

$I = \{I_1, I_2, \ldots, I_n\}$—is a set of ontology objects;

$P = \{P_1, P_2, \ldots, P_n\}$—is a set of properties of ontology classes;

$S = \{S_1, S_2, \ldots, S_n\}$—is a set of states of ontology objects of fuzzy knowledge base.

The following function is used to build a GUI based on fuzzy knowledge base:

$$F(O) = \left\{C^{T_i}, I^{T_i}, P^{T_i}, S^{T_i}, F^{T_i}, R^{T_i}\right\} \to \{L, C, I, P, S\},$$

where $\left\{C^{T_i}, I^{T_i}, P^{T_i}, S^{T_i}, F^{T_i}, R^{T_i}\right\}$ is a set of ontology entities of fuzzy knowledge base represented by expression 1 within the ith context;

$\{L, C, I, P, S\}$ is a set of GUI entities of fuzzy knowledge base represented by the expression 2.

Table 2 Correspondence between ontology elements and wiki elements

The ontology element of fuzzy knowledge base	The element of wiki-resources
Class	Category
Subclass	Subcategory
Object	Page
Class properties	The infobox elements (properties)
Object states	The infobox elements (values)
Relations	Hyperlinks

Thus, the contents of the fuzzy knowledge base are mapped to many GUI components.

This makes it easier to work with knowledge base for a user who does not have skills in ontological analysis and knowledge engineering. It also allows you to monitor the logical integrity of the user input, which leads to a reduction in the number of potential input errors.

7 Interaction of Fuzzy Knowledge Base with External Wiki-Resources

At present, wiki-technologies are used to organize corporate knowledge base [26]. It is necessary to solve the task of importing the content of such wiki-resources into fuzzy knowledge base. Table 2 contains the result of mapping the fuzzy knowledge base ontology entities to the wiki resource entities.

Thus, it becomes possible to import the content of external wiki resources for initial filling of knowledge base contents. There is also the possibility of the reverse process—generation of wiki-resources based on the contents of fuzzy knowledge base.

8 Conclusion

Thus, the use of fuzzy knowledge base stored in the Graph DBMS in the decision support process presupposes the existence of a certain set of mechanisms:

- organization of inference on the content of fuzzy knowledge base by translating SWRL-rules into Cypher-structures;
- building a graphical user interface based on the contents of fuzzy knowledge base;
- automated import of knowledge from internal and external wiki-resources.

These mechanisms allow you to automate the learning process of working with the system and simplify the work of specialists with a knowledge base.

The contexts of knowledge base allows to represent the content of the fuzzy knowledge base in space and time.

Space contexts allows to take into account the competence of the expert in the subject area. Each space context is assigned a linguistic label, for example, low, middle, high. This label determines the competence of the expert in the given subject area. The membership function is defined for each linguistic label.

The resulting ontology is formed on the basis of the values of linguistic labels.

Time contexts allows to store the history of changes of content of the knowledge base. It allows to return to a specific state of the contents of the knowledge base.

This approach allow to adapt the knowledge base to the subject area characteristics and requirements of specialists. Considered system provides them a convenient tool that is software dynamically changeable depending on the contents of the knowledge base.

References

1. J. Berant, A. Chou, R. Frostig, P. Liang, Semantic parsing on freebase from question-answer pairs, in *Proceedings of the 2013 Conference on Empirical Methods in Natural Language Processing (EMNLP)* (2013), pp. 1533–1544
2. D. Bianchini, V. De Antonellis, B. Pernici, P. Plebani, Ontology-based methodology for e-service discovery. Inf. Syst. **31**(4), 361–380 (2005)
3. D.M. Farid, M.A. Al-Mamun, B. Manderick, A. Nowe, An adaptive rule-based classifier for mining big biological data. Expert Syst. Appl. **64**, 305–316 (2016)
4. G. Guizzardi, N. Guarino, J.P.A. Almeida, Ontological considerations about the representation of events and endurants in business models, in *International Conference on Business Process Management* (2016), pp. 20–36
5. D. Rajpathak, R. Chougule, P. Bandyopadhyay, A domain-specific decision support system for knowledge discovery using association and text mining. Knowl. Inf. Syst. **31**, 405–432 (2012)
6. R.S. Renu, G. Mocko, A. Koneru, Use of big data and knowledge discovery to create data backbones for decision support systems. Procedia Comput. Sci. **20**, 446–453 (2013)
7. M. Rubiolo, M.L. Caliusco, G. Stegmayer, M. Coronel, M. G. Fabrizi, Knowledge discovery through ontology matching, an approach based on an artificial neural network model. Inf. Sci. **194**, 107–119 (2012)
8. N. Yarushkina, A. Filippov, V. Moshkin, Development of the unified technological platform for constructing the domain knowledge base through the context analysis, in *Creativity in Intelligent Technologies and Data Science* (Springer, 2017), pp. 62–72
9. T. Gruber, Ontology, in *Entry in the Encyclopedia of Database Systems*, ed. by L. Liu, M. Tamerzsu (Springer, 2008)
10. R.A. Falbo, G.K. Quirino, J.C. Nardi, M.P. Barcellos, G. Guizzardi, N. Guarino, An ontology pattern language for service modeling, in *Proceedings of the 31st Annual ACM Symposium on Applied Computing* (2016), pp. 321–326
11. N. Guarino, M.A. Musen, Ten years of applied ontology. Appl. Ontol. **10**(3–4), 169–170 (2015)
12. A. Zarubin, A. Koval, A. Filippov, V. Moshkin, Application of syntagmatic patterns to evaluate answers to open-ended questions, in *Creativity in Intelligent Technologies and Data Science* (Springer, 2017), pp. 150–162

13. D. Calvanese, B. Cogrel, S. Komla-Ebri, R. Kontchakov, D. Lanti, M. Rezk, M. Rodriguez-Muro, G.X. Ontop, *Answering SPARQL Queries over Relational Databases* [el. resource]. http://www.semantic-web-journal.net/system/files/swj1278.pdf. Accessed 11 February 2018
14. Pellet Framework. https://github.com/stardog-union/pellet. Accessed 10 January 2018
15. Neo4j. https://neo4j.com/product. Accessed 10 January 2018
16. A. Poggi, D. Lembo, D. Calvanese, G. De Giacomo, M. Lenzerini, R. Rosati, Linking data to ontologies. J. Data Semant. 133–173 (2008). https://doi.org/10.1007/978-3-540-77688-8_5
17. K. Dentler, R. Cornet, A. Ten Teije, N. de Keizer, Comparison of reasoners for large ontologies in the OWL 2 EL profile. Semant. Web **2**, 71–87 (2011)
18. A. Koukias, D. Nadoveza, D. Kiritsis, An ontology-based approach for modelling technical documentation towards ensuring asset optimization. Int. J. Prod. Lifecycle Manage. **8**(1), 24–45 (2015)
19. J. Zedlitz, J. Jorke, N. Luttenberger, From UML to OWL 2, in *Proceedings of Knowledge Technology Week 2011* (Springer, 2012)
20. N.R. Carvalho, J.J. Almeida, P.R. Henriques, M.J.V. Pereira, Conclave: ontology-driven measurement of semantic relatedness between source code elements and problem domain concepts, in *ICCSA 2014*, ed. by B. Murgante, et al. (Springer, Cham, 2014), pp. 116–131
21. A. Ferreira, A. Silva, UML to OWL mapping overview an analysis of the translation process and supporting tools, in *7th Conference of Portuguese Association of Information Systems* (2013), pp. 2536–2549
22. G. Guskov, A. Namestnikov, N. Yarushkina, Approach to the search for similar software projects based on the UML ontology. Adv. Intell. Syst. Comput. 3–10 (2018)
23. P. Wongthongtham, U. Pakdeetrakulwong, S.H. Marzooq, Ontology annotation for software engineering project management in multisite distributed software development environments, in *book: Software Project Management for Distributed Computing* (2017), pp. 66–75
24. S. Hattori, Y. Takama, Recommender system employing personal-vallue-based user model. J Adv. Comput. Intell. Intell. Inform. (JACIII) **18**(2), 157–165 (2014)
25. H. Ltifi, C. Kolski, M.B. Ayed, A.M. Alimi, A human-centred design approach for developing dynamic decision support system based on knowledge discovery in databases. J. Decis. Syst. **22**, 69–96 (2013)
26. F.B. Ruy, C.C. Reginato, V.A. Santos, R.A. Falbo, G. Guizzardi, Ontology engineering by combining ontology patterns, in *34th International Conference on Conceptual Modeling (ER2015)* (Springer, 2015), pp. 173–186

A Probabilistic Fuzzy Table Lookup Scheme with Negation Logic

Amir Samadi, Hamed Rafiei, and Mohammad-R. Akbarzadeh-T

Abstract The well known Fuzzy Table Lookup Scheme (FTLS) is extended here by introducing negation logic within a probabilistic fuzzy logic framework. The negation logic consists of contradictory negation, medium negation, and opposite negation. As shown here, this complements well the probabilistic framework with fuzzy logic by effectively utilizing the available information while also addressing higher aspects of uncertainty. In particular, it complements information in areas where there is either insufficient observations or the observations carry high ambiguity and inaccuracy. The probabilistic fuzzy framework further allows concurrent handling of both stochastic and deterministic uncertainties in the available information. The proposed approach is implemented on two classic function types, i.e. an increasing sinusoid and a pyramid step function, with fast versus slow changes as well as with few versus many observations. Results indicate the improved performance of the proposed PFTLSwNL by up to 60% at particularly higher noise levels.

Keywords Probabilistic fuzzy logic · Expert systems · Fuzzy systems · Negation logic

1 Introduction

Traditional Fuzzy Logic Systems (FLSs) consists of Fuzzifier, Fuzzy Rule Base, Inference Engine, and Defuzzifier. Fuzzy rules (IF-THEN rules) join together and form the Fuzzy Rule Base which are the most important parts of a fuzzy system. Wang et al. [6, 7] utilized I/O data pairs to generate fuzzy rules.

A. Samadi · H. Rafiei · M.-R. Akbarzadeh-T (✉)
Department of Electrical Engineering, Center of Excellence on Soft Computing and Intelligent Information Processing, Ferdowsi University of Mashhad, Mashhad, Iran
e-mail: akbazar@um.ac.ir

A. Samadi
e-mail: samadiamir@mail.um.ac.ir

H. Rafiei
e-mail: rafiei.hamed@mail.um.ac.ir

S. N. Shahbazova et al. (eds.), *Recent Developments and the New Direction in Soft-Computing Foundations and Applications*, Studies in Fuzziness and Soft Computing 393, https://doi.org/10.1007/978-3-030-47124-8_14

Fuzzy IF-THEN rules are expressed as follows:

$R^{(l)}$: IF x_1 is $A_1^{(1)}$ and $\cdots$ and x_n is $A_n^{(l)}$
THEN y is $B^{(l)}$ $(l = 1, 2, 3, \ldots, m)$

Where $x_j (j = 1, 2, \ldots, n)$ and y are I/O training data pairs respectively, A_n^l and B_l are fuzzy subsets, and m represents number of fuzzy rules.

In the real world, uncertainty in information is unavoidable.The numerous and complex information leads to a large amount of uncertainty in the system. Fuzzy uncertainty has been handled properly by using the concept of accuracy degree in fuzzy systems. Meghdadi and Akbarzadeh-T [2] brought fuzzy and probabilistic concepts together and named it probabilistic fuzzy logic system (PFLS). Tang [5] generated probabilistic fuzzy rules (PFR) where information is available in I/O pair by applying FTLS. Berg [1] used PFLS to approximate a system with uncertainty. Incompleteness in the experts' knowledge is very common. Zhang Blue [9] used three types of negation to complete fuzzy sets and cover whole domain of input and output. The rest of this paper is formed as follow. Section 2 represents probabilistic fuzzy logic and expresses concepts of contradictory, opposite, and medium negation briefly. Section 3 proposes the novel method and then illustrates through simulation studies. Finally, conclusion and discussion are given in Sect. 4.

2 Background

2.1 Probabilistic Fuzzy Logic

Probabilistic uncertainty is studied in probability area (see [2], and [10]). The probability of accuracy had been employed to represent randomness. On the other hand, fuzzy logic is capable of expressing fuzziness uncertainty whereas probability is not. Hence, PFLS had been introduced as a methodology to cope with both kinds of uncertainty which exploits the third dimension to show probability. Similar to FLS, PFLS consists of four main stages as Probabilistic Fuzzifier, Probabilistic Fuzzy Rule Base, Probabilistic Inference engine, and Probabilistic Defuzzifier (see Fig. 1). Probabilistic Fuzzy rules (the main part of PFLS) are obtained by human experts knowledge or input-output pairs analysis (see [5]). These rules that express nonlinear relations between input space $x_1 \times x_2 \times \cdots \times x_n \in \mathbb{R}^n$ and outputspace $y \in \mathbb{R}^n$ are as follows (see [6]):

$R^{(l)}$: If x_1 is A_1^l and x_2 is A_2^l and $\cdots x_n$ is A_n^l,
then
y is B_1^l with probability p_1^l and
y is B_2^l with probability p_2^l and
...

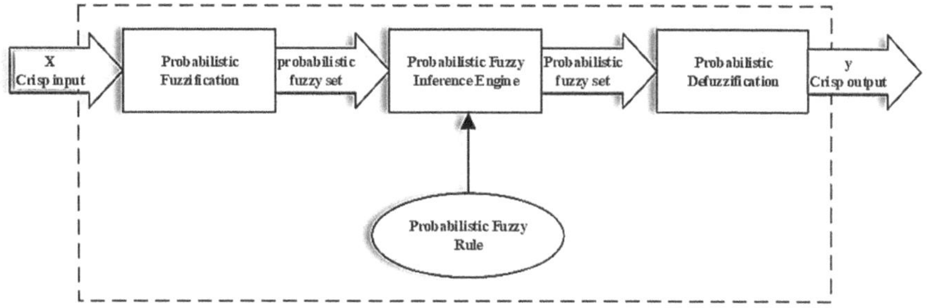

Fig. 1 PFL system contain probabilistic fuzzification, rule base, probabilistic fuzzy inference engine, and probabilistic defuzzification

$$y \text{ is } B_q^l \text{ with probability } p_q^l \ (i = 1, 2, \ldots, N)$$

where $x_j (j = 1, 2, \ldots, n)$ and y are input and output of I/O training data pairs, and A_j^l and B_k^l are membership functions respectively.
Moreover, the summation of all probabilities is equal to 1,
$\sum_{k=1}^{q} P_{ik} = 1, (i = 1, 2, \ldots, N)$.

2.2 Negation Logic

Obtaining the PFRs is the first step to define Membership Functions (MFs) as all the universe of inputs and outputs be covered. This might be so difficult due to limitation of experts' knowledge, in other words only some of fuzzy sets are available. In these circumstances, we can grasp other fuzzy sets aiding the concept of negation logic (see [4, 8] and [3]). We know traditional negation as contradictory negation that is obtained simply by applying the NOT function on linguistic variables. We use symbol '$\neg$' to represent contradictory negation, therefore $\eta\neg$ means the contradictory negation of η [9]. Contradictory negation of A is equal to $n(A(u))$ and in special case $A\neg : U \rightarrow [0, 1], A\neg(u) = n(A(u)) = 1 - A(u)$.
Opposite negation is another type of linguistic variable that creates reverse context. Negative affixes such as 'in' in inanimate, 'il' in illogical, 'un' in unbelievable, 'non' in nonsense or lexicons reverse such as beautiful and ugly, smooth and rough, glow and dim are some examples of opposite negation and shown by '$\perp$'. Opposite of A represent by $A^{\perp}$ where $A^{\perp}$: $U \rightarrow [0, 1]$ is equal to $A^{\perp}(u)$=$A(a + b - u)$ [9].
Normally, some lexical and concepts are exist between two opposite negations, for instance, wet is between dry and soak. Wet is neither complete dry nor complete soak and is named medium negation. The symbol of medium negation is '$\sim$' which refer to "to have partially". Medium negation of A shown by $A^{\sim}$ and $A^{\sim}(u)$=$T[A^{\neg}(u),((A)^{\perp})^{\neg}(u)] = T[n(A(u)), n(A(u)^{\perp})] = T[n(A(u)), n(A(a + b - u))]$ where T refer to t-norm [9].

3 Proposed Method

3.1 Advantages

The proposed method for designing a function approximation or controlling the system by using human experts' knowledge or input-output pairs has some advantages in comparison with traditional PFL and FTLS that are listed as follows:

- First, the proposed method is capable of dealing with probabilistic uncertainty and fuzzy uncertainty simultaneously.
- Second, traditional FTLS method needs fuzzy sets to be defined such that they cover the whole input and output universes, whereas this method requires fuzzy sets to cover only half of the input and output universes. For example, in Fig. 2, three membership functions of consequent parts are available and then are computed accordingly.
- Third, if we can split membership function into triplet form such as small-medium-big then we need only big or mal membership function to define membership

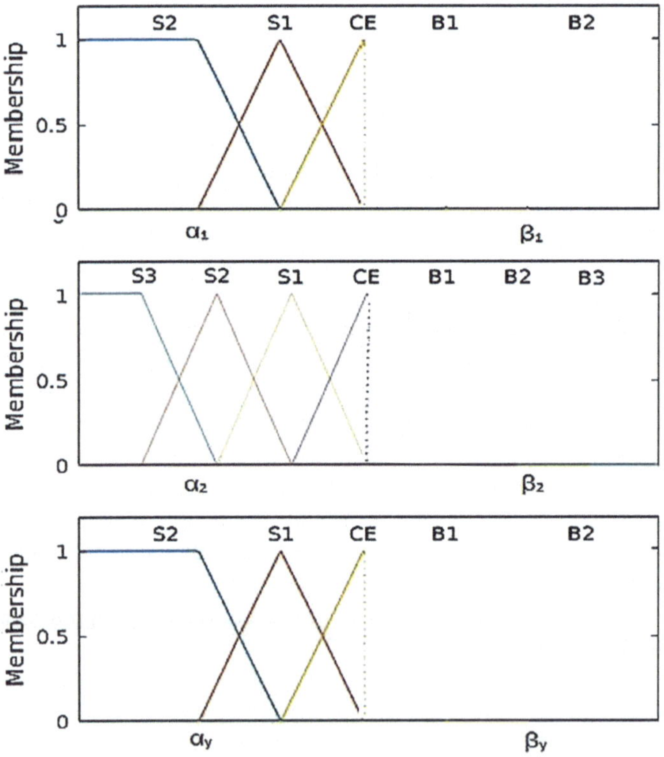

Fig. 2 Fuzzy set defined on half of input-output space

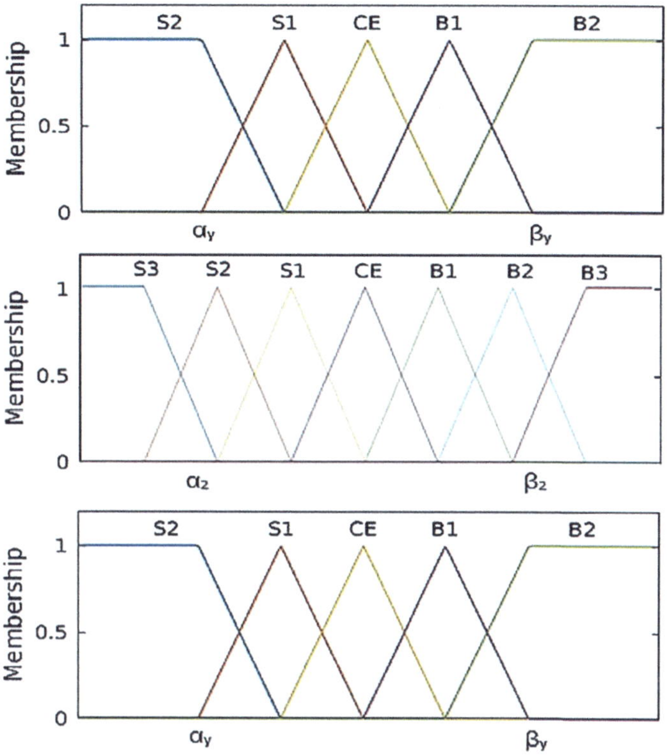

Fig. 3 Triangular fuzzy set with five membership function

function on whole input-output space. For example in Fig. 3, membership functions are ($S1$"small 1", $S2$ "small 2", CE "center", $B1$ "big 1", $B2$ "big 2") and we have only $S1$ and $S2$ membership functions then with contradictory, opposite, and medium negation (COM negation) define all membership functions (see [9]). intelligibly if we use more negation types, we have more degrees of freedom in categorization.

3.2 *The Table Lookup Scheme Based on Probabilistic Fuzzy Logic and Negation Logic*

The proposed method is represented in two forms, general case (1) and special case (2) with triangular membership function and one input, all in Algorithm 1.

Algorithm 1 The proposed method

Step 1: Insert N I/O training data.
Step 2: Define fuzzy set.
Define the typical of membership functions such as Gaussian, Triangular, Trapezoid, and Bell on half of input and output space.
The membership functions are marked as $[1, 2, \ldots, N/2]$ (Fig. 2).
Step 3: Complete the membership function.
Compute for each input $((A_i^{j_i})^{\perp})$ $(j_i = 1, 2, \ldots, [N/2], i = 1, 2, \ldots, n)$, compute for each output $((B_i^{j_i})^{\perp})$ $(j_i = 1, 2, \ldots, [N/2], i = 1, 2, \ldots, n)$, and then add them to the first half of the membership function.
Step 4: Compute firing probability of each output membership function for each rule. Compute P_{ij} the fire probability of jth output membership function for i^{th} rule as follows:
Case (1):

$$P_{ij} = \frac{\sum_{t=1}^{ki} A_{i1}(x_1^{i_t}) A_{i2}(x_2^{i_t}) \frac{B_j(y^{i_t})}{\sum_k B_k(y^{i_t})}}{\sum_{t=0}^{ki} A_{i1}(x_1^{i_t})} \tag{1}$$

Case (2):

$$P_{ij} = \frac{\sum_{t=1}^{ki} A_{i1}(x_1^{i_t}) B_j(y^{i_t})}{\sum_{t=0}^{ki} A_{i1}(x_1^{i_t})} \tag{2}$$

Step 5: Generate probabilistic fuzzy rule base.
Each row of P indicates a rule, if each row be zero then it has incompleteness in IF-THEN rules, so first find the rule that has the most neighbors. Extrapolating rules are computed as:

$$P_{ij}^c = P_{ij}^b + \frac{P_{ij}^b - P_{ij}^{b'}}{dis} \tag{3}$$

$$dis = \min_{b' \in N(b)} (\bar{x}_{ij}^b - \bar{x}_{ij}^{b'}) \tag{4}$$

Where P_{ij}^c is the result of extrapolation of P_{ij}^b the first neighbor's rule, $P_{ij}^{b'}$, $P_{ij}^{b'}$ the last neighbor, and dis is the distance between two neighbors. $N(b)$ indicates the all neighbors of the missed rule and $\bar{x}_{ij}^b$ center of input center of neighbor's rule . Repeat step 4 until all P are computed.
Step 6: Compute conditional probability
First, calculate w_i, which represents the firing rule strength. Then compute $P_r(Bj|x)$ which represents conditional probability.
w_i is computed as follows:

$$w_i = \frac{A_i(x)}{\sum_{t=1}^{N} A_t(x)} \tag{5}$$

Where $A_i(x)$=$T(A_{i1}(x_1), \cdots, A_{in}(x_n))$, and T is t-norm, in this paper "product" t-norm has been chosen. Now calculate the conditional probability as follows:

$$P_r(B_j \mid x) = \sum_{i=1}^{N} w_i(x) P_{ij} \tag{6}$$

Step 7: Construct probabilistic fuzzy system.
Case (1): First Compute conditional probability density function (pdf).

$$f(y \mid x) = \sum_{j=1}^{N} \frac{P_r(B_r(B_j \mid x)B_j(y)}{\int_{-\infty}^{+\infty} B_j(y)dy} \tag{7}$$

Now calculate average of all expected of outputs as follows:

$$y' = \int_{-\infty}^{+\infty} yf(y \mid x)dy \tag{8}$$

Where y' represents the estimation of outputs.
Case (2): Compute y' the estimation of outputs, with center average defuzzifier.

$$y' = \sum_{y} P_r(B_j \mid x)\bar{y} \tag{9}$$

Where $\bar{y}$ represent center of each output membership function.

4 Case Study

We illustrate our results in two sample continuous and discrete functions in different frequencies. We employ increasing sinusoid Eq. 10 as continuous function in fast ($w = 4$) and slow ($w = 2$) forms.

$$f(x) = \frac{x\cos(wx)}{2} + 4 + r \tag{10}$$

where r is the Guassian white noise.

In addition, we exploit pyramid step as a discrete function again in fast and slow forms.

$$f(x) = 3 + \sum_{i=0}^{3} 1.7(u(x - (6i + 5))) + r \tag{11}$$

$$f(x) = 3 + \sum_{i=0}^{11} 0.56(u(x - (2i + 1))) + r \tag{12}$$

where $u(x)$ is step function, and r is the Guassian white noise.

The proposed method and traditional FTLS function estimation are shown with blue diamond and red points respectively (Fig. 4 Slow increasing sinusoide, Fig. 5 Slow pyramid step, Fig. 6 Fast increasing sinusoid, Fig. 7 Fast pyramid step). In these examples, we can clearly see the performance of the methods in slow, fast, and abrupt changes.

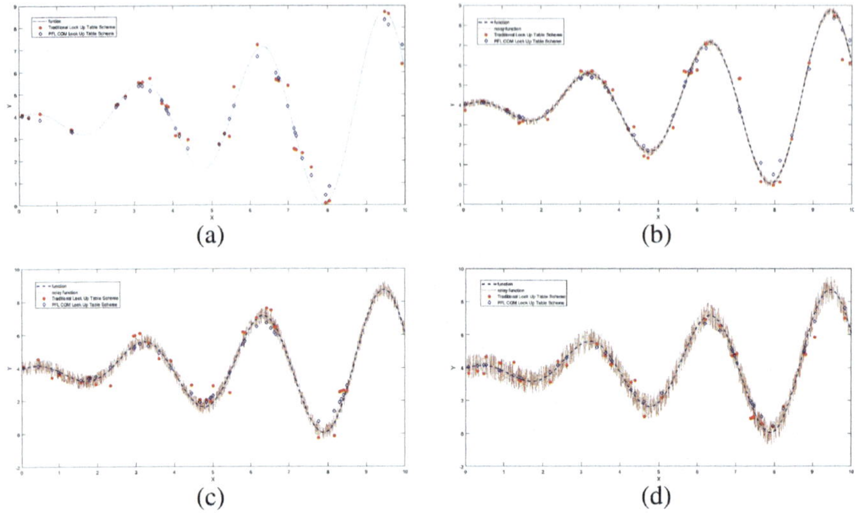

Fig. 4 Approximate slow increasing sinusoide with noise variance **a** zero **b** 0.2 **c** 0.5 **d** 0.8

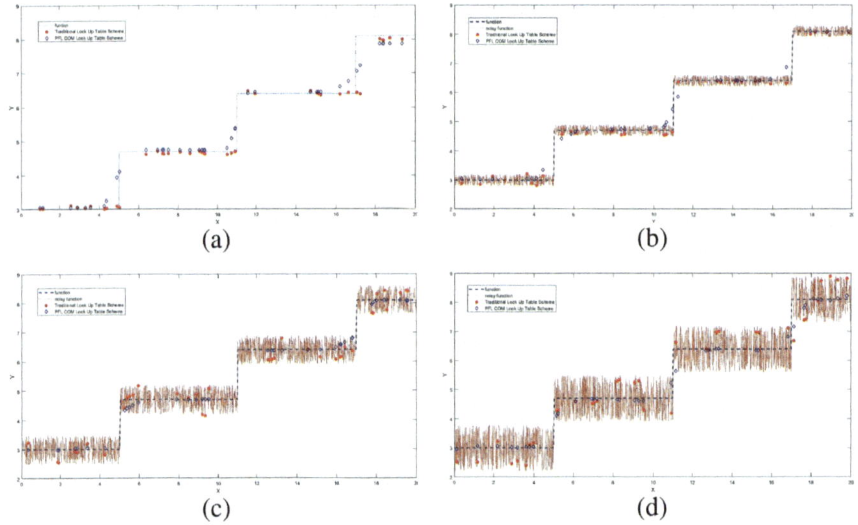

Fig. 5 Approximate slow pyramid step with noise variance **a** zero **b** 0.2 **c** 0.5 **d** 0.8

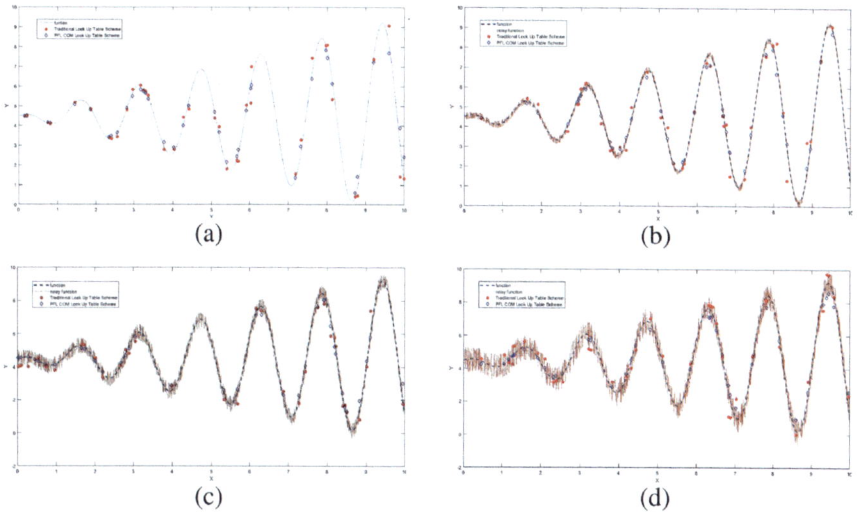

Fig. 6 Approximate fast increasing sinusoide with noise variance **a** zero **b** 0.2 **c** 0.5 **d** 0.8

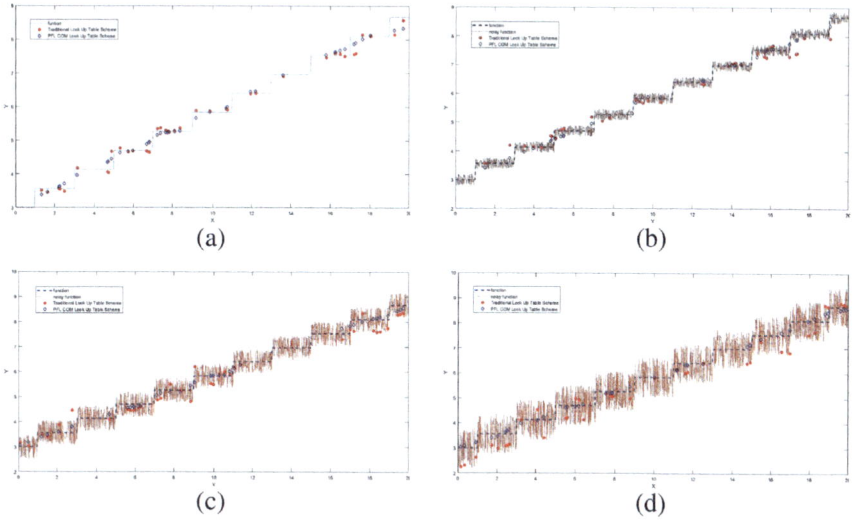

Fig. 7 Approximate fast pyramid step with noise variance **a** zero **b** 0.2 **c** 0.5 **d** 0.8

Table 1 Compare mse of traditional FTLS with PFTLSwNL

	PFTLSwNL				Traditional				Percent improvement			
Noise	S[1]	PS[2]	FS[3]	FPS[4]	S	PS	FS	FPS	S	PS	FS	FPS
0	0.0062	0.0057	0.0038	0.0063	0.0096	0.0073	0.0107	0.0046	%35.41	%21.91	%41.12	%21.73
0.2	0.0059	0.0055	0.0057	0.0053	0.0108	0.0075	0.0115	0.0046	%45.37	%26.6	%59.69	%28.26
0.5	0.0063	0.0053	0.0061	0.0033	0.0139	0.0088	0.021	0.0079	%54.67	%39.77	%49.58	%58.22
0.8	0.0063	0.0053	0.0063	0.0033	0.0155	0.0114	0.0155	0.0191	%59.35	%55.46	%59.35	%67.32

[1]Increasing sinusoid, [2]pyramid step, [3]fast increasing sinusoid, [4]fast pyramid step

5 Results

The mean squared error (MSE) of the proposed method and traditional FTLS are compared in Table 1. Clearly, in the absence of noise in the training data, there is no meaningful differences between two mentioned methods, whereas the MSE reduction in intensely noisy training data reaches %60.

6 Conclusion and Discussion

The proposed fuzzy rule generation system here uses probabilistic fuzzy sets to better cover the input-output domain within a table look up approach. The insufficiency of expert knowledge is further addressed by using negation information along with the probabilistic scheme. The approach extracts probabilistic fuzzy rules and estimates PDF. Simulation studies verify that the proposed method has a better performance when there is some randomness in the system such as noise in the training data. The proposed method has similar performance to traditional FTLS in the absence of noise. However, it creates fuzzy sets when adequate data are not available among the observation.

There are a number of future directions for this work. The first issue may be the use of the science of linguistics to exploiting more relations between linguistic variables to complete experts' knowledge. Another issue is employing the fuzzy type 2 structure instead of PFL to further address deterministic uncertainty.

References

1. R.J. Almeida, N. Basturk, U. Kaymak, V. Milea, A multi-covariate semi-parametric conditional volatility model using probabilistic fuzzy systems, in *2012 IEEE Conference on Computational Intelligence for Financial Engineering & Economics (CIFEr)* (IEEE, 2012), pp. 1–8
2. A.H. Meghdadi, M.-R. Akbarzadeh-T, Probabilistic fuzzy logic and probabilistic fuzzy systems, in *The 10th IEEE International Conference on Fuzzy Systems, 2001*, vol. 3 (IEEE, 2001), pp. 1127–1130

3. V. Novák, A comprehensive theory of trichotomous evaluative linguistic expressions. Fuzzy Sets Syst. **159**(22), 2939–2969 (2008)
4. Z. Pan, Three kinds of negation of fuzzy knowledge and their base of set. Jisuanji Xuebao (Chin. J. Comput.) **35**(7), 1421–1428 (2012)
5. M. Tang, X. Chen, H. Weidong, Y. Wenxian, Generation of a probabilistic fuzzy rule base by learning from examples. Inf. Sci. **217**, 21–30 (2012)
6. L.-X. Wang, J.M. Mendel, Generating fuzzy rules by learning from examples. IEEE Trans. Syst. Man Cybern. **22**(6), 1414–1427 (1992)
7. L.-X. Wang, J.M. Mendel, *Generating Fuzzy Rules from Numerical Data, with Applications* (Signal and Image Processing Institute, University of Southern California, Department of Electrical Engineering-Systems, 1991)
8. S. Zhang, Formal deductive system of fuzzy propositional logic with different negations. Front. Comput. Sci. Technol. **8**(4), 494–505 (2014)
9. S. Zhang, Y. Li, A novel table look-up scheme based on gfscom and its application. Soft Comput. **21**(22), 6767–6781 (2017)
10. H.-J. Zhou, A brief introduction to probabilistically quantitative logic with its applications, in *Quantitative Logic and Soft Computing 2016* (Springer, 2017), pp. 21–45

Evolutionary Algorithm for Fine-Tuning Perceptual Computer for Grading Mathematics Tests Within Bilingual Education Program

Dan Tavrov, Liudmyla Kovalchuk-Khymiuk, Olena Temnikova, and Nazar-Mykola Kaminskyi

Abstract Bilingual education implies simultaneous mastery of two major competences, linguistic one and subject one. This fact complicates the task of evaluating skills and knowledge of students, especially when it comes to grading mathematical tests. A solution to the problem can be found in using linguistic grades instead of numbers and performing grading of mathematical tests in hierarchical fashion, first by assigning grades to single tasks and then by aggregating these linguistic grades taking into consideration the relative weights of each task. This can be done within the framework of perceptual computing. However, assigning these weights is in itself a challenging and tedious problem, not free of high level of subjectivity. In this paper, we propose an evolutionary algorithm for selecting the weights for the perceptual computer based on grades already assigned by a teacher when grading works numerically.

1 Introduction

In the modern education practice, teaching foreign languages is getting replaced with the concept of bilingual education. According to [1], bilingual education is the use of two or more languages as media of instruction in subjects other than the languages

D. Tavrov (✉) · L. Kovalchuk-Khymiuk · O. Temnikova · N.-M. Kaminskyi
Applied Mathematics Department, Igor Sikorsky Kyiv Polytechnic Institute, Kyiv, Ukraine
e-mail: dtavrov@kse.org.ua

L. Kovalchuk-Khymiuk
e-mail: khl1956@gmail.com

O. Temnikova
e-mail: temnikova_elena@ukr.net

N.-M. Kaminskyi
e-mail: kaminskyi.nazar.1923@gmail.com

D. Tavrov
Kyiv School of Economics, Kyiv, Ukraine

S. N. Shahbazova et al. (eds.), *Recent Developments and the New Direction in Soft-Computing Foundations and Applications*, Studies in Fuzziness and Soft Computing 393, https://doi.org/10.1007/978-3-030-47124-8_15

themselves. Such a concept shifts perspective from learning the language to learning subjects through the media of language. Having obtained bilingual education, students can realize their academic mobility by studying in foreign universities without additional training, which enables them to integrate into international academic world and get competitive advantage on the job market.

Out of all subjects that can be taught in bilingual fashion, mathematics is of special interest. Bilingual students usually [2] switch languages during arithmetic computation and switch languages during conversation (code-switching). This means that grading their skills and knowledge cannot be boiled down to the arithmetic average of grades for solving mathematical problems and grades for language proficiency.

In [3], we proposed to divide all the relevant competences that a bilingual math student must exhibit into two groups:

(1) *Mathematical Competences*: Explaining steps of solving the problem in a foreign language (competence 1), explaining steps of solving the problem in a native language (competence 2), solving a problem (competence 3).
(2) *Linguistic Competences*: Understanding a problem definition (competence 4), ability to explain a problem in a foreign language (competence 5), ability to explain a problem in a native language (competence 6).

Assigning grades in bilingual framework is a complex problem. For instance, some students can be fluent in a foreign language, but know math relatively poorly. Other students can possess exceptional math skills but are not able to get what is being asked in the problem due to less than satisfactory language proficiency. Learning foreign language is not the main goal when studying math through the media of this language, but without knowing the language at least at some basic level, it is impossible to learn the subject. Therefore, grading math tests should consider [2] not only language proficiency as is, but rather ability to use the language in application to particular mathematical topics.

In [4], we proposed to use perceptual computer of a special kind to grade the bilingual math tests. In our approach, the test consists of a number of small problems that can be evaluated independently using words from a given set. Each test influences all competences to a certain degree that is expressed by a linguistic weight. However, the main drawback of this model is that these weights need to be assigned by the teacher who is grading the works, which is by itself highly a subjective and ill-defined problem. As experience shows, weights used when grading the test as reported by a teacher are not always in fact the weights she (subconsciously) applies in practice. Therefore, automating this process by analyzing actual (numerical) grades assigned by the teacher in order to extract the weights she uses is expedient. In other words, a method is needed to fine-tune the perceptual computer used for assigning grades. In this paper, we propose one such method that is based on evolutionary computing.

2 Perceptual Computer for Grading Math Tests

2.1 General Architecture of a Perceptual Computer

Perceptual computing is a type of computing with words (CWW) [5], in which objects of computations are words and propositions expressed in a natural language. Perceptual computing is used for making subjective judgments [6], which is why it is a natural choice for the problem discussed in this paper. Humans can interact with a perceptual computer (Per-C) using words. Each word is modeled as a type-2 fuzzy set (T2FS). Per-C processes words using the following components [7]:

(1) *Encoder*, which transforms words into T2FS representations. A vocabulary of words is called a codebook. In this paper, each word is represented using a trapezoidal interval T2FS $\tilde{A}$, for which lower $\underline{\mu}_{\tilde{A}}$ and upper $\bar{\mu}_{\tilde{A}}$ membership functions are trapezoidal:

$$\mu_{trap}(x; a, b, c, d, h) = \begin{cases} (x-a)/(b-a), & a \le x \le b \\ h, & b \le x \le c \\ (d-x)/(d-c), & c \le x \le d \\ 0, & \text{otherwise} \end{cases}. \tag{1}$$

Graphical illustration of (1) is given in Fig. 1.

(2) *CWW engine*, which processes T2FSs. In this paper, linguistic weighted average (LWA) [8, 9] is used.
(3) *Decoder*, which maps the output of the CWW engine to output of the Per-C which can be [7] a word from a codebook most similar to output of the CWW engine, a rank of competing alternatives, or a class. Here, the output of the

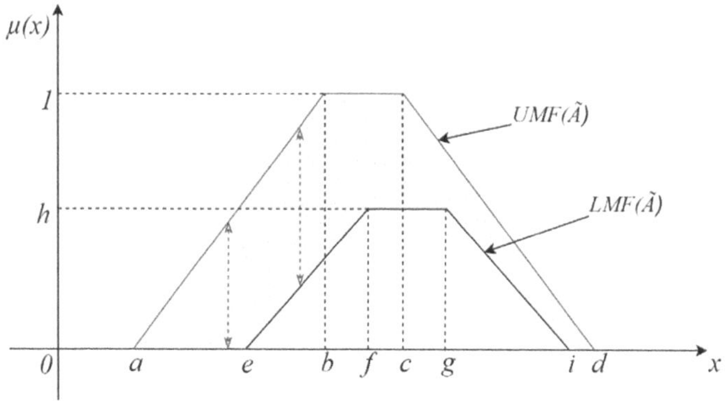

Fig. 1 Trapezoidal interval type-2 fuzzy set

Per-C is one of the four classes ("A," "B," "C," or "D") representing overall performance of a given student in a given math test. These classes are modeled as type-1 fuzzy sets with the following membership functions (scale from 1 to 12 is the standard grading scale in Ukrainian secondary educational institutions): $\mu_D(x) = \mu_{trap}(x; 1, 3, 3, 1, 1)$, $\mu_C(x) = \mu_{trap}(x; 4, 6, 6, 4, 1)$, $\mu_B(x) = \mu_{trap}(x; 7, 9, 9, 7, 1)$, $\mu_A(x) = \mu_{trap}(x; 10, 12, 12, 10, 1)$.

The decoder used in this paper is a classifier based on a Vlachos and Sergiadis's IT2 FS subsethood measure [10]:

$$ss_{VS}\left(\tilde{A}, \tilde{B}\right) = \frac{\sum_{i=1}^{N} \min\left(\underline{\mu}_{\tilde{A}}(x_i), \underline{\mu}_{\tilde{B}}(x_i)\right) + \sum_{i=1}^{N} \min\left(\bar{\mu}_{\tilde{A}}(x_i), \bar{\mu}_{\tilde{B}}(x_i)\right)}{\sum_{i=1}^{N} \underline{\mu}_{\tilde{A}}(x_i) + \sum_{i=1}^{N} \bar{\mu}_{\tilde{A}}(x_i)}$$

where x_i $(i = 1, \ldots, N)$ are equally spaced points in the universe of discourse X.

2.2 *Decision Making in a Perceptual Computer for Grading Math Tests*

The decision making process in the Per-C is depicted in Fig. 2. It is hierarchical

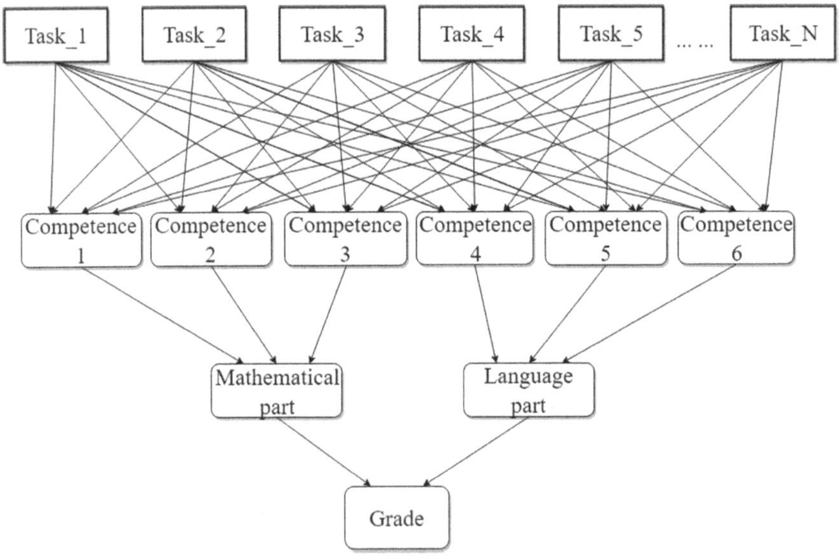

Fig. 2 Hierarchical and distributed decision making in the perceptual computer for grading math tests

and distributed [7], in that the overall grade depends on the competences in hierarchical fashion, and each task from Task_1 to Task_N influences each competence independently.

Overall grade is obtained as an LWA of the mathematical and linguistic competences. These competences are obtained as LWAs of six basic competences, each of which is an LWA of the grades assigned for each task. Grades for each task are words from the part of the codebook called grade subcodebook. Weights for each of the above LWAs are expressed as the words from the part of the codebook called weight subcodebook.

3 Evolutionary Algorithm for Setting the Weights in the Perceptual Computer

3.1 Outline of the Algorithm and Individual Representation

The evolutionary algorithm for fine-tuning the weights in the Per-C consists of the following steps:

(1) *Initialization*: Generate a population P of size μ and calculate the fitness of each individual.
(2) *Check Termination Condition*: If it holds, stop; continue otherwise.
(3) *Parent Selection*: Select $\lambda/2$ parents based on their fitness values.
(4) *Variation*: Apply crossover operator to each parent pair. Apply mutation operator to each child. Calculate fitness function for each child.
(5) *Survivor selection*: Select individuals for the new population from the combined pool of μ individuals in the current population and λ children, and put them in place of the worst individuals from the current population. Go to 2.

Each individual in the algorithm is a vector $X = (x_1, \ldots, x_{50})$ where each x_i, $i = 1, \ldots, 50$, is an integer from 1 to m, and m is the total number of words in the weight subcodebook. The first 42 elements of X correspond to the arcs connecting the tasks with the competences in the following way: the first 6 elements correspond to the arcs leading from Task 1 to each competence (from 1 to 6) the next 6 elements correspond to the arcs leading from Task 2 to each competence, and so on.

Elements $x_{43} - x_{45}$ correspond to arcs connecting the first three competences to the "Mathematical part" node, and elements $x_{46} - x_{48}$ correspond to arcs connecting the first three competences to the "Linguistic part" node. Elements x_{49} and x_{50} correspond to arcs leading from "Mathematical part" and "Linguistic part" nodes to the "Grade" node, respectively.

3.2 Fitness Function and Variation Operators

By careful consideration of the structure of the individual in the algorithm, it is evident that each individual X (genotype) fully specifies a Per-C (phenotype) for assigning grades to math tests. Therefore, in this paper, the fitness of each individual is based on comparison of the outputs of the respective Per-C to the numerical grades assigned manually by a teacher.

Therefore, the fitness function is defined as follows:

$$f(X) = \sum_{i=1}^{N} 1\left\{g_i \in {}^{0+}Y_i\right\}, \tag{2}$$

where g_i is the numerical grade assigned to the work i by a teacher, Y_i is the class determined by the Per-C corresponding to individual X, ${}^{0+}Y_i = [a_i, b_i]$ is the support of Y_i, N is the total number of students, and $1\{\cdot\}$ is an indicator function that equals 1 if the condition inside the braces holds, and 0 otherwise.

In addition to (2), we can also define another measure of Per-C performance as follows:

$$f(X) = \sum_{i=1}^{N} d\left(g_i, {}^{0+}Y_i\right), \tag{3}$$

where d is defined as

$$d(c, [a, b]) = \begin{cases} 0, & c \in [a, b] \\ a - c, & c \le a \\ c - b, & c \ge b \end{cases}$$

.

Measure (3), unlike function (2), not only captures the fact that the numerical and linguistic grades don't match, but also shows the magnitude of the mismatch. However, (3) was not used to guide the selection in the evolutionary algorithm.

3.3 Other Algorithm Elements

In this paper, we propose to use the following instances of variation and selection operators:

(1) *Crossover*: We propose to use the one-point crossover as one of the simplest and most widely used in practice.
(2) *Mutation*: We propose to use the random resetting mutation operator [11, p. 55] as the most widely used operator for integer representation.

(3) *Selection*: For both parent and survivor selection, we propose to use tournament selection [12] with the tournament size 5 as easy-to-implement and efficient.
(4) *Initialization and Termination*: Initial population is proposed to be initialized by randomly mutated (with small probability) the vector of weights as reported by the teacher (seed individual). Termination is proposed to do after reaching the prespecified number of generations.

4 Application of the Evolutionary Algorithm

4.1 Structure of a Sample Test

In this work, to validate the Per-C, we considered bilingual teaching math in French and Ukrainian, because the second author of this paper has high expertise in this field as a practicing educator. Following ideas from [4], we used the following math test:

1. Vrai ou faux (donner l'exemple) [True or False (show an example)] (2 points):
   a. Si la somme de deux nombres relatifs est positive, alors les deux nombres sont positifs [If the sum of two integers is positive, then both numbers are positive]
   b. Si deux nombres non nuls ont des signes contraires, alors ils sont opposes [If two non-zero numbers are of opposite signs, then they are opposite]
2. Complète avec les mots convenables [Complete with correct words] (2 points):
   a. Le nombre 0 est à la fois ___ et ___ [Zero is ___ and ___ at the same time]
   b. Cas de deux nombres négatifs le plus petit celui qui a la ___ ___ ___ [Out of two negative numbers, the lowest is the one that has ___ ___ ___]
3. Calcule [Calculate] (2 points):
   a. $\frac{-5{,}4 \cdot 3{,}9 \cdot (-0{,}02)}{0{,}42 \cdot (-0{,}18) \cdot (-2{,}6)} =$ b. $-1\frac{2}{9} : (-0,25 \cdot 1\frac{2}{9}) =$
4. Mettre les parenthèses oubliées pour corriger l'égalité [Insert parentheses to obtain an identity] (1 point): $D = 5 + 2 \times 3 + 2 + 4 - 1 = 18$
5. Connaissant la valeur du produit ab, calculer l'expression A [Given ab, calculate A] (1 point): $A = 4 \times (-a) \times 2 \times b$, $ab = 0{,}5$
6. Calculer pour 1,5 et –9 [Calculate for 1.5 and –9] (2 points):
   a. la somme de leurs opposes [sum of their opposites]
   b. la somme de leurs inverses [sum of their inverses]
   c. l'opposé de leur somme [opposite of their sum]
   d. l'inverse de leur produit [inverse of their product]
   e. le produit de leurs inverses [product of their inverses]
7. Résous [Solve the equations] (2 points):

a. $|3x - 1| = 5$ b. $0,02 + 0,5y = -0,18$

This test was handed out to 56 students of the Secondary bilingual school #20 of Kyiv, Ukraine, where the second author of this paper teaches math in French. Point grades assigned by the second author of this paper for each student are given in Table 4.

4.2 The Codebook

For the codebook in the Per-C, we selected the words that cover the following intervals:

- for the words from the grade subcodebook, the interval is [1; 12]. The following words were selected: *perfect* (P), *excellent* (EX) *very good* (VG), *good* (G), *sufficient* (S), *satisfactorily* (SA), *bad* (BA), *unsatisfactorily* (U), *very bad* (VB), *awful* (AW);
- for the words from the weight subcodebook, the interval is [0; 1]. The following words were selected: *not influential* (NI), *weakly influential* (WI), *moderately influential* (MI), *influential* (I), and *highly influential* (HI).

We used the interval approach [13] to determine the T2FS models for each word. Twenty experts from the Secondary bilingual school #20 of Kyiv and the Applied Mathematics Department of the Igor Sikorsky Kyiv Polytechnic Institute were asked to fill the questionnaire, in which they provided the intervals for each word. In Table 1, resulting T2FS models for each word are presented up to two significant digits, as well as interval statistics: means (m) and standard deviations (σ) of left (l) and right (r) ends of each interval.

We decided to discard words EX and U, because they either subsume or are subsumed by other words in a codebook.

4.3 Running the Evolutionary Algorithm

Initial population in the algorithm of size 50 was created by randomly mutating the seed individual given by the weights from Tables 2 and 3 with probability 0.1. Weights in both tables were assigned by the second author of this paper. The fitness of the individual defined by Tables 2 and 3 is 40, meaning that grades only 40 works (71.43%) consistently with the manual (numerical) grading.

Other parameters were picked as follows: $\lambda = 20$, probability of crossover is 0.9, probability of mutation is 0.1, maximum number of generations is 100.

We ran the algorithm for 20 times. Two best individuals from all runs had the fitness of 44 (78.57% of works graded consistently with the numerical grading). These individuals are given in Tables 4 and 5. Table 6 shows results of grading 56 tests. The "Tot." column for numerical grades was calculated as a rounded sum of

Table 1 Interval statistics for words from the codebook, and their type-2 fuzzy set models

Word	Parameters of trapezoidal T2FS									Interval statistics			
	a	b	c	d	e	f	g	i	h	m_l	σ_l	m_r	σ_r
P	9.37	11.82	12.00	12.00	10.68	11.91	12.00	12.00	1.00	10.88	0.35	12.00	0.00
EX	7.05	10.72	12.00	12.00	9.68	11.83	12.00	12.00	1.00	9.77	0.60	11.46	0.52
VG	7.59	9.00	10.00	11.41	8.79	9.50	9.50	10.21	0.59	8.80	0.42	10.20	0.42
G	6.59	8.00	9.00	10.41	7.79	8.50	8.50	9.21	0.59	7.75	0.46	9.25	0.46
S	5.38	7.00	8.00	9.62	6.79	7.50	7.50	8.21	0.59	6.67	0.52	8.33	0.53
SA	3.59	5.00	5.50	6.41	4.79	5.32	5.32	6.21	0.74	4.63	0.52	6.00	0.00
BA	1.59	3.00	4.00	5.41	2.90	3.35	3.35	3.60	0.41	2.67	0.52	4.25	0.61
U	0.38	2.50	4.25	5.41	3.19	3.58	3.58	4.21	0.37	2.44	0.98	4.25	0.46
VB	0.59	2.00	3.00	4.41	1.79	2.50	2.50	3.21	0.59	1.63	0.74	3.38	0.52
AW	0.59	1.50	2.00	3.41	0.79	1.68	1.68	2.21	0.74	0.83	0.41	2.17	0.41
NI	0.00	0.00	0.05	0.66	0.00	0.00	0.02	0.26	1.00	0.00	0.00	0.26	0.13
WI	0.01	0.15	0.25	0.46	0.08	0.18	0.18	0.22	0.59	0.07	0.04	0.27	0.07
MI	0.12	0.38	0.50	0.68	0.38	0.44	0.44	0.52	0.53	0.31	0.08	0.56	0.05
I	0.45	0.63	0.75	0.96	0.62	0.70	0.70	0.80	0.59	0.58	0.06	0.84	0.09
HI	0.47	0.96	1.00	1.00	0.74	0.98	1.00	1.00	1.00	0.73	0.08	1.00	0.01

Table 2 Weights representing relative importance of each task for assessing level of each competence in the seed individual

Task #	Competences					
	1	2	3	4	5	6
1	HI	MI	I	HI	I	NI
2	MI	MI	I	HI	I	WI
3	WI	MI	I	HI	NI	MI
4	WI	MI	I	I	NI	MI
5	MI	MI	I	I	NI	WI
6	I	MI	HI	HI	MI	WI
7	MI	MI	HI	HI	WI	WI

Table 3 Weights representing relative importance of each competence in the seed individual

Competence	1	2	3	4	5	6	Math.	Ling.
Weight	I	MI	HI	HI	I	MI	HI	I

Table 4 Weights representing relative importance of each task for assessing level of each competence in the best individuals

Task #	Competences					
	1	2	3	4	5	6
1	MI/HI	I/HI	I/WI	HI/HI	I/I	I/HI
2	MI/MI	I/I	HI/HI	MI/HI	HI/WI	HI/WI
3	WI/MI	HI/MI	HI/I	HI/HI	HI/WI	MI/I
4	WI/NI	NI/NI	HI/NI	NI/WI	NI/MI	NI/NI
5	HI/WI	MI/I	I/MI	I/I	I/NI	WI/NI
6	I/MI	HI/WI	HI/HI	I/WI	HI/MI	HI/I
7	HI/WI	WI/MI	NI/NI	WI/WI	WI/WI	WI/NI

Table 5 Weights representing relative importance of each competence in the best individuals

Competence	1	2	3	4	5	6	Math.	Ling.
Weight	WI/I	NI/WI	HI/HI	I/HI	HI/WI	HI/MI	NI/HI	I/WI

single points assigned to each task. The "Tot. 1" and "Tot. 2" columns for word grades were obtained using the two best individuals from Tables 4 and 5.

As we can see, increase from 71.43 to 78.57% is driven by two completely different forces:

- for the first individual, the mathematical competences are not influential, whereas linguistic competences are influential. This means that the teacher who initially

Table 6 Grades for the test written by students of one of the Kyiv bilingual schools

#	Point grade for each task								Word grade for each task								
	1	2	3	4	5	6	7	Tot.	1	2	3	4	5	6	7	Tot. 1	Tot. 2
1	1	1	0.75	1	1	2	2	9	G	G	SA	VG	VG	P	P	B	B
2	2	2	0.75	1	1	2	1.75	11	P	P	SA	VG	VG	P	G	A	A
3	2	2	1.75	1	1	0.5	1.5	**10***[a]	VG	VG	G	VG	VG	SA	G	**B***	**B***
4	0	0	0	1	0	0.75	1.5	**3***	BA	BA	VB	VG	VB	SA	G	**C***	**C***
5	1	1	2	0	0	1	1.5	7	G	G	VG	BA	VB	S	G	B	B
6	0	1	2	1	0.75	0.75	2	8	BA	G	VG	VG	S	SA	VG	B	B
7	1	1	1.75	1	1	0.25	2	8	G	G	G	VG	VG	BA	VG	B	B
8	2	1	1	1	1	0.25	0.75	8	VG	G	SA	VG	VG	BA	BA	B	B
9	2	1	2	1	0	0.25	2	8	VG	G	VG	VG	VB	BA	VG	B	B
10	1	1	2	1	0	0.5	2	8	G	G	VG	VG	VB	SA	VG	B	B
11	2	1	2	1	0	0.25	2	8	VG	G	VG	VG	VB	BA	VG	B	B
12	2	1	1.5	1	1	1.5	2	9	VG	G	G	VG	VG	G	VG	B	B
13	1	2	0	1	0	1	0.5	6	G	VG	VB	VG	VB	S	BA	C	C
14	1	1.75	2	0	1	0.75	1	8	G	VG	VG	BA	VG	SA	S	B	B
15	1	0	0	0	1	0	1.5	4	G	BA	VB	BA	VG	BA	G	C	C
16	1	1.5	1	0	0.5	0.25	1.75	6	G	G	SA	BA	SA	BA	G	C	C
17	1	1	2	1	1	0	1	7	G	G	VG	VG	VG	SA	G	B	B
18	1	1	2	1	1	1.25	1.25	9	G	G	VG	VG	VG	S	SA	B	B
19	1	1	2	1	0	0	1.5	7	G	G	VG	VG	VB	VB	G	B	B
20	1	0	2	1	0.5	0	0	5	G	BA	VG	VG	SA	VB	AW	C	C

(continued)

Table 6 (continued)

#	Point grade for each task								Word grade for each task								
	1	2	3	4	5	6	7	Tot.	1	2	3	4	5	6	7	Tot. 1	Tot. 2
21	0	2	2	1	0.75	0	2	8	BA	VG	VG	VG	S	VB	VG	B	B
22	1	2	2	1	0	0.25	1.5	8	G	VG	VG	VG	VB	BA	G	B	B
23	1	2	1	1	1	1.25	1.5	9	G	VG	SA	VG	VG	S	G	B	B
24	1	1	2	1	1	0.25	1.5	8	G	G	VG	VG	VG	BA	G	B	B
25	1	2	1	1	1	1.25	1.5	9	G	VG	SA	VG	VG	S	G	B	B
26	1	1	1	1	0	0	2	6	G	G	SA	VG	VB	VB	VG	C	C
27	1	0	2	1	1	0.25	1.25	7	G	BA	VG	VG	VG	BA	SA	B	B
28	1	0	1	1	0	0	0.5	4	G	BA	SA	VG	VB	VB	BA	C	C
29	0	1	2	1	0	0	1.5	6	BA	G	VG	VG	VB	BA	G	C	C
30	1	1.5	1.5	1	1	0.25	2	8	G	VG	G	VG	VG	BA	VG	B	B
31	0	1	0.5	1	0	0.5	1.75	5	BA	G	S	VG	SA	SA	VG	C	C
32	0	0	0	1	0	0	0.25	1	BA	BA	VB	VG	BA	BA	AW	D	D
33	0	1.25	0	1	1	0.5	1.5	5	BA	G	VB	VG	VG	S	VG	C	C
34	0	0	1	0	0	0	0.25	**1***	BA	BA	G	VG	VB	VB	AW	**C***	**C***
35	0	0	2	1	0	1	1.5	6	BA	BA	VG	VG	VB	S	G	C	C
36	2	0	1	1	0	0	1	5	VG	BA	G	VG	VB	VB	S	C	C
37	1	0	2	1	0.5	0	0	5	G	BA	VG	VG	SA	VB	VB	C	C
38	1	1	1	1	0	0	0.5	5	G	G	G	VG	VB	BA	VB	C	C
39	1	1	2	1	1	0.25	1.25	8	G	G	VG	VG	VG	BA	S	B	B

(continued)

Table 6 (continued)

#	Point grade for each task								Word grade for each task								
	1	2	3	4	5	6	7	Tot.	1	2	3	4	5	6	7	Tot. 1	Tot. 2
40	1	0.25	1.5	0	1	0	1.75	6	G	BA	G	BA	VG	VB	G	C	C
41	1	1.5	2	1	1	0	0	7	G	G	VG	VG	VG	BA	VB	B	B
42	1	0	2	1	1	1.5	2	9	G	BA	VG	VG	VG	G	VG	B	B
43	1	2	2	1	1	0	1.5	**4***	G	VG	VG	VG	VG	SA	G	**B***	**B***
44	1	1	2	1	1	0.25	0.5	7	G	G	VG	VG	VG	BA	BA	B	B
45	1	0	2	1	0	0	1.5	6	G	BA	VG	VG	VB	VB	S	C	C
46	1	1	2	0	1	0	1.25	**6***	G	G	VG	BA	VG	VB	G	**B***	**B***
47	1	0	2	1	1	0	1	6	G	BA	VG	VG	VG	VB	G	C	C
48	2	0.75	0	0	0	0	0	**3***	VG	G	VB	BA	VB	BA	AW	**C***	**C***
49	1	1	0	0	0	0.25	0.5	**3***	G	G	VB	BA	VB	BA	BA	**C***	**C***
50	0	1	0	0	0	0.25	0.75	**2***	BA	G	VB	BA	VB	BA	BA	**C***	**C***
51	1	0	0.75	1	0	0.25	0	**3***	G	BA	S	VG	VB	VB	AW	**C***	**C***
52	0	0	2	0	0	0.25	1	**3***	BA	BA	VG	BA	VB	VB	S	**C***	**C***
53	0	0	2	1	0	0.75	1	5	BA	BA	S	VG	VB	S	S	C	C
54	1	0	2	0	0	0	0	**3***	G	BA	VG	BA	VB	VB	AW	**C***	**C***
55	1	0	0	1	0	0.5	1	4	G	BA	VB	VG	VB	S	S	C	C
56	1	2	0	0	0	0.25	0	**3***	G	VG	VB	BA	VB	VB	AW	**C***	**C***

*Grades that don't match are marked with an asterisk

reported she values mathematical competences as highly influential in fact treats linguistic abilities much higher;
- for the second individual, we can see that the teacher indeed treats mathematical competences as highly influential, but overestimates linguistic ones.

Such discrepancies can be explained by a relatively small sample of observations. In any case, the obvious conclusion is that initial weights as reported by the teacher do not reflect her real preferences, which is why it is useful to apply the proposed method in order to enhance the accuracy of the perceptual computer.

A relatively high percentage of works that were graded not consistently (around 20%) should not be considered as the disadvantage of the method. On the contrary, if all the works were graded identically manually and with the help of Per-C, there would be no point in using the Per-C in the first place. It is the bigger freedom of choice and natural appeal of linguistic grades that makes Per-C much more fit to the task of grading math tests within the bilingual educational framework.

This conclusion is also supported by the fact that the value of (3) for both individuals is only 14, which means that, on average, the Per-C makes a mistake of 14/56 = 0.25 points in magnitude.

5 Conclusions and Further Research

In this paper, we proposed an evolutionary algorithm for assigning weights to different tasks in bilingual math tests and competences that lie at the core of the perceptual computer for grading such tests. This approach helps to reduce subjectivity when calibrating the parameters of the Per-C, and thus makes it more appealing to regular users.

Further research directions can be outlined as follows:

- expanding the number of experts and students to achieve higher accuracy of the Per-C;
- performing field research of applying the Per-C by different teachers in order to determine their degree of satisfaction with this tool.

References

1. T. Skutnabb-Kangas, T.L. McCarty, Key concepts in bilingual education: ideological, historical, epistemological, and empirical foundations, in *Encyclopedia of language and education*, ed. by N. Hornberger (Springer, US, 2008), pp. 1466–1482
2. J.N. Moschkovich, Bilingual/multilingual issues in learning mathematics, in *Encyclopedia of Mathematics Education*, ed. by S. Lerman (Springer, Netherlands, 2014), pp. 57–61
3. L.O. Kovalchuk-Khymiuk, D.Y. Tavrov, The fuzzy inference system for assessing of students' performance during the bilingual teaching of mathematics, in *System Analysis and Information*

Technologies: Materials of the 17th International Scientific and Technical Conference SAIT 2015 (in Ukrainian, 2015), pp. 245–247
4. D. Tavrov, L. Kovalchuk-Khymiuk, O. Temnikova, N.-M. Kaminskyi, Perceptual computer for grading mathematics tests within bilingual education program, in *Advances in Computer Science for Engineering and Education*, ed. by Z. Hu, S. Petoukhov, I. Dychka, M. He (Springer International Publishing, 2019), pp. 724–734
5. L.A. Zadeh, From computing with numbers to computing with words—from manipulation of measurements to manipulation of perceptions. IEEE Trans. Circ. Syst. I: Fundam. Theory Appl. **46**(1), 105–119 (1999)
6. J.M. Mendel, The perceptual computer: an architecture for computing with words, in *Proceedings of Modeling with Words Workshop in the Proceedings of FUZZ-IEEE 2001* (2001), pp. 35–38
7. D. Wu, J.M. Mendel, *Perceptual Computing. Aiding People in Making Subjective Judgments* (Wiley, Hoboken, New Jersey, 2010)
8. J.M. Mendel, The perceptual computer: an architecture for computing with words, in *Proceedings of Modeling with Words Workshop in the Proceedings of FUZZ-IEEE* (2001), pp. 35–38
9. J.M. Mendel, D. Wu, *Perceptual Computing. Aiding People in Making Subjective Judgments* (Wiley, Hoboken, New Jersey, 2010)
10. I. Vlachos, G. Sergiadis, Subsethood, entropy, and cardinality for interval-valued fuzzy sets—an algebraic derivation. Fuzzy Sets Syst. **158**, 1384–1396 (2007)
11. A.E. Eiben, J.E. Smith, *Introduction to Evolutionary Computing*, 2nd edn. (Springer, Berlin, Heidelberg, 2015)
12. A. Brindle, Genetic algorithms for function optimization, Doctoral Dissertation, Department of Computer Science, Tech. Rep. TR81-2, University of Alberta, 1981
13. F. Liu, J.M. Mendel, Encoding words into interval type-2 fuzzy sets using an interval approach. IEEE Trans. Fuzzy Syst. **16**, 1503–1521 (2008)

Soft Computing and Fuzzy Controllers

Which t-Norm Is Most Appropriate for Bellman-Zadeh Optimization

Vladik Kreinovich, Olga Kosheleva, and Shahnaz Shahbazova

Abstract In 1970, Richard Bellman and Lotfi Zadeh proposed a method for finding the maximum of a function under fuzzy constraints. The problem with this method is that it requires the knowledge of the minimum and the maximum of the objective function over the corresponding crisp set, and minor changes in this crisp set can lead to a drastic change in the resulting maximum. It is known that if we use a product "and"-operation (t-norm), the dependence on the maximum disappears. Natural questions are: what if we use other t-norms? Can we eliminate the dependence on the minimum? What if we use a different scaling in our derivation of the Bellman-Zadeh formula? In this paper, we provide answers to all these questions. It turns out that the product is the only t-norm for which there is no dependence on maximum, that it is impossible to eliminate the dependence on the minimum, and we also provide t-norms corresponding to the use of general scaling functions.

1 Formulation of the Problem

Need for optimization under constraints. In many practical problems, we need to find an optimal alternative a_{opt}, optimal in the sense that for this alternative, the value of the corresponding objective function $f(x)$ is the largest possible:

$$f(a_{\text{opt}}) = \max_{a \in P} f(a),$$

where P denotes the set of all possible alternatives.

V. Kreinovich (✉) · O. Kosheleva
University of Texas at El Paso 500 W. University, El Paso, TX 79968, USA
e-mail: vladik@utep.edu

O. Kosheleva
e-mail: olgak@utep.edu

S. Shahbazova
Azerbaijan Technical University, Baku, Azerbaijan
e-mail: shahbazova@gmail.com

S. N. Shahbazova et al. (eds.), *Recent Developments and the New Direction in Soft-Computing Foundations and Applications*, Studies in Fuzziness and Soft Computing 393, https://doi.org/10.1007/978-3-030-47124-8_16

Need for optimization under fuzzy constraints. The above formulation works well if we know the set P. In practice, however, for some alternatives a, we may not be absolutely sure that these alternatives are possible. For such alternatives, an expert can describe to what extent these alternatives are possible. This description is often made in terms of imprecise ("fuzzy") words from natural language.

To describe such knowledge, it is therefore reasonable to use techniques that Zadeh invented specifically to translate such imprecise knowledge into precise computer-understandable form—namely, the technique of fuzzy logic; see, e.g., [2–7].

Crudely speaking, we ask each expert to estimate, on a scale, say, from 0 to 10, to what extend each alternative is possible. If an expert marks 7 on a scale of 0 to 10, we say that the expert's degree of confidence that a is possible is $\mu(a) = 7/10 = 0.7$.

This way, to each alternative a, we assign a degree $\mu(a) \in [0, 1]$ to which, according to the experts, this alternative is possible. The corresponding function μ is known as a *membership function* or, alternatively, as a *fuzzy set*.

How to optimize under fuzzy constraints. So how can we optimize a function $f(a)$ under fuzzy constraints—described by a membership function $\mu(a)$? This question was raised in a joint paper that L. Zadeh wrote with Richard Bellman, a famous specialist in control [1].

Their main idea is to look for an alternative which is, to the largest extent, both possible *and* optimal. To be more precise:

- first, we need to describe the degree $\mu_{\text{opt}}(a)$ to which an alternative is optimal,
- then, for each alternative a, we need to combine the degree $\mu(a)$ to which this alternative is possible and the degree $\mu_{\text{opt}}(a)$ to which this alternative is optimal into a single degree to which a is possible *and* optimal;
- finally, we select an alternative a_{opt} for which the combined degree is the largest possible.

Let us start with the first step: finding out to what extent an alternative a is optimal.

Of course, if some alternative has 0 degree of possibility, this means that this alternative is not possible at all, so we should not consider it. So, we should consider only alternatives from the set $A \stackrel{\text{def}}{=} \{a : \mu(a) > 0\}$ of all alternatives for which there is a non-zero degree of possibility.

If two alternatives a and a' have the same value of the objective function $f(a) = f(a')$, then, intuitively, our degree of confidence that the alternative a is optimal should be the same as our degree of confidence that the alternative a' is possible. Thus, the degree $\mu_{\text{opt}}(a)$ should only depend on the value $f(a)$, i.e., we should have $\mu_{\text{opt}}(a) = F(f(a))$ for some function $F(x)$.

When the value $f(a)$ is the smallest possible, i.e., when

$$f(a) = \underline{f} \stackrel{\text{def}}{=} \min_{a \in A} f(a),$$

then we are absolutely sure that this alternative is not optimal, i.e., that $\mu_{\text{opt}}(a) = 0$. Thus, we should have $F(\underline{f}) = 0$.

On the other hand, if for the alternative a, the value of the objective function is the largest possible:

$$f(a) = \overline{f} \stackrel{\text{def}}{=} \max_{a \in A} f(a),$$

then we are absolutely sure that this alternative is optimal, i.e., that $\mu_{\text{opt}}(a) = 1$. Thus, we should have $F(\overline{f}) = 1$.

So, we need to select a function $F(x)$ for which $F(\underline{f}) = 0$ and $F(\overline{f}) = 1$. It is also reasonable to require that the function $F(f)$ increases with f. The simplest such function is linear:

$$F(f(a)) = L(f(a)) \stackrel{\text{def}}{=} \frac{f(a) - \underline{f}}{\overline{f} - \underline{f}},$$

but non-linear functions are also possible. Alternatively, we can have non-linear scaling functions

$$F(f(a)) = S(L(F(a)))$$

for some nonlinear function $S(x)$ for which $S(0) = 0$ and $S(1) = 1$.

To combine the degrees $\mu(a)$ and $F(f(a))$ of the statements "a is possible" and "a is optimal" into a single degree describing to what extent a is both possible *and* optimal, we can use an "and"-operation (t-norm) $f_\&(x, y)$. The most widely used "and"-operations are $\min(x, y)$ and $x \cdot y$.

Thus, we find the alternative a for which the value

$$d(a) = f_\&(\mu(a), F(f(a)))$$

is the largest possible. In particular, if we use a linear scaling function $F(x)$, then we select an alternative a for which the value

$$d(a) = f_\& \left(\mu(a), \frac{f(a) - \underline{f}}{\overline{f} - \underline{f}} \right)$$

is the largest possible.

When $f_\&(x, y) = \min(x, y)$, then we get

$$d(a) = \min \left(\mu(a), \frac{f(a) - \underline{f}}{\overline{f} - \underline{f}} \right).$$

When $f_\&(x, y) = x \cdot y$, then we get

$$d(a) = \mu(a) \cdot \frac{f(a) - \underline{f}}{\overline{f} - \underline{f}}.$$

Problem. The problem with this definition is that it depends on the values $\underline{f}$ and $\overline{f}$ and thus, on the exact shape of the set $A = \{a : \mu(a) > 0\}$.

In practice, experts have only approximate idea of the corresponding degrees $\mu(a)$, so when $\mu(a)$ is very small, it could be 0, or vice versa. These seemingly minor changes in the membership function can lead to huge changes in the set A and thus, to huge changes in the values $\underline{f}$ and $\overline{f}$.

There is one case when this problem is not so crucial. There is one case when the problem stops being dependent on $\overline{f}$: namely, the case of the product t-norm. Indeed, in this case, maximizing the function $d(a)$ is equivalent to maximizing the function

$$D(a) \stackrel{\text{def}}{=} (\overline{f} - \underline{f}) \cdot d(a)$$

which is equal to

$$D(a) = \mu(a) \cdot (f(a) - \underline{f}),$$

and thus, does not depend on $\overline{f}$ at all.

Natural questions and what we do in this paper. Natural questions are:

- What if we use other t-norms?
- Can we eliminate the dependence on the minimum?
- What if we use a different scaling in our derivation of the Bellman-Zadeh formula?

In this paper, we provide answers to all these questions. It turns out:

- that the product is the only t-norm for which there is no dependence on maximum,
- that it is impossible to eliminate the dependence on the minimum, and
- we also provide t-norms corresponding to the use of general scaling functions.

2 First Result: Product Is the only t-Norm for Which Optimization Under Fuzzy Constraints Does Not Depend on $\overline{f}$

Analysis of the problem. Independence on $\overline{f}$ means, in particular, that two alternatives a and a' have the same value of $d(a)$, i.e., that $d(a) = d(a')$, then the same equality holds if we change the value $\overline{f}$. In other words, we want to make sure that if

$$f_{\&}\left(\mu(a), \frac{f(a)-\underline{f}}{\overline{f}-\underline{f}}\right) = f_{\&}\left(\mu(a'), \frac{f(a')-\underline{f}}{\overline{f}-\underline{f}}\right),$$

then for a new value $\overline{f}'$, we will also have

$$f_{\&}\left(\mu(a), \frac{f(a)-\underline{f}}{\overline{f}'-\underline{f}}\right) = f_{\&}\left(\mu(a'), \frac{f(a')-\underline{f}}{\overline{f}'-\underline{f}}\right).$$

This implication must be true for any membership function $\mu(a)$, for any objective function $f(a)$, and for any values $\overline{f}$ and $\overline{f}'$. Let us denote $A \stackrel{\text{def}}{=} \mu(a)$, $A' \stackrel{\text{def}}{=} \mu(a')$,

$$b \stackrel{\text{def}}{=} \frac{f(a)-\underline{f}}{\overline{f}-\underline{f}}, \quad b' \stackrel{\text{def}}{=} \frac{f(a')-\underline{f}}{\overline{f}-\underline{f}},$$

and

$$k \stackrel{\text{def}}{=} \frac{\overline{f}-\underline{f}}{\overline{f}'-\underline{f}}.$$

In these terms, the desired implication takes the following form:

- if $f_{\&}(A, b) = f_{\&}(A', b')$,
- then for every $k > 0$, we have $f_{\&}(A, k \cdot b) = f_{\&}(A', k \cdot b')$.

Let us analyze which "and"-operations $f_{\&}(x, y)$ satisfy this property.

Main result of this section: the product $f_{\&}(x, y) = x \cdot y$is the only "and"-operation that satisfies the desired implication. We want to prove that the algebraic product $f_{\&}(x, y)$ is the only "and"-operation that satisfies the desired implication.

Indeed, by the general properties of the "and"-operation, we have

$$f_{\&}(x, 1) = f_{\&}(1, x) = x$$

for all x. Thus, the condition $f_{\&}(A, b) = f_{\&}(A', b')$ is satisfied for $A = x$, $b = 1$, $A' = 1$, and $b' = x$. So, if the desired implication holds, then, for $k = y$, we get

$$f_{\&}(x, y \cdot 1) = f_{\&}(1, y \cdot x),$$

i.e., that $f_{\&}(x, y) = f_{\&}(1, y \cdot x)$. Since $f_{\&}(1, z) = z$ for all z, we thus conclude that $f_{\&}(x, y) = x \cdot y$ for all x and y. The statement is proven.

3 Second Result: What if We Use a Non-linear Scaling Function $S(x)$?

Analysis of the problem. What if, instead of a linear scaling function, we use a non-linear function $S(x)$?

In this case, we have

$$d(a) = f_{\&}\left(\mu(a), S\left(\frac{f(a) - \underline{f}}{\overline{f} - \underline{f}}\right)\right).$$

Thus, the desired property takes the following form: if

$$f_{\&}\left(\mu(a), S\left(\frac{f(a) - \underline{f}}{\overline{f} - \underline{f}}\right)\right) = f_{\&}\left(\mu(a'), S\left(\frac{f(a') - \underline{f}}{\overline{f} - \underline{f}}\right)\right),$$

then for a new value $\overline{f}'$, we will also have

$$f_{\&}\left(\mu(a), S\left(\frac{f(a) - \underline{f}}{\overline{f}' - \underline{f}}\right)\right) = f_{\&}\left(\mu(a'), S\left(\frac{f(a') - \underline{f}}{\overline{f}' - \underline{f}}\right)\right).$$

If we use the above notations A, a', b, b', and k, then the desired implication takes the following form:

- if $f_{\&}(A, S(b)) = f_{\&}(A', S(b'))$,
- then for every $k > 0$, we have $f_{\&}(A, S(k \cdot b)) = f_{\&}(A', S(k \cdot b'))$.

Let us analyze which "and"-operations $f_{\&}(x, y)$ satisfy this property.

Let us denote $X \stackrel{\text{def}}{=} S^{-1}(A)$ and $X' \stackrel{\text{def}}{=} S^{-1}(A')$. Then $A = S(X)$, $A' = S(X')$, and the above implication takes the following form:

- if $f_{\&}(S(X), S(b)) = f_{\&}(S(X'), S(b'))$,
- then for every $k > 0$, we have $f_{\&}(S(X), S(k \cdot b)) = f_{\&}(S(X'), S(k \cdot b'))$.

It is known that for every 1-1 continuous monotonic function $S(x) : [0, 1] \to [0, 1]$ and for every "and"-operation $f_{\&}(x, y)$, the re-scaled function

$$f'_{\&}(x, y) \stackrel{\text{def}}{=} S^{-1}(f_{\&}(S(x), S(y))$$

is also an "and"-operation. In terms of this new "and"-operation,

$$f_{\&}(S(x), S(y)) = S(f'_{\&}(x, y)).$$

Thus, the desired implication takes the form:

- if $S(f'_{\&}(x, b)) = S(f'_{\&}(x', b'))$,

- then for every $k > 0$, we have $S(f'_{\&}(x, k \cdot b)) = S(f'_{\&}(x', k \cdot b'))$.

Since the scaling function $S(x)$ is increasing, $S(x) = S(y)$ is equivalent to $x = y$. Thus, the desired condition can be further simplified into the following form:

- if $f'_{\&}(x, b) = f'_{\&}(x', b')$,
- then for every $k > 0$, we have $f'_{\&}(x, k \cdot b) = f'_{\&}(x', k \cdot b')$.

In the previous section, we have proven that the only "and"-operation satisfying this condition is $f'_{\&}(x, y) = x \cdot y$. By definition of the re-scaled function $f'_{\&}$, this means that

$$S^{-1}(f_{\&}(S(x), S(y)) = x \cdot y.$$

Applying $S(x)$ to both sides, we conclude that

$$f_{\&}(S(x), S(y)) = S(x \cdot y).$$

Thus, for any $X \stackrel{\text{def}}{=} S^{-1}(x)$ and $Y \stackrel{\text{def}}{=} S^{-1}(y)$, we have $S(X) = x$, $S(y) = y$ and thus,

$$f_{\&}(X, Y) = S(x \cdot y) = S(S^{-1}(X) \cdot S^{-1}(Y)).$$

Thus, we arrive at the following conclusion.

Main result of this section: for which "and"-operation is the optimization independent on $\overline{f}$. For each scaling function $S(x)$, the only "and"-operation for which the optimization does not depend on $\overline{f}$ is the operation

$$f_{\&}(x, y) = S(S^{-1}(x) \cdot S^{-1}(y)).$$

4 Third Result: It Is Not Possible to Avoid the Dependence on $\underline{f}$

Analysis of the problem. Independence on $\underline{f}$ means, in particular, that two alternatives a and a' have the same value of $d(a)$, i.e., that if $d(a) = d(a')$, then the same equality holds if we change the value $\underline{f}$. In other words, we want to make sure that if

$$f_{\&}\left(\mu(a), \frac{f(a) - \underline{f}}{\overline{f} - \underline{f}}\right) = f_{\&}\left(\mu(a'), \frac{f(a') - \underline{f}}{\overline{f} - \underline{f}}\right),$$

then for a new value $\underline{f}'$, we will also have

$$f_{\&}\left(\mu(a), \frac{f(a) - \underline{f}'}{\overline{f} - \underline{f}'}\right) = f_{\&}\left(\mu(a'), \frac{f(a') - \underline{f}'}{\overline{f} - \underline{f}'}\right).$$

This implication must be true for any membership function $\mu(a)$, for any objective function $f(a)$, and for any values $\underline{f}$ and $\underline{f}'$. Let us take $\overline{f} = 1$ and $\underline{f} = 0$, in this case the desired condition takes the following form: if

$$f_{\&}(\mu(a), f(a)) = f_{\&}(\mu(a'), f(a')),$$

then for a new value $\underline{f}'$, we will also have

$$f_{\&}\left(\mu(a), \frac{f(a) - \underline{f}'}{1 - \underline{f}'}\right) = f_{\&}\left(\mu(a'), \frac{f(a') - \underline{f}'}{1 - \underline{f}'}\right).$$

Let us denote $A \stackrel{\text{def}}{=} \mu(a)$, $A' \stackrel{\text{def}}{=} \mu(a')$, $b \stackrel{\text{def}}{=} f(a)$, $b' \stackrel{\text{def}}{=} f(a')$, and $f_0 \stackrel{\text{def}}{=} \underline{f}'$. In these terms, the desired implication takes the following form:

- if $f_{\&}(A, b) = f_{\&}(A', b')$,
- then for every $f_0 \in (0, 1)$, we have

$$f_{\&}\left(A, \frac{b - f_0}{1 - f_0}\right) = f_{\&}\left(A', \frac{b' - f_0}{1 - f_0}\right).$$

Let us take any A and any $b < 1$. Then, for $A' = f_{\&}(A, b)$ and for $b' = 1$, we have

$$f_{\&}(A', b') = f_{\&}(A', 1) = A' = f_{\&}(A, b).$$

Thus, due to the desired property, for $f_0 = b$, we have

$$f_{\&}\left(A, \frac{b - b}{1 - b}\right) = f_{\&}\left(A', \frac{1 - b}{1 - b}\right),$$

i.e.,

$$f_{\&}(A, 0) = f_{\&}(A', 1).$$

By the properties of the "and"-operation, we have $f_{\&}(A, 0) = 0$ and $f_{\&}(A', 1) = A'$, thus we conclude that $A' = 0$. But A' is equal to $f_{\&}(A, b)$, so we get $f_{\&}(A, b) = 0$ for all A and $b < 1$—which is impossible for a continuous "and"-operation. So, we arrive at the following conclusion.

Main result of this section: for fuzzy optimization, it is not possible to get rid of the dependence on $\underline{f}$. No matter what "and"-operation we use, it is not possible to avoid the dependence of the optimization result on the value $\underline{f}$.

Acknowledgements This work was supported in part by the US National Science Foundation grant HRD-1242122.

References

1. R.E. Bellman, L.A. Zadeh, Decision making in a fuzzy environment. Manage. Sci. **17**(4), B 141–B 164 (1970)
2. R. Belohlavek, J.W. Dauben, G.J. Klir, *Fuzzy Logic and Mathematics: A Historical Perspective* (Oxford University Press, New York, 2017)
3. G. Klir, B. Yuan, *Fuzzy Sets and Fuzzy Logic* (Prentice Hall, Upper Saddle River, New Jersey, 1995)
4. J.M. Mendel, *Uncertain Rule-Based Fuzzy Systems: Introduction and New Directions* (Springer, Cham, Switzerland, 2017)
5. H.T. Nguyen, E.A. Walker, *A First Course in Fuzzy Logic* (Chapman and Hall/CRC, Boca Raton, Florida, 2006)
6. V. Novák, I. Perfilieva, J. Močkoř, *Mathematical Principles of Fuzzy Logic* (Kluwer, Boston, Dordrecht, 1999)
7. L.A. Zadeh, Fuzzy sets. Inf. Control **8**, 338–353 (1965)

Soft Computing Analytic Models for Multiplication of Asymmetrical Fuzzy Numbers

Yuriy P. Kondratenko and Nina Y. Kondratenko

Abstract This paper presents the library of analytical models of the resulting membership functions (MFs) for the arithmetic operation "multiplication" with asymmetrical fuzzy numbers (AFNs), which may have various shapes of MFs. Sixteen analytic models are included to the developed library for multiplication of the triangular AFNs. The authors discuss practical examples and applications of the developed soft computing inverse/direct models.

1 Introduction

Modern scientists have a great interest in practical applications of the fuzzy sets and fuzzy logic theory [1] in different fields: business process management, medicine, transportation, engineering, economics, etc. [2–6]. Fuzzy sets and fuzzy logic are used for the decision making tasks in uncertainty, in particular, for vehicle routing problem [7–10], evaluation of education processes [11], medical diagnostics and forecasting epidemic situations [6, 12], improving efficiency of investment in uncertainty [13], business management and financial analysis [14, 15], robotics [16–20] and critical engineering technologies [21–23], transport logistics [24, 25] and others. To solve the abovementioned problems, it is necessary to fulfill the fuzzy arithmetic operations with corresponding fuzzy sets, including such soft computing operations as addition, subtraction, multiplication and division [26–30].

Let us consider a fuzzy set $\underset{\sim}{A}$ as pairs $\left(x, \mu_{\underset{\sim}{A}}(x)\right)$, that is specified on the universal set E [1, 26, 30] and any element x, $x \in E$ of the fuzzy set $\underset{\sim}{A}$, that corresponds to the

Y. P. Kondratenko (✉)
Department of Intelligent Information Systems, Petro Mohyla Black Sea National University, Mykolaiv 54003, Ukraine
e-mail: y_kondrat2002@yahoo.com

N. Y. Kondratenko
Darla Moore School of Business, University of South Carolina, Columbia, SC 29208, USA
e-mail: nina.kondratenko@grad.moore.sc.edu

S. N. Shahbazova et al. (eds.), *Recent Developments and the New Direction in Soft-Computing Foundations and Applications*, Studies in Fuzziness and Soft Computing 393, https://doi.org/10.1007/978-3-030-47124-8_17

specific value of the membership function (MF) $\mu_{\underset{\sim}{A}}(x) \in [0, 1]$. The inverse models of the resulting fuzzy sets

$$A_\alpha = \{x \,|\, \mu_{\underset{\sim}{A}}(x) \geq \alpha\},\ \alpha \in [0, 1], x \in R$$

do not always provide high performance of such operations. It often leads to complications in [26] solving real-time decision-making problems. Prior research indicates that the computational algorithms based on α-cuts of the fuzzy sets have a high computational complexity [27–31].

The goal of this research is to provide a synthesis of the library of analytical models of the resulting MFs for arithmetic operation "multiplication" with asymmetric triangular fuzzy numbers while improving its operating speed and the accuracy of the fuzzy information processing. Triangular fuzzy numbers (TrFNs), for example, $\underset{\sim}{A} = (a_1, a_0, a_2)$ and $\underset{\sim}{B} = (b_1, b_0, b_2)$, have MFs $\mu_{\underset{\sim}{A}}(x)$ and $\mu_{\underset{\sim}{B}}(x)$ with triangular parameters $\mu_{\underset{\sim}{A}}(a_0) = \mu_{\underset{\sim}{B}}(b_0) = 1$; $\mu_{\underset{\sim}{A}}(a_1) = \mu_{\underset{\sim}{A}}(a_2) = \mu_{\underset{\sim}{B}}(b_1) = \mu_{\underset{\sim}{B}}(b_2) = 0$. The subsets A_α and B_α of fuzzy sets $\underset{\sim}{A}$ and $\underset{\sim}{B}$ determine the appropriate α-cuts and can be written as follows: $A_\alpha = [a_1(\alpha), a_2(\alpha)]$, $B_\alpha = [b_1(\alpha), b_2(\alpha)]$, where $\alpha \in [0, 1]$, $\underset{\sim}{A}, \underset{\sim}{B} \in R^+$. The inverse models A_α and B_α and direct models $\mu_{\underset{\sim}{A}}(x)$, $\mu_{\underset{\sim}{B}}(x)$ of the TrFNs $\underset{\sim}{A}, \underset{\sim}{B} \in R^+$ are determined [1, 26–31] by the corresponding dependencies (1)–(4):

$$A_\alpha = [a_1(\alpha), a_2(\alpha)] = [a_1 + \alpha(a_0 - a_1), a_2 - \alpha(a_2 - a_0)], \tag{1}$$

$$\mu_{\underset{\sim}{A}}(x) = \begin{cases} 0, \forall(x \leq a_1) \cup (x \geq a_2) \\ (x - a_1)/(a_0 - a_1), \forall(a_1 < x \leq a_0) \\ (a_2 - x)/(a_2 - a_0), \forall(a_0 < x < a_2) \end{cases}, \tag{2}$$

$$B_\alpha = [b_1(\alpha), b_2(\alpha)] = [b_1 + \alpha(b_0 - b_1), b_2 - \alpha(b_2 - b_0)], \tag{3}$$

$$\mu_{\underset{\sim}{B}}(x) = \begin{cases} 0, \forall(x \leq b_1) \cup (x \geq b_2) \\ (x - b_1)/(b_0 - b_1), \forall(b_1 < x \leq b_0) \\ (b_2 - x)/(b_2 - b_0), \forall(b_0 < x < b_2) \end{cases}. \tag{4}$$

The inverse model for arithmetic operation "multiplication" can be written as [26, 27, 30].

$$A_\alpha(\cdot)B_\alpha = [a_1(\alpha)b_1(\alpha),\ a_2(\alpha)b_2(\alpha)]. \tag{5}$$

Direct analytic model $\mu_{\underset{\sim}{C}}(x)$ of the resulting MF for the fuzzy arithmetic operation "multiplication" $\underset{\sim}{C} = \underset{\sim}{A}(\cdot)\underset{\sim}{B}$, synthesized using [27, 29, 31] for $\underset{\sim}{A}, \underset{\sim}{B} \in R^+$, can be presented as follows:

$$\forall x \in R^{+}:$$

$$\mu_{\underset{\sim}{C}}(x) = \begin{cases} 0, \text{ for } \forall(x < a_1 b_1 \cup x > a_2 b_2); \\ \left\{ \begin{array}{l} -[(a_0 - a_1)b_1 + a_1(b_0 - b_1)]+ \\ +\sqrt{[(a_0 - a_1)b_1 - a_1(b_0 - b_1)]^2 + 4(a_0 - a_1)(b_0 - b_1)x} \end{array} \right\} / \\ /\{2(a_0 - a_1)(b_0 - b_1)\}, \text{ for } \forall x \in [a_1 b_1, a_0 b_0]; \\ \left\{ \begin{array}{l} -[(a_0 - a_2)b_2 + a_2(b_0 - b_2)]- \\ -\sqrt{((a_0 - a_2)b_2 - a_2(b_0 - b_2))^2 + 4(a_0 - a_2)(b_0 - b_2)x} \end{array} \right\} / \\ /\{2(a_0 - a_2)(b_0 - b_2)\}, \text{ for } \forall x \in [a_0 b_0, a_2 b_2]. \end{cases} \quad (6)$$

The fuzzy arithmetic operations can be also realized using of max-min or min-max convolutions [26], but it often leads to more complexity and operating speed reduction, as well as, to obtaining the resulting MFs that does not meet the requirements of convexity and normality of fuzzy sets.

2 Problem Statement and Structure of Resulting Fuzzy Sets' Library Based on TrFNs Masks

The direct models $\mu_{\underset{\sim}{C}}(x)$ represented by Eq. (6) for the arithmetic operation "multiplication" is validated only for TrFNs $\underset{\sim}{A} = (a_1, a_0, a_2)$ and $\underset{\sim}{B} = (b_1, b_0, b_2)$ under the following conditions: $a_1 < a_0 < a_2, b_1 < b_0 < b_2$. At the same time a lot of real input values for decision-making and control processes can be presented as asymmetrical TrFNs with different shapes of MF, in particular, (Fig. 1) $\underset{\sim}{B} = (b_1, b_0, b_0)$ or $\underset{\sim}{G} = (g_0, g_0, g_2)$, etc. Therefore, for each special case a decision-maker should develop the separate analytic model of resulting fuzzy set for implementation of

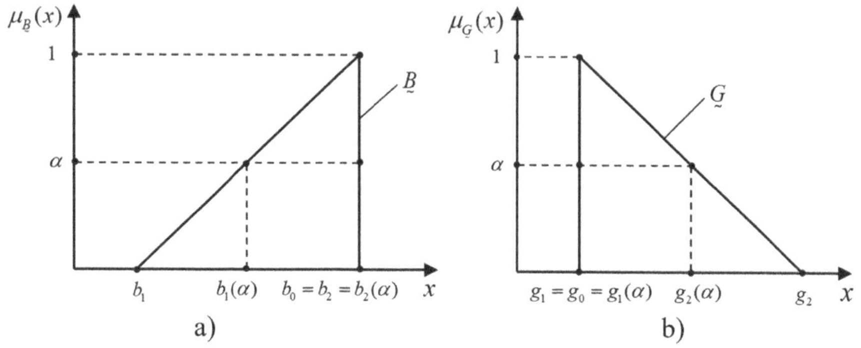

Fig. 1 TrFNs with different shapes of MF: **a)** $b_1 < b_0 = g_2$, **b)** $g_1 = g_0 < g_2$

fuzzy arithmetic "multiplication" if the asymmetrical TrFNs, $\left(\underset{\sim}{A}, \underset{\sim}{B}\right)$ have different shapes of MFs.

The authors aim to develop a library of inverse and direct analytic models of the resulting fuzzy sets $\underset{\sim}{C}$ for realization of the "multiplication" as arithmetic operation with TrFNs $\underset{\sim}{A}$ and $\underset{\sim}{B}$ taking into account various combinations of MF's shapes. Following [12, 29, 31] we can determine

$$\text{Mask}\left(\underset{\sim}{A}, \underset{\sim}{B}\right) = \{s, n, m, r\} \tag{7}$$

for any pair of the TrFNs $\underset{\sim}{A}$ and $\underset{\sim}{B}$, where indicators n, s, r and m are defined as

$$s = \begin{cases} 0, & \text{if } a_0 > a_1 \\ 1, & \text{if } a_0 = a_1 \end{cases}; n = \begin{cases} 0, & \text{if } a_0 < a_2 \\ 1, & \text{if } a_0 = a_2 \end{cases};$$
$$m = \begin{cases} 0, & \text{if } b_0 > b_1 \\ 1, & \text{if } b_0 = b_1 \end{cases}; r = \begin{cases} 0, & \text{if } b_0 < b_2 \\ 1, & \text{if } b_0 = b_2 \end{cases}.$$

The Masks (7) are the base for forming a 16-component's library of the resulting mathematical models $\{M_1 \ldots M_{16}\}$ for "multiplication" in all possible combinations of TrFNs $\left(\underset{\sim}{A}, \underset{\sim}{B}\right)$ with different shapes of MFs. The structure of the library of inverse and direct analytic models of the resulting fuzzy sets $\underset{\sim}{C}$ for realization of arithmetic operation "multiplication" is presented in Table 1.

3 Inverse and Direct Analytic Models for Multiplication of Asymmetrical TrFNs

The procedure of the inverse and direct analytic models' synthesis [27] for the arithmetic operation "multiplication" with TrFNs contains several steps.

Step 1. The formation of the inverse model C_α for α-cut of the resulting fuzzy set $\underset{\sim}{C} = \underset{\sim}{A}(\cdot)\underset{\sim}{B}$:

$$C_\alpha = A_\alpha(\cdot)B_\alpha = [c_1(\alpha), c_2(\alpha)]. \tag{8}$$

Step 2. Using the equations $c_1(\alpha) = a_1(\alpha)b_1(\alpha)$ and $c_2(\alpha) = a_2(\alpha)b_2(\alpha)$ we can find the parameter α, based on non-linear functions $F_{left}(\cdot)$ and $F_{right}(\cdot)$:

$$\alpha = F_{left}(c_1(\alpha), a_1, a_0, b_1, b_0);$$
$$\alpha = F_{right}(c_2(\alpha), a_2, a_0, b_2, b_0).$$

Table 1 Masks and models for asymmetrical TrFNs $\underset{\sim}{A}, \underset{\sim}{B} \in R^+$

$\{s, n, m, r\}$	M_i	$\underset{\sim}{A} = (a_1, a_0, a_2)$	$\underset{\sim}{B} = (b_1, b_0, b_2)$
$\{0, 0, 0, 0\}$	M_1	–	–
$\{0, 0, 0, 1\}$	M_2	–	$b_2 = b_0$
$\{0, 0, 1, 0\}$	M_3	–	$b_1 = b_0$
$\{0, 0, 1, 1\}$	M_4	–	$b_1 = b_2 = b_0$
$\{0, 1, 0, 0\}$	M_5	$a_2 = a_0$	–
$\{0, 1, 0, 1\}$	M_6	$a_2 = a_0$	$b_2 = b_0$
$\{0, 1, 1, 0\}$	M_7	$a_2 = a_0$	$b_1 = b_0$
$\{0, 1, 1, 1\}$	M_8	$a_2 = a_0$	$b_1 = b_2 = b_0$
$\{1, 0, 0, 0\}$	M_9	$a_1 = a_0$	–
$\{1, 0, 0, 1\}$	M_{10}	$a_1 = a_0$	$b_2 = b_0$
$\{1, 0, 1, 0\}$	M_{11}	$a_1 = a_0$	$b_1 = b_0$
$\{1, 0, 1, 1\}$	M_{12}	$a_1 = a_0$	$b_1 = b_2 = b_0$
$\{1, 1, 0, 0\}$	M_{13}	$a_1 = a_2 = a_0$	–
$\{1, 1, 0, 1\}$	M_{14}	$a_1 = a_2 = a_0$	$b_2 = b_0$
$\{1, 1, 1, 0\}$	M_{15}	$a_1 = a_2 = a_0$	$b_1 = b_0$
$\{1, 1, 1, 1\}$	M_{16}	$a_1 = a_2 = a_0$	$b_1 = b_2 = b_0$

Step 3. Substitution of $\alpha = \mu_{\underset{\sim}{C}}(x), x = c_1(\alpha) \in [c_1, c_0]$ and $x = c_2(\alpha) \in [c_0, c_2]$ allows to get direct models of the left and right branches of the resulting MF:

$$\forall x \in [c_1, c_0] : \mu_{\underset{\sim}{C}}(x) = F_{left}(x, a_1, a_0, b_1, b_0),$$
$$\forall x \in [c_0, c_2] : \mu_{\underset{\sim}{C}}(x) = F_{right}(x, a_2, a_0, b_2, b_0).$$

Step 4. Formation of the full direct analytical model for the resulting MF of fuzzy set $\underset{\sim}{C}$ after multiplication $\underset{\sim}{A}(\cdot)\underset{\sim}{B}$ of two TrFNs:

$$\mu_{\underset{\sim}{C}}(x) = \begin{cases} 0, \ \forall(x \leq c_1) \cup (x \geq c_2) \\ F_{left}(x, a_1, a_0, b_1, b_0), \ \forall(c_1 < x \leq c_0) \\ F_{right}(x, a_2, a_0, b_2, b_0), \forall(c_0 < x < c_2) \end{cases}. \quad (9)$$

Step 5. Calculation of the value $\mu_{\underset{\sim}{C}}(x^*)$ for any argument x^* based on (9).

The libraries of the developed inverse $C_\alpha = [c_1(\alpha), c_2(\alpha)]$ and direct $\mu_{\underset{\sim}{C}}(x)$ models $\{M_1, M_2, \ldots, M_{16}\}$ are represented in the Tables 2 and 3. The models were synthesized for the arithmetic operation "multiplication" $\underset{\sim}{C} = \underset{\sim}{A}(\cdot)\underset{\sim}{B}$ based on the above described 6-Steps procedure for the various shapes of TrFNs $\underset{\sim}{A}$ and $\underset{\sim}{B}$.

Let's consider an example with realisation of the arithmetic operation "multiplication" for the pair $\left(\underset{\sim}{A}, \underset{\sim}{B}\right)$ with the different MF's shape of TrFNs: $\underset{\sim}{A} = (7, 7, 10)$,

Table 2 Library of the inverse models C_α for resulting FN $\underset{\sim}{C} = \underset{\sim}{A}(\cdot)\underset{\sim}{B}$

M_i	$C_\alpha = A_\alpha(\cdot)B_\alpha = [c_1(\alpha), c_2(\alpha)] =$
M_1	$[a_1(\alpha), a_2(\alpha)](\cdot)[b_1(\alpha), b_2(\alpha)] = \begin{bmatrix} a_1b_1 + \alpha(a_1b_0 - 2a_1b_1 + a_0b_1) + \alpha^2(a_0 - a_1)(b_0 - b_1), \\ a_2b_2 - \alpha(2a_2b_2 - a_2b_0 - a_0b_2) + \alpha^2(a_2 - a_0)(b_2 - b_0) \end{bmatrix}$
M_2	$[a_1(\alpha), a_2(\alpha)](\cdot)[b_1(\alpha), b_0] = \begin{bmatrix} a_1b_1 + \alpha(a_1b_0 - 2a_1b_1 + a_0b_1) + \alpha^2(a_0 - a_1)(b_0 - b_1), \\ a_2b_0 - \alpha b_0(a_2 - a_0) \end{bmatrix}$
M_3	$[a_1(\alpha), a_2(\alpha)](\cdot)[b_0, b_2(\alpha)] = \begin{bmatrix} a_1b_0 + \alpha b_0(a_0 - a_1), \\ a_2b_2 - \alpha(2a_2b_2 - a_2b_0 - a_0b_2) + \alpha^2(a_2 - a_0)(b_2 - b_0) \end{bmatrix}$
M_4	$[a_1(\alpha), a_2(\alpha)](\cdot)[b_0, b_0] = [a_1b_0 + \alpha b_0(a_0 - a_1), a_2b_0 - \alpha b_0(a_2 - a_0)]$
M_5	$[a_1(\alpha), a_0](\cdot)[b_1(\alpha), b_2(\alpha)] = \begin{bmatrix} a_1b_1 + \alpha(a_1b_0 - 2a_1b_1 + a_0b_1) + \alpha^2(a_0 - a_1)(b_0 - b_1), \\ a_0b_2 - \alpha a_0(b_2 - b_0) \end{bmatrix}$
M_6	$[a_1(\alpha), a_0](\cdot)[b_1(\alpha), b_0] = \begin{bmatrix} a_1b_1 + \alpha(a_1(b_0 - b_1) + b_1(a_0 - a_1)) + \alpha^2(a_0 - a_1)(b_0 - b_1), \\ a_0b_0 \end{bmatrix}$
M_7	$[a_1(\alpha), a_0](\cdot)[b_0, b_2(\alpha)] = [a_1b_0 + \alpha b_0(a_0 - a_1), a_0b_2 - \alpha a_0(b_2 - b_0)]$
M_8	$[a_1(\alpha), a_0](\cdot)[b_0, b_0] = [a_1b_0 + \alpha b_0(a_0 - a_1), a_0b_0]$
M_9	$[a_0, a_2(\alpha)](\cdot)[b_1(\alpha), b_2(\alpha)] = \begin{bmatrix} a_0b_1 + \alpha a_0(b_0 - b_1), \\ a_2b_2 - \alpha(2a_2b_2 - a_2b_0 - a_0b_2) + \alpha^2(a_2 - a_0)(b_2 - b_0) \end{bmatrix}$
M_{10}	$[a_0, a_2(\alpha)](\cdot)[b_1(\alpha), b_0] = [a_0b_1 + \alpha a_0(b_0 - b_1), a_2b_0 - \alpha b_0(a_2 - a_0)]$
M_{11}	$[a_0, a_2(\alpha)](\cdot)[b_0, b_2(\alpha)] = \begin{bmatrix} a_0b_0, \\ a_2b_2 - \alpha(2a_2b_2 - a_2b_0 - a_0b_2) + \alpha^2(a_2 - a_0)(b_2 - b_0) \end{bmatrix}$
M_{12}	$[a_0, a_2(\alpha)](\cdot)[b_0, b_0] = [a_0b_0, a_2b_0 - \alpha b_0(a_2 - a_0)]$
M_{13}	$[a_0, a_0](\cdot)[b_1(\alpha), b_2(\alpha)] = [a_0b_1 + \alpha a_0(b_0 - b_1), a_0b_2 - \alpha a_0(b_2 - b_0)]$
M_{14}	$[a_0, a_0](\cdot)[b_1(\alpha), b_0] = [a_0b_1 + \alpha a_0(b_0 - b_1), a_0b_0]$
M_{15}	$[a_0, a_0](\cdot)[b_0, b_2(\alpha)] = [a_0b_0, a_0b_2 - \alpha a_0(b_2 - b_0)]$
M_{16}	$[a_0, a_0](\cdot)[b_0, b_0] = [a_0b_0, a_0b_0]$

Table 3 Library of the direct models $\mu_{\underset{\sim}{C}}(x)$ for resulting FN $\underset{\sim}{C} = \underset{\sim}{A}(\cdot)\underset{\sim}{B}$

M_i	$\mu_{\underset{\sim}{C}}(x)$
M_1	$0,$ for $\forall(x < a_1b_1 \cup x > a_2b_2)$; $\left\{-[(a_0-a_1)b_1+a_1(b_0-b_1)]+\sqrt{[(a_0-a_1)b_1-a_1(b_0-b_1)]^2+4(a_0-a_1)(b_0-b_1)x}\right\}/$ $/\{2(a_0-a_1)(b_0-b_1)\}$, for $\forall x \in [a_1b_1, a_0b_0]$; $\left\{-[(a_0-a_2)b_2+a_2(b_0-b_2)]-\sqrt{((a_0-a_2)b_2-a_2(b_0-b_2))^2+4(a_0-a_2)(b_0-b_2)x}\right\}/$ $/\{2(a_0-a_2)(b_0-b_2)\}$, for $\forall x \in [a_0b_0, a_2b_2]$
M_2	$0,$ for $\forall(x < a_1b_1 \cup x > a_2b_0)$; $\left\{-[(a_0-a_1)b_1+a_1(b_0-b_1)]+\sqrt{[(a_0-a_1)b_1-a_1(b_0-b_1)]^2+4(a_0-a_1)(b_0-b_1)x}\right\}/$ $/\{2(a_0-a_1)(b_0-b_1)\}$, for $\forall x \in [a_1b_1, a_0b_0]$; $(a_2b_0-x)/[b_0(a_2-a_0)]$, for $\forall x \in [a_0b_0, a_2b_0]$
M_3	$0,$ for $\forall(x < a_1b_0 \cup x > a_2b_2)$; $(x-a_1b_0)/((a_0-a_1)b_0)$, for $\forall x \in [a_1b_0, a_0b_0]$; $\left\{-[(a_0-a_2)b_2+a_2(b_0-b_2)]-\sqrt{((a_0-a_2)b_2-a_2(b_0-b_2))^2+4(a_0-a_2)(b_0-b_2)x}\right\}/$ $/\{2(a_0-a_2)(b_0-b_2)\}$, for $\forall x \in [a_0b_0, a_2b_2]$
M_4	$0,$ for $\forall(x < a_1b_0 \cup x > a_2b_0)$; $(x-a_1b_0)/((a_0-a_1)b_0)$, for $\forall x \in [a_1b_0, a_0b_0]$; $(a_2b_0-x)/(b_0(a_2-a_0))$, for $\forall x \in [a_0b_0, a_2b_0]$
M_5	$0,$ for $\forall(x < a_1b_1 \cup x > a_0b_2)$; $\left\{-[(a_0-a_1)b_1+a_1(b_0-b_1)]+\sqrt{[(a_0-a_1)b_1-a_1(b_0-b_1)]^2+4(a_0-a_1)(b_0-b_1)x}\right\}/$ $/\{2(a_0-a_1)(b_0-b_1)\}$, for $\forall x \in [a_1b_1, a_0b_0]$; $(a_0b_2-x)/(a_0(b_2-b_0))$, for $\forall x \in [a_0b_0, a_0b_2]$
M_6	$0,$ for $\forall(x < a_1b_1 \cup x > a_0b_0)$; $\left\{-[(a_0-a_1)b_1+a_1(b_0-b_1)]+\sqrt{[(a_0-a_1)b_1-a_1(b_0-b_1)]^2+4(a_0-a_1)(b_0-b_1)x}\right\}/$ $/\{2(a_0-a_1)(b_0-b_1)\}$, for $\forall x \in [a_1b_1, a_0b_0]$; $1,$ for $x = a_0b_0$.
M_7	$0,$ for $\forall(x < a_1b_0 \cup x > a_0b_2)$; $(x-a_1b_0)/((a_0-a_1)b_0)$, for $\forall x \in [a_1b_0, a_0b_0]$; $(a_0b_2-x)/(a_0(b_2-b_0))$, for $\forall x \in [a_0b_0, a_0b_2]$
M_8	$0,$ for $\forall(x < a_1b_0 \cup x > a_0b_0)$; $(x-a_1b_0)/((a_0-a_1)b_0)$, for $\forall x \in [a_1b_0, a_0b_0]$; $1,$ for $x = a_0b_0$.

(continued)

Table 3 (continued)

M_i	$\mu_{\underset{\sim}{C}}(x)$
M_9	$0,\ \text{for}\ \forall(x < a_0b_1 \cup x > a_2b_2);$ $(x - a_0b_1)/(a_0(b_0 - b_1)),\ \text{for}\ \forall x \in [a_0b_1, a_0b_0];$ $\left\{-[(a_0 - a_2)b_2 + a_2(b_0 - b_2)] - \sqrt{[(a_0 - a_2)b_2 - a_2(b_0 - b_2)]^2 + 4(a_0 - a_2)(b_0 - b_2)x}\right\}/$ $/\{2(a_0 - a_2)(b_0 - b_2)\},\ \text{for}\ \forall x \in [a_0b_0, a_2b_2]$
M_{10}	$0,\ \text{for}\ \forall(x < a_0b_1 \cup x > a_2b_0);$ $(x - a_0b_1)/(a_0(b_0 - b_1)),\ \text{for}\ \forall x \in [a_0b_1, a_0b_0];$ $(a_2b_0 - x)/(b_0(a_2 - a_0)),\ \text{for}\ \forall x \in [a_0b_0, a_2b_0]$
M_{11}	$0,\ \text{for}\ \forall(x < a_0b_0 \cup x > a_2b_2);$ $1,\ \text{for}\ x = a_0b_0;$ $\left\{-[(a_0 - a_2)b_2 + a_2(b_0 - b_2)] - \sqrt{((a_0 - a_2)b_2 - a_2(b_0 - b_2))^2 + 4(a_0 - a_2)(b_0 - b_2)x}\right\}/$ $/\{2(a_0 - a_2)(b_0 - b_2)\},\ \text{for}\ \forall x \in [a_0b_0, a_2b_2]$
M_{12}	$0,\ \text{for}\ \forall(x < a_0b_0 \cup x > a_2b_0);$ $1,\ \text{for}\ x = a_0b_0;$ $(a_2b_0 - x)/(b_0(a_2 - a_0)),\ \text{for}\ \forall x \in [a_0b_0, a_2b_0]$
M_{13}	$0,\ \text{for}\ \forall(x < a_0b_1 \cup x > a_0b_2);$ $(x - a_0b_1)/(a_0(b_0 - b_1)),\ \text{for}\ \forall x \in [a_0b_1, a_0b_0];$ $(a_0b_2 - x)/(a_0(b_2 - b_0)),\ \text{for}\ \forall x \in [a_0b_0, a_0b_2]$
M_{14}	$0,\ \text{for}\ \forall(x < a_0b_1 \cup x > a_0b_0);$ $(x - a_0b_1)/(a_0(b_0 - b_1)),\ \text{for}\ \forall x \in [a_0b_1, a_0b_0];$ $1,\ \text{for}\ x = a_0b_0.$
M_{15}	$0,\ \text{for}\ \forall(x < a_0b_0 \cup x > a_0b_2);$ $1,\ \text{for}\ x = a_0b_0;$ $(a_0b_2 - x)/(a_0(b_2 - b_0)),\ \text{for}\ \forall x \in [a_0b_0, a_0b_2]$
M_{16}	$0,\ \text{for}\ \forall(x < a_0b_0 \cup x > a_0b_0);$ $1,\ \text{for}\ x = a_0b_0.$

$\underset{\sim}{B} = (5, 9, 9)$. Using (7) and proposed algorithm we can automatically determine (a) the corresponding Mask of the pair $\left(\underset{\sim}{A}, \underset{\sim}{B}\right)$:

$$\text{Mask}\left(\underset{\sim}{A}, \underset{\sim}{B}\right) = \{s, n, m, r\} = \{1, 0, 0, 1\}$$

and (b) the corresponding model M_{10} from the library of models $\{M_1, M_2, \ldots, M_{16}\}$, presented in Table 1. Then for 10th model M_{10} we can find inverse (Table 2) and direct (Table 3) models, which for $\underset{\sim}{C} = \underset{\sim}{A}(\cdot)\underset{\sim}{B}$ can be presented as follows:

$$C_\alpha = A_\alpha(\cdot)B_\alpha = [c_1(\alpha), c_2(\alpha)] =$$
$$= [a_0b_1 + \alpha a_0(b_0 - b_1), a_2b_0 - \alpha b_0(a_2 - a_0)] = [35 + 28\alpha, 90 - 27\alpha],$$

$$\mu_{\underset{\sim}{C}}(x) = \begin{cases} 0, \text{ for } \forall(x < a_0b_1 \cup x > a_2b_0) \\ (x - a_0b_1)/(a_0(b_0 - b_1)), \text{ for } \forall x \in [a_0b_1, a_0b_0] = \\ (a_2b_0 - x)/(b_0(a_2 - a_0)), \text{ for } \forall x \in [a_0b_0, a_2b_0] \end{cases}$$
$$= \begin{cases} 0, \text{ for } \forall(x < 35) \cup (x > 90) \\ \frac{x-35}{28}, \text{ for } \forall(35 \le x \le 63) \\ \frac{90-x}{28}, \text{ for } \forall(63 < x \le 90) \end{cases}.$$

4 Implementation of Analytic Models Library for Solving Real-Life Decision-Making Problems Under Uncertainty

The implementation of the developed direct analytic models (Table 3) for calculation of the resulting MFs $\mu_{\underset{\sim}{C}}(x)$, according to TrFNs with various shapes of MFs, allows using one-step-automation-mode for the arithmetic operation "multiplication" $\underset{\sim}{C} = \underset{\sim}{A}(\cdot)\underset{\sim}{B}$. Let's consider in more details some examples with application of developed library of analytic models (Table 3) for solving real-life decision-making problems under uncertainty.

A. *Decision Making in Agriculture Sector*

In some cases [27] it is necessary to find the best solution E^* from the set of the alternatives $\{E_1, \ldots, E_m\}$ based on the prognosis of total profit after agricultural season using multiplication operation

$$\underset{\sim}{C}_i = \underset{\sim}{A}_i(\cdot)\underset{\sim}{B}_i, \tag{10}$$

where $\underset{\sim}{A}_i$ is TrFN which corresponds to future output value of i-th agricultural product, $(i = 1 \ldots m)$; $\underset{\sim}{B}_i$ is TrFN that corresponds to the future price value for a unit of i-th agricultural product, $(i = 1 \ldots m)$; $\underset{\sim}{C}_i$ is a fuzzy number that corresponds to the future profit value at the end of the agricultural season, in particular, after realization of the i-th agricultural product, $(i = 1 \ldots m)$. Such fuzzy data $\underset{\sim}{C} = \underset{\sim}{A}(\cdot)\underset{\sim}{B}$ will be very useful for building an efficient farming policy and total profit strategy for an agricultural corporation [32], taking into account peculiarities of agricultural region, customs' demands, taxes, obligatory payments and others.

B. *Ship Transportation Problem*

The transportation problem is well-known operation research problem, where it is necessary to find the best alternative solution $E^* \subset E$ for cargo transportation from several depots $N = \{N_1, N_2, \ldots, N_m\}$ to several customers $S = \{S_1, S_2, \ldots, S_n\}$. The ship transportation problem (STP) can be formulated as a classical transportation problem, which can be solved by linear programming methods with some modifications taking into account real conditions of marine environment. In general, STP deals with minimizing the cost function $Z(c, x)$ related to the transportation of various kind of cargoes (oil, coal, fuel etc.) from several (or one) supplying ports (deports) to various receiving ports (nodes):

$$\text{Min } Z(c, x) = \sum_{i=1}^{m} \sum_{j=1}^{n} c_{ij} x_{ij}, \tag{11}$$

where: x_{ij} is a quantity of cargo that can be transported from deport i to port j, $i = 1, \ldots, m; j = 1, 2, \ldots, n$; c_{ij} is associated with a transportation cost of cargo unit (for example, one ship's capacity) from deport i to port j, $i = 1, \ldots, m; j = 1, 2, \ldots, n$.

It is very important to solve STP for real marine conditions as optimization problem in fuzzy environment [33]. In reality marine environment may create superchangeable conditions and the degree of fuzziness of such environment is very high. Therefore, for the marine case each cost coefficient c_{ij} of the goal function $Z(c, x)$ is an uncertain value. It depends on the various external factors because each (ij)-separated service in STP includes transport penalty (cost) c_{ij}^{T}, unload penalty c_{ij}^{L}, penalties c_{ij}^{S} and c_{ij}^{E} to satisfy the safety and ecological requirements at the j-th port ($j = 1, 2, \ldots, n$) and others (port's sanitary penalty, custom's penalty). In this case, the goal function (11) can be transformed to a fuzzy style

$$\text{Min } \underset{\sim}{Z}(c, x) = \sum_{i=1}^{m} \sum_{j=1}^{n} \underset{\sim ij}{c} x_{ij}, \tag{12}$$

where the coefficients $\underset{\sim ij}{c}$ $(i = 1, 2, \ldots m;\ j = 1, 2, \ldots n)$ are fuzzy sets in R^{+}. Quite often the cargo demands $\sum_{j=1}^{n} \underset{\sim ij}{x}$ of several ports are also uncertain values ("about VALUE", "Approximately VALUE", "between VALUE_1 and VALUE_2") and respectively the value of total quantity of supplied cargo for each deport is also uncertain. In such conditions, the goal function (12) can be transformed to

$$\text{Min } \underset{\sim}{Z}(c, x) = \sum_{i=1}^{m} \sum_{j=1}^{n} \underset{\sim ij}{c} \underset{\sim ij}{x}, \tag{13}$$

where both components $\underset{\sim ij}{c}$, $\underset{\sim ij}{x}$ are corresponding fuzzy sets.

While solving fuzzy optimization problem (13) with corresponding restrictions [33] for the specific transportation task, it is necessary multi-timely to calculate the results

$$\underset{\sim}{c}_{ij}(\cdot)\underset{\sim}{x}_{ij} \tag{14}$$

of the multiplication operation for the different pairs of fuzzy sets $\underset{\sim}{c}_{ij}, \underset{\sim}{x}_{ij}$ $(i = 1 \ldots m;\ j = 1 \ldots n)$. In the cases of decision making process when the fuzzy sets $\underset{\sim}{c}_{ij}, \underset{\sim}{x}_{ij}$ are represented as TrFNs, it is very efficient to use analytic models $\mu_{\underset{\sim}{C}}(x) = \mu_{\underset{\sim}{A}(\cdot)\underset{\sim}{B}}(x)$, developed in the Sects. 2 and 3 (Tables 2 and 3). These analytic models can also be implemented for solving another marine problem such as optimization of the fleet size problem in uncertainty.

C. *Decision Making in Different Areas of Human Activity*

The developed resulting fuzzy models (Tables 2 and 3) may be successfully implemented for the fuzzy arithmetic "multiplication" with asymmetrical TrFNs (various shapes of MFs) in different fields of human activity, in particular, in optimization of marine bunkering processes [7, 9, 10], insurance business [34], medical service [12], social science [35], political management [36], control systems [37–40], analysis of biblio-metric information [41] and others.

5 Conclusion

Fuzzy models represented in Tables 2 and 3 for results of the soft computing operation "multiplication" with TrFNs of various shapes have a significant advantage for increasing accuracy of the calculations, decreasing time of modelling and simplifying the programming implementation in comparison with step-by-step models based on the algorithms of sorting and max-min convolutions. Series of calculation results confirm the efficient implementation of the proposed library based on the 16 analytic inverse/direct models for successful planning, evaluation and optimization of the complex systems in economics, management and engineering.

Acknowledgements Authors cordially thank the Fulbright Program (USA) and US host institutions Cleveland State University and University of South Carolina for possibility to conduct research and study in USA as well as Ukrainian Fulbright Circle and Institute of International Education for the support of this research.

References

1. L.A. Zadeh, Fuzzy sets. Inf. Control **8**, 338–353 (1965)
2. L.A. Zadeh, A.M. Abbasov, R.R. Yager, S.N. Shahbazova, M.Z. Reformat (eds.), Recent developments and new directions in soft computing. *Studies in Fuzziness and Soft Computing*, vol. 317 (Springer, Cham, 2014)

3. L.A. Zadeh, A.M. Abbasov, R.R. Yager, S.N. Shahbazova, M.Z. Reformat (eds.), Recent developments and new direction in soft-computing foundations and applications. *Studies in Fuzziness and Soft Computing*, vol. 342 (Springer, Cham, 2016). https://doi.org/10.1007/978-3-319-32229-2
4. M. Jamshidi, V. Kreinovitch, J. Kacprzyk (eds.), *Advance Trends in Soft Computing* (Springer, Germany, 2013)
5. D.E. Tamir, N.D. Rishe, A. Kandel (eds.), Fifty years of fuzzy logic and its applications. *Studies in Fuzziness and Soft Computing*, vol. 326 (Springer, Cham, 2015)
6. I. Atamanyuk, Y. Kondratenko, Calculation method for a computer's diagnostics of cardiovascular diseases based on canonical decompositions of random sequences, in *Proceedings of the 11th International Conference ICTERI-2015, CEUR-WS: ICT in Education, Research and Industrial Applications: Integration, Harmonization and Knowledge Transfer*, Lviv, Ukraine, 14–16 May 2015, ed. by S. Batsakis, et al.
7. G.V. Kondratenko, Y.P. Kondratenko, D.O. Romanov, Fuzzy models for capacitive vehicle routing problem in uncertainty, in *Proceedings of the 17th International DAAAM Symposium of Intelligent Manufacturing and Automation: Focus on Mechatronics & Robotics*, Vienna, Austria, 2006
8. P. Toth, D. Vigo (eds.), *The Vehicle Routing Problem* (SIAM, Philadelphia, 2002)
9. B. Werners, Y. Kondratenko, Alternative fuzzy approaches for efficiently solving the capacitated vehicle routing problem in conditions of uncertain demands, in *Complex Systems: Solutions and Challenges in Economics, Management and Engineering*, ed. by C. Berger-Vachon, et al. Studies in Systems, Decision and Control, vol. 125 (Springer, Heidelberg, 2018), pp. 521–543
10. M. Solesvik, Y. Kondratenko, G. Kondratenko, I. Sidenko, V. Kharchenko, A. Boyarchuk, Fuzzy decision support systems in marine practice, in *IEEE International Conference on Fuzzy Systems*, Naples, Italy, 9–12 July 2017
11. S.N. Shahbazova, Application of fuzzy sets for control of student knowledge. Int. J. Appl. Comput. Math. **10**(1), 195–208 (2011)
12. Y.B. Kotov, *New Mathematical Approaches to the Problems of Medical Diagnostics* (Editorial EPCC, Moscow, 2004). in Russian
13. J. Gil-Aluja, *Investment in Uncertainty* (Kluwer Academic Publishers, Dordrecht, Boston, London, 1999)
14. A.M. Gil-Lafuente, *Fuzzy logic in financial analysis*. Studies in Fuzziness and Soft Computing, vol. 175 (Springer, Berlin, 2005)
15. E. Anselin-Avila, A.M. Gil-Lafuente, Fuzzy logic in the strategic analysis: impact of the external factors over business. Int. J. Bus. Innov. Res. **3**(5), 515–534 (2009)
16. Y.P. Kondratenko, Robotics, automation and information systems: future perspectives and correlation with culture, sport and life science, in *Decision Making and Knowledge Decision Support Systems*, ed. by A.M. Gil-Lafuente, C. Zopounidis. Lecture Notes in Economics and Mathematical Systems, vol. 675 (Springer, Switzerland, 2015), pp. 43–56
17. Y. Kondratenko, G. Khademi, V. Azimi, D. Ebeigbe, M. Abdelhady, S.A. Fakoorian, T. Barto, A.Y. Roshanineshat, I. Atamanyuk, D. Simon, Robotics and prosthetics at Cleveland State University: modern information, communication, and modeling technologies, in *ICTERI'2016, CCIS*, vol. 783, ed. by A. Ginige, et al. (Springer, Cham, 2017)
18. A.N. Tkachenko, N.M. Brovinskaya, Y.P. Kondratenko, Evolutionary adaptation of control processes in robots operating in non-stationary environments. J. Mech. Mach. Theory **18**(4), 275–278 (1983)
19. Y. Kondratenko, O. Gerasin, A. Topalov, A simulation model for robot's slip displacement sensors. Int. J. Comput. **15**(4), 224–236 (2016)
20. Y.P. Kondratenko, J. Rudolph, O.V. Kozlov, Y.M. Zaporozhets, O.S. Gerasin, Neuro-fuzzy observers of clamping force for magnetically operated movers of mobile robots. Tech. Electrodyn. **5**, 53–61 (2017). in Ukrainian
21. J. Drozd, A. Drozd, Models, methods and means as resources for solving challenges in co-design and testing of computer systems and their components, in *Proceedings of the 9th International Conference on Digital Technologies*, Zhilina, Slovak Republic, 29–31 May 2013

22. A. Palagin, V. Opanasenko, Reconfigurable computing technology. J. Cybern. Syst. Anal. **43**(5), 675–686 (2007)
23. A.N. Trunov, An adequacy criterion in evaluating the effectiveness of a model design process. East. Eur. J. Enterp. Technol. **1**(4(73)), 36–41 (2015)
24. Y.P. Kondratenko, Ie.V. Sidenko, Decision-making based on fuzzy estimation of quality level for cargo delivery, in *Recent Developments and New Directions in Soft Computing*, ed. by L.A. Zadeh, et al. Studies in Fuzziness and Soft Computing, vol. 317 (Springer, Switzerland, 2014), pp. 331–344
25. Y.P. Kondratenko, S.B. Encheva, E.V. Sidenko, Synthesis of intelligent decision support systems for transport logistic, in *Proceeding of the 6th IEEE International Conference on Intelligent Data Acquisition and Advanced Computing Systems: Technology and Applications*, Prague, Czech Republic, 15–17 Sept 2011
26. A. Kaufmann, M. Gupta, *Introduction to Fuzzy Arithmetic: Theory and Applications* (Van Nostrand Reinhold Company, New York, 1985)
27. Y. Kondratenko, V. Kondratenko, Soft computing algorithm for arithmetic multiplication of fuzzy sets based on universal analytic models, in *ICTERI'2014: Information and Communication Technologies in Education, Research, and Industrial Application*, ed. by V. Ermolayev, et al. Communications in Computer and Information Science, vol. 469. (Springer, Switzerland, 2014)
28. S. Gao, Z. Zhang, Multiplication operation on fuzzy numbers. J. Softw. **4**(4) (2009), http://ojs.academypublisher.com/index.php/jsw/article/download/0404331338/1061
29. Y.P. Kondratenko, N.Y. Kondratenko, Soft computing analytic models for increasing efficiency of fuzzy information processing in decision support systems, in *Decision Making: Processes, Behavioral Influences and Role in Business Management*, ed. by R. Hudson (Nova Science Publishers, New York, 2015), pp. 41–78
30. A. Piegat, *Fuzzy Modeling and Control* (Springer, Heidelberg, 2001)
31. Y.P. Kondratenko, N.Y. Kondratenko, Reduced library of the soft computing analytic models for arithmetic operations with asymmetrical fuzzy numbers, in *Soft Computing: Developments, Methods and Applications*. Computer Science, Technology and Applications (Nova Science Publishers, New York, 2016), pp. 1–38
32. I. Atamanyuk, Y. Kondratenko, N. Sirenko, Management system for agricultural enterprise on the basis of its economic state forecasting, in *Complex Systems: Solutions and Challenges in Economics, Management and Engineering*, ed. by C. Berger-Vachon, et al. Studies in Systems, Decision and Control, vol. 125. (Springer, Heidelberg, 2018), pp. 453–470
33. W.A. Lodwick, J. Kacprzhyk (eds.), *Fuzzy Optimization*. Studies in Fuzziness and Soft Computing, vol. 254 (Springer, Heidelberg, 2010)
34. J.A. Sanchez, L.G.-V. Puchades, A fuzzy random variable approach to life insurance pricing, in *Soft Computing in Management and Business Economics*, vol. 287, ed. by A.M. Gil-Lafuente, et al. (Springer, Heidelberg, 2012), pp. 111–125
35. C.C. Ragin, *Redesigning Social Inquiry: Fuzzy Sets and Beyond* (The University of Chicago Press, Chicago, London, 2008)
36. A.M. Gil-Lafuente, J.M. Merigo, Decision making techniques in political management, in *Fuzzy Optimization*, ed. by W.A. Lodwick, J. Kacprzhyk. Studies in Fuzziness and Soft Computing, vol. 254 (Springer, Heidelberg, 2010), pp. 389–405
37. D. Simon, Training fuzzy systems with the extended Kalman filter. Fuzzy Sets Syst. **132**, 189–199 (2002)
38. Y. Kondratenko, E. Gordienko, Implementation of the neural networks for adaptive control system on FPGA, in *Annals DAAAM 2012 & Proceedings of the 23th International DAAAM Symposium of Intelligent Manufacturing and Automation*, Vienna, Austria, 2012, ed. by B. Katalinic
39. Z. Gomolka, E. Dudek-Dyduch, Y.P. Kondratenko, From homogeneous network to neural nets with fractional derivative mechanism, in *International Conference on Artificial Intelligence and Soft Computing*. LNCS (Springer, Cham, 2017)

40. Y.P. Kondratenko, E.Y.M. Al Zubi, The optimisation approach for increasing efficiency of digital fuzzy controllers, in *Annals DAAAM 2009 & Proceedings of the 20th International DAAAM Symposium of Intelligent Manufacturing and Automation*, Vienna, Austria, 2009
41. J.M. Merigo, A.M. Gil-Lafuente, R.R. Yager, An overview of fuzzy research with bibliometric indicators. Appl. Soft Comput. **27**, 420–433 (2015)

Combined Fuzzy Controllers with Embedded Model for Automation of Complex Industrial Plants

Yuriy P. Kondratenko and Oleksiy V. Kozlov

Abstract This paper devoted to design of the combined fuzzy controllers (CFC) with built-in model for the automatic control systems (ACS) of the complex industrial plants (CIP). The proposed CFC is designed for the ACS of the reactor temperature of the specialized pyrolysis plant (SPP) and tested in comparison with other controllers to demonstrate its advantages. The analysis of the computer simulation results confirms the high efficiency of the CFC with the proposed structure.

1 Introduction

Currently, in the industry there is a high level of automation at any stage of the material (substance) processing [1, 2]. The use of advanced ACSs of the production processes is caused by the need to increase (a) the economic indicators of plants and (b) the level of safety of the processes in general [3]. As a rule, most of the technological plants and installations, encountered today in the industry, are the complicated dynamic control objects, which can have non-linear and/or non-stationary parameters. Practice shows that automatic control of such objects based on conventional controllers in many cases does not allow achieving enough high quality of indicators of control [4]. To improve the quality indicators and total effectiveness of control systems of CIPs it is advisable to use approaches and means of advanced control theory based on artificial intelligence, in particular, intelligent fuzzy controllers (FC) [5–7].

For the effective application of FCs it is expedient to use progressive methods of their synthesis and optimization as well as specialized structures and construction

Y. P. Kondratenko (✉)
Department of Intelligent Information Systems, Petro Mohyla Black Sea National University, Mykolaiv 54003, Ukraine
e-mail: y_kondrat2002@yahoo.com

O. V. Kozlov
Department of Computerized Control Systems, Admiral Makarov National University of Shipbuilding, Mykolaiv 54025, Ukraine
e-mail: oleksiy.kozlov@nuos.edu.ua

S. N. Shahbazova et al. (eds.), *Recent Developments and the New Direction in Soft-Computing Foundations and Applications*, Studies in Fuzziness and Soft Computing 393, https://doi.org/10.1007/978-3-030-47124-8_18

principles that take into account particular features of the controlled plants. Therefore, research in the field of development and application of FCs and fuzzy control systems is conducted in two main directions: (a) development of effective methods of FCs synthesis and optimization [8–11]; (b) designing of efficient principles of FCs structural organization [12–14]. At the moment, a lot of different methods and approaches of FCs synthesis, parametric and structural optimization are developed and presented in a number of works [15–22]. Also, quite many studies are devoted to development and implementation of different types of FCs with different structural organization [23–26]. Especially effective are FCs, which combine various principles of the theory of automatic control and the theory of fuzzy logic [27, 28]. Among them PD, PI and PID FCs [29, 30], adaptive FCs [23], predictive FCs [31], hybrid FCs [26], self-organizing FCs [32] and other. Also, in the control theory one of the most effective concepts of structural organization of the control systems is the combined principle that brings together advantages of the principles of feedback, feedforward and compensation of disturbances [33]. The CFCs based on this principle can allow to significantly improve the quality indicators of control of CIPs under the action of strong disturbances. A major challenge in the implementation of these controllers is the difficulty or impossibility to measure all the existing disturbances acting on the plant. The usage of the plant model in the structure of the CFC can serve as a solution of this problem. The *aim* of this work is development and research of the CFC based on the plant model for the control systems of the CIP.

2 Structural Organization of Combined Fuzzy Controllers

The functional structure of the generalized control system of the CIP based on CFC is presented in Fig. 1.

Figure 1 has the following notations: UCL is the upper control level; SD is the setting device; FIS is the fuzzy inference system; S is the sensor of the control variable; D_1 is the disturbance, that can be measured; D_2 is the disturbance, that can not be measured; DS is the sensor of the disturbance D_1; CB is the constraint block; X_S, X_R are the set and real values of the plant control variable X; u_{SD}, u_S, u_{DS}, u_{CFC} are the corresponding output signals of SD, S, DS and CFC; ε_X is the control error (mismatch between the output signals of SD and S), $\varepsilon_X = u_{SD} - u_S$; $d\varepsilon_X/dt$ is the derivative of the control error; $\int \varepsilon_X dt$ is the integral of the control error. In turn, ε_X is the total control error caused by: changing of the set value X_S of the plant control variable, deviation of the real value X_R from the set value X_S in transients and effect of the disturbances D_1 and D_2.

As can be seen from the Fig. 1, the given control system is designed for providing automatic control and stabilization of the set value X_S of the control variable of the CIP under the influence of various disturbances. In this case, two types of disturbances effect on the CIP: the disturbance D_1, that can be measured by means of the sensor DS and the disturbance D_2, that can not be measured. In turn, the CFC calculates the

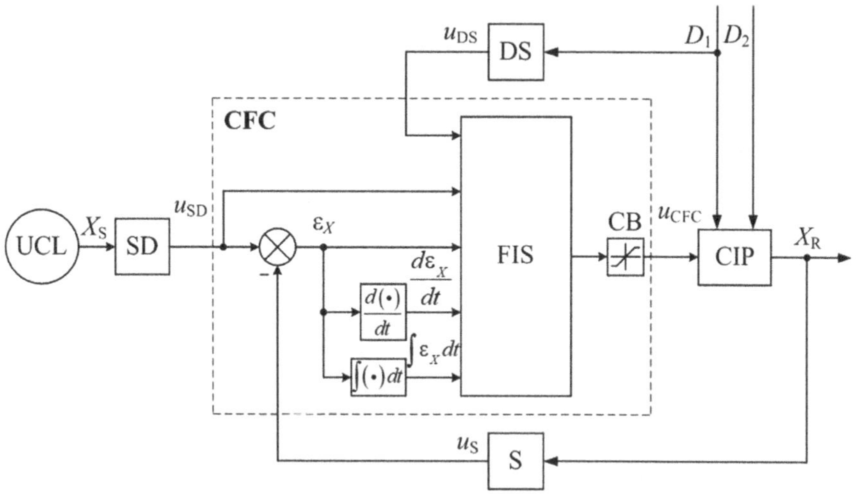

Fig. 1 Functional structure of the generalized control system of the CIP based on combined fuzzy controller

corresponding control signal u_{CFC} in accordance with the specific fuzzy control law, thereby implementing the automatic control of the variable X.

The presented control system with CFC (Fig. 1) combines together three main principles of the automatic control theory: feedback, feedforward and compensation of disturbances. The signals of control error ε_X, its derivative $d\varepsilon_X/dt$ and integral $\int \varepsilon_X dt$ are calculated and sent to the FIS first three inputs in order to implement the feedback principle in the given CFC in accordance with the proportional-integral-differential fuzzy control law. The given approach allows to realize automatic control of the plant variable X with a certain accuracy and speed, in particular, to work out the changing of its set value X_S and disturbances D_1 and D_2, that effect on the CIP. In turn, the signal u_{SD} is sent to the FIS fourth input to provide additionally the feedforward principle in the CFC. In this case, the FIS output signal u_{CFC} should depend on the signal u_{SD} taking into account the non-linear static characteristic of the CIP. Then, addition of this principle gives the opportunity to work out the changing of the set value X_S of the plant's control variable more effectively, that, in turn, allows to somewhat improve the accuracy and speed of its automatic control. The signal u_{DS}, obtained from the disturbance sensor, is sent to the FIS fifth input to provide additionally the principle of disturbances compensation in the given CFC. The additional usage of this principle allows to compensate the disturbance D_1 more effectively and, as a consequence, increase overall quality indicators of control of the CIP variable X.

Thus, the control signal of the CFC is calculated by the FIS in the following way

$$u_{CFC} = f_{FIS}\left(\varepsilon_X, \frac{d\varepsilon_X}{dt}, \int \varepsilon_X dt, u_{SD}, u_{DS}\right). \tag{1}$$

Also, the derivative du_{DS}/dt and/or integral $\int u_{DS}dt$ of the DS output signal u_{DS} can be taken as additional input signals of the FIS.

The main feature of the given CFC FIS (Fig. 1) is the large number of inputs. In turn, the total number of rules s of the rule base (RB) of any FIS can be determined by the number of all possible combinations of the linguistic terms of its inputs and calculated as follows

$$s = \prod_{i=1}^{n} m_i, \tag{2}$$

where i is the number of FIS inputs ($i = 1, \ldots n$); m_i is the number of linguistic terms for each i-th input.

Thus, on the basis of the Eq. (2) it can be concluded that the given CFC FIS (Fig. 1) will have the large number of rules of its RB and, as a consequence, the high complexity of synthesis and optimization procedures as well as further implementation.

To solve this problem, it is advisable to apply the decomposition principle for CFC modification. The structural organization of the modified CFC as a part of the generalized control system of the CIP is presented in Fig. 2.

The given modified CFC includes two FISs and the function converter. The FIS1 implements the feedback principle in the given CFC on the basis of signals ε_X, $d\varepsilon_X/dt$ and $\int \varepsilon_X dt$ in accordance with the PID fuzzy control law. In turn, the function converter implements the dependency $u_F = f(u_{SD}) = f(X_S)$, which corresponds to

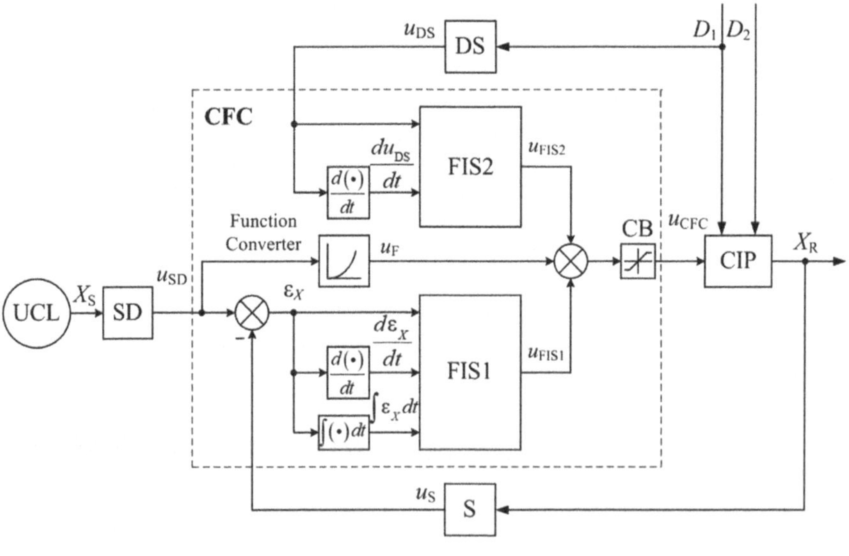

Fig. 2 Functional structure of the generalized control system of the CIP based on the modified CFC

the inverse non-linear static characteristic of the CIP, thus realizing the feedforward control principle. The FIS2, in this case, implements the principle of disturbances compensation on the basis of signal u_{DS} and its derivative. Thus, the control signal of the given modified CFC is calculated in the following way

$$\begin{aligned} u_{\text{CFC}} &= u_{\text{FIS1}} + u_{\text{F}} + u_{\text{FIS2}} \\ &= f_{\text{FIS1}}\left(\varepsilon_X, \frac{d\varepsilon_X}{dt}, \int \varepsilon_X dt\right) + f(u_{\text{SD}}) + f_{\text{FIS2}}\left(u_{\text{DS}}, \frac{du_{\text{DS}}}{dt}\right). \end{aligned} \tag{3}$$

The designing and implementation procedures of this CFC (Fig. 2) are somewhat simpler than for the previous one (Fig. 1).

In this case, it is necessary to synthesize and optimize two FISs instead of one, but the maximum number of inputs of each FIS is no more than three. So, the total number of rules of both RBs of FIS1 and FIS2 of modified CFC is significantly less than that of the FIS RB of the CFC, presented in Fig. 1.

Despite the simplification of the designing and implementation procedures of the modified CFC, the main challenge of both structural organizations of CFCs, presented in Figs. 1 and 2, is the difficulty or impossibility of measuring of all the existing disturbances, that effect on the CIP. As can be seen from the Figs. 1 and 2, only the disturbance D_1 can be measured by means of DS. In turn, the disturbance D_2 refers to the category of non-measurable disturbances. Moreover, in many cases, the unmeasured disturbances can have a much stronger influence on the plant than the measured disturbances. Also, there is a number of CIPs, whose disturbances can not be measured by any sensors at all. Thus, in the above cases the application of the CFCs with presented structures (Figs. 1 and 2) will not give a significant improvement of control quality indicators in comparison with the use of conventional fuzzy controllers.

To solve this problem the authors propose the specialized structure of the CFC with embedded CIP dynamic model (CIPDM), presented in the Fig. 3.

The given CFC (Fig. 3) also includes FIS1, FIS2 and the function converter with the inverse non-linear static characteristic of the CIP as the previous one (Fig. 2). However, instead of the channel of the disturbance D_1 measurement this CFC has the built-in CIPDM, which allows to simulate the CIP dynamics enough accurately in different modes of operation as well as to determine indirectly all the disturbances **D**, that effect on CIP.

The presented CFC with built-in CIPDM (Fig. 3) has three channels of control.

The first channel is the feedforward control channel, that is implemented on the basis of function converter with the inverse non-linear static characteristic of the CIP ($u_{\text{F}} = f(X_{\text{S}})$), as in the previous CFC (Fig. 2). The second channel is the main feedback control channel that is implemented on the basis of control variable sensor and FIS1. In turn, the total control error ε_X and its derivative $d\varepsilon_X/dt$ are the inputs of FIS1. The third channel is the additional feedback control channel, implemented on the basis of CIPDM and FIS2, which is used for the indirect determination and compensation of all the present disturbances **D**. The error e_X, caused only by the

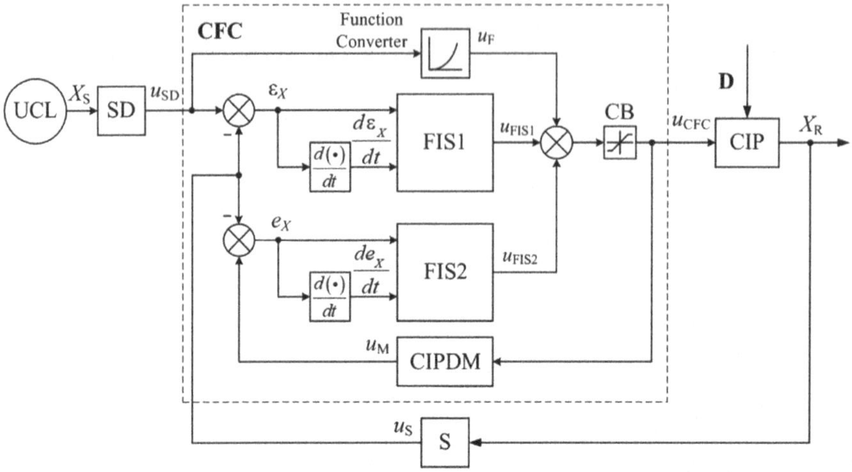

Fig. 3 Functional structure of the generalized control system based on the CFC with embedded CIP dynamic model

influence of the disturbances **D**, is determined in this channel as the deviation of the plant sensor output signal u_S from the CIPDM output signal u_M ($e_X = u_M - u_S$). Since, the same control signal u_{CFC} is sent to the both inputs of the CIP and its dynamic model, and then the presence of an error e_X indicates the effect of disturbances **D** on the CIP. Thus, the value of the disturbances **D** is determined from the error e_X value in the given control channel. To take into account the changing of the disturbances **D** in time, it is also expedient to send the derivative de_X/dt to the FIS2 s input. Thus, the control signal of the given CFC with embedded CIPDM is calculated in the following way

$$u_{CFC} = u_F + u_{FIS1} + u_{FIS2} = f(u_{SD}) + f_{FIS1}\left(\varepsilon_X, \frac{d\varepsilon_X}{dt}\right) + f_{FIS2}\left(e_X, \frac{de_X}{dt}\right). \quad (4)$$

The given CFC with the proposed structural organization (Fig. 3) has the following advantages over the CFCs, presented in Figs. 1 and 2.

1. there is no need to measure any disturbances, as they can be determined indirectly;
2. compensation of all the present disturbances **D**, that effect on the CIP, not just those that can be measured;
3. there is no need to use the integral of the error $\int \varepsilon_X dt$ in the fuzzy control law in the channel of the main feedback control due to the presence of the feedforward control channel and compensation of all the present disturbances **D** by means of third control channel.

Thus, the FIS1 and FIS2 of the given CFC have only two inputs each, so the total number of rules of their RBs will not be large and, as a consequence, their synthesis, optimization and implementation procedures will be enough simple.

The FIS1 and FIS2 of the CFC with built-in CIPDM can be of different types: Mamdani, Sugeno and others [4, 7, 25]. Furthermore, the inverse non-linear static characteristic of the CIP in the feedforward control channel can be also implemented by the additional FIS. Some other measured parameters of the CIP, that influence on its inverse non-linear static characteristic, can be the inputs of this FIS in addition to the signal u_{SD}. Also, the CIPDM can be implemented on the basis of additional FIS.

It is most effective to use the proposed CFC with embedded CIPDM for automation of CIPs which operate under conditions of strong disturbances, that cannot be measured, and for which the accurate and adequate dynamic models can be obtained. To study the effectiveness of the CFC with the proposed structure (Fig. 3) it is expedient to develop it for the ACS of the reactor temperature of the specialized pyrolysis plant [34] and compare with other existing controllers.

3 Combined Fuzzy Controller with Embedded CIPDM for the Reactor ACS of the SPP

The SPP are used for complete utilization of the polymer waste with further obtaining of alternative liquid and gaseous fuel in the environmentally friendly and energy-saving modes [3]. In turn, the SPP reactor is a CIP of the temperature control that operate under conditions of strong uncertain and unmeasurable disturbances. In addition, the accurate and adequate dynamic model of the SPP reactor as a temperature control plant is obtained and well tested in the paper [34]. Thus, it can be used in the structure of the presented CFC.

Let us consider the synthesis procedure particularities of the CFC with functional structure, presented in Fig. 3, for the reactor ACS of the SPP in detail.

The operating volume of the reactor and the maximum power of the gas burner of the chosen SPP are 100 L and 25 kW, respectively [3, 34]. In turn, the dynamic mathematical model of this reactor as a temperature control object is developed on the basis of fuzzy system for parametrical identification in the paper [34]. The given dynamic model has enough high adequacy to real processes occurring in the reactor and allows to take into account the changing of its parameters at changing of values of the reactor load level and gas burner power [34]. The function converter in the feedforward control channel of the CFC implements the SPP reactor inverse non-linear static characteristic ($u_F = f(T_S)$), obtained from its mathematical model, where T_S is the set value of the reactor temperature. The FIS1 and FIS2 are both of the Sugeno type. The temperature control error ε_T and its derivative $d\varepsilon_T/dt$ are the FIS1 inputs. In turn, signals e_T and de_T/dt are the inputs of the FIS2.

The main stages of the fuzzy logic inference of the Sugeno type are: fuzzification, aggregation, activation, accumulation and defuzzification [4]. The according linguistic meaning and degree of fuzzy set membership are determined for each input variable on the fuzzification stage [14]. For both FISs of the CFC with built-in CIPDM it is advisable to choose the following linguistic terms for the inputs, whose

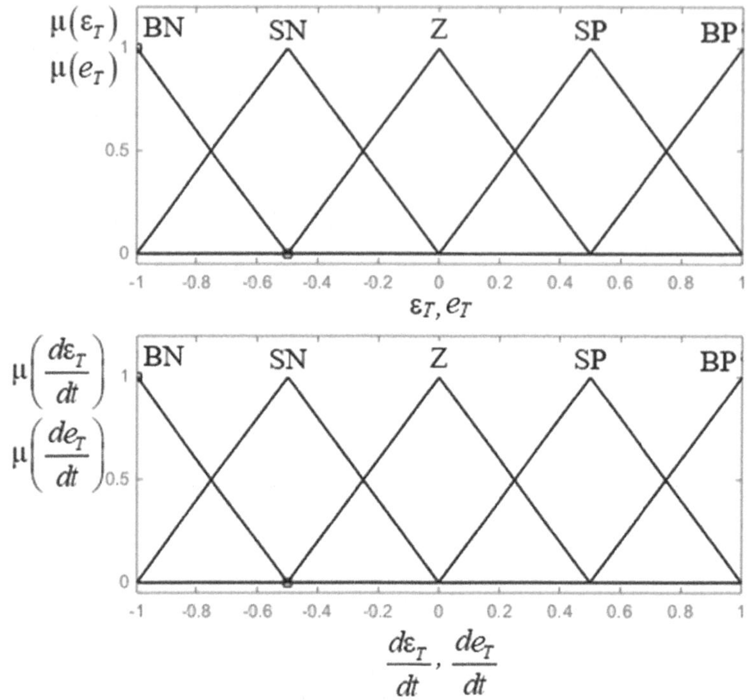

Fig. 4 Linguistic terms parameters of the CFC FISs: BN—big negative; SN—small negative; Z—zero; SP—small positive; BP—big positive

parameters are presented in Fig. 4. In turn, the parameters of the linguistic terms of the FIS1 and FIS2 inputs are presented in relative units from the maximum value.

Thereby, the formed RBs of both CFC FISs include 25 rules each, in accordance to the Eq. (2). In turn, the Sugeno's rules of the FIS1 and FIS2 can be presented in general form by the linguistic statements (5) and (6), respectively:

$$\text{IF}\,``\varepsilon_T = a\text{''}\,\text{AND}\,``\frac{d\varepsilon_T}{dt} = b\text{''}\,\text{THEN}\,``u_{\text{FIS1}} = k_{P1}\varepsilon_T + k_{D1}\frac{d\varepsilon_T}{dt}\text{''}; \tag{5}$$

$$\text{IF}\,``e_T = a\text{''}\,\text{AND}\,``\frac{de_T}{dt} = b\text{''}\,\text{THEN}\,``u_{\text{FIS2}} = k_{P2}e_T + k_{D2}\frac{de_T}{dt}\text{''}, \tag{6}$$

where a and b are according values of linguistic terms; k_{P1}, k_{D1}, k_{P2}, k_{D2} are the weights of the RBs rules.

The consequent of each rule of the RBs of both FISs is the linear combination of their inputs. The sets of consequents of the RBs of FIS1 and FIS2 are determined by the corresponding weights vector

$$\mathbf{K}_{\text{FIS}} = \{\mathbf{K}_{\text{FIS1}}, \mathbf{K}_{\text{FIS2}}\} = \{\mathbf{k}_{\text{P1}r}, \mathbf{k}_{\text{D1}r}, \mathbf{k}_{\text{P2}r}, \mathbf{k}_{\text{D2}r}\}, \tag{7}$$

where r is the rule number ($r = 1, \ldots, 25$).

In turn, weights vector $\mathbf{K}_{\text{FIS}}$ is found in the optimization process on the basis of desired transients of the reactor ACS and methods of gradient descent of numerical optimization [15, 16]. The optimization of the given FISs weights vector is implemented on the basis of the goal function I that is presented by the equation

$$I(t, \mathbf{K}_{\text{FIS}}) = \frac{1}{t_{\max}} \int_{0}^{t_{\max}} (T_{\text{D}}(t) - T_{\text{R}}(t, \mathbf{K}_{\text{FIS}}))^2 dt, \quad (8)$$

where $t_{\max}$ is the total time of the ACS transient of the reactor temperature; $T_{\text{D}}(t)$ is the desired transient characteristic of the reactor temperature; $T_{\text{R}}(t, \mathbf{K}_{\text{FIS}})$ is the real transient characteristic of the reactor temperature at the current values of the weights vector $\mathbf{K}_{\text{FIS}}$.

The desired transient characteristic $T_{\text{D}}(t)$ for the optimization process is obtained on the basis of the reference model (RM) with the differential equation

$$T_{\text{RM}} \frac{d^2 T_{\text{D}}}{dt^2} + 2T_{\text{RM}} \frac{dT_{\text{D}}}{dt} + T_{\text{D}} = T_{\text{S}}, \quad (9)$$

where T_{RM} is the time constant of the RM.

The iterative procedures of the weight vector $\mathbf{K}_{\text{FIS}}$ optimization are conducted on the basis of the vector Eq. (10) according to the method of gradient descent

$$\mathbf{K}_{\text{FIS}}[\tau + 1] = \mathbf{K}_{\text{FIS}}[\tau] - \gamma[\tau] \frac{\partial I(\mathbf{K}_{\text{FIS}})}{\partial \mathbf{K}_{\text{FIS}}} |_{\mathbf{K}_{\text{FIS}}[\tau]}, \quad (10)$$

where $\boldsymbol{\gamma}$ is the vector of steps of the gradient descent, τ is the iteration number.

Thus, the fragments of the RBs of the FIS1 and FIS2 with calculated in the optimization process rule weights are presented in Table 1.

In turn, for each developed FISs the operation "min" is selected as an aggregation operation and the operation "max" is selected for both activation and accumulation operations [25]. Moreover, the gravity center method is chosen as the defuzzification method of the given CFC FISs [14].

The computer simulation of transients and comparative analysis of the quality indicators of the ACS of the reactor temperature are carried out to verify the efficiency of the proposed by the authors CFC. The comparison is made for the following types of controllers: developed CFC with embedded CIPDM, developed CFC without CIPDM, that has only feedforward and feedback control channels, optimally tuned conventional PID-controller. In turn, the control signal u_{PID} of the conventional PID-controller is calculated as follows

$$u_{\text{PID}} = k_{\text{P}} \varepsilon_T + k_{\text{D}} \frac{d\varepsilon_T}{dt} + k_{\text{I}} \int \varepsilon_T dt, \quad (11)$$

Table 1 FISs rule bases fragments

FIS1 rule base fragment				
Rule number	Input variables	Weigh parameters		
	ε_T	$d\varepsilon_T/dt$	k_{P1}	k_{D1}
1	BN	BP	12.43	14.53
10	SN	BP	9.18	13.86
15	Z	BP	10.22	15.19
20	SP	BP	11.47	13.21
25	BP	BP	16.21	11.25
FIS2 rule base fragment				
Rule number	Input variables	Weight parameters		
	e_T	de_T/dt	k_{P2}	k_{D2}
1	BN	BP	7.22	8.46
10	SN	BP	5.63	10.05
15	Z	BP	5.24	9.44
20	SP	BP	6.94	8.27
25	BP	BP	8.29	9.12

where k_P, k_D and k_I are the corresponding gains of the controller, that are also found in the process of parametric optimization based on the desired transients using the gradient method of the fastest descent and the goal function, represented by the expression (8). In turn, $K_P = 2.189$; $K_D = 127.1$; $K_I = 0.0094$. Moreover, to eliminate the effect of integral windup, the function of the algorithmic prohibition of integration is applied in this PID-controller, when the control signal reaches the maximum possible value.

The start-up transients of the temperature ACS of the SPP reactor at the absence of disturbances are presented in Fig. 5.

The transients of the reactor ACS under conditions of strong step disturbances are presented in Fig. 6.

The quality indicators comparative analysis of the temperature ACS of the SPP reactor at the absence and at the presence of disturbances are presented in the Table 2, where t_t is the start-up transient time; σ is an overshoot; Δ_T is the static error; t_D is the time of the disturbances processing by the ACS; $\Delta_{T\max}$ is the maximum deviation caused by the disturbances effect.

As can be seen from the Figs. 5 and 6, Table 2 the quality indicators of the temperature ACS of the SPP reactor based on the developed CFC with embedded CIPDM and CFC without CIPDM are the same during the start-up transients at the absence of disturbances.

In turn, under conditions of strong step disturbances the ACS based on CFC with built-in CIPDM has considerably higher indicators than ACS based on CFC without CIPDM. In particular, the time of the disturbances processing t_D is less by 164 s and the maximum deviation caused by the disturbances effect $\Delta_{T\max}$ is also

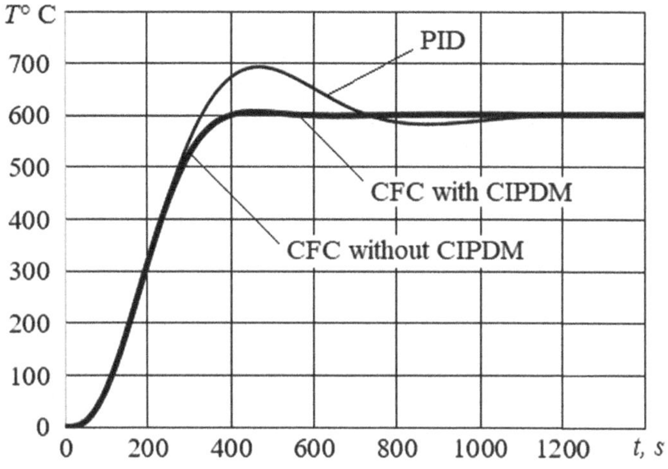

Fig. 5 Start-up transients of the temperature ACS of the SPP reactor at the absence of disturbances

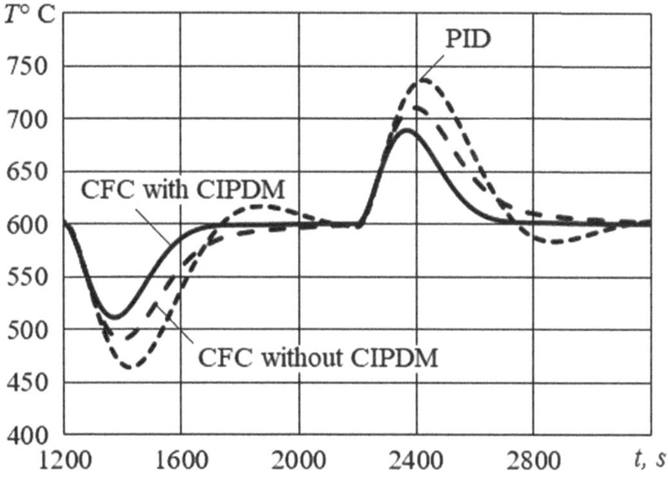

Fig. 6 Transients of the reactor ACS under conditions of strong step disturbances

Table 2 Comparative analysis of the ACS quality indicators

Controller type	t_t (s)	σ (%)	Δ_T (%)	t_D (s)	$\Delta_{T\max}$ (%)
Conventional PID	672	15.3	0	724	22.7
CFC without CIPDM	378	1.5	0	556	18.6
CFC with CIPDM	378	1.5	0	392	14.4

less by 4.2%. As for the conventional PID-controller, its application in the reactor temperature ACS gives significantly lower results in comparison with developed CFCs in both cases. All this confirms the high efficiency of the CFC with proposed structure and expediency of its application for automation of CIPs which operate under conditions of strong disturbances, that can not be measured, and for which the accurate and adequate dynamic models can be obtained.

4 Conclusions

The proposed CFC with embedded plant model gives the opportunity to significantly improve the quality indicators of CIPs, which operate under conditions of strong unmeasurable disturbances and for which the accurate and adequate dynamic models can be obtained. The CFC with the proposed structure is designed for the ACS of the reactor temperature of the SPP to study its effectiveness. The analysis of the results of computer simulation shows that the developed CFC with built-in CIPDM allows obtaining the highest quality indicators in comparison with other tested controllers, that confirms its high efficiency.

Further research should be directed towards structural and parametric optimization of the proposed CFC by means of advanced evolutionary methods and approaches.

Acknowledgements Many thanks to Fulbright Scholar Program, Institute of International Education and Ukrainian Fulbright Circle for the support of this research and for Prof. Kondratenko's possibility to conduct research at Cleveland State University.

References

1. B.R. Mehta, Y.J. Reddy, Chapter 7—SCADA systems, in *Industrial Process Automation Systems* (2015), pp. 237–300
2. Y. Kondratenko, O. Gerasin, A. Topalov, A simulation model for robot's slip displacement sensors. Int. J. Comput. **15**(4), 224–236 (2016)
3. Y.P. Kondratenko, O.V. Kozlov, O.V. Korobko, A.M. Topalov, Internet of things approach for automation of the complex industrial systems, in *Proceedings of the 13th International Conference ICTERI'2017, CEUR-WS*, vol. 1844, Kyiv, Ukraine, ed. by V. Ermolayev, et al. (2017), pp. 3–18
4. M. Pasieka, N. Grzesik, K. Kuźma, Simulation modeling of fuzzy logic controller for aircraft engines. Int. J. Comput. **16**(1), 27–33 (2017)
5. L.A. Zadeh, Fuzzy sets. Inf. Control **8**, 338–353 (1965)
6. M. Jamshidi, N. Vadiee, T.J. Ross (eds.), *Fuzzy Logic and Control: Software and Hardware Application*. Prentice Hall Series on Environmental and Intelligent Manufacturing Systems, vol. 2 (Prentice Hall, Englewood Cliffs, NJ, 1993)
7. M.T. Hayajneh, S.M. Radaideh, I.A. Smadi, Fuzzy logic controller for overhead cranes. Eng. Comput. **23**(1), 84–98 (2006)
8. W.A. Lodwick, J. Kacprzhyk (eds.), *Fuzzy Optimization*. STUDFUZ, vol. 254 (Springer, Berlin, Heidelberg, 2010)

9. L.A. Zadeh, A.M. Abbasov, R.R. Yager, S.N. Shahbazova, M.Z. Reformat (eds.), *Recent Developments and New Directions in Soft Computing*. STUDFUZ, vol. 317 (Springer, Cham, 2014)
10. Y.P. Kondratenko, L.P. Klymenko, E.Y.M. Al Zu'bi, Structural optimization of fuzzy systems' rules base and aggregation models. Kybernetes **42**(5), 831–843 (2013)
11. Y.P. Kondratenko, T.A. Altameem, E.Y.M. Al Zubi, The optimization of digital controllers for fuzzy systems design. Adv. Model. Anal. AMSE Period. Ser. A **47**, 19–29 (2010)
12. L.A. Zadeh, The role of fuzzy logic in modeling, identification and control, modeling identification and control. Model. Identif. Control **15**(3), 191–203 (1994)
13. J. Kacprzyk, R.R. Yager, S. Zadrożny, A fuzzy logic based approach to linguistic summaries of databases. Int. J. Appl. Math. Comput. Sci. **10**(4), 813–834 (2000)
14. A. Piegat, *Fuzzy Modeling and Control* (Physica-Verlag, Heidelberg, New York, 2001)
15. Y.P. Kondratenko, E.Y.M. Al Zubi, The optimization approach for increasing efficiency of digital fuzzy controllers, in *Annals of DAAAM for 2009 & Proceeding of the 20th International DAAAM Symposium on Intelligent Manufacturing and Automation* (2009), pp. 1589–1591
16. Y. Kondratenko, D. Simon, Structural and parametric optimization of fuzzy control and decision making systems, in *Recent Developments and the New Direction in Soft-Computing Foundations and Applications. Selected Papers from the 6th World Conference on Soft Computing*, May 22–25, 2016, Berkeley, USA. Series: Studies in Fuzziness and Soft Computing, vol. 361, ed. by L. Zadeh, R.R. Yager, S.N. Shahbazova, M.Z. Reformat, V. Kreinovich (Springer International Publishing, Cham, 2018) pp. 273–289. https://doi.org/10.1007/978-3-319-75408-6_22
17. Y. Kondratenko, V. Korobko, O. Korobko, G. Kondratenko, O. Kozlov, Green-IT approach to design and optimization of thermoacoustic waste heat utilization plant based on soft computing, in *Green IT Engineering: Components, Networks and Systems Implementation*. Studies in Systems, Decision and Control, vol. 105, ed. by V. Kharchenko, Y. Kondratenko, J. Kacprzyk (Springer, Cham, 2017), pp. 287–311
18. L. Wang, T.J. Kazmierski, VHDL-AMS based genetic optimisation of fuzzy logic controllers. Int. J. Comput. Math. Electric. Electron. Eng. **26**(2), 447–460 (2007)
19. D. Simon, Design and rule base reduction of a fuzzy filter for the estimation of motor currents. Int. J. Approx. Reason. **25**, 145–167 (2000)
20. R. Alcalá, J. Alcalá-Fdez, M.J. Gacto, F. Herrera, Rule base reduction and genetic tuning of fuzzy systems based on the linguistic 3-tuples representation. Soft. Comput. **11**(5), 401–419 (2007)
21. W. Pedrycz, K. Li, M. Reformat, Evolutionary reduction of fuzzy rule-based models, in *Fifty Years of Fuzzy Logic and its Applications*, STUDFUZ, vol. 326 (Springer, Cham, 2015), pp. 459–481
22. D. Simon, *Evolutionary Optimization Algorithms: Biologically Inspired and Population-Based Approaches to Computer Intelligence* (Wiley, 2013)
23. Q. Suna, R. Li, P. Zhang, Stable and optimal adaptive fuzzy control of complex systems using fuzzy dynamic model. J. Fuzzy Sets Syst. **133**, 1–17 (2003)
24. R.R. Yager, D.P. Filev, Unified structure and parameter identification of fuzzy models. Syst. Man Cybern. **23**(4) (1993)
25. R. Hampel, M. Wagenknecht, N. Chaker, *Fuzzy Control: Theory and Practice* (Physika-Verlag, Heidelberg, New York, 2000)
26. S.K. Oh, W. Pedrycz, The design of hybrid fuzzy controllers based on genetic algorithms and estimation techniques. J. Kybernetes **31**(6), 909–917 (2002)
27. K. Tanaka, H.O. Wang, *Fuzzy Control Systems Design and Analysis: A Linear Matrix Inequality Approach* (Wiley, New York, USA, 2001)
28. R.R. Yager, D.P. Filev, *Essentials of Fuzzy Modeling and Control* (Wiley, New York, NY, 1994)
29. T. Takagi, M. Sugeno, Fuzzy identification of systems and its applications to modeling and control. IEEE Trans. Syst. Man Cybern. **15**(1)
30. Y.P. Kondratenko, O.V. Kozlov, O.V. Korobko, A.M. Topalov, Synthesis and optimization of fuzzy control system for floating dock's docking operations, in *Fuzzy Control Systems: Design, Analysis and Performance Evaluation*, ed. by W. Santos (Nova Science Publishers, New York City, USA, 2017), pp. 141–215

31. I. Skrjanc, Design of fuzzy model-based predictive control for a continuous stirred-tank reactor, in *12th Zittau Fuzzy Colloquium, Proceedings of East-West Fuzzy Colloquium*, Zittau, Hochschule Zittau/Goerlitz, Germany (2005), pp. 126–139
32. C. Chen, Y.M. Chen, Self-organizing fuzzy logic controller design. J. Comput. Ind. **22**(3), 249–261 (1993)
33. A.A. Tunik, M.M. Komnatska, On structures of combined UAV flight control systems with elements of fuzzy logics. J. Electron. Control Syst. **3**(41), 20–28 (2014)
34. Y.P. Kondratenko, O.V. Kozlov, G.V. Kondratenko, I.P. Atamanyuk, Mathematical model and parametrical identification of ecopyrogenesis plant based on soft computing techniques, in *Complex Systems: Solutions and Challenges in Economics, Management and Engineering, Studies in Systems, Decision and Control*, vol. 125, ed. by C. Berger-Vachon, et al. (Springer, Berlin, Heidelberg, 2018), pp. 201–233

System Identification Using Monotonic Fuzzy Models

Petr Hušek

Abstract Fuzzy models became very popular tool for identification of nonlinear systems from input-output data. Unfortunately, their behaviour is considerably deteriorated in the regions where the data is sparse or contaminated by a noise. The paper shows that when a nonlinear function describing the systems is monotonic then a fuzzy system guaranteeing monotonicity significantly improves the performance of the fuzzy model. Furthermore, sufficient conditions for monotonicity of Takagi-Sugeno fuzzy systems with Gaussian membership functions are derived.

Keywords Monotonic fuzzy system · Fuzzy model identification · Nonlinear systems

1 Introduction

Over many decades fuzzy systems turned into very popular tool for modelling and identification of nonlinear systems. Apart from the universal approximation property the principal advantages are the possibility to include a linguistically represented knowledge about the nonlinear system and vice-versa to characterize the identified model by words and thus to better understand it.

On the other hand, when fuzzy models are created based on measured input-output data only they suffer from disadvantages typical for black-box models, especially poor interpretation and low generalization ability that is due to local character of the incoming data. Those models usually behave not adequately in the regions where the data is sparse or in the regimes in which the data was not obtained, e.g. in steady state, sometimes resulting even in instability. One possibility how to improve behaviour of data driven fuzzy models consists in including apriori information about the system.

P. Hušek (✉)
Faculty of Electrical Engineering, Department of Control Engineering, Czech Technical University in Prague, Technická 2, 166 27 Prague 6, Czech Republic
e-mail: husek@fel.cvut.cz

S. N. Shahbazova et al. (eds.), *Recent Developments and the New Direction in Soft-Computing Foundations and Applications*, Studies in Fuzziness and Soft Computing 393, https://doi.org/10.1007/978-3-030-47124-8_19

Such an information concerns typically the structure of the system. Structural information may describe behaviour of some components of the system and their interconnection. Another kind of prior knowledge is represented by partially known dynamics (typically linear coming from first principles) whereas the unknown dynamics is approximated by a nonlinear model.

Sometimes we may have some information about the nonlinear functions appearing in the description of the system. In this paper we focus on the case when the approximated nonlinear function is monotonic with respect to one or more variables. First result on monotonicity of Mamdani-type fuzzy systems was presented in [1] where sufficient conditions for triangular or trapezoidal membership functions were derived. Sufficient monotonicity conditions for Takagi-Sugeno (T-S) fuzzy systems with differentiable membership functions and membership functions that are not differentiable in finite number of points were presented by Won et al. [2]. Those conditions for monotonicity of fuzzy systems with ellipsoidal regions in the rules antecedents and linear submodels in the rules consequents were presented by Hušek [3]. In [4] procedure generating from a general fuzzy system a monotone one based on rules relabelling was introduced. More results in this topic include monotonicity of hierarchical fuzzy systems [5], interconnected single-input single-output fuzzy systems [6] or type-2 fuzzy systems [7, 8]. Properties of different defuzzification methods with respect to their ability generate monotone fuzzy systems were analyzed in [9–11]. Practical applications of monotone fuzzy systems cover detection of failures [12], decision making [13] and classification [14–16].

In this paper we present new sufficient conditions for monotonicity of Takagi-Sugeno fuzzy systems with Gaussian membership functions and first order submodels in the consequent part. As it was stated before those conditions were derived in [2] among others. However, the conditions are separated between the antecedent and consequent part that makes them very conservative. Moreover, apart from the case when the width of membership functions is the same the antecedent conditions are quite complicated. In contrast to them the presented conditions can be represented as linear matrix inequalities (LMI) on parameters of linear submodels and hence handled by efficient semidefinite programming and convex optimization techniques [17]. Next, they reduce conservatism since they consider parameters of both antecedent and consequent part at the same time.

Usefulness of taking into account the information about monotonicity of a nonlinear mapping will be demonstrated on identification of two-tank hydraulic system.

Throughout the paper the superscript $\cdot^T$ denotes transpose and $A \succ 0\,(A \succeq 0)$ stands for positive definiteness (semidefiniteness) of a square symmetric matrix A.

2 Problem Formulation

We consider Takagi-Sugeno type fuzzy model where M_i fuzzy sets $A_i^1, A_i^2 \dots, A_i^{M_i}$ are assigned to each input x_i, $i = 1, \dots, n$. The $M = \prod_{i=1}^{n} M_i$ rules describing full grid partitions can be written as

$$R_k: \quad \text{If } x_1 \text{ is } A_1^{k_1} \text{ and } \cdots \text{and } x_n \text{ is } A_n^{k_n}$$
$$\text{then } \ y = a_{\mathbf{k}}^{\mathrm{T}} x + a_{\mathbf{k}0}$$

where $x = [x_1, \ldots, x_n]^{\mathrm{T}} \in U = [U_1 \times U_2 \times \cdots \times U_n]$, $U_i = [\underline{u}_i, \overline{u}_i]$ is the input vector, y is the scalar output, $a_{\mathbf{k}} = [a_{\mathbf{k}1}, \ldots, a_{\mathbf{k}n}]^{\mathrm{T}} \in \Re^n$ and $a_{\mathbf{k}0} \in \Re$ are the parameters of the linear submodels and $\mathbf{k} = (k_1, k_2, \ldots, k_n) \in K \subset N_0^n$, $1 \le k_i \le M_i$ is the multi-index. By using the product inference engine and center of gravity defuzzification the total output of the fuzzy model is determined by

$$\begin{aligned} F(x) &= \frac{\sum_{k_1=1}^{M_1} \cdots \sum_{k_n=1}^{M_n} (a_{\mathbf{k}}^{\mathrm{T}} x + a_{\mathbf{k}0}) \prod_{i=1}^n \mu_i^{k_i}(x_i)}{\sum_{k_1=1}^{M_1} \cdots \sum_{k_n=1}^{M_n} \prod_{i=1}^n \mu_i^{k_i}(x_i)} \\ &= \frac{\sum_{\mathbf{k} \in K} (a_{\mathbf{k}}^{\mathrm{T}} x + a_{\mathbf{k}0}) \prod_{i=1}^n \mu_i^{k_i}(x_i)}{\sum_{\mathbf{k} \in K} \prod_{i=1}^n \mu_i^{k_i}(x_i)} \end{aligned} \tag{1}$$

where $\mu_i^{k_i}(x_i)$ are the membership functions characterizing input fuzzy sets $A_i^{k_i}$, $i = 1, \ldots, n$.

In this paper we assume that all the membership functions $\mu_i^{k_i}(x_i)$ are of Gaussian type, i.e.

$$\begin{aligned} &\mu_i^{k_i}(x_i) = \exp\left(-\frac{(x_i - c_{k_i})^2}{\sigma_{k_i}}\right), \ k_i = 1, \ldots M_i, \\ &i = 1, \ldots n, \ \sigma_{k_i} > 0. \end{aligned} \tag{2}$$

Definition 1 (*Monotonicity of fuzzy system*). The mapping $F : U \to \Re$ defined by (1) is said to be monotonically nondecreasing with respect to x_i if and only if $x_i^1 < x_i^2$ implies

$$F(x_1, \ldots, x_i^1, \ldots, x_n) \le F(x_1, \ldots, x_i^2, \ldots, x_n) \tag{3}$$

for any $[x_1, \ldots, x_{i-1}, x_{i+1}, \ldots, x_n]$ such that $[x_1, \ldots, x_{i-1}, x_i^1, x_{i+1}, \ldots, x_n]^{\mathrm{T}}$, $[x_1, \ldots, x_{i-1}, x_i^2, x_{i+1}, \ldots, x_n]^{\mathrm{T}} \in U$. A nonincreasing mapping is defined in a similar manner.

The question is to derive sufficient conditions for fuzzy mapping (1) to be monotonic with respect to x_i, $i = 1, \ldots, n$.

3 Monotonicity Conditions

One can easily see that the mapping (1) with membership functions (2) is differentiable in any $x \in U$ and thus it is monotonically nondecreasing with respect to x_i, $i = 1, \ldots, n$ if and only if $\frac{\partial F}{\partial x_i} \ge 0$ for all $x \in U$. The partial derivative can be expressed as

$$\frac{\partial y}{\partial x_i} = \frac{1}{(\sum_{j=1}^{M_i} \mu_j(x_i))^2} \Big\{ \sum_{p=1}^{M_i} \sum_{q=1}^{M_i} a_{\mathbf{p}i} \mu_p(x_i) \mu_q(x_i) + \sum_{p=1}^{M_i-1} \sum_{q=p+1}^{M_i} \left(a_{\mathbf{p}}^{\mathrm{T}} x + a_{\mathbf{p}0} - a_{\mathbf{q}}^{\mathrm{T}} x - a_{\mathbf{q}0} \right) \times \left(\frac{\partial \mu_p(x_i)}{\partial x_i} \mu_q(x_i) - \frac{\partial \mu_q(x_i)}{\partial x_i} \mu_p(x_i) \right) \Big\} \tag{4}$$

for all $\mathbf{p}, \mathbf{q} \in K$ where $\mathbf{p} = [k_1, \ldots, k_{i-1}, p, k_{i+1}, \ldots, M_i]$, $\mathbf{q} = [k_1, \ldots, k_{i-1}, q, k_{i+1}, \ldots, M_i]$ (i.e. for all combinations of $(k_1, \ldots, k_{i-1}, k_{i+1}, \ldots, k_n)$ where $k_i \in [1, M_i]$) such that $1 \leq p < q \leq M_i$.

Let us substitute (2) into the last term in (4) and we arrive to

$$\frac{\partial \mu_p(x_i)}{\partial x_i} \mu_q(x_i) - \frac{\partial \mu_q(x_i)}{\partial x_i} \mu_p(x_i) = 2\mu_p(x_i)\mu_q(x_i) \left(\frac{x_i - c_q}{\sigma_q^2} - \frac{x_i - c_p}{\sigma_p^2} \right). \tag{5}$$

Substituting (5) into (4) gives

$$\frac{\partial y}{\partial x_i} = \frac{1}{(\sum_{j=1}^{M_i} \mu_j(x_i))^2} \Big(\sum_{p=1}^{M_i} a_{\mathbf{p}i} \mu_p^2(x_i) + \sum_{p=1}^{M_i-1} \sum_{q=p+1}^{M_i} \mu_p(x_i)\mu_q(x_i) \big(a_{\mathbf{p}i} + a_{\mathbf{q}i} + 2(a_{\mathbf{p}}^{\mathrm{T}} x + a_{\mathbf{p}0} - a_{\mathbf{q}}^{\mathrm{T}} x - a_{\mathbf{q}0}) \left(\frac{x_i - c_q}{\sigma_q^2} - \frac{x_i - c_p}{\sigma_p^2} \right) \big) \Big). \tag{6}$$

Since the last multiplication in (6) is a quadratic function one can write

$$\frac{\partial y}{\partial x_i} = \frac{1}{(\sum_{j=1}^{M_i} \mu_j(x_i))^2} \Big(\sum_{p=1}^{M_i} a_{\mathbf{p}i} \mu_p^2(x_i) + \sum_{p=1}^{M_i-1} \sum_{q=p+1}^{M_i} \mu_p(x_i)\mu_q(x_i)(x^{\mathrm{T}} A_{pq} x + 2b_{pq}^{\mathrm{T}} x + d_{pq}) \Big) \tag{7}$$

where $A_{pq} \in \Re^{n \times n}$, $b_{pq} \in \Re^n$ and $d_{pq} \in \Re$ are given by

$$A_{pq} = \left(\frac{1}{\sigma_p^2} - \frac{1}{\sigma_q^2} \right) \begin{bmatrix} 0 & \cdots & \Delta a_1 & \cdots & 0 \\ \vdots & & \vdots & & \vdots \\ \Delta a_1 & \cdots & \Delta a_i & \cdots & \Delta a_n \\ \vdots & & \vdots & & \vdots \\ 0 & \cdots & \Delta a_n & \cdots & 0 \end{bmatrix},$$

$$b_{pq} = \left(\frac{c_p}{\sigma_p^2} - \frac{c_q}{\sigma_q^2} \right) (a_p - a_q) + \left(\frac{1}{\sigma_p^2} - \frac{1}{\sigma_q^2} \right) \begin{bmatrix} 0 \\ \vdots \\ \Delta a_0 \\ \vdots \\ 0 \end{bmatrix},$$

$$d_{pq} = 2\Delta a_0 \left(\frac{c_p}{\sigma_p^2} - \frac{c_q}{\sigma_q^2} \right) + a_{\mathbf{p}i} + a_{\mathbf{q}i} \tag{8}$$

with $\Delta a_j = a_{\mathbf{p}j} - a_{\mathbf{q}j}$, $j = 0, \dots, n$ appearing in expressions above always in j-th row or column, $1 \le p < q \le M_i$.

Since the membership functions $\mu_i^{k_i}(x_i) > 0$ for all $x_i \in U_i$ nonnegativity of the partial derivative $\frac{\partial y}{\partial x_i}$ is using (7) guaranteed if:

1. $a_{\mathbf{k}i} \ge 0$, $\mathbf{k} \in K$,
2. $\begin{bmatrix} A_{pq} & b_{pq} \\ b_{pq}^{\mathrm{T}} & d_{pq} \end{bmatrix} \succeq 0$, $p = 1, \dots, M_i - 1$,

$q = p + 1, \dots, M_i.$ (9)

Let us note that the conditions (9) are in the form of LMIs with respect to the submodel parameters $a_{\mathbf{k}i}$, $\mathbf{k} \in K$, $i = 0, \dots, n$.

4 Nonlinear Function Approximation

The key task in nonlinear dynamic system identification from input-output data is nonlinear function approximation, usually in a least-squares sense. Henceforth if monotonicity of a function needs to be enforced it is necessary to include the monotonicity conditions into the corresponding optimization task.

Suppose that the set of input samples $x^1, x^2, \dots, x^N \in \Re^n$ and the set of desired outputs $y^1, y^2, \dots, y^N \in \Re$ are given. We would like to approximate the data by a monotonic fuzzy system (1) (with respect to some or all input variables) minimizing least square error. More specifically, the aim is to find the parameters $a_{\mathbf{k}j}$, $\mathbf{k} \in K$, $j = 0, \dots, n$ minimizing the criterion

$$J = \sum_{j=1}^{N} (y^j - F(x^j))^2 \tag{10}$$

where $F(x)$ is given by (1).

At first suppose that the parameters of membership functions (2) are given. Such a task can be accomplished by different ways, the most typical are covering the whole input space uniformly or applying the fuzzy clustering techniques [18]. Then the mapping (1) can be written as

$$F(x) = \Phi(x) \cdot \theta \tag{11}$$

with

$$\Phi(x) = \frac{[\mu_{\mathbf{k}_1 0}(x), \ldots, \mu_{\mathbf{k}_M n}(x)] \otimes [1, x_1, \ldots, x_n]}{\sum_{\mathbf{k} \in K} \mu_{\mathbf{k}}(x)} \tag{12}$$

and the parameter vector

$$\theta = [a_{\mathbf{k}_1 0}, \ldots, a_{\mathbf{k}_1 n}, \ldots, a_{\mathbf{k}_M 0}, \ldots, a_{\mathbf{k}_M n}]^{\mathrm{T}} \tag{13}$$

where $\otimes$ stands for Kronecker product and $\mathbf{k}_j \in K$. The mapping (11) is linear with respect to the submodel parameters $a_{\mathbf{k}i}$.

Now assume that the regressor $\{x^l, y^l; x^l \in \Re^n, y^l \in \Re, l = 1, \ldots, N\}$ is available. When minimizing the least squares the goal is to find

$$\theta_{\mathrm{opt}} = \arg\min_{\theta} ||Z\theta - \mathbf{y}|| \tag{14}$$

where $Z = [\Phi(x^1)^{\mathrm{T}}, \ldots, \Phi(x^N)^{\mathrm{T}}]^{\mathrm{T}}$, $\mathbf{y} = [y^1, \ldots, y^N]^{\mathrm{T}}$ under the assumption that $F(x)$ is monotonically nondecreasing, i.e. under the LMI constraints (9).

The solution can be found by semidefinite programming. Write

$$||Z\theta - \mathbf{y}|| = \theta^{\mathrm{T}} Z^{\mathrm{T}} Z\theta - 2Z^{\mathrm{T}}\mathbf{y}\theta + \mathbf{y}^{\mathrm{T}}\mathbf{y} \tag{15}$$

and introduce new variable β to be minimized such that $\beta \geq ||Z\theta - \mathbf{y}||$. Then by application of Schur complements the minimum value of (15) can be determined as

$$\begin{gathered} \min \beta \\ \text{s.t.} \begin{bmatrix} I_{M(n+1)} & L^{\mathrm{T}}\theta \\ \theta^{\mathrm{T}} L & \beta + 2Z^{\mathrm{T}}\mathbf{y}\theta - \mathbf{y}^{\mathrm{T}}\mathbf{y} \end{bmatrix} \succeq 0 \end{gathered} \tag{16}$$

where L is Cholesky factor of $Z^{\mathrm{T}}Z$, $LL^{\mathrm{T}} = Z^{\mathrm{T}}Z$ and $I_{M(n+1)}$ is $M(n+1) \times M(n+1)$ identity matrix. To guarantee monotonicity the LMIs (9) will be added to (16).

5 Nonlinear Dynamical System Identification

The benefit of consideration a knowledge about monotonicity of a nonlinear function will be illustrated on identification of a nonlinear dynamical system from measured data. In that case the regressor typically contains past values of inputs and outputs and the output of the fuzzy model is one-step prediction of the output of the system.

In this paper we consider a two-tank system with the inlet to the first tank and the outlet from the second tank that is described by differential equations

$$\begin{aligned} \dot{h}_1(t) &= -k_1\sqrt{h_1(t) - h_2(t)} + k_{\mathrm{u}}u(t) \\ \dot{h}_2(t) &= k_1\sqrt{h_1(t) - h_2(t)} - k_2\sqrt{h_2(t)} \end{aligned} \tag{17}$$

where h_1 [m] and h_2 [m] are the levels in the first and second tank ($h_1(t) \geq h_2(t)$ is supposed), respectively, $u(t)$ [m^3/s] is the input inlet to the first tank and k_1, k_2 and k_u being coefficients depending on the system parameters.

State-space equations of the system may be written in the form

$$\begin{aligned} \dot{x}_1(t) &= -2k_1\sqrt{x_1(t)} + k_2\sqrt{x_2(t)} + k_u u(t) \\ \dot{x}_2(t) &= k_1\sqrt{x_1(t)} - k_2\sqrt{x_2(t)} \end{aligned} \tag{18}$$

where $x_1 = h_1 - h_2$ and $x_2 = h_2$.

Our aim is to construct and identify a corresponding state-space one-step ahead predictor taking the form

$$\begin{aligned} \Delta x_1(k) &= f_1(x_1(k), x_2(k)) + k_u u(k) \\ \Delta x_2(k) &= f_2(x_1(k), x_2(k)) \end{aligned} \tag{19}$$

where the functions f_1 and f_2 will be represented by two two-input-one-output T-S fuzzy models (1). To be able to display the nonlinear functions we will not consider dependence on the input $u(k)$ by assumption that the coefficient k_u is known. Next, we assume that we measure both levels $h_1 \in [0, 1]$ and $h_2 \in [0, 1]$. A uniformly distributed noise on the interval $[-0.005, 0.005]$ was added to both measurements. For simulations we use $k_1 = 0.0018$, $k_2 = 0.0023$ and $k_u = 400$.

Let us focus on an ideal case when we are able to set arbitrary operational conditions, i.e. we can identify both nonlinear functions for any couple $[x_1, x_2]$, $x_1 \in [0, 1]$, $x_2 \in [0, 1]$. The functions f_1 and f_2 are plotted in Figs. 1 and 2, respectively.

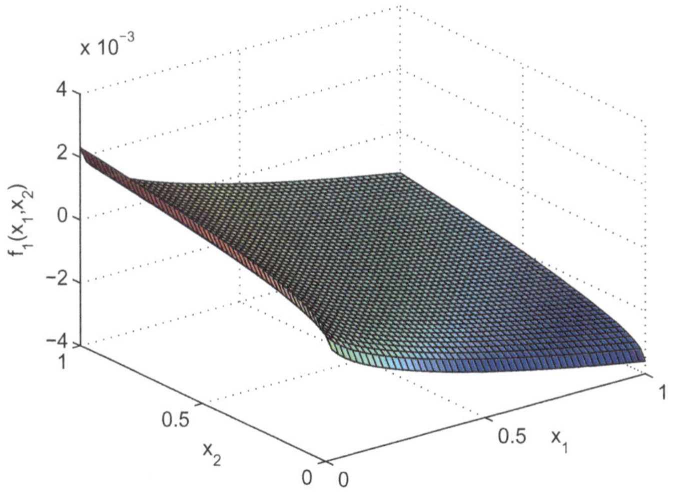

Fig. 1 Exact function f_1

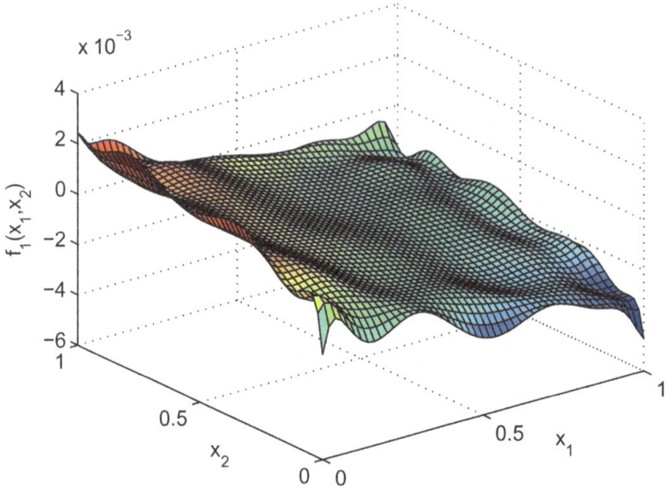

Fig. 2 Exact function f_2

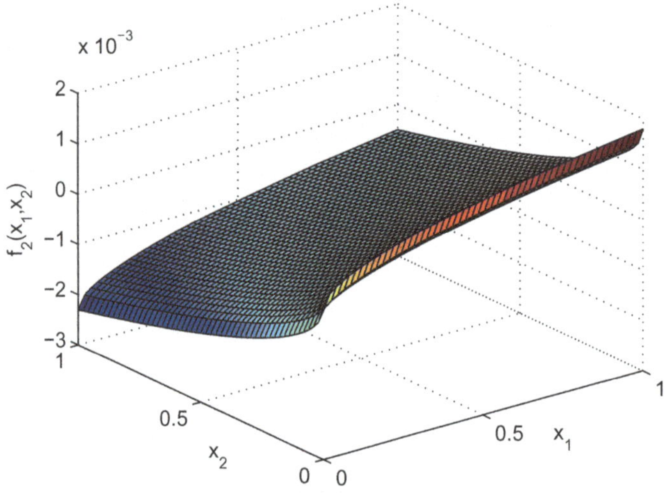

Fig. 3 Non-monotonically approximated function f_1

At first, T-S fuzzy systems with Gaussian membership functions were used for identification of the functions f_1 and f_2. We used 5 membership functions on each input x_1 and x_2 evenly distributed on the admissible interval [0, 1]. The parameter σ was chosen the same for all membership functions and such that the neighbouring functions cross at the value 0.5. The functions f_1 and f_2 were approximated in least square sense without any monotonicity constraints. The resulting approximations are depicted in Figs. 3 and 4, respectively.

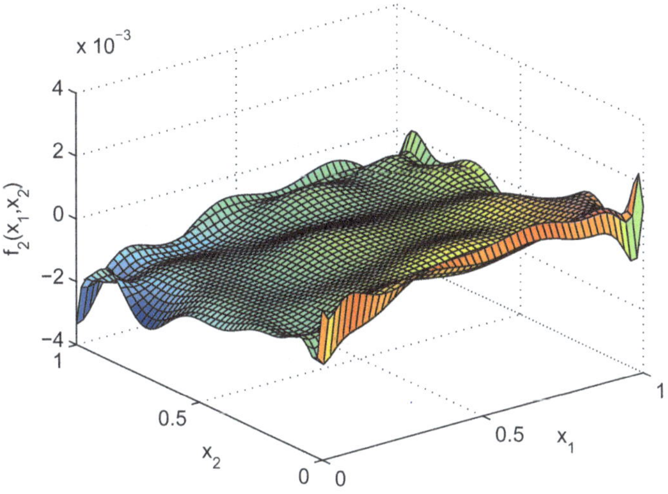

Fig. 4 Non-monotonically approximated function f_2

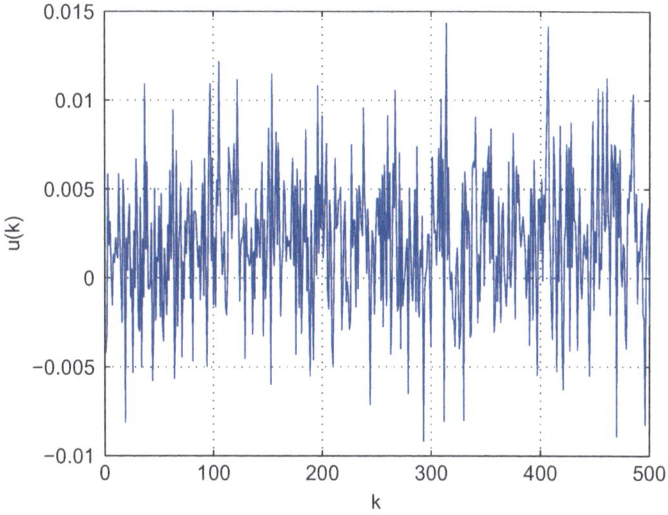

Fig. 5 Input signal

Comparison of the responses of the levels h_1 and h_2 of the corresponding one-step ahead predictor (19) and original system (17) on a random input signal $u(k)$ plotted in Fig. 5 and initial condition $h_1(0) = 0.5$, $h_2(0) = 0.306$ are shown in Figs. 6 and 7, respectively.

Next, the functions f_1 and f_2 were approximated in least square sense using (16) with monotonicity constraints (9). From the original equations (18) one can see that

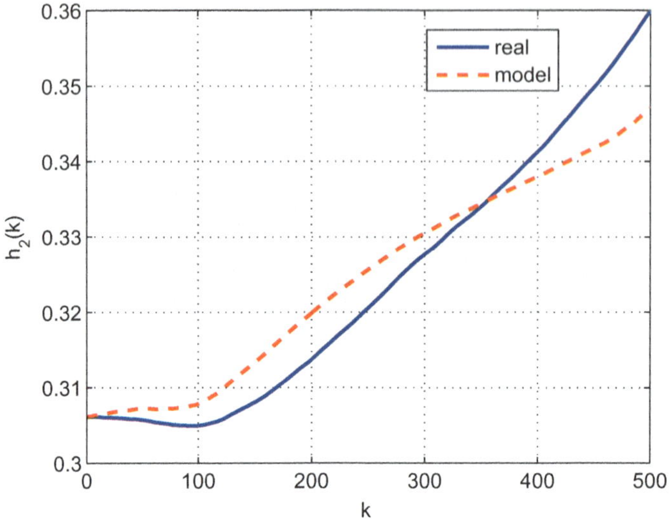

Fig. 6 h_1 for original system and non-monotonic approximation

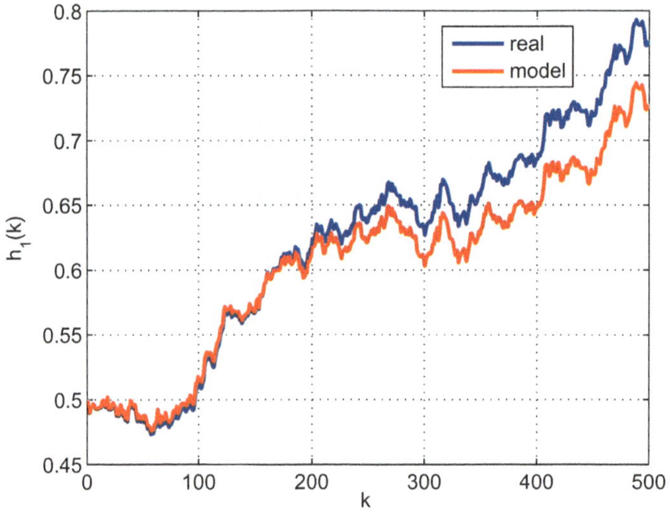

Fig. 7 h_2 for original system and non-monotonic approximation

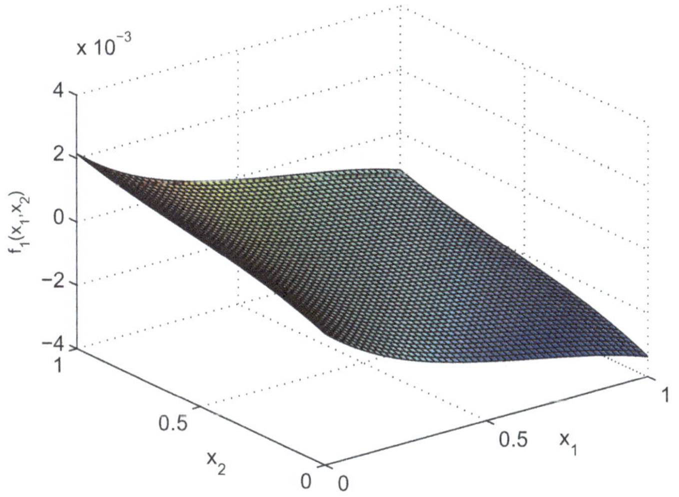

Fig. 8 Monotonically approximated function f_1

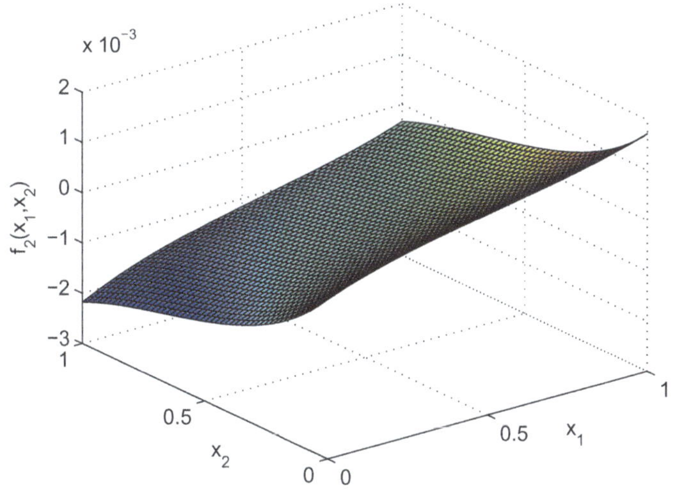

Fig. 9 Monotonically approximated function f_2

f_1 is non-increasing with respect to x_1 and non-decreasing with respect to x_2 and vice-versa for f_2. The approximated monotonic functions are depicted in Figs. 8 and 9, respectively.

The responses of the levels h_1 and h_2 of the corresponding monotonic one-step ahead predictor (19) and original system (17) on input signal $u(k)$ from Fig. 5 are compared in Figs. 10 and 11, respectively.

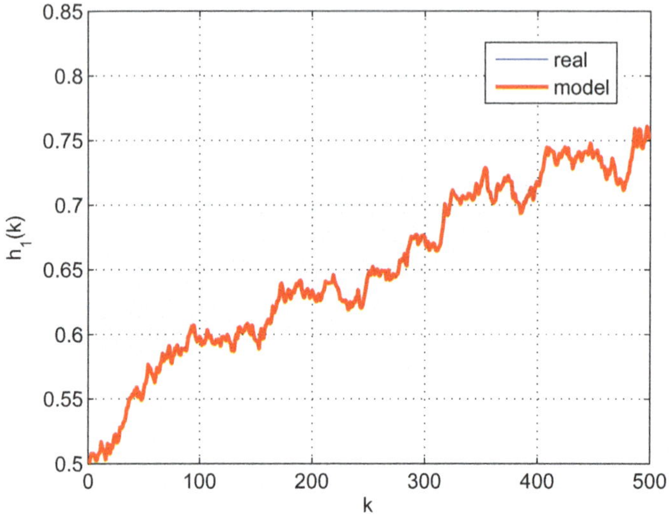

Fig. 10 h_1 for original system and monotonic approximation

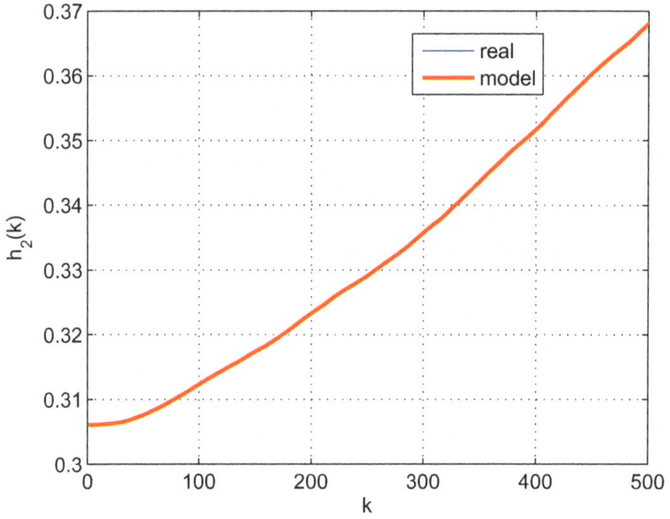

Fig. 11 h_2 for original system and monotonic approximation

6 Conclusion

The paper presents sufficient conditions for monotonicity of first order Takagi-Sugeno fuzzy systems with Gaussian membership functions. The conditions are derived in the form of LMI with respect to submodel parameters that makes it possible to be considered in optimization problems. The benefit of consideration a prior knowledge about monotonicity of a nonlinear function was illustrated on identification of two-tank dynamical system from measured data using Takagi-Sugeno fuzzy models.

Acknowledgements This work has been supported by the project 18-26278S sponsored by Grant Agency of the Czech Republic.

References

1. P. Lindskog, L. Ljung, Ensuring monotonic gain characteristics in estimated models by fuzzy model structures. Automatica **36**(2), 311–317 (2000)
2. J.M. Won, S.Y. Park, J.S. Lee, Parameter conditions for monotonic Takagi-Sugeno-kang fuzzy system. Fuzzy Sets Syst. **132**(2), 135–146 (2002)
3. P. Hušek, On monotonicity of Takag-Sugeno fuzzy systems with ellipsoidal regions. IEEE Trans. Fuzzy Syst. **24**(6), 1673–1678 (2016)
4. L.M. Pang, K.M. Tay, C.P. Lim, Monotone fuzzy rule relabeling for the zero-order TSK fuzzy inference system. IEEE Trans. Fuzzy Syst. **24**(6), 1455–1463 (2016)
5. V.S. Kouikoglou, Y.A. Phillis, On the monotonicity of hierarchical sum-product fuzzy systems. Fuzzy Sets Syst. **160**(24), 3530–3538 (2009)
6. J. Kim, J.-M. Won, K. Koo, J.S. Lee, Monotonic fuzzy systems as universal approximators for monotonic functions. Intell. Autom. Soft Comput. **18**(1), 13–31 (2012)
7. C. Li, J. Yi, G. Zhang, On the monotonicity of interval type-2 fuzzy logic systems. IEEE Trans. Fuzzy Syst. **22**(5), 1197–1212 (2014)
8. T. Wang, J. Yi, C. Li, The monotonicity and convexity of unnormalized interval type-2 TSK fuzzy logic systems, in *International Conference on Fuzzy Systems* (IEEE, Barcelona, July 2010), pp. 1–7
9. E.V. Broekhoven, B.D. Baets, Monotone Mamdan-Assilian models under mean of maxima defuzzification. Fuzzy Sets Syst. **159**(21), 2819–2844 (2008)
10. E.V. Broekhoven, B.D. Baets, Only smooth rule bases can generate monotone Mamdani-Assilian models under center-of-gravity defuzzification. IEEE Trans. Fuzzy Syst. **17**(7), 1157–1174 (2009)
11. H. Seki, H. Ishii, M. Mizumoto, On the monotonicity of fuzzy-inference methods related to T-S inference method. IEEE Trans. Fuzzy Syst. **18**(3), 629–634 (2010)
12. T.L. Jee, K.M. Tay, C.P. Lim, A new two-stage fuzzy inference system-based approach to prioritize failures in failure mode and effect analysis. IEEE Trans. Rel. **64**(3), 869–877 (2015)
13. Y. Qian, H. Xu, J. Liang, B. Liu, J. Wang, Fusing monotonic decision trees. IEEE Trans. Knowl. Data Eng. **27**(10), 2717–2728 (2015)
14. J. Alcalá-Fdez, R. Alcalá, S. González, Y. Nojima, S. García, Evolutionary fuzzy rule-based methods for monotonic classification. IEEE Trans. Fuzzy Syst. **25**(6), 1376–1390 (2017)

15. J. Garcia, A.M. AlBar, N.R. Aljohani, J.-R. Cano, S. Garcia, Hyperrectangles selection for monotonic classification by using evolutionary algorithms. Int. J. Comput. Intell. Syst. **9**(1), 184–202 (2016)
16. H. Wang, M. Zhou, K. She, Induction of ordinal classification rules from decision tables with unknown monotonicity. Eur. J. Oper. Res. **242**(1), 172–181 (2015)
17. S. Boyd, L. Vandenberghe, *Convex Optimization* (Cambridge University Press, 2004)
18. F. Hoeppner, F. Klawonn, R. Kruse, T. Runkler, *Fuzzy Cluster Analysis* (Wiley, 2000)

Conceptual Framework for Measuring Project Benefits Using Belief—Plausibility and Type 1 Fuzzy Inference System

A. Ghildyal, D. Ru, E. Chang, K. Joiner, M. Saberi, and A. Despande

Abstract We propose a combination of belief and plausibility (Bl–Pl) and Type 1 fuzzy inference system (FIS) methods to measure benefits realization in this reserach. This approach can help line managers trace the project outcomes and validate the benefits and return on investment. BI–Pl computations are centered around an expert's belief as a focal element or a power set of a classical set. This is part of Type 1 FIS, which is embraced by concepts of partial belief and fuzzy sets due to the approximate reasoning of the experts and the fuzzy rule base system. The project's output can be ranked based on the difference between Bl-Pl, while Type 1 FIS allows us to transform expert knowledge or experience and then trace the project benefits automatically. A commentary on various governing parameters in enterprise benefit management, expert classification and an illustrative example using belief and plausibility form an integral part of this Chapter.

Keywords Benefit realization · Benefit measurement · Belief · Plausibility · Fuzzy inference systems

---

A. Ghildyal (✉)
Department of Defence, Australian Government, Canberra, Australia
e-mail: amit.ghildyal@defence.gov.au

D. Ru · E. Chang · K. Joiner · M. Saberi
UNSW at Australian Defence Force Academy, Canberra, Australia
e-mail: e.chang@adfa.edu.au

K. Joiner
e-mail: k.joiner@adfa.edu.au

A. Despande
Berkeley Initiative in Soft Computing BISC, UC Berkeley, Berkeley, USA

S. N. Shahbazova et al. (eds.), *Recent Developments and the New Direction in Soft-Computing Foundations and Applications*, Studies in Fuzziness and Soft Computing 393, https://doi.org/10.1007/978-3-030-47124-8_20

1 Introduction

Measuring project benefits is a key process in benefit management and has been one of the prime objectives of policy makers. Operation automation, improved system mobility, data security are a few facets of the many project benefits. Electronic Information Management (EIM) 2025 is a digital strategy for the Australian Defence Enterprise Management Strategy delivering promising enterprise benefits guided by imperatives. There are two distinct challenges: firstly, the framework used to measure benefits at different levels of the project and secondly, the application of belief and plausibility of the Dempster-Shafer theory as well as the fuzzy approach to realize benefits. It can be argued that effectiveness, efficiency, responsiveness, compliance and interoperability are intangible yet measurable benefits [1]. A significant contribution to the analysis from the benefit management perspective was made, perhaps, by using the concept of random variable. For a long time, we have embraced two valued probability models based on Aristoteles logic. What might not be realized, is that available information may not always be in numbers but is often in the realm of linguistic description which is imprecise/ambiguous or fuzzy. Professor Lotfi Zadeh brought out this aspect of human culture and suggested the concept of fuzzy sets based on partial belief of experts and wrote a seminal paper in 1965. He coined the term fuzzy logic in 1973, which subsequently formed the concept, which was then applied in various decision making and industrial settings, computer engineering and in other areas of science and technology [1, 2]. Several books and research papers on fuzzy sets and fuzzy logic are available [3, 4] and [5]. The Dempster-Shafer Theory is based on experts' beliefs/evidence and belief and plausibility of selected experts and helps in ranking the decision on project benefit realization in linguistic terms.

2 Enterprise Benefit Management

Benefits include value, financial goal, cost, budget, performance, economic gain, and profit as illustrated in literature [6]. Benefit management is comprised of four processes, namely: benefit identification, benefit monitoring, benefit measurement, and benefit realization. "Intangible benefits", "expected benefits" and "future-oriented benefits" are important for the project success from a strategic viewpoint [7]. The reasons benefit measurement is the focus of this paper are twofold. Firstly, best practice studies do not include measures of strategic value created through present-day projects [8] and, secondly, there is a need for considerable development of the theory on the process of benefit measurement, as little theoretical work has presently been undertaken in the field of benefit management [9]. A trade-off between project benefits and organizational objectives is essential to realize project, program or even portfolio level benefits and visualize the overall organizational vision [10].

3 Challenges in Benefit Measurement for Benefit Realization

Benefits are not only about rational thinking and calculations from experts or project/program managers. Benefits frequently attribute values of all project stakeholders to individuals. Interestingly, benefits also involve human aspects such as emotions and interpretations regarding different types of benefits [11, p. 16]. While some work already exists, there is little work done at P3M levels, namely:

(1) Benefit at Portfolio level
(2) Benefit at Program level
(3) Benefit at Project level [12, 13]

P3M is defined as the essential strata or levels of enterprise or organizational projects, where different types of benefits are identified, organized, measured, delivered, and effectively realized at Portfolio, Program and Project levels [14]. This further extends to a mixed approach as practiced.

Benefit measurement for benefit realization in the dynamic business and enterprise environments involves uncertainties. Fuzzy logic embraces the idea that non-statistical data is vague. Uncertainty is classified into conventional stochastic and lexical uncertainties [15]. In most projects, it is uncertain whether benefits would be realized [16, 17] and there is ambiguity defining benefit measurements to be realized in enterprise settings [18, 19].

4 Relationship Between Portfolio Components and Organizational Objectives

Successfully overcoming the challenges is an indicator of accountability in information management for the enterprise. Project stakeholders expect trust and protected information, business-led, agile and innovative solutions. This benefit along two-dimensional table incorporating objectives and portfolio components are presented in Table 1.

Most complex projects yield benefits, outcomes and deliverables at different levels. Operation automation is important for users and the organization. It increases speed of business transactions and organizational efficiencies. It impacts the enterprise performance and the effectiveness of digital strategy. Customer engagement with the system facilitates the achievement of organizational objectives.

Time is of the essence for developing innovative business applications for the organization. This increases clear accountability towards customers and upholds standardized businesses. Systems could replace human resources in a limited way and increase overall effectiveness. While HR (Human Research) per head cost decreases, staff performance would increase as employees are deployed in other business processes.

Table 1 Relationship between portfolio components and organizational objectives

Portfolio components (PC)		Vision				
		Objective 1: innovation	Objective 2: enterprise performance	Objective 3: effectiveness	Objective 4: clear accountability	Objective 5: standardized business
Portfolio	PC 1: Staff performance			F	I	
	PC 2: Stakehloders		C			J
	PC 3: Time	A				
	PC 4: Customer engagement		D			
	PC 5: Operation automation	B	E	G		K

The RoI (Return on Investment) for projects is expected to be high and the organizational systems create value for the invested money, as an example. As an integral part of the organization, stakeholders' benefit from such initiatives generating billions of dollars in return for large organizations. What was once a tedious process is now simplified, automated and agile. The strategic mapping between the organizational vision and project benefits helps us to measure the collective contribution of projects towards achieving strategic objectives.

5 Classification of Domain Experts

The experience of users infers that the success of the application of the two possibility methods; 1. fuzzy inference system; and 2. the Dempster Shafer Theory of Evidence, to real world problems depends upon selection of the experts. No two persons think alike. Therefore, it is necessary to compute mathematical similarity within and between experts. In this chapter, the belief/evidence assignment of domain experts has been considered in estimating similarity between and within experts.

It is important to identify several experts, in this case, from Army, Air Force and Navy who are involved in the decision-making process of project benefit realization. The statistical method of proportional sampling could be used in selecting the experts for the task. Say, around 100 experts each from the Army, Navy and Air Force are identified for the estimation of their similarity.

The experts will study all the parameters of the project in detail and assign belief, based on their belief/experience in the following way: A. Project benefits very highly realized, B. Project benefits highly realized, C. Project benefits partially realized, and D. Project benefits not realized, A U B project benefits very highly realized OR highly realized. These are known as power sets (and not fuzzy sets) of a classical set.

6 Illustrative Example

6.1 The Example

We select for example, seven experts from the Defense services and they assign the degree of belief between 0 and 1 as Basic Belief or Basic Evidence Assignment (BBA/BEA). The similarity coefficients between the domain experts will be worked using expressions 7 through 13 and will be categorized in various possibility levels (α-level cut). Those experts who satisfy, for example, 0.95 possibilities will be considered in further investigations in decision research. The identified experts can be requested to draw fuzzy sets for the defined parameters based on their perception. Since the authors propose to use a Mamdani type Fuzzy Inference System, an average fuzzy set for the linguistic classes of each parameter will be used in the fuzzification and defuzzification process. Mamdani FIS is now commonly known as Type 1 FIS. Table 2 presents the Basic Belief/Evidence Assignment (BBA/BEA) of identified. For example, seven experts (out of, say 250 experts who finally agree with 0.95 possibility level). The normalized values of the evidence/belief function of experts (E1–E7) are presented.

In Table 2, we have assumed that A is Very High Benefit received, B is High Benefit received, C is an Acceptable level Benefit received, and D is Benefit Not received.

Table 2 Normalized values of the evidence/belief function of experts

Focal element no.	Basic belief/evidence assignment	E-1	E-2	E-3	E-4	E-5	E-6	E-7
1	A	0.04	0.15	0.15	0.04	0.09	0.10	0.10
2	B	0.08	0.12	0.05	0.07	0.08	0.08	0.08
3	C	0.01	0.03	0.14	0.01	0.02	0.03	0.02
4	D	0.02	0.00	0.03	0.01	0.01	0.01	0.02
5	AUB	0.09	0.14	0.07	0.09	0.10	0.10	0.10
6	AUc	0.05	0.09	0.15	0.04	0.05	0.08	0.07
7	AUD	0.05	0.08	0.05	0.06	0.06	0.06	0.04
8	BUC	0.08	0.08	0.08	0.08	0.08	0.06	0.05
9	BUD	0.09	0.06	0.02	0.08	0.04	0.05	0.05
10	CUD	0.02	0.02	0.04	0.02	0.02	0.02	0.02
11	AUBUC	0.09	0.10	0.10	0.10	0.10	0.10	0.10
12	AUBUD	0.10	0.07	0.03	0.10	0.11	0.12	0.10
13	AUCUD	0.09	0.03	0.05	0.09	0.09	0.05	0.10
14	BUCUD	0.10	0.02	0.03	0.09	0.03	0.03	0.04
15	AUBUCUD	0.10	0.02	0.01	0.10	0.11	0.12	0.10

The elements of 2 × 2 matrix in Table 2 are on two universes i.e. column as experts E1–E7 and row as BBA/BEA. To compute similarity coefficient, Cosine Amplitude algorithm will be used. Salient details of the method are explained in the next section.

6.2 Cosine Amplitude Method in Expert Classification

Fuzzy membership values for constraints are the perception of experts. So, there is the need to verify similarity between experts. This is done by Cosine Amplitude transformation. To apply Cosine Amplitude transformation, data must be probability values. Here, the data is possibility value. So, we normalize the data column wise which will become probabilities. This data must be converted into a possibility relation to find the similarity relations between them. This is done by Similarity methods in data manipulation.

$$r_{ij} = \frac{\sum_{k=1}^{n} xikxjk}{\sqrt{\left(\left(\sum_{k=1}^{n} x^2ik\right)\left(\sum_{k=1}^{n} x^2jk\right)\right)} \qquad (1)$$

Expression (1) reveals that this method is related to the dot product for the cosine function. When two vectors are colinear (most similar), their dot product is unity. When the two vectors are at right angles to one another (most dissimilar), their dot product is zero.

Similarity matrix generated from the cosine amplitude method will be invariably fuzzy tolerance relation. It is necessary to transform fuzzy tolerance relation into fuzzy equivalences relation using resemblance fuzzy operation (Table 3).

$$R_1^{n-1} = R_1 o R_1 o \cdots o R_1 = R \qquad (2)$$

Table 3 Fuzzy equivalence relation using transitivity closure

Experts'	E1	E2	E3	E4	E5	E6	E7
E1	1.00	0.80	0.64	1.00	0.94	0.92	0.93
E2	0.80	1.00	0.86	0.79	0.88	0.90	0.88
E3	0.64	0.86	1.00	0.64	0.74	0.77	0.77
E4	1.00	0.79	0.64	1.00	0.94	0.92	0.93
E5	0.94	0.88	0.74	0.94	1.00	0.98	0.99
E6	0.92	0.90	0.77	0.92	0.98	1.00	0.97
E7	0.93	0.88	0.77	0.93	0.99	0.97	1.00

(E2, E3) = 0.86 and (E3, E5) = 0.74, but (E2, E5) = 0.88 $\geq$ min [0.86, 0.74]
(E1, E5) = 0.95 and (E5, E4) = 0.94, but (E1, E4) = 1 $\geq$ min [0.95, 0.94]
(E2, E1) = 0.80 and (E1, E7) = 0.93, but (E2, E7) = 0.88 $\geq$ min[0.80, 0.93]

Fig. 1 Transformed fuzzy equivalent relations between experts

$$R^3 = R^2 \circ R = \begin{bmatrix} 1 & 0.93 & 0.88 & 0.93 & 0.71 \\ 0.93 & 1 & 0.88 & 0.97 & 0.71 \\ 0.88 & 0.88 & 1 & 0.88 & 0.71 \\ 0.93 & 0.97 & 0.88 & 1 & 0.71 \\ 0.71 & 0.71 & 0.71 & 0.71 & 1 \end{bmatrix}$$

$$\mu_{R^3}(x_3, x_2) = 0.88. \mu_{R^3} = (x_2, x_5) = 0.71$$

$$but$$

$$\mu_{R^3}(x_3, x_5) = 0.71 = \min(0.88, 0.71)$$

Now it become fuzzy equivalence relation

Fuzzy tolerance relation has been transformed to fuzzy equivalence relation using transitive closure (Expression 2).

Figure 1 below shows the transformed fuzzy equivalent relation between the experts.

Using α-cut level for defuzzification method for fuzzy to crisp converse, we will get similar experts the desired α-cut levels which is portrayed in Fig. 2 as a dendrogram.

Based on the computations, it can be inferred that except for E3, all the experts agree over 0.95 possibility levels.

7 Belief and Plausibility in Dempster-Shafer Approach for Benefit Measurement

7.1 Belief and Plausibility Model

A monotone measure [20] describes the vagueness or imprecision in the assignment of an element A to two or more crisp sets. This can also be termed as a power set in classical set. A special form associated with preconceived notions is called a belief measure. A form associated with information that is possible, or plausible is called a plausibility measure. Specific forms of belief measures and plausibility measures are known as certainty and possibility measures, respectively. A belief measure is a quantity, denoted bel(A), that expresses the degree of support, or evidence for a collection of elements defined by one or more of the crisp sets existing on the power set of a universe.

The plausibility measure of this collection A is defined as the "complement of the Belief of the complement of A," or as:

$$\mathrm{pl(A)} = 1 - \mathrm{bel(A)} \tag{3}$$

In Professor Zadeh's words, "Dempster's rule of combination may lead to counterintuitive results because of the normalization issue. The reason for this [21] is that normalization throws out evidence that asserts that the object under consideration does not exist, that is, is null or empty (Ø)". This is the reason why the computation for belief and plausibility should be carried out for each single expert and their variants.

Belief (and not combined belief) are computed for all the experts [3]. When the distance between plausibility and belief is minimum for the decision parameter such as very high benefit realization, high benefit realization and so on, then the decision is acceptable by that expert. The exercise needs to be carried out for all the Experts (in this case for 6.7 as Expert 3 does not satisfy the desired possibility level (Ref Fig. 1). The final decision will be based on maximum principle. The parameter which occurs the maximum number of times will be considered. Say, Project Benefit is highly realized, it is possible to rank the output based on belief plausibility proposed in the Dempster Shafer Theory.

7.2 *Output of Belief and Plausibility Model*

From the illustrative example, the overall minimum distance between belief and plausibility is presented below (Table 4) from the identified six experts as shown in Fig. 2, Experts E1, E2, E4, E5, E6, E7, except Expert 3.

Table 4 Minimum distance between belief and plausibility

Decision variable Description	Symbol	Min. distance between Bel and Pl	
Project benefits very highly accepted	A	0.56	
Project benefits highly accepted	B	0.48	
Project benefits partially accepted	*C*	*0.06*	*Option C—Project benefits partially accepted is the final decision from the hypothetical data*
Project benefits not accepted	D	0.13	

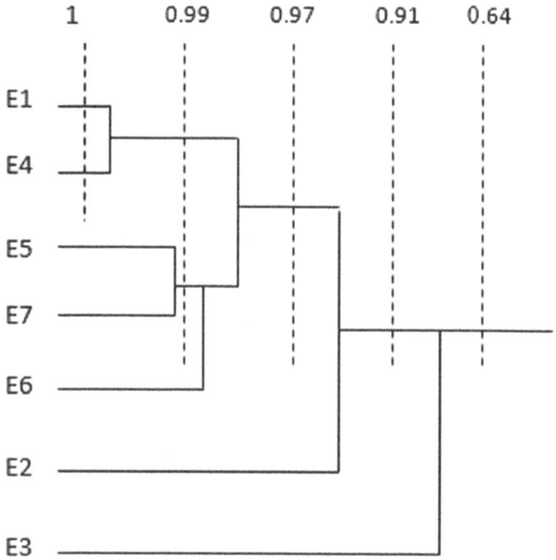

Fig. 2 Fuzzy similarity between experts for various α-cut or possibility levels

7.3 *Discussion*

After scientific examination based on the hypothetical data on belief of the identified six (E1, E2, E4, E5, E6 and E7) domain experts (Fig. 2), except E3 (Expert 3), it can be concluded that the project benefit is partially realized or accepted.

7.4 *Limitation of Dempster Shafer Theory (DST) of Evidence*

The experts, without going into sufficient details of all the governing parameters, assign their degree of belief or degree of evidence. This is one of the limitations of DST. To overcome this difficulty in decision analysis, it is proposed to use Type 1 Fuzzy Inference Method in project benefit realization.

8 Type 1 Fuzzy Inference System for Total Project Output Realization

DST is one of the approaches based on belief and plausibility/possibility. Possibility can also be effectively approached using Type 1 FIS (Fuzzy Inference Systems). Fuzzy Inference Systems have the following five vital phases of the fuzzy logic process, namely input variables, fuzzification, inference rule engines, defuzzification

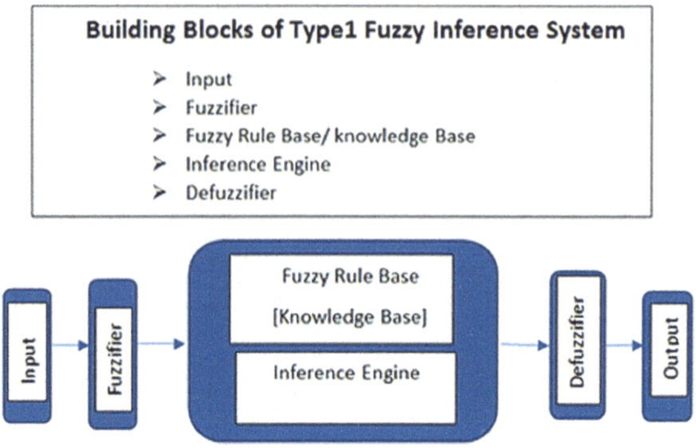

Fig. 3 Fuzzy inference systems

and outputs. Firstly, values for the input variables are entered into the model, then the rules are applied, and a qualitative output is derived for each portfolio component.

The fuzzification and application of fuzzy rules is undertaken for each portfolio component (denoted by PC variable). The contribution is determined by aggregating the qualitative outputs and then applying defuzzification to produce a crisp value. This crisp value is meaningful as it represents quantitative contributions of portfolio components aligned with organizational objectives. Qualitative/quantitative information is managed subjectively and made more meaningful objectively by FIS.

Defuzzification Process

The most important step in Type1 FIS is to transform fuzzy output into crisp output using the process of defuzzification. All the rules will fire parallel and partially in FIS. Most of the rules fire to zero degrees. Invariably, nearby rules are fired partially if the fuzzy rule-based system is well designed and with no abrupt changes in input parameters. The output, after firing of fuzzy rules using fuzzy implication rules, will be a fuzzy set and not a fuzzy number. Defuzzification transforms fuzzy output into crisp output as FIS is based on fuzzy mapping rules. The Mamdani fuzzy toolbox follows the Centroid method for defuzzification, though there are other defuzzification methods available. Figure 3 presents the salient feature of the centroid method of defuzzification.

9 Hierarchical Structure of Type 1 Fuzzy Inference System for Total Benefit Realization is Based on the 5 Categories

The linguistic term "project benefit realization" can express linguistic hedges such as Very High, High, Moderate, Poor and Very Poor. Human perception which is fuzzy or imprecise is at the center stage of any fuzzy inference system. The total benefit realization is based on the five categories which are dependent on 12 PC (Project Component) variables. Figure 4 portrays the details of Hierarchical Structure of

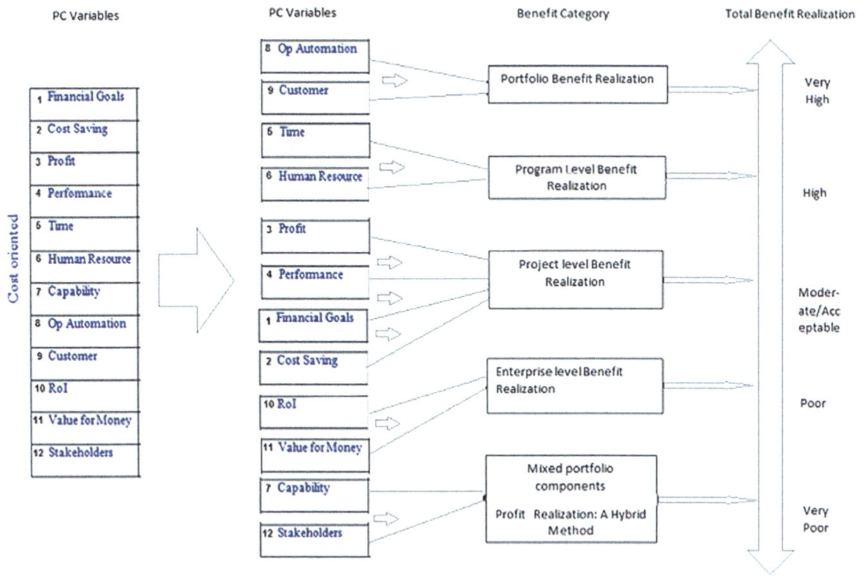

Fig. 4 Hierarchical structure of Type 1 FIS for total benefit realization is based on the 5 categories

Project Benefit Realization. Some of the features of the categories and a typical fuzzy rule base are presented in this section.

Figure 4 portrays the Hierarchical Structure of Type 1 Fuzzy Inference System for the Total Benefit realization based on the five Categories.

9.1 Portfolio and Program Level Benefits Realization

The beneficiaries are key players in driving project deliverables and user requirements. If the portfolio component is contributing to the project delivery, then it is rated 'high'. If experts intervene to improve the user requirements on behalf of the customer, then the portfolio component is rated 'medium'. If customers do not contribute ideas towards executing deliverables, then the component is rated 'low'. Using the majority of voting algorithm, we find that most experts believe that Customer engagement has "increasing" relation with benefit realization.

Thereafter, the experts in the organization will develop a set of fuzzy rules for different linguistic hedges with the following combinations and construct a knowledge base associated with fuzzy rules. Strategically managed information is the critical success factor of the public sector across the different portfolios. Using the majority of voting algorithm, we find that most experts believe that operation automation has an "increasing" relationship with benefit realization.

Table 5 Fuzzy rules in measuring portfolio level benefits

Rule 1: If PCVar8 is low AND PCVar9 is high, THEN contribution to benefits realization is sometimes
Rule 2: If PCVar8 is low AND PCVar9 is medium, THEN contribution to benefits realization is seldom
Rule 3: If PCVar8 is low AND PCVar9 is low, THEN contribution to benefits realization is never
Rule 4: If PCVar8 is medium AND PCVar9 is high, THEN contribution to benefits realization is on most occasions
Rule 5: If PCVar8 is medium AND PCVar9 is medium, THEN contribution to benefits realization is sometimes
Rule 6: If PCVar8 is medium AND PCVar9 is low, THEN contribution to benefits realization is seldom
Rule 7: If PCVar8 is high AND PCVar9 is high, THEN contribution to benefits realization is always
Rule 8: If PCVar8 is high AND PCVar9 is medium, THEN contribution to benefits realization is on most occasions
Rule 9: If PCVar8 is high AND PCVar9 is low, THEN contribution to benefits realization is sometimes

9.2 Typical Fuzzy Rule for Portfolio and Program Level Benefits Realization

Typical fuzzy rules represent customer, we use PC8 and PC9 variables to develop Fuzzy rules as shown in Table 5.

Customer engagement is an integral part of a project, program or any other work associated with portfolio components and its success.

Similarly, fuzzy rules in different linguistic hedges (Low, Medium, High) for Program level, Project level and Enterprise level benefits realization can be formulated. These fuzzy rules are formulated based on domain experts' tacit knowledge which is based on their partial belief and approximate reasoning because of (their) shallow knowledge.

The expected output of Type 1 FIS will be the total benefit realization of the project in linguistic hedges, following the standard defuzzification procedure. The outcome of these investigations will ensure the health of the project which will help the management to make a final decision on the improvement of some of the portfolios, if need arises.

10 Conclusion

In this chapter, a conceptual framework for measuring total project benefits using belief-plausibility and Type 1 Fuzzy Inference System has been proposed. We believe this is the first attempt to tackle the benefit realization measurement. It aimed to help enterprise managers or contract officers track and trace the project outcomes and measure it against committed project benefit. The conceptual model with illustrative examples was demonstrated. The next step is to apply to real world case studies such a conceptual framework and evaluate the practical significance including the measuring benefit realization at four levels namely; portfolio, program, project which leads to enterprise benefit.

Acknowledgement This research is supported by the Australian Defence Study Bank scheme, and Strategic Funds from the Rector's office of the University of New South Wales at the Australian Defence Force Academy. The authors would like to thank the many senior executives from Australian Defence (CASG, CIOG and VCDF), in particular Assistant Secretary of DoD Acquistion and Sustain Reform Chris Horscroft and the late Major General Michael Clifford of ADFA for their mentorship and encouragement to this work.

References

1. L.A. Zadeh, Stratification, target set reachability and incremental enlargement principle. Information (2016)
2. E. Cox, *Fuzzy Logic for Business and Industry* (Charles River Media, Rockland, MA, 1995)
3. T. Ross, *Fuzzy Logic with Engineering Applications* (Wiley Publication, 2003)
4. G. Klir, B. Yuan, *Fuzzy Sets and Fuzzy Logic: Theory and Applications* (Prentice Hall, Upper Saddle River, NJ, 1995)
5. Dubois, Prade, *Possibility Theory: An Approach to Computerized Processing of Uncertainty* (Plenum Press, New York, 1988) ISBN-13: 978-4684-5289-1
6. C. Ashurt, N. Doherty, J. Peppard, Improving the impact of IT development projects: the benefits realization capability model. Eur. J. Inf. Syst. **17**(4), 352–370 (2008)
7. J. Ward, P. Taylor, P. Bond, Evaluation and realisation of IS/IT benefits: an empirical study of current practice. Eur. J. Inf. Syst. **4**, 214–225 (1996)
8. M. Martinsuo, C.P. Killen, Value management in project portfolios: identifying and assessing strategic value. Proj. Manag. J. **45**(5), 56–70 (2014). https://doi.org/10.1002/pmj.21452
9. J. Braun, F. Ahlemann, G. Riempp, *Benefits Management—A Literature Review and Elements of a Research Agenda*. Wirtschaftinformatik Proceedings 2009. Paper 54 (2009), http://aisel.aisnet.org/wi2009/54
10. R. Breese, Benefits management: Lost or found in translation. Int. J. Proj. Manag. **33**(7), 1438–1451 (2015)
11. R. Young, R. Vodica, R. Bartholomeusz, *Implementing Strategy through P3M and Benefits Management* (PGCS Project and Program Management, UNSW ADFA, 2017)
12. PMI, *Implementing Organizational Project Management: A Practice Guide* (Project Management Institute, Newtown Square (PA), 2014)
13. PMI, *The Standard for Portfolio Management* (Project Management Institute, Newton Square, 2013)
14. G.M. Hill, *The Complete Project Management Office Handbook* (CRC Press LLC, Boca Raton, Florida, 2004)
15. U. Cebeci, A. Beskese, An approach to the evaluation of quality performance of the companies in Turkey. Manag. Audit. J. **17**(1), 92–100 (2002). https://doi.org/10.1108/02686900210412306
16. T. Sowell, Fuzzy logic for "Just plain folks," Fuzzy Logic Tutorial (2005), http://www.fuzzy-logic.com/Ch1.htm

17. J.M. Mendel, Fuzzy logic systems for engineering: a tutorial. Proc. IEEE **83**(3), 345–377 (1995)
18. M. Aubry, V. Sergi, S. El Boukri, Opening the black box of benefits management in the context of projects, in *IRNOP, Boston, MA* (2017)
19. C. Chen, H. Cheng, A comprehensive model for selecting information system project under fuzzy environment. Int. J. Proj. Manag. **27**(4), 389–399 (2009). https://doi.org/10.1016/j.ijproman.2008.04.001
20. Klir, Fodger, *Fuzzy Sets, Uncertainty and Information* (Prentice Hall, Englewood Cliffs, NJ 1988)
21. L. A. Zadeh, Is possibility different from probability? Hum. Sys. Manage. **3**, 253–254 (1984)
22. C.N. Enoch, L. Labuschagne, Project portfolio management: using fuzzy logic to determine the contribution of portfolio components to organizational objectives, in *Paper presented at PMI® Research and Education Conference, Limerick, Munster, Ireland* (Project Management Institute, Newtown Square, PA, 2012)
23. J. Yesn, R. Langari, *Fuzzy Logic = Intelligence, Control and Information* (Pearson Education)

An Efficient Hybrid Meta-heuristic Algorithm for Solving the Open Vehicle Routing Problem

Ali Asghar Rahmani Hosseinabadi, Atekeh Zolfagharian, and Pouya Alinezhad

Abstract Open Vehicle Routing Problem (OVRP) is categorized as one of the most important versions of Vehicle Routing Problem (VRP) attracted great interests due to its significant applications in industries and service centers. The objective of the OVRP is to minimize the number of used vehicles and total traveled distance or total travel time. Recently, some algorithms have been introduced for solving the OVRP including the Simulated Annealing (SA), Genetic Algorithm (GA), neighborhood search algorithms and etc. In this paper, a new meta-heuristic algorithm called *OVRP_SAICA* is presented for the OVRP as a kind of combinatorial optimization problem. The proposed algorithm is compared with different kinds of algorithms and the obtained results of solving some benchmark instances confirmed that it has a high efficiency and is superior to them in solving the OVRP.

Keywords OVRP · Simulated annealing · Imperialist competitive algorithm · Meta-heuristic

1 Introduction

Routing problems are a combinatorial optimization problem which aimed to find the optimal set of routes for a fleet of vehicles to traverse in order to deliver to a given set of customers [1]. The Open Vehicle Routing Problem (OVRP) involves determining the optimal routes for a set of vehicles that must deliver services to a set of customers. Every constructed route includes a sequence of customers that

A. A. Rahmani Hosseinabadi (✉) · A. Zolfagharian
Department of Computer Engineering, Islamic Azad University Behshahr Branch, Behshahr 511-48515, Iran
e-mail: A.R.Hosseinabadi@iaubeh.ac.ir

A. Zolfagharian
e-mail: At.Zolfagharian@gmail.com

P. Alinezhad
Department of Computer, University of Mazandaran Institute of Technology, Babol, Iran
e-mail: Pouya.alinezhad292@gmail.com

S. N. Shahbazova et al. (eds.), *Recent Developments and the New Direction in Soft-Computing Foundations and Applications*, Studies in Fuzziness and Soft Computing 393, https://doi.org/10.1007/978-3-030-47124-8_21

starts at the initial depot and ends in one of the customers [2–8]. On the other hand, there is no closed tour, and that's why we call it open. The general constraints of the OVRP are that all vehicles have capacity constraints, and each customer must be visited by only one vehicle to receive the required service. The total demand of all the customers in a constructed route should not exceed the total capacity of the vehicles. In some problems, vehicle usage time limitation is considered together with the traveled distance limitation: the distance traveled and the travel time spent on the vehicle for the route must not exceed the maximum allowed limits [9, 10]. Minimizing the number of vehicles besides minimizing the distance traveled through the routes, and minimizing total travel time are among the main objectives of this problem.

The OVRP and the Vehicle Routing Problem (VRP) are different in some respects and their most important difference is that there are *Hamiltonian* routes and *Hamiltonian* cycles in them, respectively. *Hamiltonian* routes begin from one node and end in another, but *Hamiltonian* cycles finally return to the starting node [9, 11]. Therefore, one of the main features of OVRP is that vehicles do not necessarily return to the initial depot after delivering service to the customers, and if they return to the initial depot, they will visit the same customers [12, 13]. The OVRP is an NP-hard problem and its solution is considered a scientific challenge. Many researchers have introduced various solution methods based on innovative and meta-heuristic methods for solving the OVRP, some important of these methods will be described here.

Brandao [12] used Tabu Search (TS) to solve the VRP by applying some constraints to vehicle capacity and on maximum travel distance in constructing the routes. In [14], a combination in the form of Genetic Algorithm and Tabu Search (GA + TS), in which the parallel computational power, the global optimization of GA, and the rapid local search of TS were used with the purpose similar to that of the above-mentioned methods in order to solve the OVRP. Then, a memory-based TS Algorithm was introduced for the OVRP with Heterogeneous Fixed Fleet Open Vehicle Routing Problem (HFOVRP) [15]. In this method, customers' demands were estimated in each route with a fixed number of vehicles having different capacities. Repoussis et al. [16] introduced a combinatorial evolution strategy for the OVRP with the aim of minimizing the number of vehicles and minimizing the traveled distance in the constructed routes. Salari et al. [17] presented a heuristic improvement procedure based on linear programming techniques for solving the OVRP. Their method can be considered as a general framework that may be expanded for solving various kinds of VRPs.

Azadeh and Farokhi-Asl [18] presented the Close–Open Mixed Vehicle Routing Problem (COMVRP) that is, a combination of routing vehicles with open and closed routes. They used a memetic algorithm for solving both problems in order to minimize the fixed and variable costs related to the performed operations in open and closed routes. They showed that their proposed method performed better than the CPLEX method in solving problems with higher dimensions. A Particle Swarm Optimization Algorithm (PSO) with self-adaptive inertia weight was developed for solving the Open Vehicle Routing problem with Time Dependent Travel Time (OVRPTD) in [19].

Also, Norouzi et al. [20] applied Multi-objective PSO (MOPSO) to solve the bi-objective OVRPC problem (which is the same as the OVRP but in competitive situations) for minimizing the travel cost and for maximizing sales in the competitive situations. The advantage of their method is lower computational costs compared to similar methods introduced for solving the problem [20]. TS Algorithms [21] and improved TS Algorithms [22] were implemented by Huang and Liu with the aim of minimizing the number of vehicles and travel costs for OVRP. Based on simulation results, the introduced algorithm can efficiently minimize the number of required vehicles and total travel costs. In [23], a PSO method was used to solve the OVRP and algorithms such as Nearest Insertion Algorithm were employed to improve the obtained solutions due to optimizing internal and external routes. Considering the slow rate of convergence and the weak search ability of the traditional GA, the combinatorial GA (which has greater convergence rate and rapid search ability) can be used to simplify the problem and improve its search efficiency [24]. Guiyun [25] presented the Open Vehicle Routing Problem with Time Windows (OVRPTW) and developed an improved GA for solving the problem with the aim of minimizing the number of vehicles and total traveled distance. The proposed algorithm improves the weakness of early convergence, and slow convergence of the traditional GA.

In [26], a Clonal Selection Algorithm (CSA) was proposed for solving the OVRP in which a new definition was developed for antibody affinity and an algorithm was proposed to preserve the diversity of the antibody. Wu and Hu [27] developed an algorithm based on genetic rules in order to upgrade the optimal performance of particle swarm and differential evolution for solving the OVRP. In their algorithm, all members had dominant and recessive characters, optimization of particle swarm took place by the dominant character and differential evolution by the recessive one, and if the proportionality of the dominant character is smaller than that of the recessive one, the recessive character will be replaced by the dominant one. In [28], the concepts of Variable Neighborhood Search (VNS) and evolutionary algorithms were used for optimization of the OVRP. This method could generate solutions with high acceptable quality.

Zachariadis and Kiranoudis [29] introduced a new search method for solving the OVRP that could investigate an extensive solution space in order to minimize the number of routes and routing costs.

Marinakis and Marinaki [30] have introduced a new version of the Bumble Bees Mating Optimization Algorithm (BBMO) for solving the OVRP problem, among the main parts of which is the replacement of outward movement by a local search that makes the proposed algorithm more efficient for solving combinatorial optimization problems. They presented a special decoding method for implementing the PSO in which a vector including customers' positions was generated in a descending order and then each customer was assigned to a specific route based on his/her position and, finally, a one-unit mutation was applied to the all constructed routes. This was an effective method for solving the problem because it allowed for studying the feasibility of the routes and for investigating the quality of the generated solutions [31].

Chen et al. [32] employed the Variable Neighborhood Descent (VND) in the framework of Iterated Local Search (ILS) and called it a Hybrid Iterated Local Search (HILS) to solve the OVRP and applied four neighborhood structures of replacement, exchange, 2-opt*, and 2-opt in the VND algorithm. Their method could find the best solutions in the shortest possible time.

In [2], an Imperialist Competitive Algorithm (ICA) was proposed to solve the OVRP problem. By applying *OVRP_ICA* algorithm and comparing with the results of different algorithms, the superiority of their proposed algorithm was proved. But in this paper, we will combine *OVRP_ICA* algorithm with SA and obtain a new algorithm, namely, *OVRP_SAICA* in order to achieve more optimal solutions, which could compete with *OVRP_ICA* algorithm and generate more optimal solutions. Because of the combination, the proposed algorithm has more run time than the previous algorithm. The differences between these two algorithms (*OVRP_ICA* and *OVRP_SAICA*) are as follow:

SA algorithm is, in fact, a robust random search method used to find a good (not necessarily optimal) solution for combinatorial problems. In contrast to the conventional search methods, in each iteration of this algorithm, besides movement towards Best Solutions (BS), solutions with better objective functions are also accepted with a non-zero probability. Computational results show that the second method, *OVRP_SAICA* converges faster than the first method by pushing smartly countries towards optimal solutions because it searches intelligently for solutions and ICA operations included reduces the probability of falling in local extremes.

The remainder of this paper is as follows; in Sect. 2 the proposed algorithm is described. Section 3 presents the computational results, and the conclusions are presented in Sect. 4.

2 The Proposed Algorithm

There are many developed algorithms/approaches in order to solve and analyze the problem and it's similar problems efficiently in the literature such as SA, GA, ACO, and so on [2, 10, 11, 33–48].

In the proposed method a Hybrid Algorithm (*OVRP_SAICA*) provided to solve the OVRP is used. The target of this algorithm is to reduce the time and traveled distance by vehicles, as well as to minimize the number of vehicles in the mentioned problem. Due to the complexity and difficulty of the problem, it would be difficult to solve it even with a small number of customers and machines. Therefore considering the characteristics of ICA and SA, and being local algorithms, they are suitable for the mentioned problem and we use it to solve the problem. As the main factors of the problem are reduced distance and number of vehicles, so ICA and SA are suitable strategies for solving an OVRP. The initial solutions for OVRP can be obtained based on SA algorithm. The proposed method works based on the combination of both ICA and SA algorithms as follow:

The proposed hybrid algorithm (*OVRP_SAICA*) starts the OVRP by defining two parameters. At first, an initial solution is generated from SA algorithm and according to the solution generated by SA algorithm, several routes are conducted, each of them is the route of trucks and their number. Then each route is considered as a TSP problem and ICA algorithm is performed on each route in order to improve the solution. This process is explained as follow: At first SA algorithm generated an initial solution and also the number of necessary vehicles to travel these routes. The time of traveling routes by using specified machines is also important and reduce the time would be our desire. After generating an initial solution and get the constructed routes, each route is considered as TSP problem and ICA algorithm is performed on them. The proposed method is fully described in the following.

2.1 Simulated Annealing Algorithm

One of the most applicable algorithms in the field of combinatorial optimization is SA [49, 50]. The main idea of SA algorithm to solve optimization problems is taken from Metropolis [51]. Authors simulated materials as a system of components. This algorithm simulated the cooling process of material as a gradual reduction of the temperature until reaching an equilibrium temperature. Later, Kirkpatrick et al. [51] applied this idea to other optimization problems. The main advantage of SA algorithm is its ability to escape from being trapped in a local optimum. This algorithm uses a random search method in which some movements that don't improve the objective function are done besides accepting the changes that don't improve the objective function. SA algorithm starts from an initial solution and finds a neighbor solution for the current solution if this neighbor solution causes to improve the objective function; the algorithm goes to the neighbor solution. There is a possibility to go to the neighborhood, even there is no improvement in the objective function. This situation prevents to being caught in a local optimum. Thermodynamics Law states that the possibility of increasing energy level of material at temperature T can be expressed as follow:

$$p(\delta e) = \exp(-\delta e/kt) \tag{1}$$

where K is Boltzmann's constant and δe is the amount of increase in energy levels.

The criteria to accept the solution in SA algorithm is as follow:

Suppose a number of changes in objective function if it goes to the neighbor point is equal to c, in minimization problems a downside movement is always accepted, (i.e. $c < 0$) and upside movement is accepted based on the following conditional probability:

$$p = \exp(-c/t) > r \tag{2}$$

Fig. 1 Solution representation

8	1	5	2	0	7	4	0	6	3

where c is the amount of change in objective function, t is the current temperature and r is a random number between 0 and 1.

2.1.1 Solution Representation

The first step in SA algorithm is considering a structure to display good solutions. Figure 1 shows a solution of a problem containing 7 customers and one depot. Zeroes indicate the end of the route and 8 represents depot. The vehicle of the first route goes from the depot to first, fifth and second customer, the second vehicle goes from the depot to forth, seventh customer and the third vehicle go from the depot to sixth and third customers.

2.1.2 Initial Solution Generation

To generate an initial solution, we generate $\left[\sum_i d_i / Q\right]$ zeroes randomly, and then we generate a random permutation of n and the number of zeroes, and finally examine the capacity constraints of vehicles, if these constraints were violated, we remove zeroes and sum customers' demands from the beginning until the capacity constraints of vehicles are violated and put the zero before the customer violating the constraint.

2.1.3 Neighborhood Generation

To generate neighbourhood for the solution, we remove depot and zeroes and utilize three methods Swap, Reversion and Insertion randomly which are described in the following: if the obtained solution isn't good, remove the zeroes and sum customers' demands from the beginning until the capacity constraints aren't violated, and then put zero before the customer violating the constraint.

Swap: select two customers randomly and exchanges their location.

Reversion: selects two customers randomly and removes the locations between these two customers and rewrite them in reverse.

Insertion: selects two customers randomly and puts the first customer after the second customer.

2.2 Imperialist Competitive Optimization Algorithm

Now we run the ICA algorithm on the obtained solutions from SA algorithm in order to improve them as the following steps:

As ICA is an evolutionary approach, so it starts with a number of countries (initial population). Therefor there is P random initial population and the values of objective functions f_i are obtained for each country $i = 1, 2, \ldots, p$. Now $m < p$ countries with less objective functions are considered as imperialist countries and constitute the initial population by exchanging indexes *1* to m. Then, an equal number of colonies are assigned to each imperialist country (The integer part of the division of p-m of m). In addition, Due to the property of the integer part function, other countries will be assigned to the most powerful empire. It must be noticed that this formula only determines the number of assigned colonies to each imperialist country, but the assignment of each colony to an Imperialist country is randomly and is equal to its probability.

Under normal condition, colonies must move toward imperialist countries culturally and socially based on absorbing function. So this algorithm uses the nearest random neighbor method in absorbing method [52]. For example, if [1 3 5 2 4] represents colony country and [3 4 1 2 5] represents the imperialist country for TSP problem with 5 nodes, the algorithm starts to move from the first colony which is 1. Then $i = 1$ is considered and the neighbors of node 1 in two countries named nodes 2, 4 and 3 which aren't visited, are put in S. if c_{ij} is the distance between nodes i and j, then nodes belonged to S are visited with probability V_j:

$$V_j = \frac{1/c_{ij}}{\sum_{j\varepsilon s}(1/c_{ij})} \quad \forall j \in S \tag{3}$$

Assume that t nodes are elected, so we have the initial population [1 2 - - -] so far. In addition, nodes 1, 4 and 5 are the neighbors of node 1. Because node 1 has already been visited, nearest neighbor between 4 and 5 which makes S, is selected randomly (if S is empty in one stage, then unvisited nodes are put the S). In this method, due to the prohibition of selecting countries that aren't visited before, obtaining one feasible solution is assured. It must be noticed that the obtained solution replaced the solution of the colony country if it has a better objective function. On the other hand, in this algorithm, the diversity of colony countries is kept due to the random structure of absorbing function, in addition to increasing the quality of the values of objective functions of colony countries.

After applying the absorbing function on all colony countries, p% of the colony countries suffer from revolution. So two-point improving method is used for this situation. This method works based on selecting to nodes and exchanges them. It must be noted that p% of countries in each empire are chosen randomly and this method is applied to them. Now the new solutions obtained from the revolution with colony countries of jth empire are considered and the BS is selected as colony countries of jth empire. After calculating the objective function for all colony countries, they might have better objective functions than their imperialist country. Therefore the best colony countries of each empire are selected and replaced imperialist country if have a better objective function. It is important to note that there are two variables in the algorithm which store the value of the BS and the value of the objective function. These variables are evaluated after updating empires. In this step, the BS and the value of the objective function of imperialist countries are selected and the triple improving algorithm would be applied to them and then if the obtained value is better than obtained values in previous iterations, the solution and the value of the objective function will be replaced.

The triple improving algorithm works based on removing three edges from the tour and reconnected them through another side. There several ways to connect edges and recreate tour, but only the one that satisfies the constraints of the problem and creates a better tour for the problem is accepted. The process of removing three arcs and reconnect them continues until there is no other triple improving movement.

In the next step, the power of each empire is measured to determine that which empire has greater power. The following formula is used for this target:

$$w_i = f_i + \lambda(s_j) \quad j = 1, \ldots, m \tag{4}$$

In the formula (4):

w_i is the power of the jth empire which consists imperialist country and colony countries created by that imperialism. s_j is the average objective function of colony countries in each empire. λ is a correlation between -1 and 0, by which we can determine the effect of the objective function of the imperialist country to the average objective function of colony countries in the imperialist country.

After determining the power of empires, each empire with higher objective function must move into decline (because the considered TSP problem is a minimization problem) and lose its power by losing its colonies. So the weakest country of weakest empire is selected and joined to the other empires. It is worthy to mention that, this insertion isn't always to the best empire, and its probability is obtained based on formula (5) for the jth empire.

$$\beta_j = \frac{{}^{1}\!/_{w_j}}{\sum_{j=1}^{m} {}^{1}\!/_{w_j}} \quad j = 1, \ldots, m \tag{5}$$

As a result, each empire that has more power take the colony country with higher probability. In addition, if the weakest empire has no colony, the empire is removed and the imperialist country transferred to the strongest empire. Otherwise, the algorithm is repeated until the termination condition is reached. Two conditions are used here as termination condition which is investigated simultaneously at the end of each iteration.

These two conditions are the number of iterations of the algorithm (t) and remain only an empire.

1- Applying SA algorithm
2- Generate initial solution w randomly and calculate the value of objective function $f(w)$
3- Select an initial temperature $t_0 > 0$, final temperature $e>0$, temperature reduction coefficient named a, and a number as a number of iterations in temperature m.
4- Repeat
5- $m=0$
6- Repeat
7- Select randomly between three methods Swap, Reversion, and Insertion and apply it on the solution and calculate w'
8- If they aren't good, remove zeroes
9- Sum customers' demands from the beginning until capacity constraint of vehicle isn't violated and then put zero before the customer violating constraint and calculate w'
10- $\delta = f(w') - f(w)$
11- If $\delta \geq 0$ $\exp(-c/t) > (1,0)$ then $w \leftarrow w'$ and $f(w) - f(w')$
12- $m=m+1$
13- Until $m=M$
14- $t = t \times a$.
15- Until $t<e$
16- w is an estimation of optimum solution
17- Applying ICA algorithm to obtained solutions from SA algorithm
18- Generate new population as a country
19- Evaluate the value of the objective function for each country
20- Approaching colony countries to the imperialist country by using a specified procedure and absorption
21- Applying revolution action to colony countries
22- Recalculating the objective function for each colony
23- If colony countries have better objective functions than imperialist c ountries, exchange these two countries
24- Stores the best solution and value for all imperialist countries as the best solution and current value of the algorithm
25- If the current objective function has better quality among obtained values so far, then the new solution and value replaced the previous solution.
26- Transfer the weakest member of weakest empire to the other empire possibly
27- If there is only one empire, the algorithm ends, otherwise go to step 4.
28- Return the best solution

Fig. 2 Steps in the proposed algorithm

By occurring each of these two conditions at any time, the algorithm ends and the best solution and best values obtained so far is introduced as the BS and the best value.

Figure 2 shows a summary of the steps in the proposed algorithm (*OVRP_SAICA*).

3 Computational Results

3.1 Benchmark Data Sets

The proposed method was tested on the famous data set of Christofide et al. [53], Fisher [54]. In Fisher's model, there are 14 problems C1–C4 [54]. The numbers of customers in both sets were in the interval of 50 to 199, with Cartesian coordinates and Euclidean distance [54]. Among them, the limitations of tour length (L) and service time (e) were considered for all customers. Table 1 lists the specifications of 16 problems introduced by Christofide et al. [53] and Fisher [54]. The maximum tour length, compared to the VRP model, was a factor of 0.9 because the lengths considered in the VRP model are not suitable. The last row, in which K_{min} (the total requests of the customers divided by the capacity of each vehicle), represents the lower limit.

Table 1 Benchmark data sets of Christofides et al. [53] and Fisher [54]

Instance	Vrpnc1	Vrpnc2	Vrpnc3	Vrpnc4	Vrpnc5	Vrpnc6	Vrpnc7	Vrpnc8	Vrpnc9	Vrpnc10	Vrpnc11	Vrpnc12	Vrpnc13	Vrpnc14	F-n72-k4	F-n135-k7
N	50	75	100	150	199	50	75	100	150	199	120	100	120	100	71	134
C	160	140	200	200	200	160	140	200	200	200	200	200	200	200	30	2210
L	N/A	N/A	N/A	N/A	N/A	180	144	207	180	180	N/A	N/A	648	936	N/A	N/A
e	N/A	N/A	N/A	N/A	N/A	10	10	10	10	10	N/A	N/A	10	10	N/A	N/A
in	5	10	8	12	6	5	10	8	12	16	7	10	7	10	4	7

N number of customer, *C* vehicle capacity, *L* maximum tour length, *e* service time, *in* minimum number of used vehicle

3.2 Computational Results

Results of simulations conducted to solve the OVRP are presented in this section. To test the efficiency of the proposed algorithm on OVRP, and to compare it with other meta-heuristic algorithms, the proposed algorithm was tested on two classes of standard examples taken from [55]. Implementation took place in the C#.NET programming language and the programs were executed on a computer with 2.06 GHz Pentium-IV processor and 6 GB RAM. Results of simulations of the proposed algorithm *(OVRP_SAICA)* were compared with 12 samples of various algorithms including: Improved Tabu Search (ITS) [21], (Clonal Selection Algorithm) [26], Broad Local Search Algorithm (BLSA) [29], Hybrid Evolution Strategy (HES) [16], Hybrid combination out of a stochastic combined Variable Neighborhood Search and a simple (1 + 1)-Evolutionary Strategy ((Hybrid (1 + 1) ES) [28], Diploid Hybrid PSO with DE (DHPD) [27], Improved Tabu Search (ITS) [22], Bumble Bees Mating Optimization for the Open Vehicle Routing Problem (BBMOOVRP) [30], Integrates a Variable Neighborhood Descent (IVND) [32], Variable Neighborhood Search [56], Open Vehicle Routing Problem Using the Gravitational Emulation Local Search Algorithm (OVRP_GELS) [33], Imperialist-Based Optimization Algorithm for the Open Vehicle Routing Problem (OVRP_ICA) [2].

Considering the mentioned steps in the previous section, results are shown in Tables 2 and 3. The performed comparisons demonstrate the advantages of the proposed algorithm (OVRP_SAICA) in terms of time, a minimum number of vehicles and minimum traveled distance.

Tables 2 and 3 demonstrate the results of the proposed algorithm (OVRP_SAICA) compared to other well-known algorithms that solved the OVRP problem in different kinds from easy to difficult problems. As highlighted in Table 3, the proposed algorithm (OVRP_SAICA) could obtain more optimum and more acceptable solutions.

Considering the mentioned steps in the previous section, results are shown in Tables 2 and 3. The performed comparisons demonstrate the advantages of the proposed algorithm (OVRP_SAICA) in terms of time, a minimum number of vehicles and minimum traveled distance.

Tables 2 and 3 demonstrate the results of the proposed algorithm (OVRP_SAICA) compared to other well-known algorithms that solved the OVRP problem in different kinds from easy to difficult problems. As highlighted in Table 3, the proposed algorithm (OVRP_SAICA) could obtain more optimum and more acceptable solutions.

No time value is calculated for two algorithms DHPD [27], Hybrid (1 + 1) ES [28] so we didn't compare them in Table 4.

Figure 3 shows the solving six OVRP problems from small to medium sizes. As you can see, the proposed algorithm (OVRP_SAICA) is able to obtain desired outputs compared to other algorithms.

Table 2 Comparison of the proposed method with ten other methods of solving OVRP based on the solutions that minimize the number of vehicles with least distance traveled in sixteen different problems

Algorithms name	vrpnc1	vrpnc2	Vrpnc3	Vrpnc4	Vrpnc5	Vrpnc6	Vrpnc7	Vrpnc8
HES [16]	416.06	567.14	639.74	733.13	894.11	412.96	584.15	644.63
IVND [32]	416.06	567.14	639.74	733.13	916.39	412.96	568.49	644.63
DHPD [27]	408.5	567.14	**617**	733.13	897.93	412.96/6	552.87/11	644.63/9
Hybrid (1 + 1) ES [28]	412.95	564.06	639.25	733.13	868.44	412.95	566.93	640.89
BBMOOVRP [30]	416.06	567.14	639.74	735.18	895.19	412.96	583.19	644.63
BLSA [29]	416.06	567.14	639.74	733.13	893.39	–	–	–
ITS [21]	451.59	649.38	928.41	1201.95	1018.56	870.06	586.64	1254.28
ITS [22]	461.37	663.28	954.34	1247.52	1059.47	890.63	602.71	1261.39
Clonal selection algorithm [26]	426.81	574.14	691.51	850.48	1070	–	–	–
OVRP_GELS [33]	406.994	543.676	630.099	729.864	841.672	418.759	**549.562**	667.384
OVRP_ICA [2]	409.68	557.14	629.74	716.73	902.45	411.87	587.76	634.52
OVRP_SAICA	**405.29**	**543.671**	621.16	**713.11**	**841.667**	**408.169**	551.134	**637.831**
Algorithms name	Vrpnc9	Vrpnc10	Vrpnc11	Vrpnc12	Vrpnc13	Vrpnc14	F-n72-k4	F-n135-k7
HES [16]	764.56	888.46	682.12	534.24	910.26	591.87	177	769.55
IVND [32]	756.38	876.38	682.12	534.24	900.32	591.87	177	770.17
DHPD [27]	752.0/14	875.67/17	682.12	534.24	909.8/11	568.87/12	–	–
Hybrid (1 + 1) ES [28]	741.44	**871.58**	678.54	534.24	**836.55**	552.64	–	–
BBMOOVRP [30]	759.35	879.38	682.12	535.28	904.04	592.16	–	–
BLSA [29]	–	–	682.12	534.24	–	–	177	769.55
ITS [21]	–	–	–	–	–	–	–	–
ITS [22]	–	–	–	–	–	–	–	–
Clonal selection algorithm [26]	–	–	–	–	–	–	–	–
OVRP_GELS [33]	765.394	873.691	672.582	**526.823**	851.937	549.372	159.725	694.419
OVRP_ICA [2]	–	–	–	–	–	–	–	–
OVRP_SAICA	**741.4**	**868.92**	**667.14**	528.86	855.03	**544.9**	**157.34**	**692.31**

Bold Indicates the solution that minimizes the number of vehicles with least distance traveled
HES Hybrid Evolution Strategy from Repoussisa et al. [16]; PIV 2.8 GHz PC using C++ language
IVND Integrates a Variable Neighborhood Descent from Chen et al. [32]; Pentium IV 2.93G with 1G of RAM using C++ language
DHPD Diploid Hybrid PSO with DE from Hu and Wu [27]; Pentium 2.8G, 1 GB of RAM using JAVA language
Hybrid (1 + 1) ES Hybrid combination out of a stochastic combined Variable Neighborhood Search and a simple (1 + 1)-Evolutionary Strategy from Schneider and Reinholz [28]
BBMOOVRP Bumble Bees Mating Optimization for the Open Vehicle Routing Problem from Marinakis and Marinaki [30]; Intel Core 2 DUO CPU T9550 at 2.66 GHz using Fortran 90 and was com- piled using the Lahey f95 compiler, Linux 9.1
BLSA Broad Local Search Algorithm from Zachariadis and Kiranoudis [29]; Intel T5500 processor 1.66 GHz using Visual C# language
ITS Improved Tabu Search from Huang and Liu [21]; 2.26 GHz processor with 1G of RAM using Matlab language
ITS Improved Tabu Search from Huang and Liu [22]; 2.26 GHz processor with 1G of RAM using Matlab language
Clonal Selection Algorithm Clonal Selection Algorithm from Pan and Fu [26]; Pentium-IV at 1.6 GHz with 256 MB of RAM using Matlab 6.5 language
OVRP_GELS Open Vehicle Routing Problem Using the Gravitational Emulation Local Search Algorithm [33]; Pentium-IV at 2.06 GHz with 6 GB of RAM using C#.Net language
OVRP_ICA Imperialist-Based Optimization Algorithm for the Open Vehicle Routing Problem [2]; Pentium-IV at 2.06 GHz with 6 GB of RAM using C#.Net language
OVRP_SAICA Open Vehicle Routing Problem Using the Simulated Annealing and Imperialist competitive algorithm; Pentium-IV at 2.06 GHz with 6 GB of RAM using C#.Net language
– No value is calculated for this case

Table 3 Comparison of the proposed method with eight state-of-the-art methods in this domain based on time used to solve the problem

OVRP_SAICA	OVRP_ICA [2]	OVRP_GELS [33]	Clonal selection algorithm [26]	ITS [22]	ITS [21]	BLSA [29]	BBMOOVRP [30]	IVND [32]	HES [16]	Instance
65.412	63.429	1.769	1.769	14.9	16. 1	25	0.08	5.78	98	vrpnc1
136.08	131.758	1.274	1.274	26.7	28.3	68	0.32	17.62	143	Vrpnc2
256.19	254.385	2.845	2.845	218.4	212.5	103	0.55	44.74	330	Vrpnc3
359.62	350.132	2.936	2.936	531.5	542.7	190	2.05	132.61	613	Vrpnc4
566.73	563.876	2.261	2.261	253.1	268.6	355	3.07	314.88	1272	Vrpnc5
89.15	85.12	1.831	–	61.4	68.3	–	0.1	8.69	215	Vrpnc6
97.41	91.67	1.382	–	17.9	18.6	–	0.22	16.04	367	Vrpnc7
583.04	576.23	1.568	–	553.1	562.8	–	0.45	33.4	510	Vrpnc8
741.98	–	2.347	–	–	–	–	1.45	83.44	933	Vrpnc9
825.43	–	2.764	–	–	–	–	3.18	228.27	1678	vrpnc10
681.65	–	1.271	–	–	–	85	1.01	84.89	318	vrpnc11
537.89	–	1.385	–	–	–	39	1.17	34.13	363	vrpnc12
839.26	–	2.937	–	–	–	–	1.15	87.08	890	vrpnc13
556.63	–	1.841	–	–	–	–	1.35	31.42	411	vrpnc14

(continued)

Table 3 (continued)

OVRP_SAICA	OVRP_ICA [2]	OVRP_GELS [33]	Clonal selection algorithm [26]	ITS [22]	ITS [21]	BLSA [29]	BBMOOVRP [30]	IVND [32]	HES [16]	Instance
4.376	–	2.124	–	–	–	93	–	12.24	264	F-n72-k4
4.816	–	2.548	–	–	–	301	–	133.9	753	F-n135-k7

HES Hybrid Evolution Strategy from Repoussisa et al. [16]; PIV 2.8 GHz PC using C++ language
IVND Integrates a Variable Neighborhood Descent from Chen et al. [32]; Pentium IV 2.93G with 1G of RAM using C++ language
BBMOOVRP Bumble Bees Mating Optimization for the Open Vehicle Routing Problem from Marinakis and Marinaki [30]; Intel Core 2 DUO CPU T9550 at 2.66 GHz using Fortran 90 and was com- piled using the Lahey f95 compiler, Linux 9.1
BLSA Broad Local Search Algorithm from Zachariadis and Kiranoudis [29]; Intel T5500 processor 1.66 GHz using Visual C# language
ITS Improved Tabu Search from Huang and Liu [21]; 2.26 GHz processor with 1G of RAM using Matlab language
ITS Improved Tabu Search from Huang and Liu [22]; 2.26 GHz processor with 1G of RAM using Matlab language
Clonal Selection Algorithm Clonal Selection Algorithm from Pan and Fu [26]; Pentium-IV at 1.6 GHz with 256 MB of RAM using Matlab 6.5 language
OVRP_GELS Open Vehicle Routing Problem Using the Gravitational Emulation Local Search Algorithm [33]; Pentium-IV at 2.06 GHz with 6 GB of RAM using C#.Net language
OVRP_ICA Imperialist-Based Optimization Algorithm for the Open Vehicle Routing Problem [2]; Pentium-IV at 2.06 GHz with 6 GB of RAM using C#.Net language
OVRP_SAICA Open Vehicle Routing Problem Using the Simulated Annealing and Imperialist competitive algorithm; Pentium-IV at 2.06 GHz with 6 GB of RAM using C#.Net language
– No value is calculated for this case

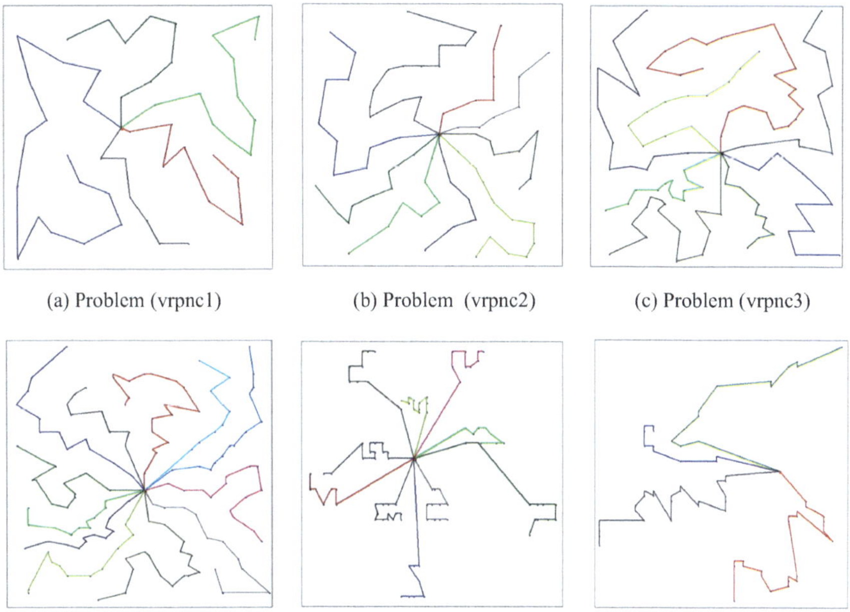

(a) Problem (vrpnc1) (b) Problem (vrpnc2) (c) Problem (vrpnc3)

(d) Problem (vrpnc4) (e) problem (vrpnc14) (f) problem (F-n72-k4)

Fig. 3 Six example outputs of the proposed algorithms for small and moderate problems

4 Conclusions

In this research, a meta-heuristic algorithm called *OVRP_SAICA* is presented for solving the OVRP. The advantages of this algorithm find the shortest tour between the customers. Its efficiency was compared with those of 12 different algorithms. This algorithm has a competitive structure compared to other meta-heuristic algorithms and could obtain good answers for standard examples of this problem. It seems used of other meta-heuristic algorithms such as Simulated Annealing or Ant Colony Optimization in combination with this algorithm can result in finding better answers for OVRP, especially in large and very large examples. On the other hand, this algorithm can be used for other VRP expansions such as the vehicle routing problem together with receiving and delivering goods. The use of such combinations of algorithms and work on these suggestions will be studied in future articles.

References

1. E.B. Tirkolaee, A. Goli, M. Bakhsi, I. Mahdavi, A robust multi-trip vehicle routing problem of perishable products with intermediate depots and time windows. Numer. Algebra Control Optim. **7**, 417–433 (2017)

2. Sh. Shamshirband, M. Shojafar, A.R. Hosseinabadi, A. Abraham, OVRP_ICA: an imperialist-based optimization algorithm for the open vehicle routing problem, in *International Conference on Hybrid Artificial Intelligence Systems (HAIS)*, LNCS, vol. 9121 (Springer, 2015), pp. 221–233
3. N. Metawaa, M. Kabir Hassana, M. Elhoseny, Genetic algorithm based model for optimizing bank lending decisions. Expert Syst. Appl. **80**, 75–82 (2017)
4. M. Elhoseny, A. Tharwat, A. Farouk, A.E. Hassanien, K-coverage model based on genetic algorithm to extend WSN lifetime. IEEE Sens. Lett. **1**(4), 1–4 (2017)
5. M. Elhoseny, A. Farouk, N. Zhou, M. Wang, S. Abdalla, J. Batle, Dynamic multi-hop clustering in a wireless sensor network: performance improvement. Wireless Pers. Commun. **95**, 3733–3753 (2017)
6. X. Yuan, M. Elhoseny, H.K. El-Minir, A.M. Riad, A genetic algorithm-based, dynamic clustering method towards improved wsn longevity. J. Netw. Syst. Manage. **25**, 21–46 (2017)
7. J. Batle, O. Ciftja, S. Abdalla, M. Elhoseny, M. Alkhambashi, A. Farouk, Equilibrium charge distribution on a finite straight one-dimensional wire. Eur. J. Phys. **38**, 1–9 (2017)
8. J. Batle, O. Ciftja, M. Naseri, M. Ghoranneviss, A. Farouk, M. Elhoseny, Equilibrium and uniform charge distribution of a classical two-dimensional system of point charges with hard-wall confinement. Phys. Scr. **92**, 34–47 (2017)
9. K. Fleszar, I.H. Osman, K.S. Hindi, A variable neighbourhood search algorithm for the open vehicle routing problem. Eur. J. Oper. Res. **195**, 803–809 (2009)
10. A.R. Hosseinabadi, M. Kardgar, M. Shojafar, Sh. Shamshirband, A. Abraham, Gravitational search algorithm to solve open vehicle routing problem, in *6th International Conference on Innovations in Bio-Inspired Computing and Applications (IBICA 2015), Chapter Advances in Intelligent Systems and Computing*, Kochi, India (Springer, 2016), pp. 93–103
11. A.R. Hosseinabadi, N.S.H. Rostami, M. Kardgar, S.S. Mirkamali, A. Abraham, A new efficient approach for solving the capacitated vehicle routing problem using the gravitational emulation local search algorithm. Appl. Math. Model. (Elsevier) **49**, 663–679 (2017)
12. D. Sariklis, S. Powell, A heuristic method for the open vehicle routing problem. J. Oper. Res. Soc. **51**, 564–573 (2000)
13. L. Tan, F. Lin, H. Wang, Adaptive comprehensive learning bacterial foraging optimization and its application on vehicle routing problem with time windows. Neurocomputing **151**, 1208–1215 (2015)
14. Sh Yu, Ch. Ding, K. Zhu, A hybrid GA–TS algorithm for open vehicle routing optimization of coal mines material. Expert Syst. Appl. **38**, 10568–10573 (2011)
15. X. Li, S.C.H. Leung, P. Tian, A multistart adaptive memory-based tabu search algorithm for the heterogeneous fixed fleet open vehicle, routing problem. Expert Syst. Appl. **39**, 365–374 (2012)
16. P.P. Repoussisa, C.D. Tarantilis, O. Braysy, G. Ioannou, A hybrid evolution strategy for the open vehicle routing problem. Comput. Oper. Res. **37**, 443–455 (2010)
17. M. Salari, P. Toth, A. Tramontani, An ILP improvement procedure for the open vehicle routing problem. Comput. Oper. Res. **37**, 2106–2120 (2010)
18. A. Azadeh, Farrokhi-Asl, The close–open mixed multi depot vehicle routing problem considering internal and external fleet of vehicles. Int. J. Transp. Res. 1–15 (2017)
19. Zh. Yanwei, W. Bin, W. Wanliang, Zh. Jingling, Particle Swarm Optimization for Open Vehicle Routing Problem with Time Dependent Travel Time (IFAC, 2008), pp. 12843–12848
20. N. Norouzi, R. Tavakkoli-Moghaddam, A. Salamatbakhsh, M. Alinaghian, Solving a novel bi-objective open vehicle routing problem in a competitive situation by multi objective particle swarm optimization. J. Appl. Oper. Res. **1**, 15–29 (2009)
21. F. Huang, C. Liu, A hybrid tabu search for open vehicle routing problem. CCTAE **1**, 132–134 (2010)
22. F. Huang, C. Liu, An improved tabu search for open vehicle routing problem, in *MASS* (2010), pp. 1–4
23. T. Zhen, Y. Zhu, Q. Zhang, A particle swarm optimization algorithm for the open vehicle routing problem. ESIAT **2**, 560–563 (2009)

24. R. Chunyu, L. Shiwei, Y. Bing, Research on the application of improved hybrid genetic algorithm in open vehicle routing problem. ISECS **1**, 532–535 (2009)
25. L. Guiyun, Research on open vehicle routing problem with time windows based on improved genetic algorithm. CiSE 1–5 (2009)
26. L. Pan, Z. Fu, A clonal selection algorithm for open vehicle routing problem, in *Third International Conference on Genetic and Evolutionary Computing* (2009), pp. 786–790
27. F. Hu, F. Wu, Diploid hybrid particle swarm optimization with differential evolution for open vehicle routing problem, in *WCICA* (2010), pp. 2692–2697
28. H. Schneider, A. Reinholz, Integrating variable neighborhood search into a hybrid evolutionary strategy for the open vehicle routing problem, in *EU/MEeting 2008*, Troyes, France, October 23–24 (2008), pp. 1–6
29. E.E. Zachariadis, ChT Kiranoudis, An open vehicle routing problem metaheuristic for examining wide solution neighborhoods. Comput. Oper. Res. **37**, 712–723 (2010)
30. Y. Marinakis, M. Marinaki, A bumble bees mating optimization algorithm for the open vehicle routing problem. Swarm Evol. Comput. **15**, 80–94 (2014)
31. S.A. MirHassani, N. Abolghasemi, A particle swarm optimization algorithm for open vehicle routing problem. Expert Syst. Appl. **38**, 11547–11551 (2011)
32. P. Chen, Y. Qu, H. Huang, X. Dong, A new hybrid iterated local search for the open vehicle routing problem, in *Pacific-Asia Workshop on Computational Intelligence and Industrial Application* (2008), pp. 891–895
33. A.R. Hosseinabadi, J. Vahidi, V.E. Balas, S.S. Mirkamali, OVRP_GELS: solving open vehicle routing problem using the gravitational emulation local search algorithm. Neural Comput. Appl. 1–14 (2016)
34. M. Alinaghian, H. Amanipour, E.B. Tirkolaee, Enhancement of inventory management approaches in vehicle routing-cross docking problems. J. Supply Chain Manag. Syst. **3**, 27–34 (2014)
35. S.H. Mirmohammadi, E.B. Tirkolaee, A. Goli, S. Dehnavi-Arani, The periodic green vehicle routing problem with considering of time-dependent urban traffic and time windows. Iran Univ. Sci. Technol. **7**, 143–156 (2017)
36. E. Babaee Tirkolaee, M. Alinaghian, M.B. Sasi, S.M. Esfahani, Solving a robust capacitated arc routing problem using a hybrid simulated annealing algorithm: a waste collection application. J. Ind. Eng. Manag. Stud. **3**, 61–76 (2016)
37. E.B. Tirkolaee, A.A.R. Hosseinabadi, M. Soltani, A.K. Sangaiah, J. Wang, A hybrid genetic algorithm for multi-trip green capacitated arc routing problem in the scope of urban services. Sustainability **10**, 1–21 (2018)
38. E.B. Tirkolaee, M. Alinaghian, A.A.R. Hosseinabadi, M.B. Sasi, A.K. Sangaiah, An improved ant colony optimization for the multi-trip capacitated arc routing problem. Comput. Electr. Eng. (2018)
39. E.B. Tirkolaee, I. Mahdavi, M.M.S. Esfahani, A robust periodic capacitated arc routing problem for urban waste collection considering drivers and crew's working time. Waste Manag. (2018)
40. P. Khosravi, M. Alinaghian, S.M. Sajadi, E. Babaee, The periodic capacitated arc routing problem with mobile disposal sites specified for waste collection. J. Appl. Res. Ind. Eng. **2**, 64–76 (2015)
41. A. Goli, M. Bakhshi, E.B. Tirkolaee, A review on main challenges of disaster relief supply chain to reduce casualties in case of natural disasters. J. Appl. Res. Ind. Eng. **4**, 77–88 (2017)
42. A. Mostafaeipour, M. Qolipour, M. Rezaei, E. Babaee-Tirkolaee, Investigation of off-grid photovoltaic systems for a reverse osmosis desalination system: a case study. Desalination (2018)
43. Sh. Shamshirband, M. Shojafar, A.R. Hosseinabadi, A. Abraham, A solution for multi-objective commodity vehicle routing problem by NSGA-II, in *IEEE International Conference on Hybrid Intelligent Systems (HIS)* (2014), pp. 12–17
44. A.R. Hosseinabadi, M. Kardgar, M. Shojafar, Sh. Shamshirband, A. Abraham, GELS-GA: hybrid metaheuristic algorithm for solving multiple travelling salesman problem, in *IEEE International Conference on Intelligent Systems Design and Applications (ISDA)* (2014), pp. 76–81

45. A.R. Hosseinabadi, F. Alavipour, Sh. Shamshirbnd, V.E. Balas, *A Novel Meta-Heuristic Combinatory Method for Solving Capacitated Vehicle Location-Routing Problem with Hard Time Windows* (Springer International Publishing Switzerland, 2017); Transp. Syst. Adv. Intell. Syst. Comput. (China, Springer) **454**, 707–728 (2017)
46. A.R. Hosseinabadi, J. Vahidi, B. Saemi, A.K. Sangaiah, M. Elhoseny, Extended genetic algorithm for solving open-shop scheduling problem. Soft Comput. (Springer) 1–18 (2018)
47. N. Metawa, M. Elhoseny, M. Kabir Hassan, A.E. Hassanien, Loan portfolio optimization using genetic algorithm: a case of credit constraints, in *Proceedings of 12th International Computer Engineering Conference (ICENCO)* (IEEE, 2016), pp. 59–64
48. M. Elhoseny, A. Hosny, A.E. Hassanien, K. Muhammad, A.K. Sangaiah, Secure automated forensic investigation for sustainable critical infrastructures compliant with green computing requirements. IEEE Trans. Sustain. Comput. 1–1 (2017)
49. A.R. Hosseinabadi et al., Using the gravitational emulation local search algorithm to solve the multi-objective flexible dynamic job shop scheduling problem in small and medium enterprises. Ann. Oper. Res. **229**(1), 451–474 (2015)
50. A.R. Hosseinabadi et al., An ameliorative hybrid meta-heuristic algorithm for solving the capacitated vehicle routing problem. IEEE Access **7**, 175454–175465 (2019)
51. N. Metropolis, A.W. Rosenbluth, M.N. Rosenbluth, A.H. Teller, E. Teller, Equation of state calculation by east computing machines. J. Chem. Phys. **21**, 1087–1091 (1953)
52. P. Pellegrini, Application of two nearest neighbor approaches to a rich vehicle routing problem, in *TR/IRIDIA/2005-15, IRIDIA* (Universite Librede Bruxelles, Brussels, Belgium, 2005)
53. N. Christofides, A. Mingozzi, P. Toth, The vehicle routing problem, in *Combinatorial Optimization,* ed. by N. Christofides, A. Mingozzi, P. Toth, C. Sandi (Wiley, Chichester, UK, 1979), p. 31538
54. M. Fisher, Optimal solution of vehicle routing problems using minimum k-trees. Oper. Res. (1994), pp. 626–642
55. http://neo.lcc.uma.es/vrp/vrp-instances/capacitated-vrp-instances/
56. B. Güler, A.Z. Şevkli, *Oscillated Variable Neighborhood Search for Open Vehicle Routing Problem* (Springer International Publishing Switzerland 2015); in *ICONIP 2015*, Part III, LNCS, vol. 9491 (2015), pp. 182–189

Fuzzy Regression and Modeling

Modeling Objects and Processes in GIS by Fuzzy Temporal Graphs

Alexander Bozhenyuk, Stanislav Belyakov, and Margarita Knyazeva

Abstract In this paper the concepts of a strongly connected fuzzy temporal graph and a fuzzy set of strong connectivity are introduced. Fuzzy temporal graph is generalization of a fuzzy graph on the one hand, and a temporal graph on the other hand. In this graph the incidence of graph vertices is changed in the discrete time. Fuzzy temporal graph is offered to use as a model in GIS. A method is proposed for finding a fuzzy set of strong connectivity. An example of finding a fuzzy set of strong connection of a fuzzy temporal graph is considered too.

Keywords Fuzzy temporal graph · Geoinformation system · Reachability degree · Fuzzy path · Fuzzy temporal set

1 Introduction

Geoinformation technologies are aimed at creating information models of objects, events and phenomena of the real world. Visualization is of great importance in geoinformation technologies. Maps and charts play simultaneously the role of the source data, the modeling environment and the presentation of the simulation result, the means of dialogue with the user. The analytical apparatus used by geoinformation systems (GIS) for designing information models includes discrete and continuous methods of representing and manipulating objects [1, 2]. One of the important classes of such objects is graphs. Graphs and methods of graph theory are used for visualization, analysis and synthesis of geoinformation models [3, 4] as a universal means of relationship modeling.

A. Bozhenyuk (✉) · S. Belyakov · M. Knyazeva
Southern Federal University, Nekrasovsky 44, 347922 Taganrog, Russia
e-mail: avb002@yandex.ru

S. Belyakov
e-mail: beliacov@yandex.ru

M. Knyazeva
e-mail: margarita.knyazeva@gmail.com

S. N. Shahbazova et al. (eds.), *Recent Developments and the New Direction in Soft-Computing Foundations and Applications*, Studies in Fuzziness and Soft Computing 393, https://doi.org/10.1007/978-3-030-47124-8_22

Geoinformation models are not static. The change in spatial coordinates, time marks, semantic content of cartographic objects and changes in the real world are interdependent and provide for the dynamics of geoinformation models. As the analysis shows, traditionally used graph models in some cases become inadequate to the formulation of tasks under conditions of dynamism. The reason for this can be:

- incompleteness of the data, which is the cartographic basis of the GIS. The process of production of maps involves the use of a set of several sources of cartographic information and geodatabases. Coordination of information leads to a "renegotiation", i.e. the distortion of map. In addition, the construction of the map is accompanied by cartographic generalization, that is, the purposeful allocation of the main and generalization. Since this procedure is not devoid of subjectivity, the construction of graph models of processes and phenomena on electronic maps should introduce categories of fuzziness [5, 6]. Without them, the model is simplified, which will lead to the complexity of the subsequent interpretation of the simulation results;
- using of data from unreliable sources outside GIS. The solution of many tasks in the GIS environment assumes knowledge of the current state of the displayed area of space. For example, the analysis of the transport situation in the city assumes the availability of data on the carrying capacity of road sections, the intensity of traffic flows, the ways of daily population migration, and the logistics schemes for the movement of goods and products produced by urban enterprises. However, it is impossible to obtain such data with absolute certainty. As a consequence, mechanisms are required for extracting indirectly information, for example, from the Internet. The completeness, reliability and consistency of such information is assessed by experts and is described by fuzzy;
- when using graphs, it is traditionally assumed that the reachability of vertices is uniquely determined by a given variant of the graph. In some cases this is not so [7]. The weights of the individual edges of the graph can change in time, in the process of modeling, which forces us to take a different look at the procedures for analyzing and synthesizing graphs, algorithms for solving optimization problems on graphs.

2 Fuzzy Temporal Graph and GIS

Let us consider an example illustrating the listed features of statement of tasks. In Fig. 1 shows a map of the railways, on the basis of which the scheme is constructed.

According to the scheme, a graph of the road network can be constructed, which will solve the problems of finding the shortest path, evaluate flows, find service centers, determine connectivity, and solve other problems known from graph theory [3, 4]. At the same time, for a given train schedules, it is impossible to solve the task of transporting an additional (out-of-schedule) composition between a given pair of stations in the minimum time in the traditional way. Individual sections of the path at

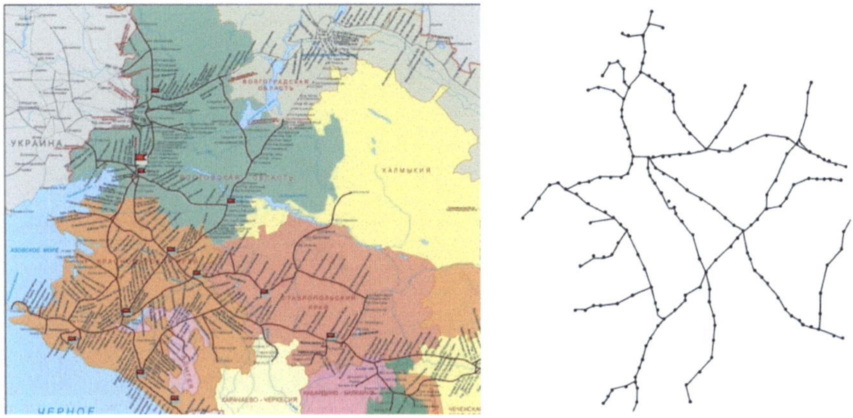

Fig. 1 Fragment of the railway network and road map

some time intervals in the network graph are not included. Here the following tasks can be posed:

- what is the shortest path from the vertex *A* to the vertex *B* if the time interval of the beginning of the motion from the vertex *A* is given;
- what is the shortest path from the vertex *A* to the vertex *B* if the time interval of arrival at the vertex *B* is given.

The above statements are in practice related to the uncertainty of the time and metric characteristics of the transport network.

The complexity of graph models grows in the case of solving logistics problems. In Fig. 2 shows an example of a logistics network describing possible directions for the movement of material flows. The weight of each arc is determined in a complex manner, based on the properties of the object being moved, the spatial environment of the transport network, and the dynamics of processes affecting transportation.

Thus, the solution of a number of problems using geoinformation models requires a specific-temporal description of the graph. The behavior is described by a function of time. We can specify the following models for describing behavior:

- analytical model. A well-known drawback of this approach is the need to apply simplifications that make it possible to apply mathematical formalisms. Simplifications result in model inadequacy;
- statistical model. This model is applicable in cases where the researcher has the necessary means of observation under unchanging conditions. In practice, this is difficult to achieve. But improving the technical means of Earth observation makes it possible to talk about the prospects of such an approach;
- expert logical model. This model is built on the basis of individual experience of experts. Its drawback is subjectivism. Nevertheless, expert knowledge in certain cases gives a more adequate result than a complete formalization [8];

Fig. 2 Logistics network

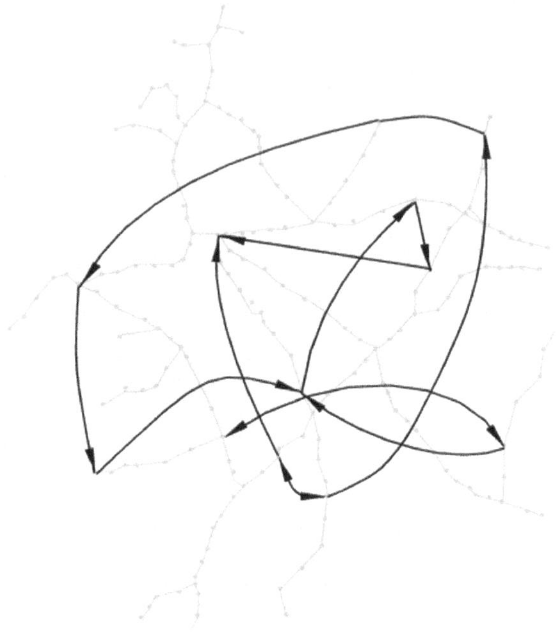

- cart metric model. A measurement on a map is a generalized operation that involves estimating the relationships between objects on the map. An electronic map is considered as a source of metric and non-metric data. The modern structural organization of GIS makes it possible to associate with any cartographic objects any complex behavior, maintains communication with any external data sources of local and global networks. There is an idea of models of a combined nature, which includes the collection of real information, the accumulation of experience and the use of the logic of experts.

3 Basic Concepts and Definitions

Definition 1 [9] A temporal fuzzy graph is a triple $\tilde{G} = (X, \tilde{U}_t, T)$, where X is the set of vertices of the graph c with the number of vertices $|X| = n$; $T = \{1, 2, \ldots, N\}$ is the set of natural numbers defining (discrete) time; $\tilde{U}_t = \{<\mu_t(x_i, x_j)/(x_i, x_j)>\}$—fuzzy set of edges, where x_i, $x_j \in X$; $\mu_t(x_i, x_j) \in [0, 1]$ is the value of the membership function μ_t for the edge (x_i, x_j) at time $t \in T$. Moreover, for different instants of time for the same edge (x_i, x_i), the values of the membership function are, in general, different.

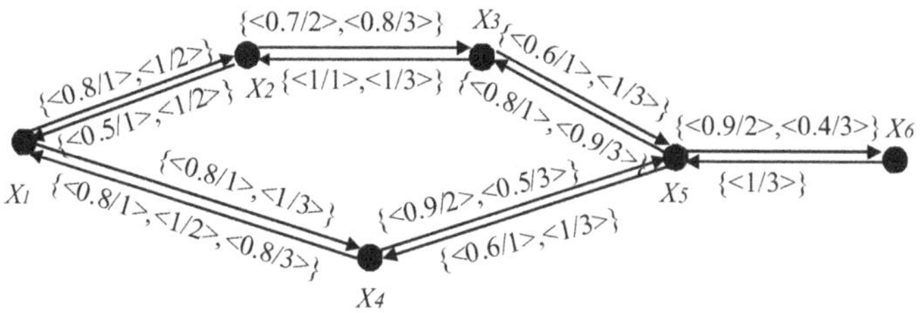

Fig. 3 Example of fuzzy temporal graph for $T = \{1, 2, 3\}$

Example 1 Consider a fuzzy temporal graph $\tilde{G} = (X, \tilde{U}_t, T)$ for which the set of vertices $X = \{x_1, x_2, xx_3, x_4, {}_5, x_6\}$, time $T = \{1, 2, 3\}$, $n = 6$, $N = 3$, and fuzzy set of edges is given in the form:

$$\tilde{U}_t = \{<0,8_1/(x_1, x_2>, <1_2/(x_1, x_2)>, <0,5_1/(x_2, x_1)>, <1_2/(x_2, x_1)>,$$
$$<0,7_2/(x_2, x_3)>, <0,8_3/(x_2, x_3)>, <1_1/(x_3, x_2)>, <1_3/(x_3, x_2)>,$$
$$<0,6_1/(x_3, x_5)>, <1_3/(x_3, x_5)>, <0,8_1/(x_5, x_3)>, <0,9_3/(x_5, x_3)>,$$
$$<0,9_2/(x_5, x_6)>, <1_3/(x_6, x_5)>, <0,8_1/(x_1, x_4)>, <1_3/(x_1, x_4)>,$$
$$<0,8_1/(x_4, x_1)>, <1_2/(x_4, x_1)>, <0,9_2/(x_4, x_5)>,$$
$$<0,6_1/(x_5, x_4)>, <1_3/(x_5, x_4)>\}.$$

The graphically fuzzy temporal graph can be defined as a fuzzy oriented graph, on the edges of which the values of the membership function μ_t at the instants $t \in T$ are indicated.

The graph considered in Example 1 has the form shown in Fig. 3.

Definition 2 [9] The vertex x_j is a fuzzy adjacent vertex x_i with respect to the time instant $t \in T$ if the following condition is satisfied: $\mu_t(x_i, x_j) > 0$.

The directed fuzzy path $\tilde{L}(x_i, x_k)$ of the fuzzy temporal graph is called the directed sequence of fuzzy edges leading from the vertex x_i to the vertex x_k, in which the finite vertex of any edge different from the last one is the initial vertex of the next edge:

$$\begin{aligned}\tilde{L}(x_i, x_k) = &<\mu_{t_1}(x_i, x_1)/(x_i, x_1)>,\\ &<\mu_{t_2}(x_1, x_2)/(x_1, x_2)>,\\ &<\mu_{t_k}(x_{k-1}, x_k)/(x_{k-1}, x_k)>\end{aligned} \tag{1}$$

for which the following conditions are satisfied:

$$\mu_{t_1}(x_i, x_1) > 0, \mu_{t_2}(x_2, x_3) > 0, \ldots, \mu_{t_k}(x_{k-1}, x_k) > 0 \tag{2}$$

and for the instants of time $t_1, t_2, \ldots, t_k \in T$ the following inequality holds:

$$t_1 \le t_2 \le \cdots \le t_k \tag{3}$$

In other words, if in a fuzzy path (1) each successive vertex is an adjacent previous vertex by a moment of time no less than the moments at which all the previous vertices in this sequence are fuzzy adjacent.

The conjunctive strength of the path $\tilde{L}(x_i, x_k)$ is determined by the expression:

$$\mu_{\&}(\tilde{L}(x_{\mathrm{i}}, \mathrm{x_k})) = \underset{t_1,t_2,\ldots,t_k}{\&} \mu_{\mathrm{t_j}}<\mathrm{x_i}, \mathrm{x_j}>.$$

Definition 3 A fuzzy path $\tilde{L}(x_i, x_k)$ will be called a simple path between the vertices x_i and x_k if its part is not any other fuzzy path between the same vertices x_i and x_k.

It is obvious that this definition coincides in form with the definition for a crisp graph.

The vertex x_k is fuzzy achievable from the vertex x_i in the fuzzy temporal graph if there is an oriented fuzzy path $\tilde{L}(x_i, x_k)$ from the vertex x_i to the vertex x_k.

Definition 4 The value t_k is called the reachability time of the vertex x_k from the vertex x_i, and the value $\mu_{\&}(\tilde{L}(x_{\mathrm{i}}, \mathrm{x_k}))$ is called the reachability degree by the path $\tilde{L}(x_i, x_k)$.

Let there exist several sequences $\tilde{L}$ of the form (1) from the vertex x_1 to the vertex x_k, then the values $i_{k-1}^{(j)}$ for each sequence can be different. The smallest of these values is called a minimal time reachability $t_{\min}(x_1, x_k)$ vertex x_k from the vertex x_1, that is:

$$t_{\min}(x_1, x_k) = \min_{j=\overline{1,L}}\{i_{k-1}^{(j)}\}, \tag{4}$$

and the corresponding value $\mu(t_{\min})$ will be called the degree of attainability at the minimum time.

Example 2 In the fuzzy temporal graph shown in Fig. 3, the vertex x_3 is reachable from the vertex x_1 by the sequence $seq_1= (x_1, x_2, x_3)$ with a reachability degree 0.7 with reachability time $t = 2$, and with a reachability degree 0.8 with reachability time $t = 3$; is achievable with the sequence $seq_2= (x_1, x_4, x_5, x_3)$ with a reachability degree 0.8 for $t = 1$ and with a reachability degree 0.9 for $t = 3$. Therefore, the value $t_{\min}(x_1, x_3) = 1$, and value $\mu(t_{\min}) = 0.8$.

Let $\{\tilde{L}(x_i, x_j)\}$ be a family of fuzzy paths with the help of which the vertex x_j is accessible from the vertex x_i. We denote by $\alpha_t^{i,j}$ the greatest degree of attainability of some path from vertex x_i to vertex x_j with time $t \in \overline{1, T}$.

Definition 5 A fuzzy set $\tilde{A}(x_i, x_j) = \{< \alpha_t^{i,j}/t > | t \in \overline{1, n}\}$ is called a fuzzy temporal set of reachability of the vertex x_j from the vertex x_i.

We will assume that each vertex is attainable from itself with degree 1 at any instant of time $t \in \overline{1, T}$. That is, the following is true:

$$(\forall i \in \overline{1,n})[\tilde{A}(x_i, x_i) = \{<1/t>|t \in \overline{1,T}\}].$$

Definition 6 A fuzzy set

$$\tilde{B} = \bigcap_{i=\overline{1,n}} \bigcap_{j=\overline{1,n}} \tilde{A}(x_i, x_j) = \{<\beta_t/t>|t \in \overline{1,T}\}$$

is called a fuzzy set of strong connection of the fuzzy temporal graph $\tilde{G}$.

For a fuzzy set of strong connection, the following property holds:

$$\tilde{B} = \bigcap_{i=\overline{1,n}} \tilde{A}(x_k, x_i) \cap \bigcap_{j=\overline{1,n}} \tilde{A}(x_j, x_k).$$

Here x_k is an arbitrary vertex from the set X.

If for some time $t \in \overline{1,T}$ the value $\beta_t \neq 0$, then it means that between any vertices of the temporal graph $\tilde{G}$ at time t there is a fuzzy path with a degree not less than β_t.

Definition 7 A fuzzy temporal graph $\tilde{G}$ is said to be strongly connected if $\tilde{B} \neq \emptyset$.

We introduce the operation of temporal intersection of fuzzy sets to calculate a fuzzy temporal set of reachability.

Let $\tilde{C}_1 = \{<\mu_1(t)/t>\}$ and $\tilde{C}_2 = \{<\mu_2(t)/t>\}$ be fuzzy sets, $t \in \overline{1,T}$.

Definition 8 The temporal intersection of the sets $\tilde{C}_1$ and $\tilde{C}_2$ is the set $\tilde{C}_1 \cap_t \tilde{C}_2 = \{<\mu_{\cap t}(t)/t>\}$ in which the membership function $\mu_{\cap t}$ is defined as: $\mu_{\cap t}(t) = \min\{\max_{\tau=\overline{1,t}}\{\mu_1(\tau)\}, \mu_2(t)\}$.

Property $\tilde{C}_1 \cap_t \tilde{C}_2 \neq \tilde{C}_2 \cap_t \tilde{C}_1$.

Example 3 Let $\tilde{C}_1 = \{<0.8/1>, <0.6/2>, <0.4/3>\}$ and $\tilde{C}_2 = \{<0.3/1>, <0.7/2>, <0.9/3>\}$. Then $\tilde{C}_1 \cap_t \tilde{C}_2 = \{<0.3/1>, <0.7/2>, <0.8/3>\}$ and $\tilde{C}_2 \cap_t \tilde{C}_1 = \{<0.3/1>, <0.6/2>, <0.4/3>\}$.

4 Method for Determining the Strong Connectivity of Fuzzy Temporal Graph

To answer the question "whether the fuzzy temporal graph $\tilde{G}$ is strongly connected" we consider the method for finding the fuzzy set of strong connection $\tilde{B}$.

To do this, we take an arbitrary vertex, for example x_1, and find for it a fuzzy set $\bigcap_{i=\overline{1,n}} \tilde{A}(x_1, x_i)$, and then we find a fuzzy set of $\bigcap_{j=\overline{1,n}} \tilde{A}(x_j, x_1)$. Then we find their intersection: $\tilde{B} = \bigcap_{i=\overline{1,n}} \tilde{A}(x_1, x_i) \cap \bigcap_{j=\overline{1,n}} \tilde{A}(x_j, x_1)$.

	x_1	x_2	x_3	x_4	x_5	x_6
x_1	{1/1,1/2,1/3}	{0.8/1,1/2}	∅	{0.8/1,1/3}	∅	∅
x_2	{0.5/1,1/2}	{1/1,1/2,1/3}	{0.7/1,0.8/2}	∅	∅	∅
$R=x_3$	∅	{1/1,1/3}	{1/1,1/2,1/3}	∅	{0.6/1,1/3}	∅
x_4	{0.8/1,1/2,0.8/3}	∅	∅	{1/1,1/2,1/3}	{0.9/2,0.5/3}	∅
x_5	∅	∅	{0.8/1,0.9/3}	{0.6/1,1/3}	{1/1,1/2,1/3}	{0.9/2,0.4/3}
x_6	∅	∅	∅	∅	{1/3}	{1/1,1/2,1/3}

Fig. 4 Incidence matrix of graph $\tilde{G}$

If $\tilde{B} \neq \emptyset$, then the fuzzy temporal graph $\tilde{G}$ is strongly connected.

Let us consider this approach using the example of the graph presented in Fig. 3. The incidence matrix of this graph is shown in Fig. 4.

We introduce three columns and three rows. The first column A^+ corresponds to the elements of $\tilde{A}(x_1, x_i)$ sets. Elements of the second column Pr^+ indicate that the line has already been viewed (1), or not yet (0). The elements of the third column Y_+ contain the number of the previous vertex. The first row A^- corresponds to the elements of $\tilde{A}(x_j, x_1)$ sets. Elements of the second row Pr^- indicate that the column has already been viewed (1), or not yet (0). The elements of the third row Y_- contain the number of the previous vertex.

The initial values of these columns and rows are shown in Fig. 5.

- According to the algorithm, we select an arbitrary vertex, for example x_I. Assign value $A^+(1) := \{<1/1>, <1/2>, <1/3>\}$, $Y_+(1): = 1$.
- We select the first row of the matrix R for which $Pr^+(\mathrm{i}) = 0$ (the row has not yet been viewed) and $A^+(i) \neq \emptyset$ (there is some path from the vertex x_I). At this step,

Fig. 5 Initial values of columns and rows

A^+	Pr^+	Y_+
∅	0	0
∅	0	0
∅	0	0
∅	0	0
∅	0	0
∅	0	0

A^-	∅	∅	∅	∅	∅	∅
Pr^-	0	0	0	0	0	0
Y_-	0	0	0	0	0	0

this is the first row (i = 1). In the first row of the matrix R we find the elements (r_{12}, r_{14}), which are not equal Ø. The number of the previous vertex is written in the corresponding elements of the column Y_+: $Y_+(2)$: = 1; $Y_+(4)$: = 1. Then, in the corresponding elements of column A^+, we write down of reachability degree of the vertex from the vertex x_1:

$$A^+(2) := A^+(2) \cup A^+(Y_+(2)) \cap_t r_{12} = \{<0.8/1>, <1/2>\},$$
$$A^+(4) := A^+(4) \cup A^+(Y(4)) \cap_t r_{14} = \{<0.8/1>, <1/3>\}.$$

- Assign the value of $Pr^+(1)$: = 1 (row 1 of the matrix has been viewed).
- Then we choose the first row of the matrix R for which $Pr^+(i) = 0$ (the row has not yet been viewed) and $A^+(i) \neq \emptyset$ (there is some path from the vertex x_1). At this step, this is the second row (i = 2). In the second row of the matrix R we find the elements, which are not equal Ø. This is the element r_{23}. The number of the previous vertex is written in the corresponding element of the column Y_+: $Y_+(3)$: = 2. Then, in the corresponding element of column A^+, we write down of reachability degree of the vertex from the vertex x_1:

$$A^+(3) := A^+(3) \cup A^+(Y_+(3)) \cap_t r_{23} = \{<0.7/2>, <0.8/3>\}.$$

- Assign the value of $Pr^+(2)$: = 1 (row 2 of the matrix has been viewed).
- The process continues until all the elements of the vector Pr^+ become equal to 1 (all rows were viewed).
- Make the same steps and for all columns of the matrix R.

As a result, we get the column A^+ and row A^- values, which are shown in Fig. 6. Hence it follows that the fuzzy set of strong connectivity of the graph in question is $\tilde{B} = \{<0.8/3>\}$. In other words, the graph is strongly connected, and the degree of accessibility between any pair of vertices of at least 0.8.

A^+	Pr^+	Y_+
{<1/1>,<1/2>,<1/3>}	1	0
{<0.8/1>,<1/2>,<0.8/3>}	1	1
{<0.7/2>,<0.8/3>}	1	2
{<0.8/1>,<1/3>}	1	1
{<0.8/2>,<0.8/3>}	1	2
{<0.8/2>,<0.8/3>}	1	3

A^-	{<1/1>,<1/2>,<1/3>}	{<0.5/1>,<1/2>,<0.8/3>}	{<0.6/1>,<1/2>,<1/3>}	{<0.8/1>,<1/2>,<0.8/3>}	{<0.6/2>,<1/3>}	{<1/3>}
Pr^-	1	1	1	1	1	1
Y_-	0	1	2	1	2	3

Fig. 6 Values of columns and rows after the operation of algorithm

5 Conclusions

The introduced concepts of strong connectivity of the fuzzy temporal graph and fuzzy set of strong connectivity can serve as a basis for modeling complex processes in GIS, in which the elements have fuzzy relationships that vary in discrete time. It should be noted that the definition of a fuzzy set of strong connectivity refers to the problem of analysis. Another, more complex task is the problem of synthesis, which in general form can be formulated as follows: the degree of which edges of the graph, and at what time it is necessary to increase, so that the strong connectivity of the fuzzy graph under consideration reaches a certain, predetermined value.

Acknowledgements The reported study was funded by the Russian Foundation for Basic Research according to the research projects N 18-01-00023, N 20-01-00197.

References

1. J. Malczewski, *GIS and Multicriteria Decision Analysis* (Wiley, New York, 1999)
2. P. Longley, M. Goodchild, D. Maguire, D. Rhind, *Geographic Information Systems and Science* (Wiley, New York, 2001)
3. A. Kaufmann, *Introduction a la theorie des sous-ensemles flous* (Masson, Paris, 1977)
4. N. Christofides, *Graph Theory. An Algorithmic Approach* (Academic Press, London, 1976)
5. M. Goodchild, Modelling error in objects and fields, in *Accuracy of Spatial Databases*, ed. by M.F. Goodchild, S. Gopal (Taylor & Francis, Basingstoke, 1989), pp. 107–113
6. J. Zhang, M. Goodchild, *Uncertainty in Geographical Information* (Taylor & Francis, New York, 2002)
7. Ia.M. Erusalimskiy, Graph with attenuation on arcs and amplification in vertices and routing in information networks. Eng. J. Don **1** (2015). ivdon.ru/ru/magazine/archive/n1y2015/2782
8. L. Bershtein, A. Bozhenyuk, The using of temporal graphs as the models of complicity systems. Izvestiya UFY. Technicheskie nayuki. TTI UFY, Taganrog **4**(105), 198–203 (2010)
9. L. Bershtein, S. Belyakov, A. Bozhenyuk, The using of fuzzy temporal graphs for modeling in GIS. Izvestiya UFY. Technicheskie nayuki. TTI UFY, Taganrog **1**(126), 121–127 (2012)

Method for Object Allocation to the Intermediate Resource Center Based on Fuzzy Estimation

Alexander Bozhenyuk, Olesiya Kosenko, and Margarita Knyazeva

Abstract This paper focuses on developing the method of object allocation to the intermediate centers. The goal is to search the optimal solutions/strategy of allocations of number of objects according to some critical parameters in the way they move from producers to customers through intermediate centers. The paper introduces a method for fixing demand area under a certain intermediate center. It takes into account the distance between the grouping objects and demand level in a certain area as well as capacity of such intermediate centers. Allocation and distribution problems are always treated as uncertain variables to meet the practice-oriented situations due to some expert estimation. Here we treat this uncertainty as a fuzzy variable estimation process. Special software was developed to solve this problem.

Keywords Distribution of resources · Intermediate centers · Demand · Grouping · Uncertainty of initial parameters · Interaction potentials

1 Introduction

The development of production relations and resource allocation technologies led to actual problem of creating intermediate distribution centers. Such intermediate distribution centers can be presented as pool of technological equipment, pool of some specific resource centers, logistic centers, warehouses and stocks, transmission networks servers etc. A common feature of distribution centers is the accumulation and storage of resources, and if it is necessary—solving the problem of efficient management of resource consumption at other levels. Among the distribution problems

A. Bozhenyuk (✉) · O. Kosenko · M. Knyazeva
Southern Federal University, Nekrasovsky 44, 347922 Taganrog, Russia
e-mail: avb002@yandex.ru

O. Kosenko
e-mail: o_kosenko@mail.ru

M. Knyazeva
e-mail: margarita.knyazeva@gmail.com

S. N. Shahbazova et al. (eds.), *Recent Developments and the New Direction in Soft-Computing Foundations and Applications*, Studies in Fuzziness and Soft Computing 393, https://doi.org/10.1007/978-3-030-47124-8_23

a special cases are problems of operational planning and transportation management, the mathematical model of which is the transportation problem, that is usually devoted to finding the minimum cost of transporting a single commodity from a given number of sources (e.g. factories) to a given number of destinations (e.g. warehouses) [1]. In practice it is often impossible to solve effectively the problem of transporting inhomogeneous objects by different transportation modes (transmitting elements) [2–4], since the initial uncertain data of the problem is not fully taken into account.

One of the efficient approaches to optimizing the interaction of several transportation modes and implementation of advanced business technologies is the creation of intermediate network centers. The goal is to minimize temporal costs and the total costs of senders (producers) and consumers of resources.

Resource transferring in modern market conditions is impossible without the concentrating resources (objects) in certain places in certain amounts, storages of such resources represents intermediate distribution centers. The amount of transportation costs varies significantly depending on the location of the intermediate centers. In papers [5, 6] was shown, that optimal location of the center allows to reduce the cost of allocating resources to 30%.

Scheduling distribution processes are of particular importance in the case of managing multi-stage technological processes for processing a resource, coming from its initial source to a warehouse and then to the consumer. From the point of view of organization of distribution procedure, it is relevant to analyze the whole process of transportation as a whole from producer to consumer [7].

Analysis of the practical implementation of transport and distribution systems shows that distribution centers are usually created for the following reasons [8]:

- losses of material and labor resources;
- weak information support of marketing activities;
- lack of experience in managing material flows;
- losses associated with the dispersal of resources.

Changing the number and location of intermediate centers allows to reduce total costs of delivering resources to consumers so the question of the rational location and number of intermediate centers is actual. This problem becomes especially urgent when creating a new transportation and distribution system or expanding the existing one.

2 Distribution Problem with Intermediate Centers

Analysis of the classification measures aimed at reducing the costs of resource allocation allows us to conclude that the indicators that significantly affect the robustness of solution are [5]: multifactor formalization of transport-distribution problem parameters in the stage "producer—intermediate resource center—consumer"; effective allocation of distribution centers.

If it is necessary to realize the condition of fixing specific consumers to the center, the problem of allocation intermediate centers is considered and rational number and location for the considered demand area is determined. The supply of resources through the center to the customer requires the construction of a model for a multi-stage transportation and distribution problem, in which the minimum value of the total resource allocation costs as an optimality criterion is considered [2, 3, 7]. The distribution problem with intermediate centers was formulated in [8] as a method of reduction high dimensionality of a given problem and an effective way to reduce the overall transport costs. Figure 1 presents a diagram of the interaction between elements of the system "producers—intermediate centers consumers".

Formal problem definition can be stated as follows in [1]. There are *I* numbers of sources (producers) of resources, where a_i *(i* = 1, 2, …, *I)* is a set of resource distribution centers and J is the set of consumers, where b_j *(j* = 1, 2, …, *J)* is the set of resource level demand. Let's consider K as a number of intermediate centers and matrix of costs C = (c_{ijk}), where c_{ijk}—is the cost of moving for one resource unit from the *i*-source to *j*-consumer points through the *k*-center. Matrix $X = (x_{ijk})$ determines the vector for the number of resources intended for transportation from the *i*-source to *j*-consumer points through the *k*-center. The objective functions is minimizing total costs of transportation and searching the matrix $X = (x_{ijk})$: $F(X) = \sum_{i=1}^{I} \sum_{j=1}^{J} \sum_{k=1}^{K} c_{ijk} x_{ijk} \to \min$
with respect to:

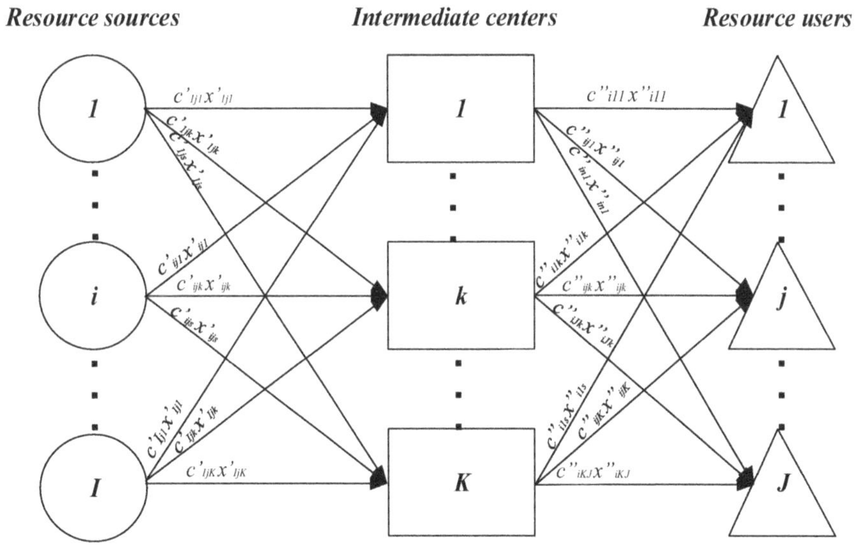

Fig. 1 Interaction between elements of the system "producers—intermediate centers consumers"

$$\sum_{k=1}^{K}\sum_{j=1}^{J} x_{ijk} = a_i, i = 1, 2, \ldots, I$$

,

$$\sum_{k=1}^{K}\sum_{i=1}^{J} x_{ijk} = b_i, j = 1, 2, \ldots, J$$

,

$$\sum_{i=1}^{I}\sum_{j=1}^{J} x_{ijk} = d_k, k = 1, 2, \ldots, k$$

,

$$x_{ijk} \geq 0, i = 1, 2, \ldots, I; j = 1, 2, \ldots, J; k = 1, 2, \ldots, K$$

where d_k is the capacity of the following centers.

There are many analytical and expert methods considering the rational allocation of intermediate centers, they have a number of advantages and disadvantages. As the centers providing distribution of resources for the consumers are assigned to them, the so-called intermediate distribution centers are being investigated in the following works [9, 10] and it was shown that clustering is the most effective method for combining several homogeneous elements, which can be considered as an independent unit with certain properties. When solving the problem of determining the rational location of the intermediate center under the certain cluster, we mean that a set of homogeneous resource consumers are assigned to one center. Clustering involves the allocation of objects that are far from each other, determining the partitioning into the cluster areas of objects. It is used when the initial data are presented in the form of matrices whose elements are the distances between objects or considered as points in a multidimensional space. The idea of constructing the simplest clustering algorithm is represented by the following sequence of steps.

Step 1. Selection of grouping centers.

Step 2. Calculating the distance for all cluster objects to each of the grouping centers.

Step 3. Attaching the processed object to the group whose distance to the center of the selected metric is minimal.

Grouping algorithms differ a lot from each other in the way they select the grouping centers, in the metric in which the distance is calculated, and in the way objects are attached to the formed groups. Depending on the rules for the clusters formation clustering methods for attaching objects (points) can be classified into the following groups [10, 11]:

(1) the first group of methods supposes to associate the next point to the grouping center with the help of calculating the distance from the desired point to each

of the grouping centers; the point joins the center of the grouping class, the distance to which is minimal;

(2) the second group of methods supposes the sum of the distances to all points of each of the clusters that is calculated for each point, and then considers the average distance to the clusters, and the point joins the certain center of the grouping, the mean distance to which is minimal.
(3) the third group of methods establishes a rational order of attaching points, and the first one joins that point and a cluster for which the mean distance is the minimum possible.

The disadvantage of these methods is that distance is the only parameter of determining the belonging to the grouping center. Distance cannot fully characterize the adequacy of the formation of clusters when constructing transport-distribution structure. So it is necessary to develop a method that takes into account the demand of consumption areas and the capacity of intermediate centers to determine the rational association and location of intermediate centers. It should also be taken into account that transportation and distribution systems are characterized by a large number of elements and connections between them, a high degree of dynamism. This leads to the fact that the system operates under conditions of uncertainty of external and internal environment. The uncertainty of the initial parameters of transport-distribution problem is determined by the fluctuations in demand, changes in the costs of delivering resources, distribution conditions, inaccuracy of the data, and can significantly affect the final result. This circumstance predetermines the use of intelligent technologies, such as the theory of fuzzy sets.

3 Method for Object Allocation Based on Fuzzy Variables Estimation

Taking into account the uncertainty of demand while estimating and processing a real data, we can get a solution that minimizes the total cost of moving, and can plan the distribution of resources needed to meet the demand. To determine the rational allocation of the intermediate distribution centers under the requirements to specify the problem variables as fuzzy intervals, the following algorithm is proposed.

Step 1. Set the areas of demand.

Step 2. Set the number of intermediate centers.

Step 3. Determine (by expert way) the initial location of the intermediate centers and the distribution of the resource (power) between the centers.

Step 4. Calculate the potentials of the relationship between demand areas and intermediate centers, based on which consolidate the areas of demand for the centers.

Step 5. Based on the results of the steps, adjust the area of rational allocation of intermediate centers.

Step 6. Solve the problem of determining the cost of transportation for resources from the centers to the areas of consumption assigned to them.

The execution of steps 1–6 corresponds to the first iteration of solving the problem of the rational arrangement of intermediate centers. The subsequent iterations begin with the implementation of the third step, taking into account the decision made at the previous iteration of the location of the intermediate centers and the amount distribution of resources between them. The decision is repeated for a new number of centers, the search procedure provides the problem solution.

The obtained values allow on the next step of the main algorithm to determine for the ith iteration the value $\tilde{F}_i$—as total cost of allocation resources from intermediate centers to areas of demand while consolidating their position. The cycle is completed if the results of grouping (fixing the areas for the centers) before the adjustment and after adjusting coordinates of the location of the intermediate centers coincide.

The set of potentials calculated on the basis of the proposed method allows to determine the degree of assessment of the relationship between resource consumption areas characterized by the demand value with intermediate centers considering capacity and taking into account the distance between them.

Application of the theory of fuzzy sets to determine the areas of grouping, in contrast to deterministic methods, allows one and the same qth area of resource consumption to belong simultaneously to several centers of grouping, but with different degrees of belonging.

When specifying the parameters of the grouping task in the form of crisp values, the matrix of belonging of the qth resource consumption area to the kth clustering center W is defined as follows:

$$W = \left[\begin{array}{c|cccccc} k/q & 1 & . & . & . & . & m \\ \hline 1 & 0 & 0 & 0 & 1 & 0 & 0 \\ \vdots & 0 & 0 & 1 & 0 & 0 & 0 \\ K & 0 & 0 & 0 & 0 & 0 & 1 \end{array} \right],$$

$$W = [\eta_{kq}],\ \eta_{kq} \in \{0,1\}\ k = 1,\ldots,K,\ \ q = 1,\ldots,m,$$

$$\sum_{k=1}^{K} \eta_{qk} = 1,\ k = 1,\ldots,K,\ 0 < \sum_{q=1}^{m} \eta_{qk} < m,\ \ q = 1,\ldots,m, \tag{1}$$

where η_{kq} is a crisp value (0 or 1), that denotes membership of qth resource consumption area to the kth grouping center. So that matrix W consider values η_{kq}, according to which it is possible to say exactly whether the region belongs to the grouping center or not. If the domain is assigned to a center ($\eta_{kq} = 1$), then this area can not belong to another grouping center ($\eta_{kq} = 0$).

When specifying the parameters of the grouping problem in the form of fuzzy values, the membership matrix of the qth resource consumption area to the kth clustering center is defined as follows:

$$\tilde{W} = \left[\begin{array}{c|cccccc} k/q & 1 & . & . & . & . & m \\ \hline 1 & \mu_{11} & \mu_{1q} & \mu_{1q} & \mu_{1q} & \mu_{1q} & \mu_{1m} \\ \vdots & \mu_{k1} & \mu_{kq} & \mu_{kq} & \mu_{kq} & \mu_{kq} & \mu_{km} \\ K & \mu_{K1} & \mu_{Kq} & \mu_{Kq} & \mu_{Kq} & \mu_{Kq} & \mu_{Km} \end{array}\right], \tag{2}$$

$$\tilde{W} = [\mu_{kq}],\ \mu_{kq} \in [0,1],\ k = 1,\ldots,K,\ \ q = 1,\ldots,m,$$

$$\sum_{k=1}^{K} \mu_{kq} = 1,\ k = 1,\ldots,K,\ 0 < \sum_{q=1}^{m} \eta_{qk} < m,\ \ q = 1,\ldots,m,$$

where μ_{kq}—is a fuzzy value that determines if the qth resource consumption area belongs to the kth grouping center or not. Matrix $\tilde{W}$ contains values of the region's belonging to the centers of grouping. Within a fuzzy estimation the consumption area can belong to one grouping center with some degree μ_{kq} and with a degree 1-μ_{kq} can belong to another grouping center.

In the problem of allocating the areas of demand to a specific intermediate distribution center, the distance between the center and the grouping object is important, but we also need to take into account the demand of the consumption area, and the so-called capacity of the distribution center that characterizes its throughput. Considering this, to determine the membership values of belonging resource consumption areas to distribution centers, it is proposed to use the average value of the potentials P_{kq}^{cp}, which is defined as the center of gravity of the trapezoid [12], constructed from the corresponding values P_{kq}. Then the membership function of belonging to the qth resource consumption area to the kth grouping center can be defined as follows:

$$\mu_{kq} = \frac{P_{kq}^{cp}}{\sum_{k=1}^{K} P_{kq}^{cp}}, q = 1, \ldots, m; k = 1, \ldots, K. \tag{3}$$

The implementation of potential values avoids the complexity of the fuzzy partition with respect to regions remote from all clustering centers. The result of determining the values of the membership of consumption areas to the grouping centers is shown in Fig. 2, where:

(a) defines a clear specification of the parameters, according to Eq. 1;
(b) specification of parameters in the form of fuzzy intervals, according to Eq. 2.

Specifying and estimating parameters of the problem in fuzzy values gives us opportunity to describe the initial data and take into account the uncertainties associated with incompleteness and inaccuracy.

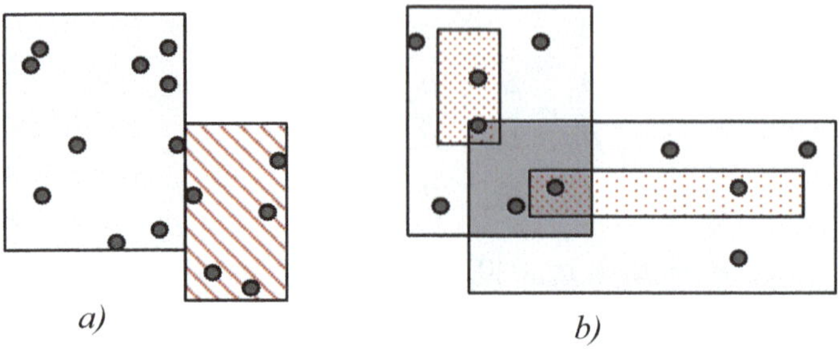

Fig. 2 Result of determining the membership values of consumption areas to the grouping centers

4 Software Realization

A software application was developed in the Delphi 10 Seattle environment that allows input initial data in the form of fuzzy intervals based on which geographic location (coordinates) of the intermediate centers and the total the cost of delivering resources through the centers is estimated (Fig. 3).

The result of solving the problem of allocation intermediate centres is considered as a set of fuzzy intervals—the coordinates of their location, depending on the fuzzy demand of each sub-area.

The allocation of the intermediate distribution centres is determined not by specific numerical values, but by a fuzzy interval area (Fig. 4), which determines the best possible allocation of the intermediate centre.

The developed software application is based on the proposed method and algorithm. It calculates and presents to the user the following analytical parameters:

the distance from each intermediate center to the center of the demand area (region);

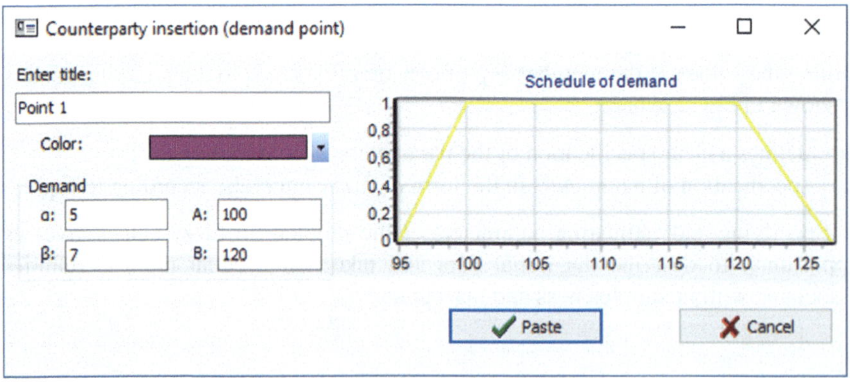

Fig. 3 Specifying demand parameters in the form of a fuzzy interval

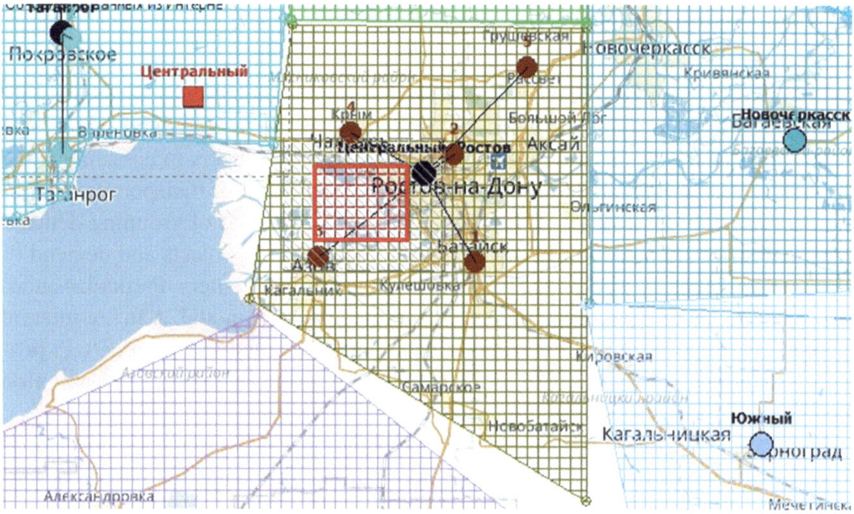

Fig. 4 Area of the best location of the intermediate center

the value of the potentials for relations between the centers of resources accumulation and the regions of demand, based on which the regions of demand are grouped;
the value of membership degree of belonging of the demand area (area) to the intermediate centers.

Using Eq. (3) for each region of demand, one can determine membership degree according to which demand areas belong to each of the intermediate centers. Figure 5 illustrates an example of matrix, where columns from left to right shows distance, potentials and membership function for each object allocation. The left green button on the panel allows to re-calculate coordinates of these objects.

For example, for the region of demand "Novocherkassk" the degree of belonging to the center "Central" is 0.51, and to the center "Vostochny" is 0.49; so the decision should be to fix and assign the area to the intermediate center named "Central".

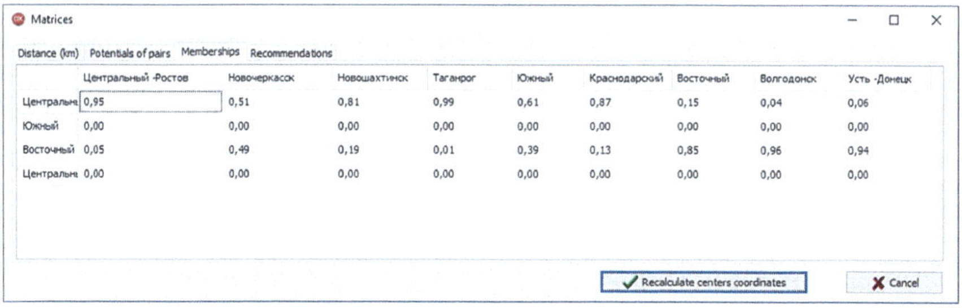

	Центральный -Ростов	Новочеркасск	Новошахтинск	Таганрог	Южный	Краснодарский	Восточный	Волгодонск	Усть -Донецк
Центральны	0,95	0,51	0,81	0,99	0,61	0,87	0,15	0,04	0,06
Южный	0,00	0,00	0,00	0,00	0,00	0,00	0,00	0,00	0,00
Восточный	0,05	0,49	0,19	0,01	0,39	0,13	0,85	0,96	0,94
Центральны	0,00	0,00	0,00	0,00	0,00	0,00	0,00	0,00	0,00

Fig. 5 Values of membership function of belonging the demand area to the intermediate centre

5 Conclusions

An effective way to reduce the cost and time of resource allocation is the availability to introduce and determine intermediate centers. Since the intermediate center is an integral part of the chain of delivering resources from producers to consumers, the problem of its rational allocation is an important component of transportation problem and distribution systems. The paper introduced a method of grouping demand areas that takes into account distance between the grouping objects and demand of consumption areas as well as the capacity of intermediate centers. Particular attention is paid to the uncertainty of the initial parameters of the model. A fuzzy-interval approach is proposed that allows obtaining adequate results without a certain possibility to indicate the magnitude of the demand for a certain resource consumption area.

A software realization was developed. It includes a set of system elements (modules) that allows to input raw data with the help of fuzzy intervals based on geographic location (coordinates) for resource accumulation centers. The total cost of transportation for resources through these centers was calculated. Solution obtained with the proposed method is much more informative comparing with classical methods for solving resource allocation problems. Fuzzy sets theory was implemented for modeling distributive problems and its extensions.

Acknowledgements The reported study was funded by the Russian Foundation for Basic Research according to the research projects N 18-01-00023, N 20-01-00197.

References

1. C. Villani, Topics in optimal transportation. Am. Math. Soc. (2003)
2. Mohamed, S., Mahmoud, M. El-S., Hamdeen, A., Hanan, H.: A Fuzzy Technique for Solving Rough Interval Multiobjective Transportation Problem. Int. J. Comput. Appl. **147**(10), 49–57 (2016)
3. P. Kundu, S. Kar, M. Maiti, Multi-objective multi-item solid transportation problem in fuzzy environment. Appl. Math. Model. **4**(37), 2028–2038 (2013)
4. S. Kikuchi, A method to defuzzify the fuzzy number: transportation problem application. Fuzzy Sets and Sys. **1**(116), 3–9 (2000)
5. O. Kosenko, E. Sinyavskaya, E. Shestova, E. Kosenko, S. Antipin, Method of rational placement of intermediate centers with setting parameters in the form of the fuzzy intervals, in *XIX IEEE International Conference on Soft Computing and Measurements*, (SCM, St. Petersburg, 2016), pp. 186–189
6. O. Kosenko, E. Shestova, E. Sinyavskaya, E. Kosenko, A. Nomerchuk, A. Bozhenyuk, Development of information support for the rational placement of intermediate distribution centers of fuel and energy resources under conditions of partial uncertainty, in *XX IEEE International Conference on Soft Computing and Measurements*, (SCM 2017), pp. 224–227
7. T. Gudehus, H. Kotzab, *Comprehensive Logistics* (Springer, Heidelberg, 2012)
8. G. Ghiani, G. Laporte, R. Musmanno, *Introduction to Logistics Systems Planning and Control* (Wiley, West Sussex, England, 2004)
9. A. Corban, A multidimential transportation problem. Rev. Roum. Appl. **8**(9), 14–27 (1967)

10. B.S. Everitt, S. Landau, M. Leese, D. Stahl, Cluster Analysis. West Sussex, England, 5th edn. (Wiley, 2011)
11. G. Ritter, *Robust Cluster Analysis and Variable Selection* (Chapman and Hall/CRC, N.-Y., 2014)
12. F. Höppner, F. Klawonn, R. Kruse, T. Runkler, *Fuzzy Cluster Analysis* (Wiley, West Sussex, England, 2014)

Outlier Detection in Fuzzy Regressions

Jalal Chachi and S. Mahmoud Taheri

Abstract Existence of outliers among the observation data leads to inaccurate results in modeling. Either elimination or reduction of the outlier data influence is two ways to prevent their negative effect on the modeling. The approach of impact reduction is taken into account in dealing with the mentioned problem in fuzzy regression, where the input is crisp and the output data is fuzzy. The main idea is based on optimizing a weighted target function into fuzzy regression. Some experiments and simulation studies are designed to show its performance in the presence of different kinds of outliers in the data set. The experimental results suggest that the proposed model is capable of dealing with the data set contaminated by outliers and has high prediction accuracy. The proposed fuzzy regression method is capable of determining the weigh of the outlying data points.

1 Introduction

Many revisions have been proposed on fuzzy regression models since it was introduced by Tanaka et al. [1]. Linear programming method [2], and the least-squares [3, 4] are the two classes of fuzzy regression methods. The provided relationships can be led to inaccurate models with the existence of outlier data. Detection and omission of outlier data is an important process that may prevent from obtaining untrustworthy models [5, 6]. Outlier affect both on parameters estimates and on the fit of the model to the majority of the data [10–12]. Therefore, an assessment of potential outliers is important in any analysis especially for formalizing a regression model in a fuzzy

J. Chachi
Faculty of Mathematical Sciences and Computer, Shahid Chamran University of Ahvaz, 61355-83151 Ahvaz, Iran
e-mail: jalal.chachi@gmail.com; jalal.chachi@scu.ac.ir

S. M. Taheri (✉)
School of Engineering Science, College of Engineering, University of Tehran, 1417613131 Tehran, Iran
e-mail: sm_taheri@ut.ac.ir

S. N. Shahbazova et al. (eds.), *Recent Developments and the New Direction in Soft-Computing Foundations and Applications*, Studies in Fuzziness and Soft Computing 393, https://doi.org/10.1007/978-3-030-47124-8_24

domain, in which model parameters and/or data are fuzzy (imprecise or vague) [13]. Outliers could be arisen in a fuzzy regression framework with respect to [14]:

- one or more crisp explanatory variables X;
- the centers and/or the spreads of the fuzzy dependent variable $\widetilde{y}$;
- the fuzzy regression model (but not with respect to X and $\widetilde{y}$);
- both X and centers of $\widetilde{y}$ (but not with respect to the model);
- a combination of the above aspects.

This paper deals with outlier data problem of the non-fuzzy input and fuzzy output models. Outlier data are determined by applying a weighted least-squares estimation method in fuzzy regression to detect, down-wight, omit or lessen their effects. Finally, the results of the numerical examples show that weighted approach is able to neutralize and/or smooth the disruptive effects of possible crisp/fuzzy outliers in the estimation process.

The organization of this paper is as follows. In the next section, preliminary definitions of fuzzy numbers are considered. Proposed method will be introduced in Sect. 3. Numerical examples will be reported in Sect. 4 in order to demonstrate the ability of proposed approach. Conclusion remarks will be pointed out in the last section.

2 Fuzzy Sets and Fuzzy Arithmetic

A fuzzy set $\widetilde{A}$ on the universal set $\mathbb{X}$ is described by its membership function $\widetilde{A}(x) : \mathbb{X} \to [0, 1]$. Through this paper it is assumed that $\mathbb{X} = \mathbb{R}$ (the real line number). A specific type of fuzzy sets is the so-called LR-fuzzy number denoted by $\widetilde{N} = (n, l, r)_{LR}$ with central value $n \in \mathbb{R}$, left and right spread values $l, r \in \mathbb{R}^+$, decreasing left and right shape functions $L, R : \mathbb{R}^+ \to [0, 1]$, with $L(0) = R(0) = 1$. Such LR-fuzzy numbers have the following membership function [15]

$$\widetilde{N}(x) = \begin{cases} L(\frac{n-x}{l}) & \text{if } x \leq n, \\ R(\frac{x-n}{r}) & \text{if } x > n. \end{cases}$$

A special type of LR-fuzzy numbers is the so-called triangular fuzzy numbers $\widetilde{N} = (n, l, r)_T$, with the following membership function

$$\widetilde{N}(x) = \frac{x - (n - l)}{l} \mathcal{I}_{[n-l,n]}(x) + \frac{(n + r) - x}{r} \mathcal{I}_{(n,n+r]}(x),$$

where $x \in \mathbb{R}$ and $\mathcal{I}_A$ stands for the characteristic function of a crisp set A. When $l = r$, the triangular fuzzy number $\widetilde{N}$ is called symmetric and is abbreviated by $\widetilde{N} = (n, l)_T$.

For the algebraic operations on $LR-$fuzzy numbers, we have the following proposition on the basis of Zadeh's Extension Principle (for more details, see [15]).

Proposition 2.1 *Let* $\widetilde{M} = (m, l_m, r_m)_{LR}$ *and* $\widetilde{N} = (n, l_n, r_n)_{LR}$ *be two LR–fuzzy numbers, and* λ *be a real number. Then*

$$\lambda \otimes \widetilde{M} = \begin{cases} (\lambda m, \lambda l_m, \lambda r_m)_{LR} & \text{if } \lambda > 0, \\ \mathcal{I}_{\{0\}} & \text{if } \lambda = 0, \\ (\lambda m, |\lambda| r_m, |\lambda| l_m)_{RL} & \text{if } \lambda < 0, \end{cases}$$
$$\widetilde{M} \oplus \widetilde{N} = (m + n, l_m + l_n, r_m + r_n)_{LR}.$$

3 The New Approach

Suppose we have a data set consists of n observations $(\widetilde{y}_1, \boldsymbol{x}_1), \ldots, (\widetilde{y}_n, \boldsymbol{x}_n)$, where $\widetilde{y}_i = (y_i, l_i, r_i)_{LR}$ $(i = 1, \ldots, n)$ determines the fuzzy observed of the output variable and

$$\boldsymbol{x}_i = [x_{i0}, x_{i1}, x_{i2}, \ldots, x_{ik}] \in \mathbb{R}^{k+1}, \ x_{i0} = 1, \quad i = 1, \ldots, n,$$

are k input real-valued variables. The relationship between $\widetilde{y}$ and $\boldsymbol{x}$ is formulated as follows

$$\begin{aligned} (y, g(l), h(r)) &= \widetilde{\beta}_0 \oplus (\widetilde{\beta}_1 \otimes x_1) \oplus \ldots \oplus (\widetilde{\beta}_k \otimes x_k) \\ &= (\beta_0, \sigma_0, \theta_0) \oplus ((\beta_1, \sigma_1, \theta_1) \otimes x) \oplus \ldots \\ &\qquad \ldots \oplus ((\beta_k, \sigma_k, \theta_k) \otimes x_k) \\ &= (\beta_0 + \beta_1 x + \cdots + \beta_k x_k, \sigma_0 + \sigma_1 |x_1| + \\ &= \ldots + \sigma_k |x_k|, \theta_0 + \theta_1 |x_1| + \cdots + \theta_k |x_k|). \end{aligned}$$

where $g : \mathbb{R}^+ \to \mathbb{R}$ and $h : \mathbb{R}^+ \to \mathbb{R}$ are invertible functions. In this model we do not face the non-negativity constraints of the spreads of the response variable because of introducing the invertible functions g and h [16]. Indeed, using the invertibility of the functions g and h, the estimated fuzzy response variable is obtained as the following model

$$\begin{aligned} \widehat{\widetilde{y}} &= (\widehat{y}, \widehat{l}, \widehat{r})_{LR} \\ &= (\widehat{\beta}_0 + \widehat{\beta}_1 x_1 + \cdots + \widehat{\beta}_k x_k, \\ &\quad g^{-1}\{\widehat{\sigma}_0 + \widehat{\sigma}_1 |x|_1 + \cdots + \widehat{\sigma}_k |x_k|\}, \\ &\quad h^{-1}\{\widehat{\theta}_0 + \widehat{\theta}_1 |x|_1 + \cdots + \widehat{\theta}_k |x_k|\})_{LR}. \end{aligned}$$

Theorem 3.1 *In the problem of estimating the fuzzy parameter* $\widetilde{\boldsymbol{\beta}}$ *in the model*

$$(y, g(l), h(r)) = \widetilde{\beta}_0 \oplus (\widetilde{\beta}_1 \otimes x_1) \oplus \ldots \oplus (\widetilde{\beta}_k \otimes x_k),$$

the weighted least-squares estimator of $\widetilde{\beta}$ is obtained by minimizing the following objective function

$$\begin{aligned} E &= \sum_{i=1}^{n} w_i\, \mathcal{D}^2\Big((y_i, g(l_i), h(r_i)),\\ &\qquad \widetilde{\beta}_0 \oplus (\widetilde{\beta}_1 \otimes x_1) \oplus \ldots \oplus (\widetilde{\beta}_k \otimes x_k)\Big)\\ &= \sum_{i=1}^{n} w_i \left[\left(y_i - \sum_{j=0}^{k} x_{ij}\, \beta_j\right)^2 \right.\\ &\qquad + c_1 \left(g(l_i) - \sum_{j=0}^{k} |x_{ij}|\, \sigma_j\right)^2\\ &\qquad\quad \left. + c_2 \left(h(r_i) - \sum_{j=0}^{k} |x_{ij}|\, \theta_j\right)^2 \right]. \end{aligned}$$

In the above objective function, $w_i \geq 0\,(i = 1, \ldots, n)$ is the weight of the ith residual and must be optimally determined and the constants c_1 and c_2 are defined as $c_1 = \int_0^1 L^{-1}(\alpha)d\alpha$, and $c_2 = \int_0^1 R^{-1}(\alpha)d\alpha$.

4 Outlier Detection: An Example

This section will exhibit the capability of detecting and down-weighting outliers by the proposed approach, and demonstrate how outliers can grossly deteriorate the least-squares estimators of a fuzzy regression model by considering scatter plot. Consider the data set in Table 1 which consist of crisp input x and triangular fuzzy output $\widetilde{y} = (y, l, r)_T$ [17]. Suppose that the theoretic model between variables x and $\widetilde{y}$ is

$$(y, g(l), h(r)) = \widetilde{\beta}_0 \oplus (\widetilde{\beta}_1 \otimes x) \oplus (\widetilde{\beta}_2 \otimes x^2).$$

Consider $g(\cdot) = h(\cdot) = Ln(\cdot)$ in the solved numerical example. By applying the proposed approach to the data set in Table 1, the following weighted fuzzy regression model is obtained

$$\begin{aligned} \widehat{\widetilde{y}}_w &= (15.03 + 54.90\,x - 7.27\,x^2,\\ &\qquad \exp\{1.20 + 1.45\,x - 0.32\,x^2\},\\ &\qquad\quad \exp\{1.15 + 1.54\,x - 0.33\,x^2\}). \end{aligned}$$

Table 1 Data set

i	$(x_i, \widetilde{y}_i)$	i	$(x_i, \widetilde{y}_i)$
1	$(0.0; (11.5, 3, 2.5)_T)$	9	$(1.6; (84, 15, 16)_T)$
2	$(0.2; (24.8, 4.5, 4)_T)$	10	$(1.8; (82, 15, 16)_T)$
3	$(0.4; (40, 6, 7)_T)$	11	$(2.0; (\mathbf{33, 3, 5})_T)$
4	$(0.6; (45.2, 7, 7)_T)$	12	$(2.2; (\mathbf{150, 80, 30})_T)$
5	$(0.8; (49.1, 9, 9)_T)$	13	$(2.4; (103.1, 16, 17)_T)$
6	$(1.0; (70, 11, 12)_T)$	14	$(2.6; (111, 17, 19)_T)$
7	$(1.2; (70.9, 12, 12)_T)$	15	$(2.8; (109.1, 17, 19)_T)$
8	$(1.4; (80.1, 14, 15)_T)$	16	$(3.0; (\mathbf{160, 8, 30})_T)$

The above model is shown in Fig. 1, where the horizontal axis represents the value of the independent variable x and the vertical axis represents the upper value $(y + r)$, the center value (y) and the lower value $(y - l)$ of the triangular fuzzy number $\widetilde{y} = (y, l, r)_T$. As shown in Fig. 1, the proposed estimation procedure can determine the optimal weight of any observation which are obtained and shown for each point in Fig. 1.

From the results given in Fig. 1 it is clear that the $\widetilde{y}_{11} = (33, 3, 5)$, $\widetilde{y}_{12} = (150, 80, 30)$ and $\widetilde{y}_{16} = (160, 8, 30)$ are detected as outliers. According to the estimation procedure the smallest estimated optimal weight values are obtained for these points which are $w_{11} = 0.01$, $w_{12} = 0.01$ and $w_{16} = 0.007$.

4.1 A Competitive Study

In the sequel, in order to provide a comparative study and to investigate the performance of the weighted fuzzy regression model in the presence of outliers a least-squares fuzzy regression model will be considered (called as LS method). In the LS method all of the observations have the same weight $w_i = 1$ for $i = 1, \ldots, 16$ in the iteratively reweighted algorithm. Indeed, for the LS method the iteratively reweighted algorithm is an algorithm with fixed values of $w_i = 1$ for $i = 1, \ldots, 16$. Now, by applying the LS method to the given data set the following least-squares fuzzy regression model is obtained

$$\widehat{\widetilde{y}}_{LS} = (21.37 + 33.69\,x + 1.47\,x^2,\\ \exp\{1.16 + 1.53\,x - 0.36\,x^2\},\\ \exp\{1.24 + 1.19\,x - 0.19\,x^2\})$$

The above model is shown in Fig. 2.

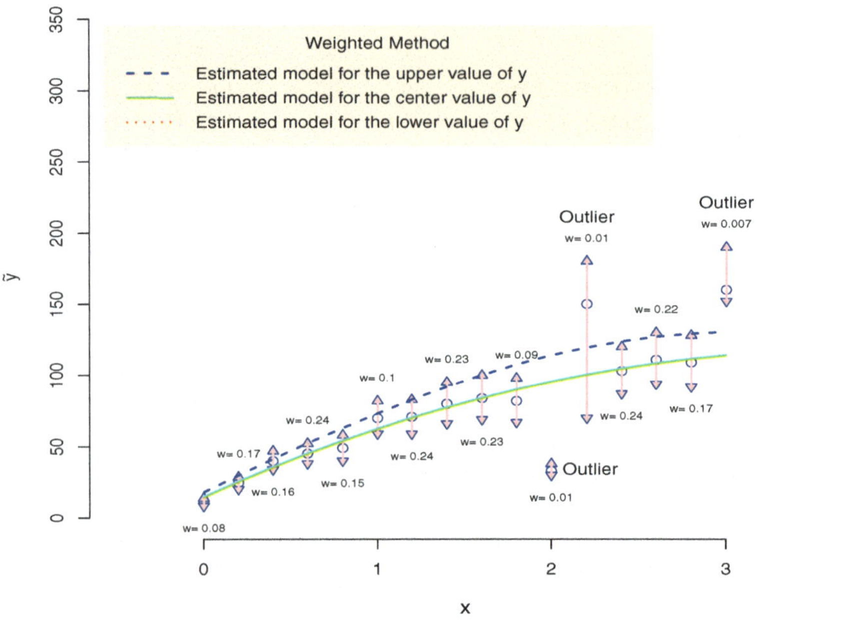

Fig. 1 The estimated weighted fuzzy regression model for the data set in Table 1

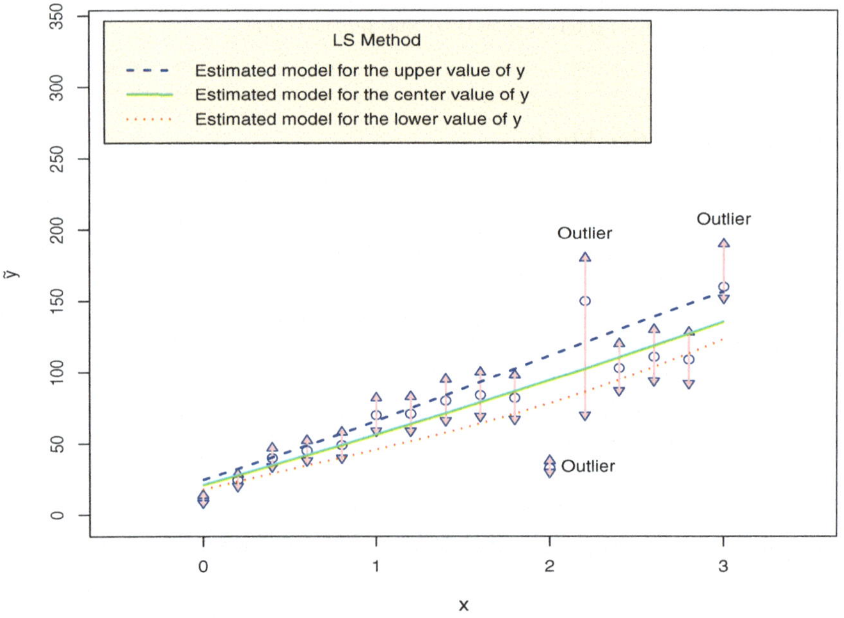

Fig. 2 The estimated LS model for the data set in Table 1

Table 2 Comparison between fuzzy regression models $\widehat{\widetilde{y}}_w$ and $\widehat{\widetilde{y}}_{LS}$

Models	*MSM*	*MAE*
$\widehat{\widetilde{y}}_w$	0.50	11.09
$\widehat{\widetilde{y}}_{LS}$	0.28	16.02

We employ two criteria known as Mean of Similarity Measures (*MSM*) and Mean of Absolutes Errors (*MAE*) which have been widely used for evaluation of fuzzy regression models [3, 5, 18]. These criteria are defined as follows:

$$MSM = \frac{1}{n}\sum_{i=1}^{n}\frac{\int \min\{\widetilde{y}_i(t), \widehat{\widetilde{y}}_i(t)\}\,dt}{\int \max\{\widetilde{y}_i(t), \widehat{\widetilde{y}}_i(t)\}\,dt},$$

$$MAE = \frac{1}{n}\sum_{i=1}^{n}\int |\widetilde{y}_i(t) - \widehat{\widetilde{y}}_i(t)|\,dt.$$

To compare the performances of fuzzy regression models $\widehat{\widetilde{y}}_w$ and $\widehat{\widetilde{y}}_{LS}$, the criteria *MSM* and *MAE* are adopted to calculate the accuracy of the models in estimating the observed responses. The corresponding values of these criteria are listed in Table 2 which are almost the same for the two models $\widehat{\widetilde{y}}_w$ and $\widehat{\widetilde{y}}_{LS}$. By these results, the *MSM* value for model $\widehat{\widetilde{y}}_w$ is 0.50, which is greater than 0.28, the *MSM* value for model $\widehat{\widetilde{y}}_{LS}$. On the other hand, the *MAE* value for the proposed method is 11.09 which is clearly smaller than that of 16.02 calculated from the *LS* method. Therefore, like to the conclusion taken by the scatter plot criteria, it can be concluded the estimated weighted fuzzy regression model $\widehat{\widetilde{y}}_w$ fits much better than the estimated least-squares fuzzy regression model $\widehat{\widetilde{y}}_{LS}$ to the data set.

After removing the down-weighted observations $\widetilde{y}_{11} = (33, 3, 5)$, $\widetilde{y}_{12} = (150, 80, 30)$ and $\widetilde{y}_{16} = (160, 8, 30)$ from the data set, the proposed weighted model and least-squares model to this data set are obtained as follows

$$\begin{aligned}\widehat{\widetilde{y}}_w &= (13.91 + 57.76x - 8.44x^2,\\ &\quad \exp\{1.20 + 1.47x - 0.32x^2\},\\ &\quad \exp\{1.10 + 1.63x - 0.36x^2\})\\ \widehat{\widetilde{y}}_{LS} &= (14.22 + 56.60x - 8.03x^2,\\ &\quad \exp\{1.19 + 1.47x - 0.32x^2\},\\ &\quad \exp\{1.09 + 1.65x - 0.36x^2\}).\end{aligned}$$

The goodness-of-fit criteria for the above models are calculated in Table 3. The results show that there is a much improvement of the fitness of the fuzzy regression models $\widehat{\widetilde{y}}_w$ and $\widehat{\widetilde{y}}_{LS}$ when the down-weighted observations are omitted.

Table 3 Comparison between fuzzy regression models $\widehat{\tilde{y}}_w$ and $\widehat{\tilde{y}}_{LS}$ when the down-weighted observations are omitted

Models	*MSM*	*MAE*
$\widehat{\tilde{y}}_w$	0.6318	5.1217
$\widehat{\tilde{y}}_{LS}$	0.6335	5.0736

5 Conclusion

The existence of outliers in a set of experimental data can cause incorrect interpretation of the fuzzy linear regression results. The present investigation focused on this problem in a fuzzy regression model for the crisp input and fuzzy output data type and proposed weighted approach to handle the outlier problem. In this regard, a weighted robust fuzzy regression analysis have been developed as an improvement to least-squares estimation in the presence of outliers and to provide us information about what a valid observation is and whether this should be thrown out or down-weighted. The advantages of the proposed approach in contrast with least-squares estimation were compared and discussed in the presence of outliers. This approach not only can detect abnormal values, it can also reduce the impact of these points on the estimation problem by down-weighting them.

References

1. H. Tanaka, S. Uegima, K. Asai, Linear regression analysis with fuzzy model. IEEE Trans. Sys Man Cybern **12**, 903–907 (1982)
2. F. Chen, Y. Chen, J. Zhou, Y. Liu, Optimizing h value for fuzzy linear regression with asymmetric triangular fuzzy coefficients. Eng. Appl. Artif. Intel. **47**, 16–24 (2016)
3. J. Chachi, S.M. Taheri, H.R. Pazhand, Suspended load estimation using l_1-fuzzy regression, l_2-fuzzy regression and mars-fuzzy regression models. Hydrol. Sci. J. **61**(8), 1489–1502 (2016)
4. J. Chachi, S.M. Taheri, S. Fattahi, S.A.H. Ravandi, Two robust fuzzy regression models and their application in predicting imperfections of cotton yarn. J. Text. Polym. **4**(2), 60–68 (2016)
5. J. Chachi, M. Roozbeh, A fuzzy robust regression approach applied to bedload transport data. Commun. Stat. Simul. Comput. **47**(3), 1703–1714 (2017)
6. H. Shakouri, R. Nadimi, Outlier detection in fuzzy linear regression with crisp input-output by linguistic variable view. Appl. Soft Comput. **13**, 734–742 (2013)
7. R. Andersen, *Modern Methods for Robust Regression* (Sage, Thousand Oaks, CA, 2007)
8. P.J. Rousseeuw, A.M. Leroy, *Robust Regression and Outlier Detection* (Wiley, Hoboken, NJ, 1987)
9. S. Varga, Robust estimations in classical regression models versus robust estimations in fuzzy regression models. Kybernetika **43**, 503–508 (2007)
10. Y.S. Chen, Outliers detection and confidence interval modification in fuzzy regression. Fuzzy Sets Syst. **119**, 259–272 (2001)
11. W.L. Hung, M.S. Yang, An omission approach for detecting outliers in fuzzy regressions models. Fuzzy Sets Syst. **157**, 3109–3122 (2006)
12. T.D. Nguyen, R. Welsch, Outlier detection and least trimmed squares approximation using semi-definite programming. Comput. Stat. Data Anal **54**, 3212–3226 (2010)

13. J.M. Leski, M. Kotas, On robust fuzzy c-regression models. Fuzzy Sets Syst. **279**, 112–129 (2015)
14. P. D'Urso, R. Massari, A. Santoro, Robust fuzzy regression analysis. Inf. Sci. **181**, 4154–4174 (2011)
15. G. Klir, B. Yuan, *Fuzzy Sets and Fuzzy Logic: Theory and Applications* (Prentice-Hall, Taiwan, 2002)
16. P. D'Urso, R. Massari, Weighted least squares and least median squares estimation for the fuzzy linear regression analysis. Metron **71**, 279–306 (2013)
17. W. Zeng, Q. Feng, J. Lia, Fuzzy least absolute linear regression. Appl. Soft Comput. **52**, 1009–1019 (2016)
18. M. Arefi, S.M. Taheri, Least squares regression based on atanassov's intuitionistic fuzzy inputs-outputs and atanassov's intuitionistic fuzzy parameters. IEEE Trans. Fuzzy Syst. **23**, 1142–1154 (2015)

A Robust Variable-Spread Fuzzy Regression Model

S. Mahmoud Taheri and Jalal Chachi

Abstract A novel regression model with variable spreads fuzzy error is proposed for crisp explanatory and fuzzy response observations to tackle the spreads increasing problem. It can also cope with the situations of increasing, decreasing, constant or variable spreads. The coefficients of the model are estimated by using a least absolutes method. Then, based on minimizing the difference of membership values between the observed and estimated response variable, the fuzzy error terms are estimated. The results from comparative examples, based on some well-known data sets, show effectiveness of the proposed model.

1 Introduction

Most of researches on the topic of fuzzy regression analysis can be classified in two categories: (I) the possibilistic regression, and (II) the fuzzy least-squares (LS) regression. The first one, using the possibilistic concepts, formulates the problem of estimating the parameters as a linear/non-linear programming problem (e.g. [5, 16, 19]). The second one, which is a generalization of the LS regression modeling, is based on a metric on the space of fuzzy numbers (e.g. [1, 3, 4, 7, 9–12, 14, 17]).

Several researches, however, have pointed to some drawbacks of these methods (see [6, 15]), such as

S. M. Taheri (✉)
School of Engineering Science, College of Engineering, University of Tehran, 1417613131 Tehran, Iran
e-mail: sm_taheri@ut.ac.ir

J. Chachi
Department of Mathematical Sciences and Computer, Shahid Chamran University of Ahvaz, 61355-83151 Ahvaz, Iran
e-mail: jalal.chachi@gmail.com; jalal.chachi@scu.ac.ir

S. N. Shahbazova et al. (eds.), *Recent Developments and the New Direction in Soft-Computing Foundations and Applications*, Studies in Fuzziness and Soft Computing 393, https://doi.org/10.1007/978-3-030-47124-8_25

1. both possibilistic and fuzzy LS methods are sensitive to outliers,
2. they cannot avoid spread increasing problem, i.e. the spreads of the estimated dependent variable become wider as the magnitudes of the independent variables increase.

There have been a lot of attempts to avoid the above problems. Especially, to avoid the spread increasing problem, some methods were investigated in which the coefficients of the model were crisp rather than fuzzy numbers and/or a fuzzy error term was added to the model in order to handle the fuzziness of the model (see, for example, [2, 6, 10, 11, 15]).

In the present article, we propose a robust model to avoid both two problems mentioned above. The method, which is somehow similar to the method proposed by Chen and Dang [6], is a two-phase method to obtain the crisp coefficients and the fuzzy error terms. In the first phase, the coefficients are estimated by applying the LAD (Least Absolute Deviations) method. In the second phase, the fuzzy error term is estimated for each instance, by minimizing the area between the membership functions of the estimated and observed response variable, provided that their spread values are equal. Briefly, the proposed method has two advantages simultaneously:

1. It is not sensitive to outliers, and
2. It resolves the spreads increasing problem.

In the next section, some concepts of fuzzy set theory will be recalled. In Sect. 3, we introduce a two-phase optimization method which provides a variable-spread fuzzy multiple regression model for crisp input-fuzzy output data. In Sect. 4, we explain the method of forecasting of the proposed model via a fuzzy inference system. Based on a couple of data sets, comparative studies are provided with some well-known recent methods in Sect. 5. Section 6 gives some concluding remarks.

2 Fuzzy Sets and Fuzzy Arithmetic

A fuzzy set $\widetilde{A}$ on the universal set $\mathbb{X}$ is described by its membership function $\widetilde{A}(x) : \mathbb{X} \to [0, 1]$. Through this paper it is assumed that $\mathbb{X} = \mathbb{R}$ (the real numbers). A specific type of fuzzy set is the so-called LR-fuzzy number denoted by $\widetilde{N} = (n, l, r)_{LR}$ with central value $n \in \mathbb{R}$, left and right spread values $l, r \in \mathbb{R}^+$, decreasing left and right shape functions $L, R : \mathbb{R}^+ \to [0, 1]$, with $L(0) = R(0) = 1$. Such LR-fuzzy numbers have the following membership function [13]

$$\widetilde{N}(x) = \begin{cases} L(\frac{n-x}{l}) & \text{if } x \leq n, \\ R(\frac{x-n}{r}) & \text{if } x > n. \end{cases}$$

A special type of such numbers is the so-called triangular fuzzy numbers $\widetilde{N} = (n, l, r)_T$, with the following membership function

$$\widetilde{N}(x) = \frac{x-(n-l)}{l}\mathcal{I}_{[n-l,n]}(x) + \frac{(n+r)-x}{r}\mathcal{I}_{(n,n+r]}(x),$$

where $x \in \mathbb{R}$ and $\mathcal{I}_A$ stands for the characteristic function of a crisp set A. When $l = r$, the triangular fuzzy number $\widetilde{N}$ is called symmetric and is abbreviated by $\widetilde{N} = (n, l)_T$.

For algebraic operations on $LR-$fuzzy numbers, we have the following proposition on the basis of Zadeh's Extension Principle (for more details, see [13]).

Proposition 2.1 *Let $\widetilde{M} = (m, l_m, r_m)_{LR}$ and $\widetilde{N} = (n, l_n, r_n)_{LR}$ be two $LR-$fuzzy numbers, and λ be a real number. Then*

$$\lambda \otimes \widetilde{M} = \begin{cases} (\lambda m, \lambda l_m, \lambda r_m)_{LR} & \text{if } \lambda > 0, \\ \mathcal{I}_{\{0\}} & \text{if } \lambda = 0, \\ (\lambda m, |\lambda| r_m, |\lambda| l_m)_{RL} & \text{if } \lambda < 0, \end{cases}$$

$$\widetilde{M} \oplus \widetilde{N} = (m+n, l_m + l_n, r_m + r_n)_{LR}.$$

3 The Proposed Fuzzy Regression Model

The considered model in this paper is expressed as follows

$$\widetilde{y}_i = \beta_0 + \beta_1 x_{1i} + \cdots + \beta_k x_{ki} \oplus \widetilde{\varepsilon}_i, \qquad i = 1, \ldots, n,$$

where $\widetilde{y}_i = (y_i, y_i^l, y_i^r)_{LR}$ is LR-type fuzzy response, β_j, $j = 0, 1, \ldots, k$, are the crisp regression coefficients, $\mathbf{x}_i = [x_{0i}, x_{1i}, \ldots, x_{ki}] \in \mathbb{R}^{k+1}$ $(i = 1, \ldots, n; k < n;$ $x_{0i} = 1)$ forms the $(k+1)$-dimensional vector of the crisp independent variables, and $\widetilde{\varepsilon}_i = (0, l_i, r_i)_{LR}$ is the fuzzy error term associated with the ith observation.

In order to estimate the above fuzzy regression model, a two-phase procedure is introduced in this paper. In the first phase, the crisp regression coefficients are estimated, and in the second phase, the fuzzy error terms are determined by a mathematical programming problem.

Phase 1. The regression coefficients β_j, $j = 0, 1, \ldots, k$, are estimated by modeling the centers of fuzzy dependent variable on the independent variable $\mathbf{x}$, through the following model

$$y_i = \beta_0 + \beta_1 x_{1i} + \cdots + \beta_k x_{ki} + \varepsilon_i, \qquad i = 1, \ldots, n.$$

By using the LAD method, to estimate the regression coefficient $\boldsymbol{\beta} = [\beta_0, \beta_1, \ldots, \beta_k]^t$, we need to minimize the sum of absolute values of residuals, i.e.

$$\min_{\beta} \sum_{i=1}^{n} |e_i| = \min_{\beta} \sum_{i=1}^{n} \left| y_i - \sum_{j=0}^{k} x_{ji}\beta_j \right|.$$

To make such a minimization problem easy, we employ non negative variables e_i^+ and e_i^-, $i = 1, \ldots, n$, such that

$$\left| y_i - \sum_{j=0}^{k} x_{ji}\beta_j \right| = e_i^+ + e_i^-, \quad y_i - \sum_{j=0}^{k} x_{ji}\beta_j = e_i^+ - e_i^-.$$

Let $\boldsymbol{e}_{n\times 1}^+ = [e_1^+, \ldots, e_n^+]^t$, $\boldsymbol{e}_{n\times 1}^- = [e_1^-, \ldots, e_n^-]^t$, and

$$\begin{aligned}
\boldsymbol{e}_{(k+1+2n)\times 1} &= [(\boldsymbol{\beta}_{(k+1)\times 1})^t \; (\boldsymbol{e}_{n\times 1}^+)^t \; (\boldsymbol{e}_{n\times 1}^-)^t]^t, \\
\mathbf{H}_{n\times(k+1+2n)} &= [\mathbf{X}_{n\times(k+1)} \; \mathbf{I}_{n\times n} \; -\mathbf{I}_{n\times n}], \\
\boldsymbol{h}_{(k+1+2n)\times 1} &= [\mathbf{0}_{1\times(k+1)} \; \boldsymbol{J}_{1\times 2n}]^t,
\end{aligned}$$

where $\mathbf{I}_{n\times n}$ is the identity matrix of order n, and $\boldsymbol{J}$ denotes a vector of ones. Now, the above minimization problem could be formulated by the following linear programming problem

$$\begin{aligned}
\min_{\boldsymbol{e}_{(k+1+2n)\times 1}} \quad & \boldsymbol{h}^t_{(k+1+2n)\times 1}\boldsymbol{e}_{(k+1+2n)\times 1} \\
\text{s.t.} \quad & \mathbf{H}_{n\times(k+1+2n)}\boldsymbol{e}_{(k+1+2n)\times 1} = \mathbf{y}_{n\times 1}, \\
& \boldsymbol{e}_{n\times 1}^+ \in \mathbb{R}^{+^n}, \; \boldsymbol{e}_{n\times 1}^- \in \mathbb{R}^{+^n}, \; \boldsymbol{\beta}_{(k+1)\times 1} \in \mathbb{R}^{k+1}.
\end{aligned}$$

By solving this problem, the crisp coefficients of the model can be estimated, and so the centers of the fuzzy responses are estimated by

$$\widehat{y} = \widehat{\beta}_0 + \widehat{\beta}_1 x_1 + \cdots + \widehat{\beta}_k x_k.$$

Phase 2. To provide a model with variable spread, we estimate the fuzzy error term $\widetilde{\varepsilon}_i = (0, l_i, r_i)_{LR}$ for each i, $i = 1, \ldots, n$. Each fuzzy error term $\widetilde{\varepsilon}_i$ should be estimated such that the spread of $\widetilde{\varepsilon}_i$, i.e. $l_i + r_i$, be equal to the observed spread of $\widetilde{y}_i$, i.e. $y_i^l + y_i^r$. The spreads of $\widetilde{\varepsilon}_i$ are estimated by minimizing the total errors subject to the equality of spreads.

The optimization problem Based on the above explanation, we are led to an optimization problem with the following objective function

$$\sum_{i=1}^{n} \int |\widetilde{y}_i(x) - \widehat{\widetilde{y}}_i(x)|\, dx.$$

The constraints are such that the spreads of each estimated response variable are equal to those of the observed response variable, i.e. $l_i + r_i = y_i^l + y_i^r$, $i = 1, \ldots, n$. To make such constraints, let y_m^l and y_m^r be the smallest left and right spreads of the observed responses, respectively, i.e. $y_m^l = \min\{y_1^l, \ldots, y_n^l\}$, and

$y^r_m = \min\{y^r_1, \ldots, y^r_n\}$. Then, we consider part of the estimated spread of the response variable as the fixed value $y^l_m + y^r_m$. Note that

$$D_i = (y^l_i + y^r_i) - (y^l_m + y^r_m) \geq 0, \qquad i = 1, \ldots, n.$$

Therefore, this minimal spread needs to be expanded in both sides to become equal to the spread of the observed fuzzy response. To do this, the value of D_i is divided into two values d_i and $D_i - d_i$, where $0 \leq d_i \leq D_i$. Consequently, the optimization problem which yields the optimal values of $\widehat{d}_1, \ldots, \widehat{d}_n$, is obtained by solving the following problem

$$\begin{aligned}
\min_{d_1,\ldots,d_n} \quad & \sum_{i=1}^{n} \int |\widetilde{y}_i(x) - \widehat{\widetilde{y}}_i(x)|\, dx, \\
\text{s.t.} \quad & \widehat{\widetilde{y}}_i = \widehat{\beta}_0 + \widehat{\beta}_1 x_{1i} + \cdots + \widehat{\beta}_k x_{ki} \oplus \widetilde{\varepsilon}_i, \\
& \widetilde{\varepsilon}_i = (0, y^l_m + d_i, y^r_m + D_i - d_i)_{LR}, \\
& 0 \leq d_i \leq D_i, \qquad i = 1, \ldots, n.
\end{aligned}$$

On the basis of the optimal values of $\widehat{d}_1, \ldots, \widehat{d}_n$, the spreads of each estimated response is derived and so the optimal fuzzy regression model is obtained as follows

$$\begin{aligned}
\widehat{\widetilde{y}}_i &= \widehat{\beta}_0 + \widehat{\beta}_1 x_{1i} + \cdots + \widehat{\beta}_k x_{ki} \oplus \widehat{\widetilde{\varepsilon}}_i \\
&= \widehat{\beta}_0 + \widehat{\beta}_1 x_{1i} + \cdots + \widehat{\beta}_k x_{ki} \\
&\quad \oplus (0, y^l_m + \widehat{d}_i, y^r_m + D_i - \widehat{d}_i)_{LR}, \quad i = 1, \ldots, n.
\end{aligned}$$

Evaluating the model In the present study, we use two well-known criteria to evaluate the performances of the proposed method.

(I) The error in estimation: This criterion is defined as follows [12]

$$E_{1i} = \int |\widetilde{y}_i(t) - \widehat{\widetilde{y}}_i(t)|\, dt, \qquad i = 1, \ldots, n.$$

This index measures the difference between the estimated and observed membership values of response variable.

(II) The similarity measure: The similarity of fuzzy numbers is also used as a measurement to evaluate the effectiveness of fuzzy regression model. This index is defined as follows

$$S_i = \frac{\int \min\{\widetilde{y}_i(t), \widehat{\widetilde{y}}_i(t)\}\, dt}{\int \max\{\widetilde{y}_i(t), \widehat{\widetilde{y}}_i(t)\}\, dt}, \qquad i = 1, \ldots, n.$$

4 Forecasting via Fuzzy Inference System

By estimating $\widehat{\beta}$ and $\widehat{d}_1, \ldots, \widehat{d}_n$, the proposed fuzzy regression model becomes

$$\widehat{\widetilde{y}}_i = \widehat{\beta}_0 + \widehat{\beta}_1 x_{1i} + \cdots + \widehat{\beta}_k x_{ki} \oplus (0, y^l_m + \widehat{d}_i, y^r_m + D_i - \widehat{d}_i)_{LR},$$

which can be adopted to forecast the associated response for the new observations of the independent variables. By inception of the procedure introduced in [6], we introduce our forecasting method via fuzzy inference system (see also Fig. 1). Here, the simplest case of the Mamdani fuzzy model with one input-one output is adopted for deriving the predicted fuzzy error terms [13].

Let $\mathbf{x}^* = [1, x^*_1, \ldots, x^*_k]$ denotes the new specific value of the independent variable, $Y = \widehat{\beta}_0 + \widehat{\beta}_1 x^*_1 + \cdots + \widehat{\beta}_k x^*_k$ denotes the estimated center of the response obtained from the model, and $\widehat{\widetilde{\epsilon}}$ denotes its estimated error term. Suppose that the observed responses and their associated error terms activated by $\mathbf{x}^*$ are $(\widetilde{y}_a, \widehat{\widetilde{\varepsilon}}_a)$, $a \in A$, where $A = \{i : \widetilde{y}_i(Y) > 0,\ i = 1, \ldots, n\}$. The form of the ath if-then rule is represented as follows

$$R^a : \text{IF } Y \text{ is } \widetilde{y}_a \text{ THEN } \widehat{\widetilde{\epsilon}} \text{ is } \widehat{\widetilde{\varepsilon}}_a,$$

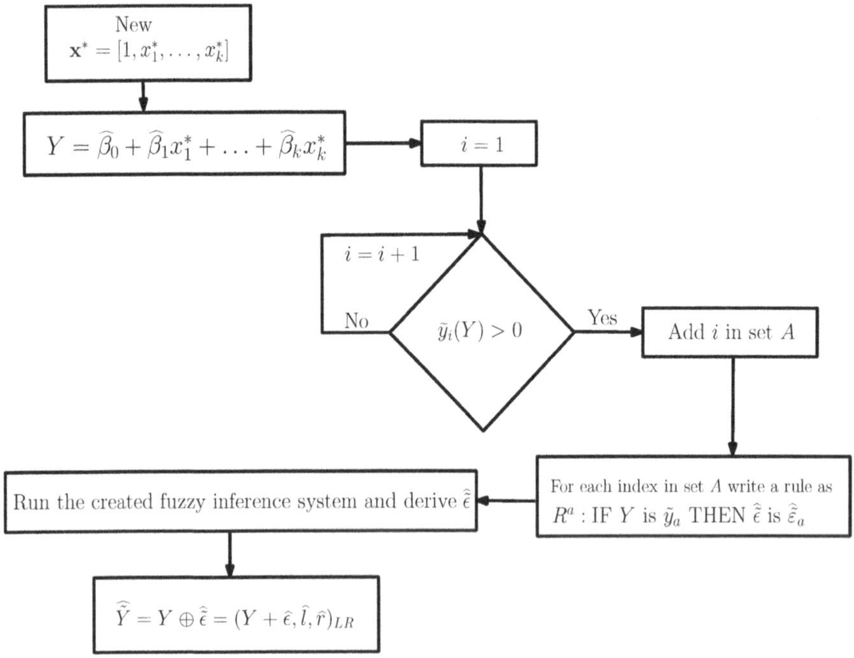

Fig. 1 The flowchart of the forecasting procedure

where R^a is the ath rule for $a \in A$. For the value Y, each output membership function is activated to arrive at the membership grade $\widetilde{y}_a(Y)$, and the predicted error term associated to Y, the fuzzy set created by the superposition of individual rule actions, can be obtained.

Note that, while all observations are LR-fuzzy numbers, the fuzzy set of the predicted error term could be any irregular shapes. But, it is suitable that the predicted error term is a LR-fuzzy number which makes the obtained estimated response a LR-fuzzy number, too. To achieve this, the obtained predicted error term $\widehat{\widetilde{\epsilon}}$ should be transformed into a LR-fuzzy number. Finally, we consider the transformed predicted error term as $\widehat{\widetilde{\epsilon}} = (\widehat{\epsilon}, \widehat{l}, \widehat{r})_{LR}$, where $\widehat{\epsilon}$ is the defuzzified value of the predicted error term calculated by center of gravity, $\widehat{l}$ and $\widehat{r}$ are set as the minimum and maximum of the possible values of the predicted error term, respectively (see also [6]).

Thus, the estimated response for new independent variable $\mathbf{x}^*$ becomes

$$\widehat{\widetilde{Y}} = Y \oplus \widehat{\widetilde{\epsilon}} = (Y + \widehat{\epsilon}, \widehat{l}, \widehat{r})_{LR}.$$

5 Competitive Studies

In this section, the proposed method is applied on a couple of data sets to show its performances. The numerical examples of some simple and multiple fuzzy regression models provide comparative studies with the model proposed by Chen and Dang [6] which will be indicated by VS(LS) model.

Note that, they compared the performances of their method with several common methods in fuzzy regression analysis to show that their model provided higher explanatory power than the compared methods. In the sequel, we compare our model (indicated by VS(LA)) with their model to provide comparative studies and to show that our proposed model provides higher explanatory power than the model in [6].

Example 5.1 Consider the crisp input-fuzzy output data set in Table 1, given by Tanaka et al. [19]. The observations of the dependent variable $\widetilde{y}$ are symmetric triangular fuzzy numbers. By applying the proposed approach described in Sect. 2, as well as VS(LS), the following fuzzy regression models are derived

$$\begin{aligned}\widehat{\widetilde{y}}_i^{\text{VS(LA)}} &= 6.44 + 1.31\,x_i \oplus (0, 1.8 + \widehat{d}_i, 1.8 + D_i - \widehat{d}_i)_T,\\ \widehat{\widetilde{y}}_i^{\text{VS(LS)}} &= 4.95 + 1.71\,x_i \oplus \widehat{\widetilde{\varepsilon}}_i,\end{aligned}$$

where, the values of $\widehat{d}_i$ and $\widehat{\widetilde{\varepsilon}}_i$, $i = 1, \ldots, 5$, for the proposed model and VS(LS) model are shown in Table 1.

The sum of similarity measures for the proposed model is $S = 3.00$, which is greater than that of VS(LS) model, i.e. $S = 2.59$. The total error in estimation for the proposed model is $E = 7.02$, which is smaller than $E = 7.85$ for VS(LS) model. The results are summarized in Table 2.

Table 1 Data set and the values of $\widehat{d}_i$ and $\widehat{\varepsilon}_i$ in Example 5.1

i	$(\widetilde{y}_i; x_i)$	$\widehat{d}_i$	$\widehat{\varepsilon}_i$
1	$((8.0, 1.8)_T; 1)$	0.0	$(0, 1.8, 1.8)_T$
2	$((6.4, 2.2)_T; 2)$	0.8	$(0, 2.6, 1.8)_T$
3	$((9.5, 2.6)_T; 3)$	1.6	$(0, 3.4, 1.8)_T$
4	$((13.5, 2.6)_T; 4)$	0.0	$(0, 1.8, 3.4)_T$
5	$((13.0, 2.4)_T; 5)$	0.6	$(0, 3.0, 1.8)_T$

Table 2 The similarity measures and estimated errors for the models in Example 5.1

	Proposed method		Chen and Dang's method [6]	
i	S_i	E_i	S_i	E_i
1	0.76	0.47	0.24	2.18
2	0.12	3.45	0.23	2.73
3	0.72	0.82	0.84	0.42
4	0.39	2.27	0.41	2.13
5	1.00	0.00	0.85	0.38
Sum	3.00	7.02	2.59	7.85

Example 5.2 (Example V.1 Continued: Forecasting accuracy) Here, we exactly follow a method similar to that of Chen and Dang [6] to illustrate the forecasting accuracy of the proposed model. To illustrate the forecasting accuracy of the two methods, suppose that the fourth pair of observations, i.e. $(\widetilde{y}_4; x_4) = ((13.5, 2.6)_T; 4)$, is unknown and we would like to use the other four pairs of observations to predict it. By applying the proposed method, the constructed model is

$$\widehat{\widetilde{y}}_i^{\mathrm{VS(LA)}} = 3.5 + 2x_i \oplus (0, 1.8 + \widehat{d}_i, 1.8 + D_i - \widehat{d}_i)_T,$$

where, $i = 1, 2, 3, 5$, $\widehat{d}_1 = 0$, $\widehat{d}_2 = 0.8$, $\widehat{d}_3 = 0.8$, and $\widehat{d}_5 = 1.2$. For the independent observation $x_4 = 4$, the corresponding estimated center is $\widehat{y}^4 = 3.5 + (2 \times 4) = 11.5$. Since

$$\begin{aligned} &\widetilde{y}_1(11.5) = 0, && \widetilde{y}_2(11.5) = 0, \\ &\widetilde{y}_3(11.5) = 0.23, && \widetilde{y}_5(11.5) = 0.37, \end{aligned}$$

therefore, $A = \{3, 5\}$. Thus, in the Mamdani fuzzy inference system, the following two error terms are activated

$$\widehat{\varepsilon}_3 = (0, 1.8 + \widehat{d}_3, 1.8 + D_3 - \widehat{d}_3)_T = (0, 2.6, 2.6)_T,$$
$$\widehat{\varepsilon}_5 = (0, 1.8 + \widehat{d}_5, 1.8 + D_5 - \widehat{d}_5)_T = (0, 3, 1.8)_T.$$

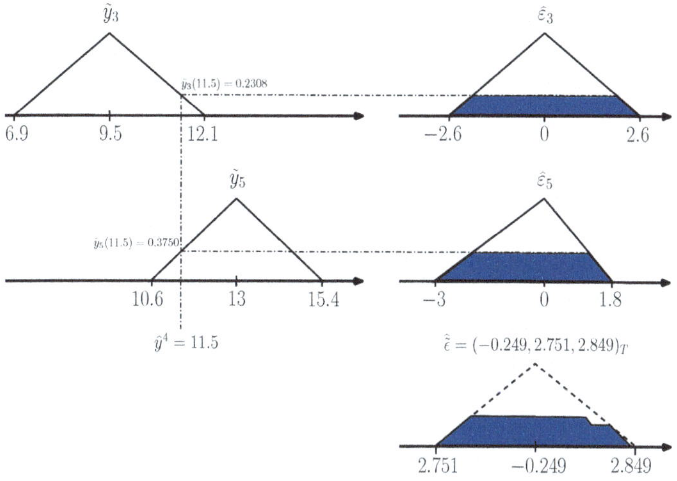

Fig. 2 The predicted error term $\widehat{\varepsilon}$ by fuzzy inference system in Example 5.1 (continued)

So, the following if-then rules

$$R^3 \;: \text{IF } 11.5 \text{ is } (9.5, 2.6)_T \text{ THEN } \widehat{\epsilon} \text{ is } (0, 2.6, 2.6)_T,$$
$$R^5 \;: \text{IF } 11.5 \text{ is } (13, 2.4)_T \text{ THEN } \widehat{\epsilon} \text{ is } (0, 3, 1.8)_T,$$

are represented to derive the transformed predicted error term as $\widehat{\widehat{\epsilon}} = (-0.25, 2.75, 2.85)_T$ (see Fig. 2). Consequently, based on the constructed model, the estimated response for $x_4 = 4$ is obtained as

$$\widehat{\widetilde{y}}^4_{\text{VS(LA)}} = 3.5 + 2\,x_4 \oplus \widehat{\widehat{\epsilon}} = (11.25, 2.75, 2.85)_T.$$

On the other hand, by applying VS(LS) method, the fuzzy regression model is

$$\widehat{\widetilde{y}}_i^{\text{VS(LS)}} = 5.19 + 1.46x_i \oplus \widehat{\widetilde{\varepsilon}}_i, \qquad i = 1, 2, 3, 5,$$

where $\widehat{\widetilde{\varepsilon}}_1 = (0, 1.8, 1.8)_T$, $\widehat{\widetilde{\varepsilon}}_2 = (0, 2.6, 1.8)_T$, $\widehat{\widetilde{\varepsilon}}_3 = (0, 2.73, 2.47)_T$, and $\widehat{\widetilde{\varepsilon}}_5 = (0, 1.8, 3)_T$. For the independent observation $x_4 = 4$, the corresponding estimated center is $\widehat{y}^4_{\text{VS(LS)}} = 5.19 + (1.46 \times 4) = 11.05$. Here, two error terms $\widehat{\widetilde{\varepsilon}}_3$ and $\widehat{\widetilde{\varepsilon}}_5$ are activated, which can be used to obtain the transformed predicted error term as $\widehat{\widehat{\epsilon}} = (0.02, 2.75, 2.97)_T$. Therefore, the estimated response for $x_4 = 4$ based on VS(LS) method is $\widehat{\widetilde{y}}^4_{\text{VS(LS)}} = (11.08, 2.75, 2.97)_T$.

Referring to data set, the real response for x_4 is $\widetilde{y}_4 = (13.5, 2.6)_T$. Using the error in estimation and the relative error in estimation as measures of the forecasting performances, the errors in estimations for the proposed model is $E(\widetilde{y}_4, \widehat{\widetilde{y}}^4_{\text{VS(LA)}}) = 3.52$

which is smaller than that of VS(LS) model, i.e. $E(\widetilde{y}_4, \widehat{\widetilde{y}}^4_{VS(LS)}) = 3.67$. Moreover, the similarity measures between the real response for x_4 and the estimated value are $S(\widetilde{y}_4, \widehat{\widetilde{y}}^4_{VS(LA)}) = 0.21$ and $S(\widetilde{y}_4, \widehat{\widetilde{y}}^4_{VS(LS)}) = 0.19$ for the proposed model and for VS(LS) model, respectively, in favor of the proposed model (see Table 3).

Example 5.3 Consider the data set in Table 4, with three crisp explanatory variables and one fuzzy response variable, designed by Kim and Bishu [12]. This data has been considered by several authors (e.g. [11, 18]). Choi and Buckley [8] claimed that to model this data set the least absolutes fuzzy regression works better than the Kim and Bishu's approach (which is the least squares method) since there is one outlier point (No. 3). By applying the proposed method and VS(LS) method the regression models are

$$\begin{aligned}\widehat{\widetilde{y}}_i^{VS(LA)} &= -2.82 - 0.38\,x_{1i} - 1.01\,x_{2i} + 0.61\,x_{3i}\\ &\quad \oplus(0, 0.52 + \widehat{d}_i, 0.52 + D_i - \widehat{d}_i)_T,\\ \widehat{\widetilde{y}}_i^{VS(LS)} &= -16.79 - 1.10\,x_{1i} - 1.18\,x_{2i} + 1.86\,x_{3i} \oplus \widehat{\widetilde{\varepsilon}}_i,\end{aligned}$$

in which, the values of $\widehat{d}_i$ (for the proposed model), and $\widehat{\widetilde{\varepsilon}}_i$ (for VS(LS) model), $i = 1, \ldots, 8$, are given in Table 4.

The total errors in estimation for the proposed model is $E = 15.00$; for VS(LS) model is $E = 15.77$; which is in favor of the proposed model. In addition, the sum of similarity measures for the proposed model is $S = 6.11$, while for VS(LS) model is

Table 3 Forecasting accuracy in estimation in Example 5.1 (continued)

The model	$S(\widetilde{y}_4, \widehat{\widetilde{y}}^4)$	$E(\widetilde{y}_4, \widehat{\widetilde{y}}^4)$
The proposed model	0.21	3.52
Chen and Dang's model [6]	0.19	3.67

Table 4 Data set and the values of $\widehat{d}_i$ and $\widehat{\widetilde{\varepsilon}}_i$ in Example 5.3

i	$\widetilde{y}_i = (y_i, s_i)_T$	x_{1i}	x_{2i}	x_{3i}	$\widehat{d}_i$	$\widehat{\widetilde{\varepsilon}}_i$
1	$(5.83, 3.56)_T$	2.00	0.00	15.25	3.040	$(0, 6.60, 0.52)_T$
2	$(0.85, 0.52)_T$	0.00	5.00	14.13	0.000	$(0, 0.52, 0.52)_T$
3	$(13.93, 8.50)_T$	1.13	1.50	14.13	0.000	$(0, 0.52, 16.48)_T$
4	$(4.00, 2.44)_T$	2.00	1.25	13.63	1.303	$(0, 3.60, 1.28)_T$
5	$(1.65, 1.01)_T$	2.19	3.75	14.75	0.490	$(0, 1.50, 0.52)_T$
6	$(1.58, 0.96)_T$	0.25	3.50	13.75	0.880	$(0, 1.40, 0.52)_T$
7	$(8.18, 4.99)_T$	0.75	5.25	15.25	0.000	$(0, 0.52, 9.46)_T$
8	$(1.85, 1.13)_T$	4.25	2.00	13.50	0.610	$(0, 0.52, 1.74)_T$

Table 5 The similarity measures and estimated errors for the models in Example 5.3

	Proposed method		Chen and Dang's method [6]	
i	S_i	E_i	S_i	E_i
1	1.00	0.00	0.44	2.72
2	1.00	0.00	0.00	1.04
3	0.36	7.98	0.55	4.84
4	0.87	0.32	0.79	0.57
5	1.00	0.00	0.01	1.95
6	0.66	0.38	0.00	1.92
7	0.22	6.32	0.63	2.24
8	1.00	0.00	0.63	0.49
Sum	6.11	15.00	3.05	15.77

$S = 3.05$, in favor of the proposed method (Table 5). It is noticeable VS(LA) model yields more efficient results than those of the models based on least squares methods in the present of outliers.

6 Conclusion Remarks

A two-phase method was proposed to construct a fuzzy regression model for crisp input-fuzzy output observations. In the first phase, using the LAD method, the coefficients of the model were calculated as crisp values. In the second phase, the fuzzy error terms were estimated as variable values by setting each estimated spread equals to its corresponding observed spread in an optimization problem. This method eliminates the problem of increasing spreads for the estimated fuzzy responses as the magnitude of the independent variable increases. In the proposed fuzzy model, the centers of the fuzzy dependent variable are estimated by adopting the LAD approach, which represents a robust estimation technique. Considering two goodness-of-fit criteria, the investigated examples indicate that the method proposed in this paper has a better explanatory power and higher forecasting accuracy than some current methods. The extension of the proposed approach to fuzzy input-fuzzy output data is a potential subject for future research.

References

1. A. Bargiela, W. Pedrycz, T. Nakashima, Multiple regression with fuzzy data. Fuzzy Sets Syst. **158**, 2169–2188 (2007)
2. J. Chachi, S.M. Taheri, N.R. Arghami, A hybrid fuzzy regression model and its application in hydrology engineering. Appl. Soft Comput. **25**, 149–158 (2014)

3. J. Chachi, S.M. Taheri, Multiple fuzzy regression model for fuzzy input-output data. Iran. J. Fuzzy Syst. **13**, 63–78 (2016)
4. J. Chachi, S.M. Taheri, H. Rezaee Pazhand, Suspended load estimation using L_1-Fuzzy regression, L_2-Fuzzy regression and MARS-Fuzzy regression models. Hydrol. Sci. J. **61**, 1489–1502, (2016)
5. F. Chen, Y. Chen, J. Zhou, Y. Liu, Optimizing h value for fuzzy linear regression with asymmetric triangular fuzzy coefficients. Eng. Appl. Artif. Intell. **47**, 16–24 (2016)
6. S.P. Chen, J.F. Dang, A variable spread fuzzy linear regression model with higher explanatory power and forecasting accuracy. Inform. Sci. **178**, 3973–3988 (2008)
7. L.H. Chen, C.C. Hsueh, Fuzzy regression models using the least squares method based on the concept of distance. IEEE Trans. Fuzzy Syst. **17**, 1259–1272 (2009)
8. S.H. Choi, J.J. Buckley, Fuzzy regression using least absolute deviation estimators. Soft Comput. **12**, 257–263 (2008)
9. P. D'Urso, R. Massari, A. Santoro, Robust fuzzy regression analysis. Inform. Sci. **181**, 4154–4174 (2011)
10. C. Kao, C.L. Chyu, A fuzzy linear regression model with better explanatory power. Fuzzy Sets Syst. **126**, 401–409 (2002)
11. C. Kao, C.L. Chyu, Least-squares estimates in fuzzy regression analysis. Eur. J. Oper. Res. **148**, 426–435 (2003)
12. B. Kim, R.R. Bishu, Evaluation of fuzzy linear regression models by comparison membership function. Fuzzy Sets Syst. **100**, 343–352 (1998)
13. G. Klir, B. Yuan, *Fuzzy Sets and Fuzzy Logic-Theory and Applications* (Prentice Hall, New Jersey, 2005)
14. J. Li, W. Zeng, J. Xie, Q. Yin, A new fuzzy regression model based on least absolute deviation. Eng. Appl. Artif. Intell. **52**, 54–64 (2016)
15. J. Lu, R. Wang, An enhanced fuzzy linear regression model with more flexible spreads. Fuzzy Sets Syst. **160**, 2505–2523 (2009)
16. S. Pourahmad, S.M.T. Ayatollahi, S.M. Taheri, Fuzzy logistic regression: A new possibilistic model and its application in clinical vague status. Iran. J. Fuzzy Syst. **8**, 1–17 (2011)
17. M.R. Rabiei, N.R. Arghami, S.M. Taheri, B.S. Gildeh, Least squares approach to regression modeling in full interval valued fuzzy environment. Soft Comput. **18**, 2043–2059 (2014)
18. S.M. Taheri, M. Kelkinnama, Fuzzy linear regression based on least absolute deviations. Iran. J. Fuzzy Syst. **9**, 121–140 (2011)
19. H. Tanaka, I. Hayashi, J. Watada, Possibilistic linear regression analysis for fuzzy data. Eur. J. Oper. Res. **40**, 389–396 (1989)

Robust Modeling of Acceptance Control Chart to Specify Best Design Parameters

Samrad Jafarian-Namin, Mohammad Saber Fallah Nezhad, Reza Tavakkoli-Moghaddam, and Mehrdad Mirzabaghi

Abstract Control charts are one of the most important techniques in statistical process control. Acceptance control chart (ACC) is an effective tool for monitoring highly capable processes in which 6-sigma spread of the process is smaller than that of the specification limits. Accordingly, the fluctuation of the process mean is permitted or inevitable. For designing ACC, an economic-statistical model is developed in this paper. In some practical processes, the parameters are not exactly defined or estimated. Such uncertainty could bring difficulty in getting the best design. Thus, a robust optimization approach is proposed to investigate them through a solution procedure utilizing the Differential Evolution (DE) algorithm. Some numerical examples are given for illustration. Moreover, comparisons are made to assess the performances of economic design (ED) versus economic-statistical design (ESD). The results indicate the better performance of the proposed approach for designing ACC and more reliable solutions for practitioners.

1 Introduction

Statistical Process Control (SPC) is a method of quality control (QC) that employs statistical techniques for controlling and improving the quality of manufacturing and service enterprises. Control chart is the most popular tool in SPC that graphically plots measurable or countable data obtained from a process in specific time order.

S. Jafarian-Namin · M. S. Fallah Nezhad (✉)
Department of Industrial Engineering, Faculty of Engineering, Yazd University, Yazd, Iran
e-mail: fallahnezhad@yazd.ac.ir

S. Jafarian-Namin
e-mail: samrad.jafarian@stu.yazd.ac.ir

R. Tavakkoli-Moghaddam · M. Mirzabaghi
School of Industrial Engineering, College of Engineering, University of Tehran, Tehran, Iran
e-mail: tavakoli@ut.ac.ir

M. Mirzabaghi
e-mail: m.mirzabaghi@ut.ac.ir

S. N. Shahbazova et al. (eds.), *Recent Developments and the New Direction in Soft-Computing Foundations and Applications*, Studies in Fuzziness and Soft Computing 393, https://doi.org/10.1007/978-3-030-47124-8_26

These charts are widely used to stabilize and monitor to determine whether a process stays in-control or out-of-control.

Among various types of control charts, acceptance control charts (ACCs) are applied to monitor the mean of high capable processes distributed according to a Normal distribution [1]. In order to describe the principles of this chart, let *USL* and *LSL* be the upper and lower specification limits, respectively. The simplest capability index is defined as $C_p = (USL\text{-}LSL)/6\sigma$ where σ is known standard deviation. Considering processes with $C_p \geq 1$, it is clear that the difference between specification limits can be 6-sigma or more. Hence, it may be suitable to allow the process mean to vary over a specific range and the nonconforming fraction not exceeding a target value. In fact, the aim of ACC is over the classic control charts in which only verifying the stability of process mean is considered.

In this paper, the design of ACC is considered via parameters related to them. Sample size (n), sampling interval (h), and control limits width (k) are three main variables of control chart design. Mohammadian and Paynabar [2] proposed an economic model for ACC to determine optimal design variables. Nevertheless, designs only based on pure economic models have been criticized for unsatisfactory statistical features. Thus, Saniga [3] presented an economic-statistical model, which minimized the Duncan's [4] cost function subject to statistical constraints. This approach was implemented with some changes for ACC [5]. In this field of study, the certainty of parameters has been assumed for ACC during the years.

In some practical processes, the parameters are not exactly defined or estimated. Such uncertainty could bring difficulty in getting the best design. Thus, the parameters are investigated by the robust optimization approach in [6] which is built on the notion of the budget of uncertainty. In fact, range estimates of some imprecise parameters are proposed instead of point estimates. Accordingly, Safaei et al. [7] proposed a robust economic-statistical design (RESD) of X-bar control chart considering the cost function introduced by Lorenzen and Vance [8]. Recently, the robust design of ACC was proposed in [9] and [10] to minimize Duncan's [4] cost model subject to some statistical constraints. Optimization procedures of Both studies in [7] and [9] are based on the Genetic algorithm (for more information on metaheuristic algorithms, refer to [11]).

In this paper, a similar model for ACC is provided in which minimizing the cost function of Lorenzen and Vance [8] is considered. Moreover, the solution procedure according to the Differential Evolution (DE) algorithm is proposed. The remainder of the article is organized as follows. In Sect. 2, ACC is briefly reviewed. The proposed robust model for ACC, assuming uncertainty in its parameters, is presented in Sect. 3. Then, a solution procedure utilizing DE algorithm is illustrated for robust optimization. Section 5 reports the numerical results of an example in the investigation of the best designs. Finally, we conclude the paper.

2 Acceptance Control Chart

Freund [1] introduced ACC in 1957 for monitoring high capable processes where the process standard deviation is small enough compared to the spread of specification limits. In this condition, the fluctuations of process mean are permitted to certain amounts which are called acceptance process levels.

Commonly, the range of expected variation in a control chart is shown by *UCL* and *LCL*. Table 1 shows notations of ACC (note that the distribution of quality characteristic is assumed Normal throughout the paper). The control limits of ACC can be statistically computed using either: (1) α, *APL*, and specified sample size (n), (2) β, *RPL*, and n, and (3) α, β, *APL*, and *RPL* [12], which are calculated as follows:

$$\begin{aligned} APL_U &= USL - Z_{AQL}\sigma, \\ APL_L &= LSL + Z_{AQL}\sigma, \end{aligned} \tag{1}$$

$$\begin{aligned} RPL_U &= USL - Z_{RQL}\sigma, \\ RPL_L &= LSL + Z_{RQL}\sigma, \end{aligned} \tag{2}$$

$$\alpha = P(\bar{X} \leq LCL|\mu = APL_L) = \int_{-\infty}^{-Z_\alpha} f_z(z)dz, \tag{3}$$

Table 1 Notations of ACC

APL	Acceptance process level
RPL	rejectable process level
AQL	maximum of the rejected product proportions, when process mean lies within the APL
RQL	maximum of rejected product proportions that can be tolerated
Z_{AQL}	standard normal value associated with the probability *AQL*
Z_{RQL}	standard normal value associated with the probability *RQL*
USL	upper specification limit
LSL	lower specification limit
UCL	upper control limit
LCL	lower control limit
$f_z(z)$	probability destiny function of the standard normal distribution
α	type I error, probability of indicating an unacceptable status when the process mean is at *APL*
β	type II error, probability of indicating an acceptable status when the process mean is at *RPL*
σ	standard deviation corresponding to inherent process variability

$$B = 1 - \int_{-\infty}^{Z_{AQL}\sqrt{n} - Z_{\alpha} - Z_{RQL}\sqrt{n}} f_z(z)dz. \quad (4)$$

Using the first method in [12], the control limits are computed as follows:

$$UCL_{\alpha} = APL_U + Z_{\alpha}\frac{\sigma}{\sqrt{n}},$$
$$LCL_{\alpha} = APL_L - Z_{\alpha}\frac{\sigma}{\sqrt{n}}. \quad (5)$$

In practice, the process is permanently sampled with size n in every definite h hours. Then, these sampled data, via ACC with k coefficient of standard deviation, are inspected and the results are interpreted to monitor the state of the process. In the next section, a model is proposed for optimally selecting n, h, and k.

3 Economic-Statistical Design of ACC

In this section, a model is developed for the acceptance control chart considering uncertainty for related parameters. In this regard, some assumptions are first stated. Next, economic cost function is introduced. The proposed model is presented in the last subsection.

3.1 Assumptions of the Model

Some assumptions, for simplifying the mathematical modeling, are listed to be held as follows:

1. The quality characteristic follows a Normal distribution with known σ,
2. Only the lower specification limit is active, and considered for calculations,
3. The control limits based on desired α, APL, and specified sample size are applied for monitoring,
4. The process is accepted when mean alters inside acceptable process level, i.e. $APL_L \leq \mu \leq APL_U$,
5. Occurring a random assignable cause leads to upward or downward shifts in the process mean. Thus, the process is rejected when $\mu \leq RPL_L$ or $\mu \geq RPL_U$,
6. The occurrence of an assignable cause possesses an exponential distribution with mean time $1/\lambda$,
7. The process is allowed to continue during the search but is ceased during the repair.

3.2 Economic Cost Function

With the aim of finding the best design variables of ACC through a constrained nonlinear model, we present the traditional definition of the cost function in this subsection. This function is based on a control cycle of monitoring the production process introduced in [8]. Accordingly, the expected hourly cost (EHC), adapted to ACC is as follows (list of variables and parameters are defined in Table 2).

$$EHC = \frac{1}{T_c}(C_S + C_{NC} + C_{OoC}), \quad (6)$$

where:

Table 2 List of parameters and variables

Design variables	n	sample size
	h	sampling interval
	k	coefficient of std. dev
Process parameters	λ	Poisson distribution parameter for failure rate
	δ	shift in std. dev. in out-of-control (OoC) state
	C_0	quality cost/hour while producing in-control (IC)
	C_1	quality cost/hour while producing Ooc
	d_1	1(0) if production continues (ceases) during search
	d_2	1(0) if production continues (ceases) during repair
Sample parameters	a	fixed cost of sampling
	b	variable cost of sampling
OoC parameters	E	expected time to sample and chart one item
	W	cost to locate and repair the assignable cause
	Y	cost per false alarm
Calculated variables	T_0	expected time to search for a false alarm
	T_1	time to detect the assignable cause
	T_2	time to repair the system
	ξ	mean number of sample while IC
	τ	mean time between last sample while IC and drift
	T_c	length of a control cycle
	C_S	direct sampling cost for a cycle
	C_{NC}	non-conformities costs for a cycle
	C_{Ooc}	false alarm and detection costs for a cycle
	ATS_0	average time to signal when a false alarm occurs
	ATS_1	average time to signal when an assignable cause occurs

$$T_c = \frac{1}{\lambda} + ATS_1 - \tau + nE + T_1 + T_2 + \frac{\xi(1-d_1)T_0 h}{ATS_0},$$

$$C_S = (a+bn)\frac{(1/\lambda + ATS_1 - \tau + nE + d_1T_1 + d_2T_2)}{h},$$

$$C_{NC} = \frac{C_0}{\lambda} + C_1(ATS_1 - \tau + nE + d_1T_1 + d_2T_2),$$

$$C_{OoC} = W + \frac{Y\xi h}{ATS_0},$$

$$\xi = \frac{e^{-\lambda h}}{(1-e^{-\lambda h})},$$

$$\tau = \frac{(1-(1+\lambda h)\xi)}{\lambda}.$$

For different sets of n, h, and k, the optimal setting in the ED of ACC is chosen for design variables with the lowest value of cost function. The optimality obtained here is based on a single-point estimate of model parameters. In fact, this is a single scenario when all parameters are known. But, these estimates are often not accurately known in reality.

That is aimed to find a robust setting given the set of scenarios that could potentially be realized within the bounds of the uncertain parameters in the proposed model. Uncertainty in the model means that some of the parameters in the planning phase are not exactly defined. Suppose that a vector of cost parameters $C^s = \{C_0^s, C_1^s, a^s, b^s, W^s, Y^s\}$ and process parameters $P^s = \{\lambda^s, \delta^s\}$ are defined for each scenario $s \in S$. Thus, different values of *EHC* is found for various scenarios of these parameters.

3.3 The Proposed Model for ACC

In order to develop an ESD for ACC, we used the traditional objective function introduced by Lorenzen and Vance [8] and employing some statistical constraints. According to [7], these statistical constraints under scenario s include: (1) a minimum value for ATS_0^s which maintains a reasonable time to signal when a false alarm occurs, and (2) a maximum value to limit ATS_1^s when an assignable cause occurs. These extreme values, i.e. ATS_0^{min}, and ATS_1^{max}, may be determined as desired bounds by decision-maker (DM) or quality engineers. Therefore, the proposed economic-statistical model used to design the ACC is as follows:

$$\min_{x\in X} \max_{s\in S} EHC^s$$

$$s.t.$$

$$\begin{aligned} ATS_0^s &\geq ATS_0^{\min} \quad \forall s \in S \\ ATS_1^s &\leq ATS_1^{\max} \quad \forall s \in S \\ n_{\min} &\leq n \leq n_{\max} \\ h_{\min} &\leq h \leq h_{\max} \\ k_{\min} &\leq k \leq k_{\max}, \end{aligned} \tag{7}$$

where the design parameters n, h, and k are set between lower and upper bounds as well. Briefly speaking, we altered the model presented in [9] by (1) changing the cost function according to the definition in [8], and (2) considering some other statistical constraints. Two classes of parameters are considered including the class with uncertainty and the class with nominal value or point estimation. Therefore, different scenarios can be generated by putting lower and upper bounds on uncertain parameters.

Through optimization of a robust counterpart of the proposed model, three decision variables, i.e. n, h and k are determined. In the next section, a methodology to tackle the uncertainty in the model and a procedure to optimize are provided.

4 The Solution Algorithm

A robust optimization methodology to tackle the uncertainty in mathematical models is proposed in [13]. In such cases, the uncertainty set is defined in which possible values of the uncertain parameters are determined. Indeed, uncertain parameters are unknown but bounded. Furthermore, in order to avoid having an infinite number of constraints, the set of uncertain values separated into a limited set of scenarios.

The concept of "budget of uncertainty" is developed in [14] to avoid excessive conservatism that enables the designer to trade-off between risk (robustness) and performance. Parameter Γ, as budget of uncertainty, is an integer number ($\Gamma \in [0, m]$) which is equal to a maximum number of uncertain parameters in each scenario, and m is the number of uncertain parameters in the proposed model. If $\Gamma = 0$, all parameters take nominal values and no uncertainty is considered in the model. In contrast, if $\Gamma = \text{m}$, all uncertain parameters take values between their lower and upper bounds and the results are quite conservative. If $\Gamma \in [0, m]$, a trade-off is made between performance and the degree of conservatism of the solution. Due to the non-linear nature of our model, a metaheuristic approach based on DE algorithm is developed in this section.

4.1 Scenario Generation

The first step to solve the problem is scenario generation according to the value of the budget of uncertainty Γ that is equal to a maximum number of uncertain parameters in each scenario. The scenarios are randomly generated and in each scenario, Γ numbers of uncertain parameters are selected and take values within their given bounds. The description of this procedure is summarized below:

Procedure 1: The scenario generation
Step 1: Γ of uncertain parameters take value within their bounds randomly and (*m*-Γ) of them take their nominal value. Hence, "random scenarios" are generated as follows:

$$Sc\binom{m}{\Gamma}$$

Step 2: Γ of uncertain parameters take an extreme value of their bounds and (*m*-Γ) of them take their nominal value. Thus, "extreme scenarios" are generated as follows:

$$2^{\Gamma}\binom{m}{\Gamma}$$

Step 3: Sum of steps 1 and 2 forms the total scenarios.

4.2 Proposed Differential Evolution Algorithm

DE is a parallel direct search method that uses *n* (population size) vectors to search the solution space in which each vector is equivalent to a point of solution space. An initial population of vectors is generated randomly. In each iteration, the current population is considered as parents and offsprings are generated using mutation and crossover operators. At first, equivalent to each parent, a trial vector is generated using the mutation operator, and next, the offspring vector is generated using the crossover operator. Then, to generate the next population, all vectors (parents and offsprings) are assessed and the next population is formed by the selection procedure. All steps of proposed DE are briefly mentioned here.

Procedure 2: Main steps of proposed DE algorithm
Begin
Step1: *Initialization*
1.1. Load model parameters: Cost, Process and Control chart Parameters, lower and upper bounds of decision variables also constraints
1.2. Set DE parameters: Number of iterations (*imax*), Population size (*PopSize*), Crossover rate (*Pc*), Scale coefficient (*B*)

1.3. Set scenario parameters include Budget of uncertainty (Γ), Indicator of the number of random scenarios (*Sc*), and the number of uncertain parameters (m*)* and generate scenarios

1.4. Generate initial random population, $i = 0$

Repeat

Step2: *Mutation* (create trial vectors using procedure 3)

Step3: *Crossover* (create offsprings using procedure 4)

Step4: *Evaluation* (evaluation of all parent and offspring vectors using procedure 5)

Step5: *Selection* (select next generation using selection procedure, $i = i+1$)

Until $i = imax$

Procedure 3: Mutation operator

```
Begin
    For i=1:popsize
        Choose 3 different vectors from the current population
        (a(i),b(i),c(i)) which are different from vector i and
        create trial vector u(i)=a(i)+B(b(i)-c(i))
    End
End
```

Procedure 4: Crossover operator

```
End
```

```
Begin
    For i=1:popsize
        For j=1:number_of_decision_variables
            If rand<Pc , Offspring(i,j)=u(i,j)
            Else Offspring (i,j)=parent(i,j)
            End
        End
    End
End
```

In each iteration, evaluation of all vectors (parents and offspring) is done to select the next generation according to the following procedure.

Procedure 5: Evaluation

Step 1: Calculate the objective function of each vector (parents and offspring) in all scenarios ($EHC_{X,S}$). The fitness of each vector is equal to *Maximum* ($EHC_{X,S}$).

Step 2: According to constraints, if a vector is infeasible (in one or more scenarios), penalize its fitness by 10 times.

After evaluating all vectors, between each parent and its offspring, the best vector-based on its fitness is selected for the next generation. According to this procedure, the best vector of the next generation is not worse than the best vector of the current generation.

5 An Illustrative Example

In order to illustrate the design of ACC by optimization of robust counterpart, an example is investigated in this section. Table 3 shows the point estimates or nominal values of parameters adapted from [8] by matching them to ACC.

Uncertainty in various parameters can lead to a loss for manufacturer and customer. Thus, a range estimate of cost parameters $C^s = \{C_0^s, C_1^s, a^s, b^s, W^s, Y^s\}$ and process parameters $P^s = \{\lambda^s, \delta^s\}$ are considered as uncertain ones for each scenario $s \in S$. Deviation of these parameters from their nominal values includes two scenarios: (1) 10% shift scenarios from nominal estimates, and (2) 20% shift scenarios from the nominal estimates. The maximum number of uncertain parameters (m) is set to eight according to eight uncertain parameters. Hence, we can consider all combinations of expected uncertain parameters, i.e. $\Gamma = 0, \ldots, 8$.

Feasible space is determined by considering $1 \leq n \leq 20$, $0.1 \leq h \leq 8$, and $1 \leq k \leq 4$. The right-hand-side values of constraints in Eq. (7) are respectively defined as 105.31, and 3.29. All calculations have been facilitated under a program coded in the MATLAB (version R2016b) environment in this study.

In Table 4, the results for different scenarios based on budgets $\Gamma = 0, \ldots, 8$ are shown. Moreover, the trend of cost function under different shift scenarios and budgets of uncertainty are depicted in Fig. 1. For the nominal scenario when $\Gamma = 0$, the best plan for ACC is obtained as $n = 7$, $h = 120$, and $k = 2.24$. The expected hourly cost is 189.34. For different scenarios, it is clear that design variables are nearly unchanged. In fact, sampling time interval changes to some extent about 25 min, for example when $\Gamma = 0$ and $\Gamma = 4$ are compared for 20% shift scenario. This value for h decreases to 15 min under 10% shift scenario.

Interesting results can be obtained from Fig. 1 when comparing the various RESD with nominal design. The least *EHC* is computed for the nominal design. It leads to 4% cost less than the least expensive RESD in 10% shift scenario and 10% in 20% shift scenario. Besides, a consistent increase in cost is detected as the budget of uncertainty becomes larger. In fact, it is expected that incorporating more uncertainty in the model causes an increase in optimal cost. Moreover, the most conservative robust design, i.e. $\Gamma = 8$, has the most *EHC*.

Table 3 Nominal values of parameters

$a = 5$	$C_0 = 114.24$	$E = 0.0833$	$\lambda = 0.02$	AQL = 0.001
$b = 4.22$	$C_1 = 949.2$	$T_0 = 0.0833$	$\delta = 0.86$	RQL = 0.025
$W = 977.4$	$d_1 = 1$	$T_1 = 0.0833$	$\mu = 10$	USL = 10.5
$Y = 977.4$	$d_2 = 0$	$T_2 = 0.75$	$\sigma = 0.01$	LSL = 9.5

Table 4 The robust solution of the proposed model under different shift scenarios and different budgets

Shift scenario	Γ	n	h	k	EHC
10%	0	7	120	2.24	189.34
	1	7	110	2.29	194.14
	2	7	115	2.24	204.85
	3	7	105	2.26	209.05
	4	7	105	2.26	211.08
	5	6	105	2.19	212.61
	6	7	110	2.26	213.27
	7	7	110	2.26	213.54
	8	7	115	2.24	213.54
20%	0	7	120	2.24	189.34
	1	7	120	2.25	210.89
	2	7	110	2.25	220.14
	3	7	100	2.24	228.90
	4	7	95	2.27	233.32
	5	6	95	2.20	236.38
	6	6	95	2.23	237.93
	7	7	110	2.24	238.52
	8	7	110	2.24	238.52

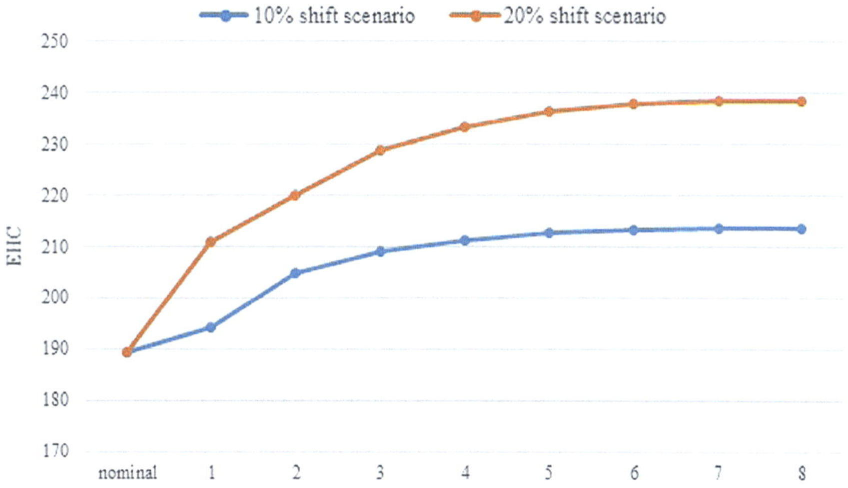

Fig. 1 Changes in cost under different scenarios (10 and 20%) and budgets of uncertainty

6 Conclusion and Future Researches

In designing a control chart, some parameters of some processes are uncertain due to measurement errors or other technical obstacles. With the aim of getting the best design in such circumstances, the concept of budgets of uncertainty was considered in the robust framework to take into account the uncertainty of parameters in a systematic routine. Thus, the robust design of the acceptance control chart was presented in this paper. The provided model included an economic cost function subject to some statistical constraints. For this reason, a solution procedure utilizing DE algorithm was proposed.

Then, a numerical example was presented to explain the proposed model under different shift scenarios and different budgets of uncertainty. Such results could help the practitioners to select the best design according to excellence in cost regarding statistical performance while providing the most protection against uncertainties in parameters. The results indicated that a trivial increase in costs would be suitable to get a design for ACC with the desired statistical performance.

Acknowledgements This work was supported by the Iran National Science Foundation (INSF) [grant number 97015328]. The initial version of this paper was presented in the 7th World Conference on Soft Computing (WConSC) held in Baku, Azerbaijan.

References

1. R.A. Freund, Acceptance control charts. J. Ind. Qua. Control **14**, 13–23 (1957)
2. F. Mohammadian, K. Paynabar, Economic design of acceptance control charts, in *Proceedings of the 2008 IEEE and IEEM*, (Singapore, 2008), pp. 2132–2136
3. E.M. Saniga, Economic statistical control chart designs with an application to X and R charts. Technometrics **31**, 313–320 (1989)
4. A.J. Duncan, The economic design of x charts used to maintain current control of a process. J. Am. Stat. Assoc. **51**, 228–242 (1956)
5. F. Mohammadian, A. Amiri, Economic-statistical design of acceptance control chart. Qual. Reliab. Eng. Int. **29**, 53–61 (2012)
6. D. Bertsimas, M. Sim, The price of robustness. Oper. Res. **52**, 35–53 (2004)
7. A.S. Safaei, R.B. Kazemzadeh, H.-S. Gan, Robust economic-statistical design of X-bar control chart. Int. J. Prod. Res. **53**, 4446–4458 (2015)
8. T.J. Lorenzen, L.C. Vance, The economic design of control charts: a unified approach. Technometrics **28**, 3–10 (1986)
9. S. Jafarian-Namin, M. Mirzabaghi, M.S. Fallah Nezhad, R. Tavakkoli-Moghaddam, Economic-statistical design of acceptance control chart under uncertainty conditions, in *14*th *International Industrial Engineering Conference (IIEC 2018)*, (Iran, 2018), pp. 3384–3393
10. S. Jafarian-Namin, M.S. Fallah Nezhad, R. Tavakkoli-Moghaddam, and M. Mirzabaghi, Robust Economic-statistical design of acceptance control chart. J. Qual. Eng. Prod. Optim. **4**, 55–72
11. R. Tavakkoli-Moghaddam, N. Norouzi, S-M. kalami, A. Salamat-bakhsh, *Meta-heuristic Algorithms Theory and Implementation in MATLAB (In Persian)*, (Tehran: Islamic Azad University-South Tehran Branch, 2013)
12. E.L. Grant, R.S. Leavenworth, *Statistical Quality Control*, 6th edn. (McGraw-Hill, New York, 1988)
13. A. Ben-Tal, S. Boyd, A. Nemirovski, Extending scope of robust optimization: comprehensive robust counterparts of uncertain problems. Math. Program. **107**, 63–89 (2006)
14. D. Bertsimas, D.B. Brown, C. Caramanis, Theory and applications of robust optimization. SIAM Rev. **53**, 464–501 (2011)

Fuzzy Modeling of Dynamic Multistage Process with Non-linear Stages

Alexander Galkin, Pavel Saraev, Natalia Zhbanova, and Sam Blyumin

Abstract Class of difference neuro-fuzzy switched models which can describe complex multistage processes is discussed in the article. Two-dimensional fuzzy sets which allow to reduce model parameter's amount are considered. Numerical algorithm for identifying parameters of two-dimensional fuzzy sets with non-linear centres is proposed.

1 Introduction

Process control often requires computation of significant parameters. In many cases computation is based on process data handling. Modern complex industrial processes are characterized by measure errors, complex relations between factors, and multistaging. Multistaging means that stage changing may to modificate relations between process parameters. So, process data handling is a difficult problem, which needs for using of artificial intelligence methods, such as fuzzy or neuro-fuzzy modelling [1].

Fuzzy and neuro-fuzzy models is an extensively investigated problem. The investigation of this theme arouses much interest, and there are a lot of articles with analysys fuzzy, neuro and neuro-fuzzy modelling theoretical aspects.

A. Galkin · P. Saraev (✉)
Faculty of Automation and Computer Science,
Lipetsk State Technical University, Lipetsk, Russia
e-mail: psaraev@yandex.ru

A. Galkin
e-mail: avgalkin82@mail.ru

N. Zhbanova · S. Blyumin
Applied Mathematics Department, Lipetsk State Technical University,
Lipetsk, Russia
e-mail: zbanoid@gmail.com

S. Blyumin
e-mail: sabl@lipetsk.ru

S. N. Shahbazova et al. (eds.), *Recent Developments and the New Direction in Soft-Computing Foundations and Applications*, Studies in Fuzziness and Soft Computing 393, https://doi.org/10.1007/978-3-030-47124-8_27

In [2, 3] Babuska studied fuzzy and neuro-fuzzy modeling for nonlinear system identification, and given some industrial applications. Saraev [4] investigated neural networks training methods and proposed neurostructural approach.

A lot of researches are about ANFIS and their applications. Zhou [5] analysed neuro-fuzzy modeling in chemical process. Rubinstein [6] used combination of subtractive clustering and fuzzy inference neural networks for crude oil consumption forecasting. In [7] Enke applied neuro-fuzzy models for inflation forecasting.

Difference fuzzy models have a side benefit due to larger quantity of prerequisites, and used for dynamic processes modelling. Kudinov [8] had been studied using of difference fuzzy models in metallurgy.

Switching systems shows considerable promise, because of a large amount of real objects, which may be described by such systems. In [9], Liberzon described switched systems features and some of their applications. Lubentzov [10] studied application of fuzzy interference system with switchable rule database in the systems of automatic control.

In this article we consider class of difference neuro-fuzzy switched models [11], which combine features of difference fuzzy models [8], neuro-fuzzy models [2], and switching systems [9]. Besides, we consider special type of fuzzy sets, designed for foregoing model class, and algorithms of their identification.

So, the purpose of this article is the investigation and development of difference neuro-fuzzy switched models, assigned for multistage dynamic processes modelling, and parametric identification methods for this models.

2 Modeling of Multistage Dynamic Process

2.1 Difference Neuro-fuzzy Switched Model

Multistage process data handling means that data observed by the sensors, arrives at the input of the neuro-fuzzy model. It is supposed that input process have several stages, in other words it represents piecewise-linear function (Fig. 1). As the result of process data handling, the model computes output values, that are of explorer's or engineer's interest. Let us notice that one of a number of difference neuro-fuzzy switched model advantages is possibility to take into account dynamic of input process. It is possible due to inputs of the model, which receive values for a series of consecutive time points.

Including the switching to the difference neuro-fuzzy model's structure we may describe complex process made of several stages with accuracy. Difference neuro-fuzzy switched model (DNFSM) consists of several difference neuro-fuzzy submodels and may be described by the rule database (1).

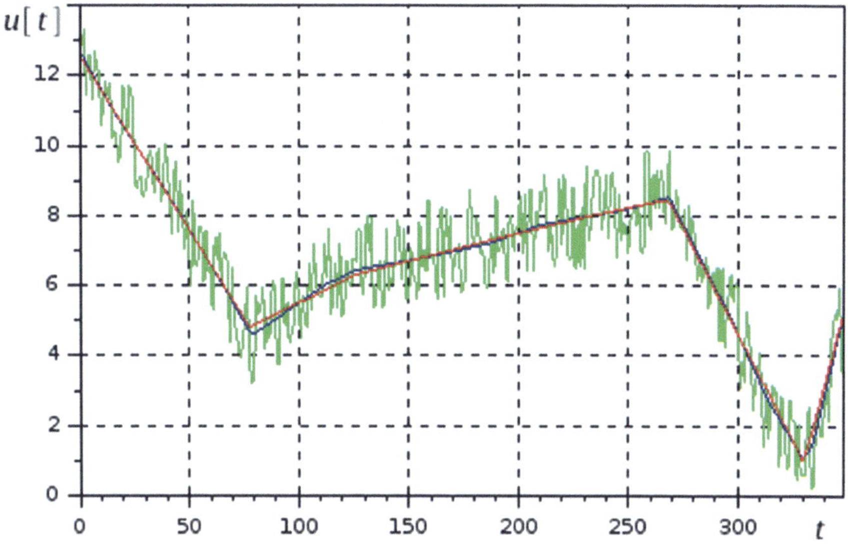

Fig. 1 Multistage process with five stages

$$\begin{aligned} R_s^l : If\quad & u_{s1}[t]\quad is\quad A_{s11}^l[t], \dots, \\ & u_{sm}[t-n+1]\quad is\quad A_{smn}^l[t], \\ then\quad & y_s^l[t+1] = \overline{a}_s^l \overline{U}_s[t]^T + b_s^l. \end{aligned} \tag{1}$$

Here $u_{si}[t], \dots, u_{si}[t-n+1]$—values of the input multistage processes, $A_{sij}^l[t]$—input prerequisite (antecedent of the rule) fuzzy sets, $y_s^l[t+1] \in R^1$—output of the each rule, $\overline{a}_s^l$, b_s^l—parameters of the difference equations in the consequents. Parameter $l = 1, \dots L_s$ determines the number of rules in each submodel.

Vector $\overline{U}_s[t] = [\overline{u}_{s1}[t], \dots \overline{u}_{sm}[t]]$ consists of a multistage input process' values, where $\overline{u}_{si}[t] = [u_{si}[t], \dots u_{si}[t-n+1]]$—values of the each input, $i = 1, \dots m$ is a number of input processes, n is the DNFSM memory depth.

The base (1) also includes a switching signal $s \in S = \{1, 2, \dots |S|\}$, which determines active submodel at each time point. Each submodel is represented as a difference neuro-fuzzy model. The value of $|S|$ is determined by the number of input process' stages.

Output of the DNFSM submodel at the fixed value s is determined by the formula for Takagi-Sugeno fuzzy models [1] and ANFIS [2]. Output values of the DNFSM (1) at each time point are determined by outputs of the active submodel.

Due to switchings DNFSM (1) allows effective modelling of multistage processes characterized by abrupt changes in structure or parameters. Becides, DNFSM is sensitive to features of input multistage process. This advantage evolves from a large amount of inputs, and leads to a disadvantage of DNFSM: its parameters increase in number, and curse of the dimension appears.

2.2 Two-Dimensional Fuzzy Sets

In order to reduce the effect of the curse, a mechanism for fuzzification of the input values of the DNFSM is considered, which is based on two-dimensional fuzzy sets [11]. Let us note that the input values $\overline{u}_{si}[t] = [u_{si}[t], \ldots u_{si}[t-n+1]]$ of the DNFSM submodel (1) at fixed s after fuzzification with standard fuzzy sets $A^l_{sij}[t]$ can be considered as fuzzy processes [12]. Evidently, input fuzzy processes identification expects evaluation of the input fuzzy sets parameters, and represents a part of the problem of parametric identification of DNFSM. Big memory depth of the DNFSM input process leads to increasing of its parameters number.

But it is possible to decrease parameters amount if we describe each step $\overline{u}_{si}[t]$ of multistage fuzzy input processes with one two-dimensional fuzzy set instead of several one-dimensional fuzzy sets $A^l_{sij}[t]$. Two-dimensional fuzzy set based on Gaussian fuzzy set, because it has handy analytical form.

The two-dimensional Gaussian fuzzy set $A^l_{si}[t]$ with a linear centre and constant width is shown in Fig. 2. The two-dimensional Gaussian fuzzy set with a linear centre and linear width is shown in Fig. 3. The sections represent one-dimensional fuzzy sets $A^l_{sij}[t]$.

So, with single two-dimensional fuzzy set $B^l_{si}[t]$ we can fuzzificate the whole step of the input multistage dynamic process.

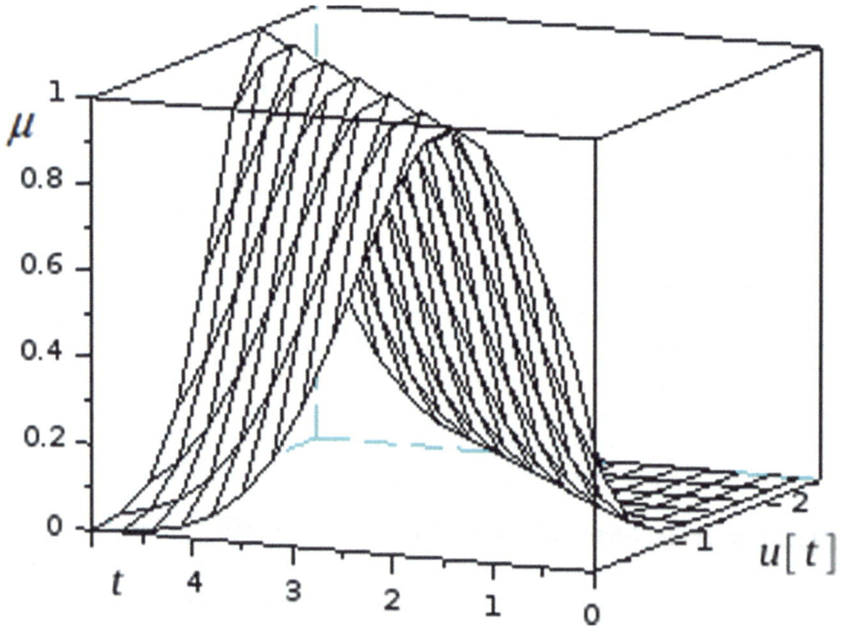

Fig. 2 Two-dimensional Gaussian fuzzy set with constant width

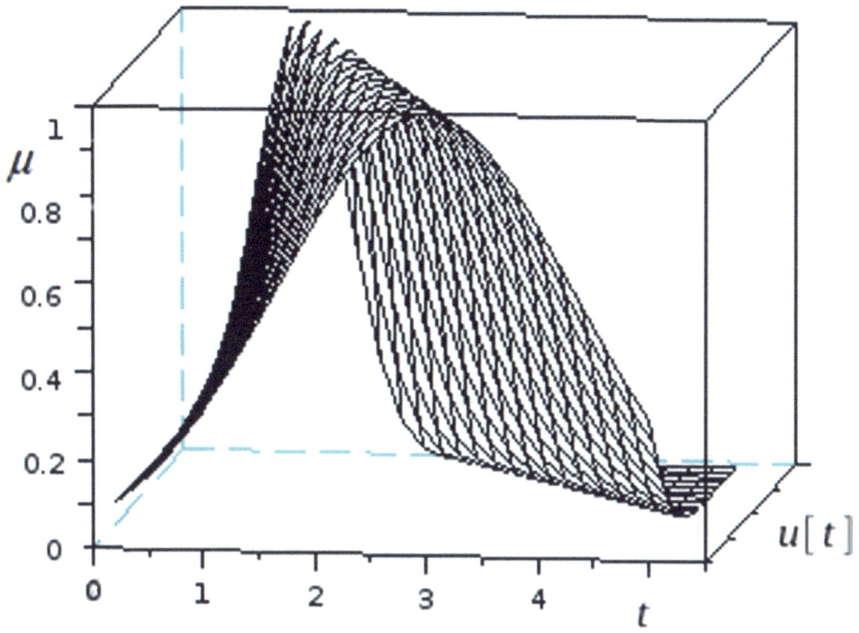

Fig. 3 Two-dimensional Gaussian fuzzy set with linear width

For an arbitrary DNFSM submodel, the membership degree of process $\overline{u}_{si}[t]$ in the two-dimensional fuzzy set $B^l_{si}[t]$ is calculated by the formula

$$\mu^l_{si} = \sum_{j=1}^{n-1} exp\left(c^l_{s1i}(t-j) + c^l_{s0i} - u_{si}(t-j)\right)^2 / w^l_{si}, \tag{2}$$

where c^l_{s0i} and c^l_{s1i} are the linear centre parameters of a two-dimensional fuzzy set, w^l_{si} is a parameter defining set width, which may be constant or function.

DNFSM rules with two-dimensional fuzzy sets in antecedents will be as follows:

$$\begin{array}{l} R^l_s : If \quad \overline{u}_{s1}[t] \quad is \quad B^l_{s1}[t], \ldots, \overline{u}_{sm}[t] \quad is \quad B^l_{sm}[t], \\ \qquad then \quad y^l_s[t+1] = \overline{a}^l_s \overline{U}_s[t]^T + b^l_s. \end{array} \tag{3}$$

After comparing DNFSM rule base (1) with DNFSM rule base (3) we can see that last variant demands lesser amount of the fuzzy sets. Every stage of the multistage input process may be described with one two-dimensional fuzzy set instead of n one-dimensional fuzzy sets.

So, two-dimensional fuzzy sets are a convenient supporting handle for neuro-fuzzy modelling of multistage technological processes with linear stages, and allows to reduce a number of DNFSM parameters. This fact profits from using DNFSM especially in cases of big memory depth of the input processes.

2.3 Modeling of the Processes with Non-linear Stages

Fuzzification of the DNFSM input multistage process' steps $\overline{u}_{si}[t]$ based on calculation of the average nearness between level of the process and the linear centre of the fuzzy set (2). This fact puts restrictions on two-dimensional fuzzy sets using. In form (2) it may be used only for fuzzification of the piecewise-linear input multistage processes $\overline{u}_{si}[t]$.

But real systems' modelling often leads to a necessity of working with input multistage processes with non-linear stages. Two-dimensional fuzzy sets in form (2) are not suitable to fuzzification of such processes, but it is possible to adapt form (2) for this problem.

Let us suppose that one stage of the DNFSM input multistage process looks like a non-linear power function. We can use two-dimensional fuzzy set with non-linear center and parameters c_{s0i}^{l} and c_{s1i}^{l} to fuzzificate such process [13].

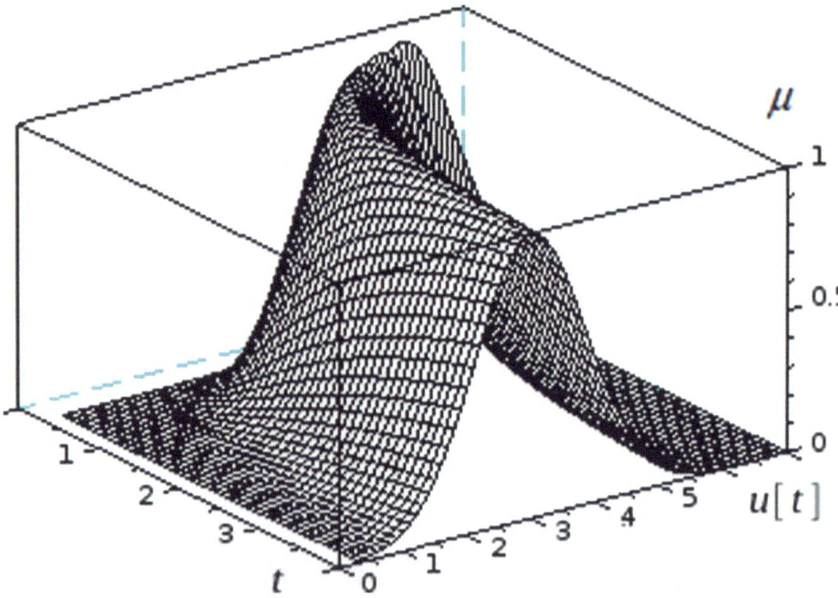

Fig. 4 Two-dimensional Gaussian fuzzy set with non-linear centre

Membership function of such fuzzy set will be as follows:

$$\mu_{si}^{l} = \sum_{j=1}^{n-1} \exp\left(c_{s0i}^{l}(t-j)^{c_{s1i}^{l}} - u_{si}(t-j)\right)^{2} / w_{si}^{l}. \quad (4)$$

Besides, two-dimensional fuzzy sets may be adapt for any type of non-linear process—logarithmic, parabolic, and other. Two-dimensional Gaussian fuzzy set with non-linear centre and constant width is shown in Fig. 4.

3 Parameter Identification of DNFSM

For DNFSM which multistage fuzzy input processes art described by the sets (2) and (4), we will provide algorithms of parameter configuration.

One of the most convenient method of neuro-fuzzy models identification comprising individual configuration of conclusions and antecedents. Parameters of the linear functions in the rule conclusions are configured through optimization. We should minimize sum-squared error function of the model. Fuzzy sets parameters in the rule antecedents are determined by the training set. So, optimization problem does not include antecedent parameters, which means decrease of its dimension and exact determination of parameters in rule conclusions.

When using this fuzzification mechanism, based on two-dimensional Gaussian fuzzy sets $B_s^l[t]$, we should determine parameters of the membership function (2). We can do it with composite method includes two phases.

On the first phase multistage input process falls into different stages, and on the second phase each stage we can describe by two-dimensional fuzzy set (2).

The following algorithm [12] is developed for identification of fuzzy sets (2). Notice that all stages of multistage input process are linear.

Algorithm 1. Numerical algorithm for evaluation of the linear centre's parameters for two-dimensional fuzzy sets (2).

Step 1. Suppose the k-th element of the DNFSM submodel's training set at a fixed value s has the form of $\{\overline{U}_s^k[t], y_s^k[t+1]\}$, and $k = 1, \ldots, K$. Here $\{\overline{U}_s^k[t] = \left[\overline{u}_{s1}^k[t], \ldots, \overline{u}_{sm}^k[t]\right]$ is a vector, where $\overline{u}_{si}^k[t] = \left[u_{si}^k[t], \ldots, u_{si}^k[t-n+1]\right]$ is the implementation of the stage with fixed number s of the i-th multistage input process, and m is a number of inputs.

Step 2. On the implementations of $\overline{u}_{si}^k[t]$ (for $k = 1, \ldots, K$) with the least squares method regression lines of the form $u = c_1 t + c_0$ are built. The result is a set of parameters $\left\{\left(c_1^1, c_0^1\right), \ldots, \left(c_1^K, c_0^K\right)\right\}$.

Step 3. Suppose $t_0 = t - n + 1$. Let us calculate $u_0^k = c_1^k t_0 + c_0^k$. Let us choose $u_0^{\max}$ and $u_0^{\min}$.

Step 4. Let us divide segment $\left[u_0^{\min}, u_0^{\max}\right]$ into Q parts. Here Q is the number of two-dimensional fuzzy sets $B_{si}^1[t], \ldots, B_{si}^Q[t]$ required for the fuzzification of the process $\overline{u}_{si}^k[t]$. As a result, domain of the process $\overline{u}_{si}^k[t]$ will be broken into Q parts.

Step 5. Coefficients c_0^k, c_1^k of the straight lines entering into fragments are averaged over each fragment. Averaged values $\left(c_{1a}^1, c_{0a}^1\right), \ldots, \left(c_{1a}^K, c_{0a}^K\right)$ are taken as the parameters of centres of input fuzzy sets $B_{si}^1[t], \ldots, B_{si}^Q[t]$ of the input number i of DNFSM submodel at fixed s.

Step 6. Steps 1–5 are repeated for each input $i = 1, \ldots, m$ of each DNFSM submodel, $s = 1, \ldots, |S|$.

For case when multistage input process includes non-linear stages, let us consider modification of the first algorithm [13]. Modification ensues from substitution for the step 2.

Algorithm 2. Numerical algorithm for configuring non-linear centres of two-dimensional fuzzy sets (4).

Step 2.1. According to the common method, values t and $u_{si}^k[t]$ (absciss and ordinate) of the input process are linearized, and then replaced by $T = \ln(t)$ and $U_{si}^k[t] = \ln\left(u_{si}^k[t]\right)$.

Step 2.2. On the values T and $U_{si}^k[t]$ (for $k = 1, \ldots, K$) using the least square method, regression lines of the form $u = c_1 t + c_0$ are built. The result is a set of parameters $\left\{(c_1^1, c_0^1), \ldots, (c_1^K, c_0^K)\right\}$ for non-linear centres of the two-dimensional fuzzy sets.

In order to identify the width parameters of fuzzy processes, which are described by two-dimensional fuzzy sets with non-linear centres (4), an algorithm that provides for fulfilment of the unity partition condition is proposed. Condition of the unity partition is fulfilled when for each input value u the sum of degrees of membership in all fuzzy sets A^l, $l = 1, \ldots, L$, equals to 1: $\sum_{l=1}^{L} \mu_{A^l}(u) = 1$. Fuzzy and neuro-fuzzy models show the best results, when the condition of unity partition for input fuzzy sets is fulfilled.

Algorithm for configuring width of two-dimensional fuzzy sets with linear centres based on the condition of unity partition was proposed in [12]. Here we propose a algorithm for fuzzy sets with non-linear centres (4).

Algorithm 3. Numerical algorithm for configuring width of two-dimensional fuzzy sets with non-linear centres.

Step 1. Suppose we are given the adjacent two-dimensional fuzzy sets $B^1[t]$ and $B^1[t]$ with the non-linear centres configured by Algorithmg 2, which surfaces are determined by the following expressions $exp\left[-\left(c_0^1 t^{c_1^1} - u(t)\right)^2 / w^1\right]$, $\exp\left[-\left(c_0^2 t^{c_1^2} - u(t)\right)^2 / w^2\right]$, and level $z = 0.5$, at which sets must intersect. Requirement $z = 0.5$ express fulfilment of the unity partition condition.

Step 2. Suppose we require that the line of fuzzy sets intersection (critical line) is lying on the level desired. Let us equate $exp\left[-\left(c_0^l t^{c_1^l} - u(t)\right)^2 / w^l\right] = 0.5$.

Step 3. Let us determine type of the function that assigns the width value. It is obvious that for the critical line to be non-linear power function, widths w^l must be assigned as $w^l(t) = w_0^2 t^{w_1^2}$, $l = 1, 2$. The equation of critical line will be as follows $u = (c_0^l - w_0^l \sqrt{\ln 2}) t^{c_1^l}$, $l = 1, 2$.

Step 4. Let us require that the critical line in OTU plane projection shall divide distance between the projections of fuzzy sets' centres in half. For the two adjacent two-dimensional sets with non-linear centres $B^1[t]$ and $B^2[t]$, the line extending between the centres is defined by the formula $u = c_0^{cr} t^{c_1^{cr}}$, where $c_1^{cr} = c_1^1 = c_1^2$, $c_0^{cr} = \left(c_0^1 + c_0^2\right)/2$.

Step 5. By equating the corresponding coefficients of power functions obtained in steps 3 and 4, let us calculate the unknown parameters of width w_1^l and w_0^l of two-dimensional fuzzy sets with non-linear centres $B^l[t]$: $w_1^l = \left(c_1^l - c_1^{cr}\right)/\sqrt{\ln 2}$, $w_0^l = \left(c_0^l - c_0^{cr}\right)/\sqrt{\ln 2}$, $l = 1, 2$.

4 Computational Experiment

Let us provide an example of two-dimensional fuzzy sets building based on Algorithm 2 and 3 for non-linear dynamic input process identifying. In this example were used data on ball pen lines discoloration [14]. It is known that a pigment amount decrease in the course of time, and this dependance may be described by power function.

Dynamic discoloration process has single stage and may be fuzzificated by two-dimensional fuzzy set with non-linear centre (4). Notice that measurements with mistakes is a reason to represent discoloration process in fuzzy form.

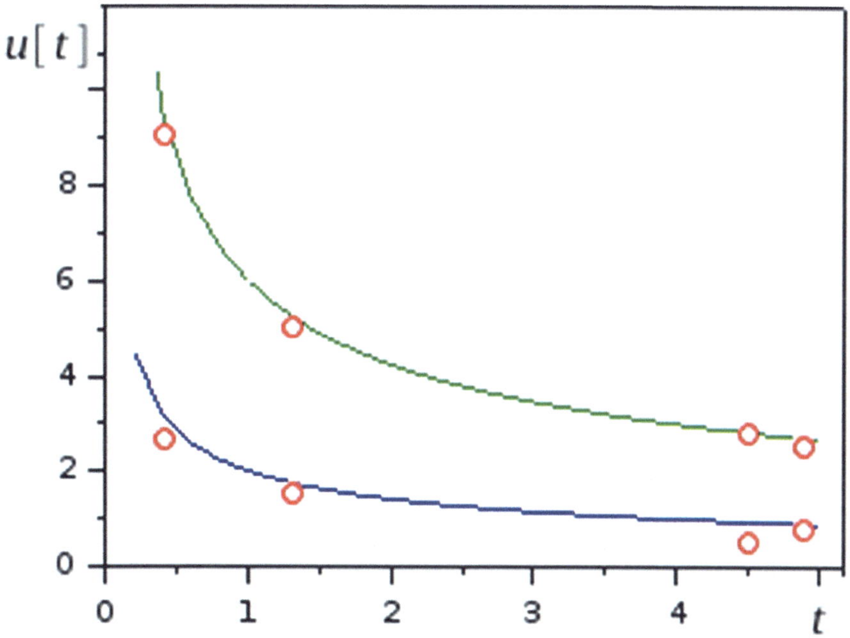

Fig. 5 Non-linear discoloration processes

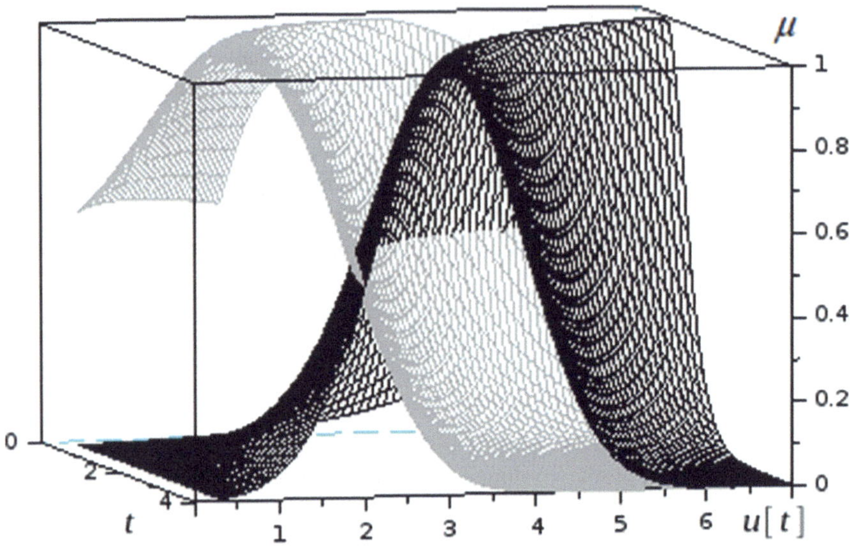

Fig. 6 Two-dimensional fuzzy sets for discoloration processes

Fuzzification includes configuring of the centre c and width w parameters. Considering fuzzy set has non-linear width, which should be defined with the condition of unity partition.

Values of the non-linear input dynamic process and lines represent non-linear centres of the fuzzy sets is shown in Fig. 5 (it was chosen $Q = 2$). Let us note that small amount of data for regression lines is due to features of the experiment on the pigment amount estimation.

Two-dimensional fuzzy sets for discoloration processes is shown in Fig. 6, where we can see that the condition of unity partition for input fuzzy sets is fulfilled. Parameters of non-linear centres and widths were evaluated according to Algorithms 2 and 3.

5 Conclusion

Necessity of DNFSM with two-dimensional fuzzy sets using follows from great amount of dynamic technological processes with complex and varying functioning laws. For example, in the articles [13, 15] DNFSM and two-dimensional fuzzy sets using for sugar boiling modelling was investigated. Main advantages of DNFSM are ability to work with large amount of input data and ability to model complex technological processes with variable structure.

The main results of the article: class of DNFSM for multistage dynamic process modelling is described; two-dimensional fuzzy sets with non-linear centres and nonlinear width to simplify DNFSM parameter identification are proposed; algorithm for configuring parameters of two-dimensional fuzzy sets with non-linear centres with the unity partition condition is proposed.

The results show us that two-dimensional fuzzy set is a convenient tool for fuzzy dynamic input processes identification. Two-dimensional fuzzy sets simplify DNFSM training and allow to identify multistage piecewise-linear and piecewise-nonlinear input processes. This fact confirm efficacy of DNFSM for technological processes modelling.

Acknowledgements The reported study was supported by the Russian Foundation For Basic Research within the framework of the scientific research 17-47-480305 r_a.

References

1. A. Piegat, *Fuzzy Modeling and Control* (Binom, Moscow, 2009)
2. R. Babuska, Neuro-fuzzy methods for modeling and identification, in *Recent Advances in Intelligent Paradigms and Applications* (2000), pp. 161–186
3. R. Babuska, Z. Lendek, T. Guerra, B. Schutter, *Stability Analysis and Nonlinear Observer Design Using Takagi-Sugeno Fuzzy Models* (Springer, Berlin, Heidelberg, 2011)
4. P. Saraev, M. Zhuralyova, O. Nazarkin, V. Alexeev, ANFIS training and refining based on clustering of training data set. Chernozem Sci. Bull. **2**, 54–60 (2015)
5. Q. Zhou, Y. Wu, C. Chan, From neural network to neuro-fuzzy modeling: applications to the carbon dioxide capture process. Energy Procedia **4**, 2066–2073 (2011)
6. S. Rubinstein, A. Goor, A. Rotshtein, Time series forecasting of crude oil consumption using neuro-fuzzy inference. J. Ind. Intell. Inf. **3**(2), 84–90 (2015)
7. D. Enke, N. Mehdiyev, A hybrid neuro-fuzzy model to forecast inflation. Procedia Comput. Sci. **36**, 254–260 (2014)
8. Yu. Kudinov, I. Kudinov, S. Suslova, *Nechetkie modeli dinamicheskih protsessov* [Fuzzy models of dynamic processes] (Nauchnaja kniga [Science book], Moscow, 2007)
9. D. Liberzon, *Switchings in Systems and Control* (Birkhauser, 2003)
10. V. Lubentsov, Intellectual control system with variable structure on the basis of fuzzy logic. Fundam. Res. **12**, 252–257 (2014)
11. N. Zhbanova, S. Blyumin, A. Shmyrin, Identifikatsia parametrov vkhodnikh nechetkikh protsessov raznostnikh neironechetkikh perekljuchaemikh modelei. Sistemi upravlenija i informatsionnie tekhnologii [Control Syst. Inf. Technol.] **1**(55), 8–12 (2014)
12. N. Zhbanova, O. Kravets, M. Grigoriev, L. Babich, Neuro-fuzzy modeling and control of multistage dynamic processes that depend on inputs with uncertainty elements. JTAIT-LLS J. Theor. Appl. Inf. Technol. **80**(1), 2–12 (2015)
13. N. Zhbanova, S. Blyumin, Parametricheskaya identifikatsia kusochno-lineinikh i kusochno-nelineinikh nechetkih protsessov. Vestnik IrGTU [Proc. IrSTU] **11**, 84–93 (2016)
14. B. Sitnikov, A. Venevtsev, N. Zhbanova, S. Zhbanov, Osobennosti postroenija dinamicheskoi modeli starenija shtrikhov rekvisitiov dokumentov. Vestnik LGTU [Proc. LSTU] **2**(28), 35–42 (2016)
15. N. Zhbanova, Osobennosti identifikatsii nechetkikh mnogoetapnikh protsessov [Features of identifying fuzzy multistage processes], in *Materialy III Mezhdunarodnoi konferentsii 'Intellektualnye tekhnologii obrabotki informatsii I upravleniia'* [Materials of the 3rd International Conference on Intelligent Technologies for Information Processing and Management], no. 1 (2015), pp. 154–158

Fuzzy Algorithms and Fuzzy Cognitive Map

The Effects of Preprocessing on Colorectal Polyp Detecting by Fuzzy Algorithm

Brigita Sziová, Szilvia Nagy, and László T. Kóczy

Abstract In the following study the effects of two image preprocessing methods, namely Gaussian filtering and Wiener filtering, is studied on the results of a fuzzy inference method previously developed by the authors, for determining whether a colonoscopy picture segment contains any colorectal polyp. As earlier results show that less blurry, less compressed and less noisy images tend to be better classifiable, the effects of noise suppression with a Gaussian filter, which makes the images also blurrier, was beneficial on noisy, compressed images, and rather maleficent in good quality pictures. The effects of the Wiener filter, which both decreases noise and enhances edges, did not really manifest in classification improvement.

1 Introduction

Medical image processing can provide help for the medical staff in many types of diagnosis. Human diagnosis will not be automated, only visual aid can be given to the trained medical personnel.

One of the most dangerous cancer types is the colorectal cancer, the main reason for its dangerousity is that it is usually diagnosed in an advanced state. The tumours

B. Sziová (✉)
Department of Information Technology, Szechenyi Istvan University, Győr, Hungary
e-mail: szi.brigitta@sze.hu

S. Nagy
Department of Telecommunications, Szechenyi Istvan University, Győr, Hungary
e-mail: nagysz@sze.hu

L. T. Kóczy
Department of Information Technology, Szechenyi Istvan University,
Budapest University of Technology and Economics, Győr, Hungary
e-mail: koczy@sze.hu

S. N. Shahbazova et al. (eds.), *Recent Developments and the New Direction in Soft-Computing Foundations and Applications*, Studies in Fuzziness and Soft Computing 393, https://doi.org/10.1007/978-3-030-47124-8_28

in the colorectal tract of the bowel start as polyps, little warts on the bowel wall. Not all the colorectal polyps develop into cancer, some types are completely beneficent only about 10% is dangerous. The most reliable method for determining, whether a polyp is maleficent or beneficent is biopsy, however, in many cases the pit pattern of the polyp can give high reliability results as well [1]. The shape of the polyp [2] is also important. Pit pattern based classification was developed for polyps protruding into the bowel, while flat and depressed polyps can not be classified with this method. Our main goal is to develop a method for classification of polyps in the colorectum, however, as a first step we aim for no full classification, just an identification of protruding polyps.

We used three databases [3–5] of colonoscopic images consisting of pictures of a bowel part with polyp and their masks (the masks are black and white pictures with the white domain indicating the polyp area). The first database [3] is built from 380 pictures, 18 series, where the resolution is medium, 574 × 500, this database was used in our previous works. The pictures in the first database [5] are small, 384 × 288 pixels, and there are long series of images from the same polyp, but their quality is the lowest among the samples. There are all together 29 series of pictures in this database with 612 pictures. The third group [4] contains large images and short sequences, the total number of pictures is 196, the number of the series is 35, and the resolution is 1225 × 966. These images are of the best quality, and no compression artifacts are visible on them, moreover, the focus is also almost always correct, thus the patterns of the polyps and the bowel wall are almost always well distinguishable.

Previously, the group of the authors developed a fuzzy classification method and tested it on the second database [6, 7]. The algorithm is the following: the images are cut into tiles, and to each of the tiles the polyp percentage of its area is calculated from the masks provided in the database. Also 25–75 antecedent parameters were calculated for all of the tiles. Based on every second tile, i.e., a uniformly distributed 50% of the tiles, a two-output (polyp–no polyp) fuzzy classification algorithm is parameterized. The remaining tiles (the ones that did not take part in training the classification algorithm, also 50% of the total tiles) were used for testing the method.

The idea behind cutting the images into tiles is, that the thus arising grid with polyp-no polyp markers in the tiles can provide a good guide of the eye for the diagnosist, moreover it can be a basis of a more advanced image clustering technique applied afterwards, like active contour methods [8, 9]. As in the previous approaches we did not use any image preprocessing, in the present work we study the effect of noise filtering on the results of the classification. The main motivation is that the new databases are of different quality, and the results in the second, low-quality database are much worse than in the case of the old, medium sized images, and the large images perform even better than our previous results.

As a first step, in Sect. 2 we summarise the effects of noise and the methods for noise filtering. Next, in Sect. 3 the fuzzy classification scheme together with the applied stabilized KH rule interpolation and [10, 11] the used antecedent dimensions are reviewed. The effects of the different types of filtering are summarized in Sect. 4, and the conclusions are given in the last section.

2 Image Filtering

As the image qualities of the databases are not uniform, and the classification method's applicability is also different for the different databases (especially as the small images give extremely bad results, while the large images practically almost always lead to excellent results), it is natural to test, whether image quality enhancements, like noise filtering, influences the performance of the classification scheme.

2.1 *On Noise and Distortions in Pictures*

Image noise is a random fluctuation in the pixel information (both in brightness and in colour) [12]. Noise represents unsought, misleading information decreasing image quality. The source of this noise can be diverse, from point disturbances in the CCD-s acquiring the image to image compression artifacts. The types of noise are amplifier noise, speckle noise, salt and pepper noise, shot noise. In our case the images contain additive noise, thus we selected filters that can reduce additive noise, such as Gaussian and Wiener filters.

2.2 *Gaussian Filtering*

Gaussian filters change the pixel intensity into the weighted average of the surrounding pixel intensities. The weights are larger around the center, i.e., the pixel to be substituted. This property is favourable because it does not increase the blur in the image too much. It is easy to calculate the weights as Gaussians are simple linear low-pass convolutional filters, and the matrix elements are computed as

$$g(i,j) = \frac{1}{2\pi\sigma^2} e^{-\frac{i^2+j^2}{\sigma^2}} \tag{1}$$

with σ determining the width of the Gaussian, i.e., it indicates how rapidly the matrix elements decrease to 0 with the distance from the center. This means, that if σ is large, the effect of the Gaussian filter is very similar to that of the mean filter, it makes the edges blurry, while the small σ belongs to very rapidly decreasing filter coefficients.

In the case of 2D images, Gaussian filters are mostly implemented as combinations of 1D filters, as Gaussian filters are separable. Gaussian filters are usually also direction-independent.

A Gaussian filtered image and its original can be seen in the first two subplots of Fig. 1.

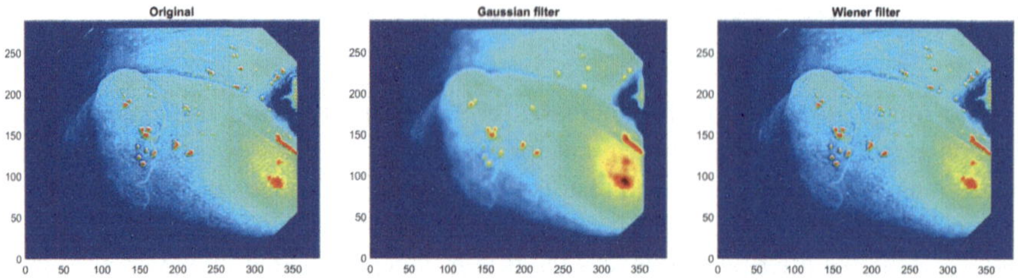

Fig. 1 Original, Gaussian ($\sigma = 2$) filtered and Wiener (5 by 5) filtered version of the blue colour channel of image 540 from database [5]

2.3 *Wiener Filtering*

Although the noise in the pictures can be decreased or eliminated with Gaussian filters, at the same time, the image becomes blurrier. As we would like to use the patterns of the bowel or polyp surface, blurring can cause decrease in the classification performance. The filter type which can decrease the noise level and simultaneously does not increase the blur significantly, or even enhance the edges, is the Wiener filter. The Wiener filter has a statistical background, it builds on the a-priori knowledge about the noise. The mathematical formula of the Fourier transform of the filter is

$$W(i, j) = \frac{H^*(i, j)}{|H(i, j)|^2 + \frac{S_n(i,j)}{S_s(i,j)}}, \tag{2}$$

where the power spectrum of the signal, i.e., the image and the noise are S_s, and S_n, respectively, and the filter that decreases the noise and increases the blur is represented by H. The sign * stands for complex conjugation.

The effect of Wiener filtering can also be seen in Fig. 1. It is visible that while the Gaussian filter does decrease the fine structure of the image, the Wiener filter does not, however, extra edges may appear as well.

3 Fuzzy Rule Interpolation and Fuzzy Classification Scheme

In order to determine, whether the studied tile contains any polyp, we apply fuzzy classification.

3.1 *On Fuzzy Classification*

The concept of sets with non-crisp boundaries was introduced by Zadeh [13, 14]. This concept is in accordance with the natural way of human thinking, thus it is excellently suitable for modelling biomedical problems and diagnosis support. Usually there is no rigorous boundary between two extreme properties, like e.g., the weather being warm or cold, i.e., belonging to the set warm does not necessarily exclude belonging to the set cold at the same time. In our case, most of the image tiles consist only parts of polyps, thus it is logical to use fuzzy classification for them. Our method uses Mamdani's fuzzy inference system [15]. This algorithm checks the membership values of the match of the measured data in the fuzzy antecedent sets of the rules. Each rule consists of fuzzy sets of the antecedents, i.e., the input parameters, and corresponding consequent sets (or, in some cases, values). For each rule, the conclusion result is calculated as the t-norm (minimum, in our case) of the weights obtained from the membership values of the degrees of matching between measured values and antecedent sets, and the consequent sets (or values), and the overall conclusion is the s-norm (in our case, maximum) of the resulting consequents of each rule. This may be still defuzzified by some method (like Center of Gravity, Center of Area, or Mean of Maxima, e.g.).

3.2 Fuzzy Rule Interpolation

The classification scheme, however, typically has sparse rulebases, with no matching between the measured values and any of the rule antecedents, where the Mamdani-algorithm cannot be applied. In such a case (which is the usual case for the colorectal images) interpolation of the rules is needed. While many fuzzy rule interpolation approaches are known from the literature, for its straightforwardness and mathematical stability, for overcoming the problem of sparseness, in this application we selected the Stabilized KH Interpolation algorithm [10, 11, 16], which can be described by

$$\inf\{B_\alpha^*\} = \frac{\sum_{i=1}^{2n}\left(\frac{1}{d_{\alpha L}(A^*,A_i)}\right)^k \inf\{B_{i\alpha}\}}{\sum_{i=1}^{2n}\left(\frac{1}{d_{\alpha L}(A^*,A_i)}\right)^k} \tag{3}$$

$$\sup\{B_\alpha^*\} = \frac{\sum_{i=1}^{2n}\left(\frac{1}{d_{\alpha U}(A^*,A_i)}\right)^k \sup\{B_{i\alpha}\}}{\sum_{i=1}^{2n}\left(\frac{1}{d_{\alpha U}(A^*,A_i)}\right)^k}. \tag{4}$$

Here, B_α means the α-cut of the consequent, A_i is the ith rule's antecedent fuzzy set, B_i is its consequent, and A^* is the measured antecedent parameter. The distance $d_{\alpha L}$ is the lower bound and $d_{\alpha U}$ is the upper bound of the Euclidean distance between the α-cuts of its variables. Previously, we have successfully applied a similar complex method for the classification of telecommunication network lines [17].

3.3 On the Selected Antecedent Dimensions

In our previous research [6, 7, 18], we first determined possible antecedent dimensions. We treated the three colour channels (red, green, blue) separately, and as the most straightforward step, we calculated the mean and the standard deviation of the intensity value in the tile for each. Next, after Canny edge filtering [19], we also calculated the mean of the edge intensity within the tile, thus receiving an edge density parameter. The structural entropies [20–23] are two parameters characterising the general shape of the distribution, they both originated from Rényi entropies. Also, the gradients of the image are used, the magnitude, the direction, and the x and y components of the gradients, as well. These properties are calculated not only for the image segment to be evaluated, but for its wavelet transforms, too [24].

4 Results

Using the above mentioned 99 input parameters we implemented a fuzzy classification method. The tile sizes were different for the various databases in order to achieve approximately the same rate between the tile sizes and the picture sizes. The first database [3] of medium-sized images were cut into tiles of size 60×60 and 100 × 100, the second database [5] with small images had 30 × 30 and 50 × 50 tile sizes, and the third database, that contained the largest images, had 120 × 120 and 200 × 200 tile sizes.

In order to test, why the different image sizes result in different classifiability, we used Gaussian filtering with a not too wide filter width in order to eliminate the noise and the artifacts from the image, and Wiener filter to both for suppressing the noise and enhancing the edges. The rates of false classification can be seen in Fig. 2.

As it can be seen in Fig. 1, Gaussian filtering smoothens the edges, but makes the pictures blurry. This effect is beneficial in the case of the smallest-sized images, as they contain the most noise. This fact is visible in Fig. 3: in some cases the rate of the wrong classifications decreased in the middle part of the image group number (image group number 19–44) domain. In the other cases, Gaussian filtering neither improves nor worsens the results. It can also be concluded, that the larger tile sizes give better results, thus later we will present only the 100 × 100, 50 × 50, 200 × 200 tile-size series results. The other, 60 × 60, 30 × 30, 120 × 120 tile size series produce in the case of the Wiener filter also worse results, than the larger tile sizes.

In the case of the Wiener filtering, the edges remain intact while the noise is suppressed. These effects do not influence the performance of the classification, as it can be seen in Fig. 4.

As it can be seen form the previous figures, the rate of the false negative classification is very high for almost all the image groups. The reason for this is partially the fact that there are many of image segments that contain only a small area of polyp. In order to demonstrate this, we introduced the area corrected (fuzzy)rate of wrong

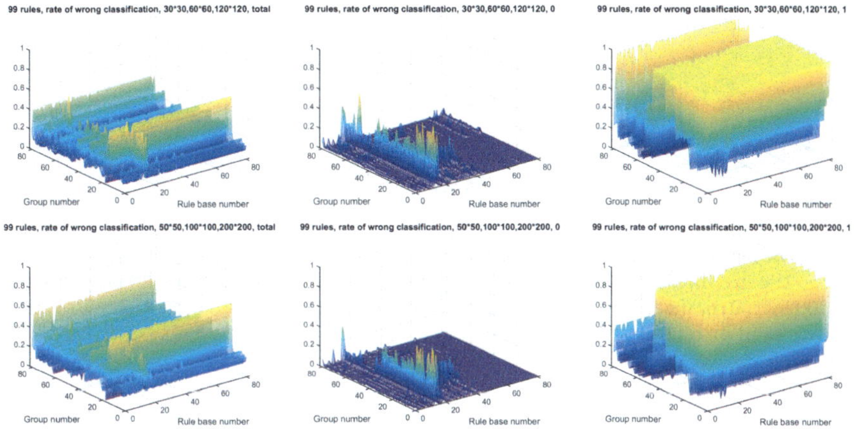

Fig. 2 The rate of false classification for each rulebase and each image group for the original images, namely, tile size 60 by 60 for the first 18 image groups, 30 by 30 for the second 25 image groups, and 120 by 120 for the last 29 image groups. The first subplot is: total rate, the second subplot: rate of false positive, and the third subplot: rate of false negative classification

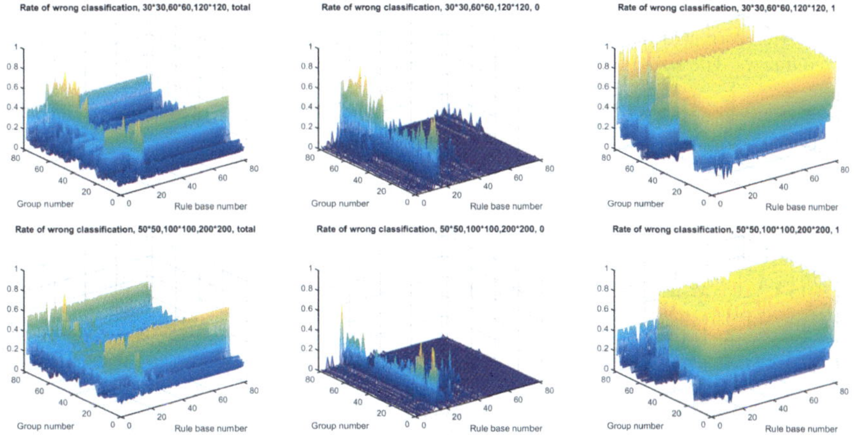

Fig. 3 The rate of false classification for each rulebase and each image group for Gaussian filtered images. First row tile size: 60 by 60 for the first 18 image groups, 30 by 30 for the second 25 image groups, and 120 by 120 for the last 29 image groups, second row tile size: 100 by 100 for the first 18 image groups, 50 by 50 for the second 25 image groups, and 200 by 200 for the last 29 image groups. First column: total rate of false classification, second column: rate of false positive, third column: rate of false negative classification

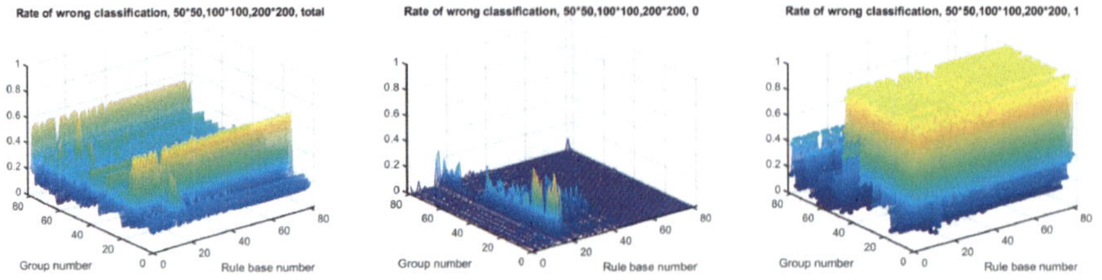

Fig. 4 The rate of false classification for each rulebase and each image group for Wiener filtered images. Tile size 100 by 100 for the first 18 image groups, 50 by 50 for the second 25 image groups, and 200 by 200 for the last 29 image groups. First subplot: total rate. Second subplot: rate of false positive. Third subplot: rate of false negative classification

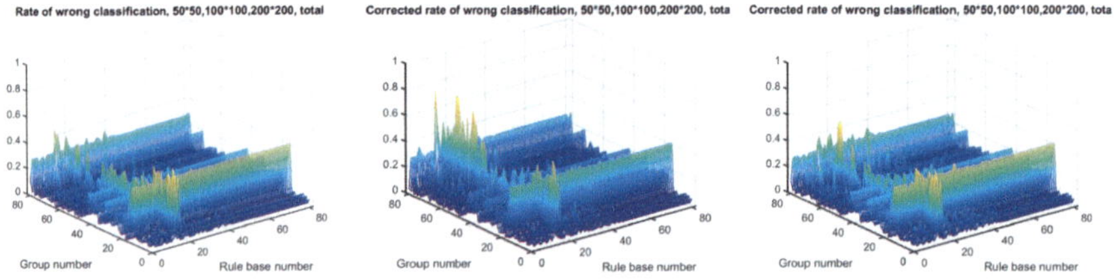

Fig. 5 Area corrected rate of false classification for the original, the Gaussian and the Wiener filtered images. Tile sizes: 100, 50, and 200. Total rate

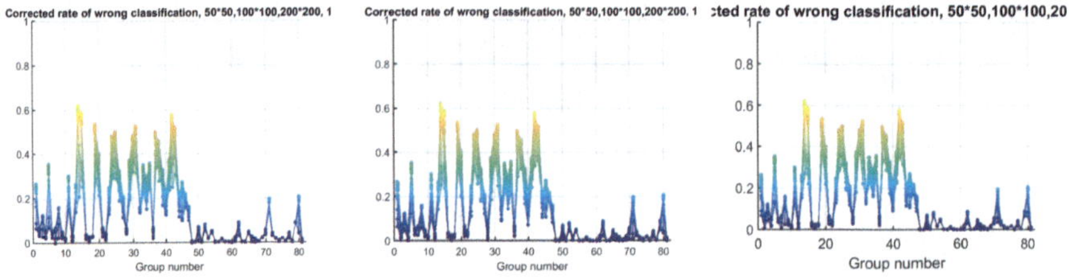

Fig. 6 Area corrected rate of false negative classification for original, Gaussian and Wiener filtered images. The classification success rate does not depend much on the rulebase, this is the reason for plotting the 2D projection of the 3D figures similar to the ones in Fig. 5

classification, i.e., if the area of the tile is covered by polyp in only e.g. 30% of the total area, then the tile is only 30% wrong, not 100%. This way the results became much better, as it can be seen in Fig. 5.

For a better view, we plotted the area corrected rate of false negative classifications in Fig. 6.

ROC plots give another type of visualization for the results, as they give the rate of true positive classifications as the function of false positive classifications. As we have 82 image groups and 82 different rulebases, the ROC plot is extremely crowded, as it can be seen in Fig. 7. It can be concluded, however, that most of the rulebases give acceptable false positive rates.

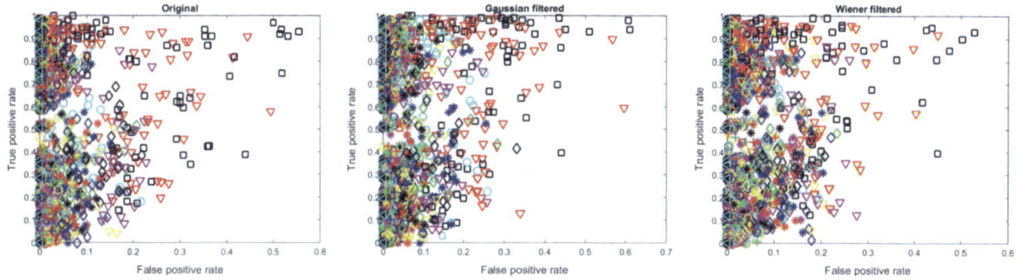

Fig. 7 ROC plot for original, Gaussian and Wiener filtered images. The colours mean different rulebases

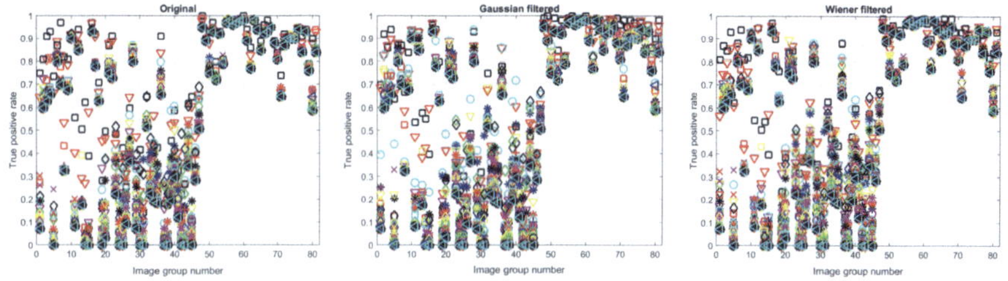

Fig. 8 True positive rates for different image groups and different rulebases for original, Gaussian and Wiener filtered images. The image group number is the horizontal axis plot. The colours indicate different rulebases

Focusing on the true positive rate, gives more visible and interpretable results. In Fig. 8 we present the true positive rate as a function of the image groups. It can be seen that the first 18 image groups perform differently, the best results arise in the case of images with visible vein structure, or different, very contrasty colour. In the second, low resolution image group the results are originally rather bad, but the improvement after Gaussian filtering is visible. The last 29 image groups, i.e., the ones with the highest resolution and best quality, perform very well, however, they are not improved by filtering.

5 Conclusion and Future Work

A method for classifying colonoscopy image tiles according to their polyp content was proposed. The method uses 99 antecedent dimensions, including the mean intensities, the standard deviations, the edge densities, the structural entropies and the gradients of the 3 colour channels of the original and wavelet transformed pictures. A fuzzy inference system was used with stabilized KH rule interpolation, because of the sparsity of the available knowledge.

The method was tested on 3 colonoscopy picture databases of 3 different image sizes and qualities. Two filtering methods were tested in order to decide how image edge enhancing, blurring and noise suppression influenced the performance of the classification method. It could be concluded, that only in the case of low-resolution, noisy images the Gaussian filtering improved the classification rates only slightly, whereas the Wiener filtering did not influence the results in any positive way.

In the future, various other preprocessing techniques will be tested, based on combined filtering and edge detection, with and without applying wavelet transform on the images. An efficient feature extraction method applied first might reduce the computational complexity and thus provide more opportunity to more complex preprocessing and clustering technologies.

Acknowledgements The authors would like to thank the financial support of the project EFOP-3.6.2-16-2017-00015 HU MATHS-IN-Intensification of the activity of the Hungarian Industrial Innovation Mathematical Service Network and the ÚNKP-17-4 New National Excellence Programme of the Ministry of Human Capacities of Hungary. This work was supported by National Research, Development and Innovation Office (NKFIH) K108405, K124055 and GINOP-2.3.4-15-2016-00003.

References

1. S. Kudo, S. Hirota, T. Nakajima et al., Colorectal tumours and pit pattern. J. Clin. Pathol. **47**, 880–885 (1994)
2. J.R. Jass, Classification of colorectal cancer based on correlation of clinical, morphological and molecular features. Histopathology (Wiley) **50**, 113–130 (2006)
3. J. Bernal, F.J. Sanchez, F. Vilariño, Towards automatic polyp detection with a polyp appearance model. Pattern Recognit. **45**, 3166–3182 (2012)
4. J.S. Silva, A. Histace, O. Romain, X. Dray, B. Granado, Towards embedded detection of polyps in WCE images for early diagnosis of colorectal cancer. Int. J. Comput. Assist. Radiol. Surg. **9**, 283–293 (2014)
5. J. Bernal, F. J. Sanchez, G. Fernández-Esparrach, D. Gil, C. Rodrígez, F. Vilariño, WM-DOVA maps for accurate polyp highlighting in colonoscopy: validation vs. saliency maps from physicians. Comput. Med. Imaging Gr. **43**, 99–111 (2015)
6. Sz. Nagy, F. Lilik, L. T. Kóczy, Entropy based fuzzy classification and detection aid for colorectal polyps, in *IEEE Africon 2017*, Cape Town, South Africa, 15–17 September 2017 (2017)
7. Sz. Nagy, F. Lilik, L. T. Kóczy, On fuzzy classification with interpolation of the sparse rule bases, in *ESCIM 2017*, Faro, Portugal, 6–8 October 2017 (2017)
8. M. Kass, A. Witkin, D. Terzopoulos, Snakes: active contour models. Int. J. Comput. Vis. **1**, 321–331 (1988). https://doi.org/10.1007/BF00133570
9. V.M. Georgieva, S.G. Vassilev, *Kidney segmantation in ultrasound images via active contours, in 11th International Conference on Communications* (Electromagnetics and Medical Applications, Athens, Greece, 2016)
10. L.T. Kóczy, K. Hirota, Approximate reasoning by linear rule interpolation and general approximation. Int. J. Approx. Reason. **9**, 197–225 (1993)
11. D. Tikk, I. Joó, L.T. Kóczy, P. Várlaki, B. Moser, T.D. Gedeon, Stability of interpolative fuzzy KH-controllers. Fuzzy Sets Syst. **125**, 105–119 (2002). https://doi.org/10.1016/S0165-0114(00)00104-4
12. V. Govindaraj, G. Sengottaiyan, Survaey of image denoising using different filters. Int. J. Sci. Eng. Technol. Res. **2**, 344–351 (2013)
13. L.A. Zadeh, Fuzzy sets. Inf. Control **8**, 338–353 (1965)
14. L.A. Zadeh, Fuzzy algorithms. Inf. Control **12**, 94–102 (1968)
15. E.H. Mamdani, S. Assilian, An experiment in linguistic synthesis with a fuzzy logic controller. Int. J. Man-Mach. Stud. **7**, 1–13 (1975)

16. L.T. Kóczy, K. Hirota, Interpolative reasoning with insufficient evidence in sparse fuzzy rule bases. Inf. Sci. **71**, 169–201 (1993)
17. F. Lilik, J. Botzheim, Fuzzy based prequalification methods for EoSHDSL technology. Acta Technica Jaurinensis **4**(1), 135–144 (2011)
18. V.M. Georgieva, Sz. Nagy, E. Kamenova, A. Horváth, *An approach for pit pattern recognition in colonoscopy images*. Egypt. Comput. Sci. J. **39**, 72–82 (2015)
19. J. Canny, A computational approach to edge detection. IEEE Trans Pattern Anal. Mach. Intell. **8**, 679–698 (1986)
20. J. Pipek, I. Varga, Universal classification scheme for the spatial localization properties of one-particle states in finite d-dimensional systems. Phys. Rev. A **46**, 3148–3164 (APS, Ridge NY-Washington DC) (1992)
21. Cs. Dominkovics, I. Mojzes, et al., *Heat treatment parameters effecting the fractal dimensions of AuGe metallization on GaAs*. Appl. Phys. Lett. **91**, 073107 (2007)
22. L.M. Molnár, Sz. Nagy, I. Mojzes, Structural entropy in detecting background patterns of AFM images. Vacuum **84**, 179–183 (Elsevier, Amsterdam) (2010)
23. A. Bonyár, AFM characterization of the shape of surface structures with localization factor. Micron **87**, 1–9 (2016)
24. I. Daubechies, *Ten Lectures on Wavelets, CBMS-NSF Regional Conference Series in Applied Mathematics*, vol. 61 (SIAM, Philadelphia, 1992)

Stability of Fixed-Point Values in Reduced Fuzzy Cognitive Map Models

Miklós F. Hatwágner and László T. Kóczy

Abstract The authors have already presented their method for reducing oversized FCM models, and also have analyzed the prediction error of the reduced models. These investigations assumed that models have a single fixed-point attractor. The novelty of this paper is that it deals with the stability behavior of the fixed-point attractor value of original-reduced model pairs and compares the number of fixed-point attractors found, the asymptotic values of the concepts, and also checks if any limit cycles or chaotic behavior occur. The method of comparison and also the first results made with two real-life and one synthetic model are presented and some conclusions are taken.

1 Introduction

Fuzzy Cognitive Maps (FCMs) are widely applied in practice to support various decision or policy making tasks [1–5]. The technique has gained great popularity mainly because it can be easily used by experts of different fields, the explored causal connections among different factors of the investigated systems can be described in a straightforward, clear way, and simulations can be performed to predict the consecutive states of systems.

One drawback of the method is that the increasing number of modeled factors quadratically increases the possible connections among them. If the model is created by using the experiences and knowledge of experts, the iterative refinement of inaccurate models not only needs much valuable human resources and is rather

M. F. Hatwágner (✉)
Department of Information Technology, Széchenyi István University,
Egyetem Tér 1, Győr, Hungary
e-mail: miklos.hatwagner@sze.hu

L. T. Kóczy
Department of Telecommunication and Media Informatics,
Budapest University of Technology and Economics, Budapest, Hungary
e-mail: koczy@sze.hu; koczy@tmit.bme.hu

S. N. Shahbazova et al. (eds.), *Recent Developments and the New Direction in Soft-Computing Foundations and Applications*, Studies in Fuzziness and Soft Computing 393, https://doi.org/10.1007/978-3-030-47124-8_29

time consuming, but it may also lead to unmanageably large models. The handling difficulties of large models has led to various model reduction techniques [6–8]. A proper reduction technique allows experts to create a single, even rather large model that obviously meets the accuracy requirements, then to reduce this original model to varying degrees in an automated way until they find the best balance between model accuracy and complexity. The effect of a specific model reduction method on FCM models prediction capabilities has been already investigated in [9]. The prediction errors of reduced models was measured, and also recommendations were given for parameter values and algorithm selection. This investigation concentrated on models that have a single stable outcome of simulations, however.

This paper is the first attempt to examine the effect of model reduction on behavioral properties. The reduction may cause more or less stable states, repeating sequences of the same state vectors or chaotic behavior. First, a short overview of FCMs is given, followed by a brief description of the applied model reduction technique. In the next section the selected models for behavioral examination and their most important properties are presented. It is followed by the summary of simulation results, and finally some conclusions are taken.

2 Brief Introduction to the Applied Methods

2.1 Fuzzy Cognitive Maps

The idea of Cognitive Maps originates from Robert Axelrod [10], who used this method to describe the causal connections among political elites and their possible actions, in the '70s. His technique was further developed to its fuzzy extension by Bart Kosko [11], during the '80s. During the next decades, FCM became more and more popular among researchers, and several improvements of the original method were suggested [12].

FCMs are signed, weighted, bipolar fuzzy graphs [13]. The nodes of the graph represent the important factors of the model, and are called 'concepts' in the literature. Concepts will be denoted by C_i. The states of concepts are expressed by a value usually normalized to [0, 1], in some rare cases to the $[-1, +1]$ interval. The current value of the concept at a specified discrete time step t is often referred to as the 'activation value', and is denoted by A_i^t. For example, if the concept is associated with the fluid level of a tank, the value 0 expresses the empty tank, while 1 means that the tank is full of liquid [14].

The edges of the graph express the direction and strength of causal relationships among concepts by their weights. The weight of an edge between concepts C_i and C_j is usually denoted by w_{ij}, and its value falls in the $[-1, +1]$ interval. The sign of the value expresses the amplifying or inhibiting nature of this connection, and the absolute value of the weight defines the strength of it. The lack of causal relationship is expressed by a zero-weight edge.

If the model contains n concepts, their relationships can be collected in the so called 'connection matrix', denoted by $W = [w_{ij}]$. According to Kosko's original definition, the main diagonal contains only zeros. With other words, the concepts cannot influence their own state directly, they do not remember their earlier states. It is easy to see however, that some real life systems behave exactly this way. Modifications were suggested in [15] to the original FCM in order to compensate this deficiency. The models investigated in our study contain '1'-s in the main diagonal of their connection matrices.

Probably the most important ability of FCMs is that simulations can be executed and the future state of concepts can be predicted. If the initial states of concepts (also called 'scenario') and the connection matrix are given, the next state vector of concepts can be calculated by the activation function: $A : [W, A_i^{(t)}] \rightarrow A_i^{t+1}$. In order to guarantee that activation values will remain in the allowed interval, the 'threshold function' f is used in the activation function. The authors used Eq. 1 during their investigations.

$$A_i^{t+1} = f\left(\sum_{j=1}^{N} w_{ji} A_j^t\right) \tag{1}$$

Several widely applied versions of the threshold function exist [16], but the authors used the rather popular 'sigmoid' function defined by Eq. 2. Its parameter λ defines the steepness of the function. The value of λ can be any positive number, but usually $\lambda = 5$ is applied in the literature. With smaller values the function approximates a linear function better, and may change the model behavior significantly.

$$f(x) = \frac{1}{1 + e^{-\lambda x}} \tag{2}$$

According to [16], the FCM with continuous concept values can behave in three different ways during simulations: (1) it may converge to one of the stable, equilibrium states called *fixed-point attractors* (FPs). (2) If a sequence of state vectors appear repeatedly, a *limit cycle* is found. (3) *Chaotic behavior* can also be observed, when the activation values never stabilize.

Sometimes its useful to classify the concepts into three groups: the *input concepts* are not influenced by other concepts, but they influence others. The consequence of the lack of incoming edges is that they preserve their initial values during simulations. *Intermediate concepts* have both incoming and outgoing edges. The *output concepts* are influenced by other concepts, but they do not have any outgoing edges. Decision makers are often interested in the fixed-point values of only these latter concepts.

Algorithm 1 The *buildCluster* function

```
function BUILDCLUSTER(initialConcept, ε)
    c ← {initialConcept}
    for i ← 0; i < n; i + + do
        if i ≠ initialConcept then
            member ← true
            while member and HASNEXTELEMENT(c) do
                j ← NEXTELEMENT(c)
                member ← ISNEAR(j, i, ε)
                if member then
                    c ← c + {i}
                end if
            end while
        end if
    end for
    return c
end function
```

2.2 Model Reduction

The authors presented their model reduction method in [17, 18], and later further developed it in order to handle models with differentiated input, intermediate and output concepts [9]. In order to save space, only the original method is presented here briefly.

The reduction method can be considered as generalization of the state reduction of finite state machines and sequential systems with partially specified states [19]. The main idea of the method is similar to clustering and based on fuzzy tolerance relations. The state reduction creates clusters of similar concepts and uses these clusters as concepts in the reduced model. The similarity of concepts can be measured with several distance metrics, but in this paper only normalized, squared Euclidean distance is used.

Function *buildCluster* (see Algorithm 1) is the most important part of the method. It starts the creation of (the fuzzy) cluster K_i by putting C_i in it. Later it compares the similarity of other concepts with all the current cluster members using the similarity metrics *isNearC* defined in [17] and decides about the expansion of the cluster with the concept under investigation. Experts can influence the reduction rate of models by defining the value of the design parameter ε, which defines the threshold of similarity. Another function, called *buildAllClusters* initiates the creation of n clusters by *buildCluster*, where n is the number of concepts. If the same cluster appears more than once, only one of them will be kept.

Several methods were proposed to calculate the weights of connections among clusters in [9], but only two of them are used and thus presented in this paper. The method 'average' (Eq. 3) calculates the arithmetic mean weight of connections between concepts included in clusters a and b:

$$w_a(a, b) = \frac{\sum_{i \in a} \sum_{j \in b} a(i, j) \times w_{ij}}{\sum_{i \in a} \sum_{j \in b} a(i, j)} \tag{3}$$

where $a(i, j)$ is defined as

$$a(i, j) = \begin{cases} 1 \text{ if } w_{ij} \neq 0, \\ 0 \text{ otherwise.} \end{cases} \tag{4}$$

The equation of method 'sum' is the following:

$$w_s(a, b) = \begin{cases} 1 & \text{if } s(a, b) > 1 \\ -1 & \text{if } s(a, b) < -1 \\ s(a, b) & \text{otherwise.} \end{cases} \tag{5}$$

$s(a, b)$ is calculated as

$$s(a, b) = \sum_{i \in a} \sum_{j \in b} w_{ij} \tag{6}$$

3 Short Description of the Selected Models

Three prototype models were selected for behavioral analysis, based on their different, but typical properties. Each one of them has been already published in e.g. [9, 18, 20].

3.1 *Waste Management Model*

The first selected model was the 'Waste management' system. This model can be found, e.g. in [18, 21], and describes the connections among the factors of solid waste management systems in Hungary. The refined model developed earlier by our team is rather large: it contains 33 concepts (C1.1, C1.2, …, C1.5, C2.1, …, C2.6, …, C6.1, …, C6.5), which were originally clustered into one of the 6 groups of 'main factors'. There are no distinguished inputs, intermediate or output concepts. The number of really existing connections is 638 (the theoretical maximum would be $33 \times 32 = 1056$), therefore the density of the model is 60.4%, taking into account that the main diagonal of the connection matrix has to contain '1'-s in contrary to Kosko's definition [11]. This way the previous states of the concepts are taken into account at the next iteration (FCM of Type III, [15]).

The reduced model has only 23 clusters (K1, K2, …, K23), but the density has increased to 76.5%. Here, the expert-defined value of the design parameter ε was 0.024, and the 'average' weight calculation method was applied.

Table 1 Clusters of the 'Waste management' model

Concepts	Cluster	Concepts	Cluster
C1.1 + C1.4	K1	C3.3 + C3.4 + C3.5	K13
C1.2 + C6.4	K2	C3.6	K14
C1.3 + C4.7	K3	C4.1	K15
C1.1 + C1.5	K4	C4.2 + C4.4	K16
C2.1	K5	C6.1 + C6.2 + C6.3 + C6.4 + C4.3	K17
C2.2	K6	C4.4 + C4.5 + C4.6	K18
C2.3	K7	C5.1	K19
C2.4	K8	C5.2	K20
C2.5	K9	C5.3	K21
C2.6	K10	C5.4	K22
C3.1	K11	C1.1 + C6.5	K23
C3.2	K12		

Neither the description and the interpretation of clusters, nor the connection matrices are included here for reasons of economizing with space, but can be found in [18]. Only the elements of clusters are provided by Table 1 for easy comparison of the behavior of the original and the reduced models. Simulations were performed with $\lambda = 5$ and 0.25.

3.2 Synthetic Model

The second examined model was first published in [9]. As its name implies, the model was generated with a special tool designed for this particular purpose. This FCM contains 6 input, 4 output and 15 intermediate concepts. Tables 2 and 3 contain only the final fixed-point values of the output concepts (C22–C25). The density is $d = 10\%$, according to Eq. 7.

Table 2 'Synthetic' model, $\lambda = 5$

	FP1		FP2		FP3	
Conc.	Orig.	Redu.	Orig.	Redu.	Orig.	Redu.
C22	0.99	**0.95**	0.98	**0.94**	0.95	**0.93**
C23	1.00	1.00	1.00	1.00	1.00	1.00
C24	0.99	0.99	0.99	0.99	0.99	0.99
C25	1.00	1.00	1.00	1.00	1.00	1.00
Freq. (%)	34.8	12.5	3.3	2.0	61.9	85.5

Table 3 'Synthetic' model

	$\lambda = 1.7$			$\lambda = 1$	
	FP1		FP2	FP1	FP1
Conc.	Orig.	Redu.	Orig.	Orig.	Redu.
C22	0.69	**0.65**	0.68	0.59	0.59
C23	0.84	0.84	0.84	0.69	0.69
C24	0.80	0.80	0.80	0.66	**0.67**
C25	0.97	**0.98**	0.97	0.88	**0.89**
Freq. (%)	96.0	100.0	4.0	100.0	100.0

$$d = 100 \frac{n}{((c - i)(c - o) - (c - i - o))} \tag{7}$$

where n denotes the number of non-zero matrix elements, c is the number of concepts, i is the number of input concepts and o is the number of output concepts.

Only the intermediate concepts were grouped into 10 clusters, the input and output concepts were kept untouched. This way the results of simulations with original and reduced models could be compared easily. The reduction rate was $\varepsilon = 0.024$, the weights of the new connections were calculated with the 'sum' method. The density of the reduced version was 17.3%.

The behavior of the model was analyzed with $\lambda = 5$, 1.7 and 1 parameter values.

3.3 New BPS Model

The third model was published in [20, 22], and similarly to the first one, it dealt with a real problem.

The concepts collected and briefly described in Table 4 represent the most important aspects of the everyday operation of a bank. The connection matrix is also provided in Table 5. The model has 13 concepts, but one of them (C4) is an input concept and that is why it is left out from Tables 6 and 11. The density of the model is 42.4%.

The reduced model had 9 clusters (Table 6). It is interesting, that in this case the clusters do not overlap: every concept appears in only one cluster. The density increased again, to 56.3%. Parameter ε of the reduction was set to 0.22, the weights of the reduced model were calculated by the 'average' method.

Table 4 Concept IDs, names and categories of the 'New BPS' model

Concept ID	Concept name	Category
C1	Clients	Assets
C2	Rules & regulations	
C3	New IT solutions	
C4	Funding	Money
C5	Cost reduction	
C6	Profit/loss	Financials
C7	Investments	
C8	Staff	Human resources
C9	New services	Product and process development
C10	Quality	
C11	Client development	Output measures
C12	Service development	
C13	Productivity	

Table 5 Connection matrix of the 'New BPS' model

	C1	C2	C3	C4	C5	C6	C7	C8	C9	C10	C11	C12	C13
C1	0	0	0.5	0	0	0.5	1	0.5	0	0.5	1	0.5	0
C2	1	0	0.5	1	0	0	1	1	0.5	0	1	1	0
C3	1	0.5	0	0	0	−1	0	−1	1	0	1	1	1
C4	0	0	0	0	0	0	0	0	0	0	0	0	0
C5	0	0	1	−0.5	0	0	0	−1	0	0	0	1	0
C6	0	0	0	0	−0.5	0	0	0	0	0	0	0	0
C7	0.5	0	0.5	1	0	0.5	0	0	0	−0.5	0	0	0
C8	0	0	0	0	0	−0.5	0	0	0	0.5	0	0	−0.5
C9	0	0	0	1	0	0.5	0.5	0.5	0	−0.5	0	0.5	0
C10	0.5	0	0	0	0	0	0.5	0.5	0.5	0	1	0	0
C11	0	0	0.5	0.5	0	0	0	0	0.5	0.5	0	0	1
C12	0	0	0.5	0.5	0	0	1	0	0.5	0	0.5	0	−0.5
C13	0	0	1	0	0	0.5	0	0	0	0	1	0	0

Table 6 Clusters of the 'New BPS' model

Concepts	Cluster	Concept	Cluster
C1 + C2 + C12	K1	C7 + C9	K5
C3	K2	C10	K6
C5	K3	C11	K7
C6 + C8	K4	C13	K8

Table 7 'Waste management' original model, $\lambda = 5$

Conc.	FP1	Conc.	FP1	Conc.	FP1
C1.1	1.00	C3.1	1.00	C4.6	1.00
C1.2	1.00	C3.2	1.00	C4.7	1.00
C1.3	1.00	C3.3	1.00	C5.1	1.00
C1.4	1.00	C3.4	1.00	C5.2	1.00
C1.5	1.00	C3.5	1.00	C5.3	1.00
C2.1	1.00	C3.6	1.00	C5.4	1.00
C2.2	1.00	C4.1	1.00	C6.1	1.00
C2.3	1.00	C4.2	1.00	C6.2	1.00
C2.4	**0.06**	C4.3	1.00	C6.3	1.00
C2.5	1.00	C4.4	1.00	C6.4	1.00
C2.6	1.00	C4.5	1.00	C6.5	1.00

4 Results

Simulations were executed with both the original and the reduced versions of all the three models, with various λ values, and the results were compared. The statistical data gathered were based on 1000 elements, evenly distributed, random-generated scenarios (initial state vectors of the concepts).

4.1 Waste Management Model

The simulation with $\lambda = 5$ resulted in a single fixed-point attractor (FP1, see Table 7). The final values of the concepts were 1.00, except for C2.4, which was 0.06. The reduced model behaved the same way: it had also a single FP, and almost every cluster had the same final fixed-point value, 1.00. The only exception was K8, which is the equivalent cluster of concept C2.4 of the original model. Its value decreased to exactly 0.00, but such a small change usually has no importance in practical applications. (See Table 8).

Next, the simulations were repeated with setting λ to 0.25. The lower λ resulted in various lower concept values, but there was still one FP and C2.4 had the lowest fixed-point value here as well (0.53, see Table 9). The reduced model preserved this property, because K8 has also the lowest value among all clusters (0.50, see Table 10). The order of magnitude of final concept/cluster fixed-point values were the same in both cases.

Table 8 'Waste management' reduced model, $\lambda = 5$

Clus.	FP1	Clus.	FP1	Clus.	FP1
K1	1.00	K9	1.00	K17	1.00
K2	1.00	K10	1.00	K18	1.00
K3	1.00	K11	1.00	K19	1.00
K4	1.00	K12	1.00	K20	1.00
K5	1.00	K13	1.00	K21	1.00
K6	1.00	K14	1.00	K22	1.00
K7	1.00	K15	1.00	K23	1.00
K8	**0.00**	K16	1.00		

Table 9 'Waste management' original model, $\lambda = 0.25$

Conc.	FP1	Conc.	FP1	Conc.	FP1
C1.1	0.86	C3.1	0.87	C4.6	0.90
C1.2	0.92	C3.2	0.80	C4.7	0.85
C1.3	0.80	C3.3	0.82	C5.1	0.70
C1.4	0.93	C3.4	0.84	C5.2	0.96
C1.5	0.92	C3.5	0.82	C5.3	0.92
C2.1	0.75	C3.6	0.86	C5.4	0.90
C2.2	0.76	C4.1	0.97	C6.1	0.92
C2.3	0.62	C4.2	0.96	C6.2	0.83
C2.4	**0.53**	C4.3	0.89	C6.3	0.89
C2.5	0.72	C4.4	0.89	C6.4	0.87
C2.6	0.78	C4.5	0.91	C6.5	0.95

Table 10 'Waste management' reduced model, $\lambda = 0.25$

Clus.	FP1	Clus.	FP1	Clus.	FP1
K1	0.86	K9	0.70	K17	0.81
K2	0.84	K10	0.75	K18	0.84
K3	0.78	K11	0.76	K19	0.67
K4	0.86	K12	0.75	K20	0.94
K5	0.73	K13	0.81	K21	0.90
K6	0.70	K14	0.84	K22	0.87
K7	0.59	K15	0.91	K23	0.87
K8	**0.50**	K16	0.89		

4.2 Synthetic Model

The first simulation was performed with $\lambda = 5$ setting again, because this value is widely applied in the literature. The original version of the model had 3 FPs, but FP2 occured very rarely, only in 3.3% of all cases. The output concepts C23–C25 had the same final FP values in all FPs, only C22 had different values, and these values are also very similar (0.99, 0.98 and 0.95).

Despite the reduction, the model behaved in a very similar way: the number of FPs was still three, the final values of concepts C23-C25 had not changed, only the values of C22 decreased slightly. Even the frequency of FPs was very similar (See Table 2).

The repeated simulations with $\lambda = 1.7$ led to different results (see Table 3). The original model had 2 FPs, even if one them occurred only in 4% of all cases, and only the final value of C22 differed slightly (0.69 vs. 0.68).

The reduced model had only one FP, however, and the final value of C22 decreased to 0.65 and C25 increased slightly to 0.98.

$\lambda = 1$ was applied during the last simulations. In this case only one FP was found, and the values of C24 and C25 increased by 0.01, which is negligible.

4.3 New BPS Model

At first, $\lambda = 5$ was chosen during the simulation of the New BPS model. The model had 2 FPs, but only the values of C6 and C8 differed from 1.00. FP1 occurred in the majority of the scenarios (76.2%, see Table 11).

During the simulations of the reduced model a third FP (FP1) emerged, but it appeared only in 0.3% of the scenarios. The final value of cluster K4, which is a merger of C6 and C8, is 0.50. The values of K4 in FP2 and FP3 were 0.86 and 0.15, respectively. Exactly the same values can be observed in FP1 and FP2 of the original model.

The repeated simulations with the original model and $\lambda = 2.8$ resulted in only one FP. Again, the values of most concepts were 1.00, but C8, C6 and this time C5 was also somewhat different (0.84, 0.51 and 0.99, respectively).

The reduced model behaved similarly: the value of K4 was 0.50 (very close to C6), the value of K3, which is the equivalent cluster of C5, was 1.00. Nevertheless, the 0.01 difference seems negligible.

Simulations with $\lambda = 0.7$ led to various concept values in case of the original model, but most of them were still close to 1, except C6 and C8, which were closer to 0.50. C5 with 0.78 was somewhere in the middle.

The concepts of the reduced model behaved very similarly. The value of K3 was the same as C5 of the original model, and K4 had the lowest value among all (0.52) (Table 12).

Table 11 'New BPS' original model

	$\lambda = 5$		$\lambda = 2.8$	$\lambda = 0.7$
Conc.	FP1	FP2	FP1	FP1
C1	1.00	1.00	1.00	0.96
C2	1.00	1.00	1.00	0.99
C3	1.00	1.00	1.00	0.97
C5	1.00	1.00	0.99	0.78
C6	0.86	0.15	0.51	0.52
C7	1.00	1.00	1.00	0.82
C8	0.92	0.99	0.84	0.55
C9	1.00	1.00	1.00	0.84
C10	1.00	1.00	1.00	0.92
C11	1.00	1.00	1.00	0.92
C12	1.00	1.00	1.00	0.89
C13	1.00	1.00	1.00	0.89
Freq. (%)	76.2	23.8	100.0	100.0

Table 12 'New BPS' reduced model

	$\lambda = 5$			$\lambda = 2.8$	$\lambda = 0.7$
Clus.	FP1	FP2	FP3	FP1	FP1
K1	1.00	1.00	1.00	1.00	0.93
K2	1.00	1.00	1.00	1.00	0.94
K3	1.00	1.00	1.00	1.00	0.78
K4	**0.50**	**0.86**	**0.15**	**0.50**	**0.52**
K5	1.00	1.00	1.00	1.00	0.82
K6	1.00	1.00	1.00	1.00	0.89
K7	1.00	1.00	1.00	1.00	0.92
K8	1.00	1.00	1.00	1.00	0.89
Freq. (%)	0.3	72.7	27.0	100.0	100.0

5 Conclusion

The number of concepts were reduced to $\approx$70% in case of the Waste management model, to 80% in case of the Synthetic model and to $\approx$69% in case of the New BPS model. The reduction rate can be specified by design parameter ε of the state reduction method between 0 (no reduction) and 1 (the model collapses into a single concept). It is hard to define the appropriate value of ε, which provides the best balance between model size and acceptable behavioral similarity. Even these early simulation results show that a reasonable reduction rate does not affect the models behavior seriously. The investigated models never behaved chaotically or produced limit cycles, at most they had different numbers of fixed-point attractors (FPs). In most cases, the reduced models had the same number of FPs. In only two cases number of FPs changed by one. The corresponding FP had very low occurrence frequencies, so its disappearing might be the result of the limited number of experiments. Even the final concept values were very similar.

This paper reports the first results of the behavioral comparison of original and reduced FCM models. The method of comparison needs further development in order to characterize the effect of different reduction rates on behavioral similarity. The applied reduction technique essentially preserved the model behavior under reasonable reduction rates, and thus can be used for real-life problems.

Acknowledgements This research was supported by the ÚNKP-17-4 New National Excellence Program of the Ministry of Human Capacities, by National Research, Development and Innovation Office (NKFIH) K108405, K124055, by EFOP-3.6.2-16-2017-00016 "Dynamics and control of autonomous vehicles meeting the synergy demands of automated transport systems" and by EFOP-3.6.1-16-2016-00017 "Internationalization, initiatives to establish a new source of researchers and graduates as instruments of intelligent specializations at Szechenyi University".

References

1. S. Ahmadi, E. Papageorgiou, C.-H. Yeh, R. Martin, Managing readiness-relevant activities for the organizational dimension of ERP implementation. Comput. Ind. **68**, 89–104 (2015)
2. B.F. Hobbs, S.A. Ludsin, R.L. Knight, P.A. Ryan, J. Biberhofer, J.J.H. Ciborowski, Fuzzy cognitive mapping as a tool to define management objectives for complex ecosystems. Ecol. Appl. **12**, 1548–1565 (2002)
3. E.I. Papageorgiou, ed., *Fuzzy Cognitive Maps for Applied Sciences and Engineering - From Fundamentals to Extensions and Learning Algorithms*, Intelligent Systems Reference Library, vol. 54 (Springer, Berlin, 2014)
4. E.I. Papageorgiou, J.L. Salmeron, Methods and algorithms for fuzzy cognitive map-based decision support, in *Fuzzy Cognitive Maps for Applied Sciences and Engineering* ed. by E.I. Papageorgiou (2013)
5. M. van Vliet, K. Kok, T. Veldkamp, Linking stakeholders and modellers in scenario studies: the use of fuzzy cognitive maps as a communication and learning tool. Futures **42**(1), 1–14 (2010)
6. S. Alizadeh, M. Ghazanfari, M. Fathian, Using data mining for learning and clustering FCM. Int. J. Comput. Intell. **4**(2), 118–125 (2008)
7. W. Homenda, A. Jastrzebska, W. Pedrycz, Time series modeling with fuzzy cognitive maps: simplification strategies, in *Computer Information Systems and Industrial Management: 13th IFIP TC8 International Conference, CISIM 2014, Ho Chi Minh City, Vietnam, November 5-7, 2014. Proceedings*, (Springer Berlin Heidelberg, Berlin, Heidelberg, 2014), pp. 409–420
8. G. Nápoles, I. Grau, R. Bello, R. Grau, Two-steps learning of fuzzy cognitive maps for prediction and knowledge discovery on the HIV-1 drug resistance. Expert. Syst. Appl. **41**(3), 821–830 (2014). Methods and Applications of Artificial and Computational Intelligence
9. M.F. Hatwagner, E. Yesil, F. Dodurka, E.I. Papageorgiou, L. Urbas, L.T. Koczy, Two-stage learning based fuzzy cognitive maps reduction approach. IEEE Trans. Fuzzy Syst. (2018)
10. R. Axelrod, *Structure of Decision: The Cognitive Maps of Political Elites* (Princeton University Press, 1976)
11. B. Kosko, Fuzzy cognitive maps. Int. J. Man-Mach. Stud. **24**, 65–75 (1986)

12. E.I. Papageorgiou, J.L. Salmeron, A review of fuzzy cognitive maps research during the last decade. IEEE Trans. Fuzzy Syst. **21**(1), 66–79 (2013)
13. W.-R. Zhang, Bipolar fuzzy sets, in *The 1998 IEEE International Conference on Fuzzy Systems Proceedings, 1998. IEEE World Congress on Computational Intelligence*, vol. 1 (IEEE, 1998), pp. 835–840
14. K.E Parsopoulos, E.I. Papageorgiou, P.P. Groumpos, M.N. Vrahatis, A first study of fuzzy cognitive maps learning using particle swarm optimization, in *CEC'03. The 2003 Congress on Evolutionary Computation, 2003*, vol. 2 (IEEE, 2003), pp. 1440–1447
15. C.D. Stylios, P.P. Groumpos, Mathematical formulation of fuzzy cognitive maps, in *Proceedings of the 7th Mediterranean Conference on Control and Automation* (1999), pp. 2251–2261
16. A.K. Tsadiras, Comparing the inference capabilities of binary, trivalent and sigmoid fuzzy cognitive maps. Inf. Sci. **178**(20), 3880–3894 (2008)
17. M.F. Hatwagner, L.T. Koczy, Parameterization and concept optimization of fcm models, in *2015 IEEE International Conference on Fuzzy Systems (FUZZ-IEEE)* (IEEE, Istanbul, 2015), pp. 1–8
18. E.I. Papageorgiou, M.F. Hatwágner, A. Buruzs, L.T. Kóczy, A concept reduction approach for fuzzy cognitive map models in decision making and management. Neurocomputing **232**, 16–33 (2017)
19. Z. Kohavi, N.K. Jha, *Switching and Finite Automata Theory*, 5th edn. (Cambridge University Press, 2009)
20. M.F. Hatwágner, G. Vastag, V.A. Niskanen, L.T. Kóczy, Banking applications of FCM models, in *9th European Symposium on Computational Intelligence and Mathematics* (2017), pp. 60–68. http://escim2017.uca.es/wp-content/uploads/2015/02/OralCommunications.pdf
21. M.F. Hatwágner, A. Buruzs, P. Földesi, L.T. Kóczy. A new state reduction approach for fuzzy cognitive map with case studies for waste management systems, in *Computational Intelligence in Information Systems* (Springer, Berlin, 2015), pp. 119–127
22. M.F. Hatwágner, G. Vastag, V.A. Niskanen, L.T. Kóczy, Improved behavioral analysis of fuzzy cognitive map models, in *Submitted to the 17th International Conference on Artificial Intelligence and Soft Computing (ICAISC 2018)* (Zakopane, Poland, 2018)

FUZIPS: Fuzzy v2 Based Algorithm for Automatic Switching from GPS Based Location Services to the Indoor Positioning Service

Farid Orujov, Rytis Maskeliūnas, and Robertas Damaševičius

Abstract We address the problem of seamless positioning/navigation and ubiquitous localization. The problem is relevant for commercial service providers who are interested that their positioning services will be able to work seamlessly in all environments, including, indoor, urban, and rural. We present an algorithm capable of automatic switching from traditional location services (GPS, A-GPS) to indoor positioning systems (WiFi based triangulation, Bluetooth LE beacons). The proposed fuzzy logic algorithm based on the signal strength and the distance showed a better result than the Proximity algorithm based only on the signal strength, reaching an error rate of 12%.

1 Introduction

Location-Based Services (LBS) and navigation/guidance systems have become increasingly important for a large variety of services such as for health, entertainment, work, personal life, real-time tracking, security alerts, informational services, etc. [1]. Outdoor positioning services usually rely on Global Positioning System (GPS) technology or the Global System for Mobile communications (GSM) radio signal delay from the cell-phone towers [2]. As GPS and GSM do not work very well indoors, the indoor localization may depend on Wi-Fi, Bluetooth Low Energy (BLE) beacons, or Near Field Communication (NFC) technology [3, 4].

A great variety of current positioning technologies, each having very different features, combined with different environments wherein they are to be used, requires from system developers to consider how positioning services could be integrated to ensure seamless provision of LBS services. The problem is known as seamless positioning/navigation [5], or ubiquitous localization [6]. The problem is especially relevant for commercial service providers who are interested that their positioning

F. Orujov (✉) · R. Maskeliūnas · R. Damaševičius
Multimedia Engineering Department, Faculty of Informatics, Kaunas University of Technology, Studentu 50-401, Kaunas, Lithuania
e-mail: farid.orujov@ktu.edu

S. N. Shahbazova et al. (eds.), *Recent Developments and the New Direction in Soft-Computing Foundations and Applications*, Studies in Fuzziness and Soft Computing 393, https://doi.org/10.1007/978-3-030-47124-8_30

services will be able to work seamlessly in all environments, including, indoor, urban, and rural [5].

An integration of positioning technologies is especially required in hybrid environments such as outdoor-to-indoor environments such as pedestrian tracking in navigation systems [7]. The main requirements for such integration is scalability, i.e. the ability to adapt seamlessly to any change in the environment, e.g., transfer from outdoor to indoor, switching automatically while using the information provided by the available positioning technology. The problem of service switching has attracted considerable attention from the researchers in recent years. For example, Cheng et al. [8] proposes a system based on GPS, BLE beacons, and NFC technology that is able to switch automatically from an outdoor mode to an indoor mode and to guide the user to requested target destination. De Angelis et al. [9] proposed a hybrid positioning scheme for urban scenarios that uses both a Global Navigation Satellite System (GNSS) and a Magnetic Positioning System (MPS), while a Kalman filter algorithm is used for data integration. He [10] proposed a layered indoor 3D positioning technology based on WiFi positioning, pedestrian dead reckoning (PDR) and architecture information for the pedestrian location service using individual intelligent terminal. Jia et al. [11] developed an Android APP for both outdoor positioning using GPS and Baidu Map, and indoor positioning using Wi-Fi fingerprint positioning, contextual (indoor/outdoor) and floor detection, while employing information from smartphone sensors and functional configuration. Kulshrestha et al. [12] proposed a human identification, monitoring, and location tracking system, called SmartITS, which continuously tracks MAC IDs of personal smart devices (smartphones, BLE tags, and Bluetooth devices) and provides a Google map-based visualization of their trajectories to be used for navigation in dense urban areas. Li et al. [13] used WiFi and light sensor data, and semi-CRF as machine learning method to exploit the inner property of IOS transition. Owing to the semi-CRF (Markov conditional random fields) algorithm to model interdependency among indoor/outdoor/semi-open (IOS) states and improve the localization performance. Li et al. [14] presented a hybrid seamless image-based positioning system for LBS in both outdoor and indoor environments. The system uses visual sensor input to align with geo-referenced images for image-based positioning resolution, while employing smartphone's sensors (built-in GPS receiver and digital compass) to improve positioning accuracy in indoor environments as well as in complex outdoor environments such as urban canyons. Li et al. [15] integrated PDR/GPS/Bluetooth to realize seamless indoor/outdoor localization of the pedestrian, while smartphone's magnetometer and gyroscope sensors are used to detect context. Liu et al. [16] proposed advanced integration of Wi-Fi, MARG (magnetic, angular rate, and gravity sensors) and GPS using an adaptive integration structure based on Kalman Filter (KF) and Particle Filter (PF) for applications in real indoor/outdoor environments. Maghdid [17] proposed three localization schemes including Wireless Access Point (WAP) synchronization/localization scheme, and indoor localization schemes. The schemes are based on hybridizing GNSS with WiFi, Bluetooth and inertial-sensors measurements using combined localization techniques including time-of-arrival (TOA) and dead-reckoning (DR). Molina et al. [18] developed and indoor/outdoor localization application aimed at complex environment of airports. The system is based on Fingerprinting algorithm and uses Wi-Fi and BLE measurements, while the position is found using the Nearest Neignbour (NN) algorithm applied on the Fingerprinting database (radio map). Moon et al. [19] used a terrestrial digital multimedia broadcasting (T-DMB) signal in Korea to obtain the mobile user's position in case of GPS signal failure in indoor/outdoor urban environments. Takatsuka et al. [20] proposes a unified locating service, KULOCS, which horizontally data and operations of the existing heterogeneous locating services (BLE Locating Service and GPS Locating Service). Wang et al. [21] proposed a data fusion scheme for the accurate localization and tracking in mixed GPS-friendly, GPS-challenging, and GPS-denied scenario, which provides for intelligently switching between GPS and inertial sensor-based schemes end deals with related problems such as error accumulation and GPS faults for robust indoor/outdoor seamless tracking. Zhu et al. [22] presented the Chinese national standard of Indoor Multi-Dimensional Location GML for use in ubiquitous indoor location intelligent applications. The standard is intended as an indoor multi-dimensional location information model and exchange data format standard for integration of different indoor positioning and navigation technologies. Zou et al. [23] propose an accurate, fast response and energy-efficient scheme for IO detection and seamless LBS running on the mobile device based on iBeacon technology. The scheme enables the implementation of seamless positioning and navigation services in a semi-outdoor environment.

The main goal of this article is a presentation of algorithm capable of automatic switching from traditional location services (GPS, A-GPS, etc.) to indoor positioning systems (WiFi based triangulation, Bluetooth LE beacons, etc.).

The structure of the paper is as follows. Section 2 described the proposed model. Section 3 describes the implementation of the proposed model. Section 4 described the experimental settings. Section 5 presents the results of experiments. Finally, Sect. 6 presents the conclusions.

2 Model

This section describes model of the environment in which proposal positioning algorithm are used. Main components used are a typical smartphone with built in GPS and Bluetooth modules and a set of beacons. The concept is to detect, identify and get the data from various beacon modules or other sources of identification (e.g. WiFi access points). These should be detectable near the entrances to buildings, walkways, and other locations with pedestrian traffic, close to the target building itself. The concept is illustrated in Fig. 1.

Such system allows a potential user to initiate triggering the switch either by detecting a Bluetooth based beacon (in blue)—these should be positioned near the pedestrian crossing in car parks and near the entrance and would work most accurately or by calculating an approximate position via detectable WiFi signal strength based triangulation (in red), and while this in most cases is often quite accurate, but not all

Fig. 1 A concept of our system

access points are in available location databases (e.g. Google and Apple gather and use such data to improve outdoor location services).

The model of a working principle is displayed in Fig. 2. Here a smartphone user is guided to his destination via a traditional, GPS based indoor positioning method. Once a phone detects a stop or significant reduction of motion, a service based on our algorithm initiates and starts monitoring the available WiFi and Bluetooth access point data. Then the system checks if the signals are disappearing over time or are of significantly variable strength over a short period of time as this indicated access points on other users phones, if not—a presumption is made that there is a building nearby. Once a positive signal is found the fuzzy trigger algorithm is started (see Sect. 3).

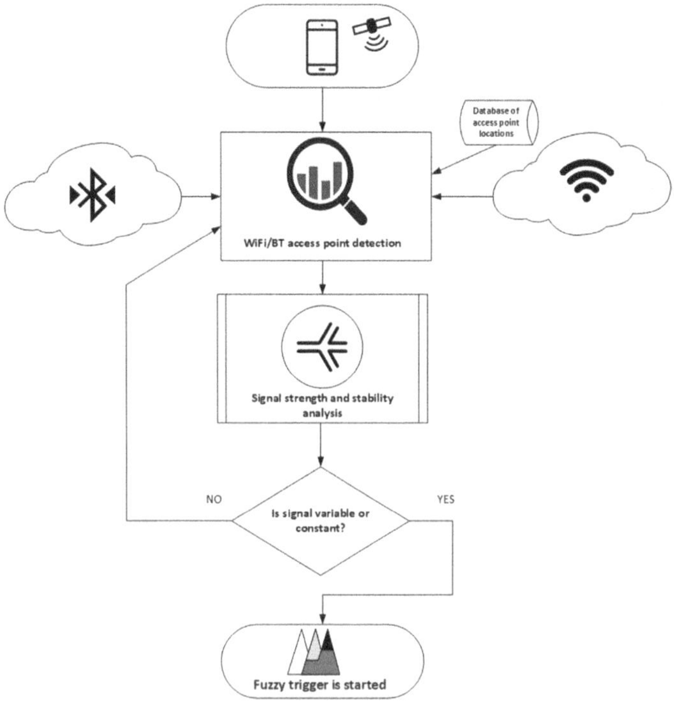

Fig. 2 Model of the system

3 Implementation

3.1 Proximity Localization

The proximity algorithm [24] is assigned to the receiver that coordinates the beacon which emits the greatest power signal. All beacons must be tied to both the local coordinate system and the global GPS coordinate system for the correct operation of the algorithm.

For instance, if three beacons are located in different entrances of building and the highest power signal has been received from, then the receiver is assigned coordinates of that beacon which will trigger IPS on the smartphone for indoor environments.

However, this method has some major drawbacks with signal interference, which can lead to the launch of IPS for another building or room

3.2 *Fuzzy Trigger*

We propose to use a hybrid technique based on fuzzy logic to avoid mistakes in choosing a beacon. We calculate the distance between the location of the smart-phone and the beacon. To calculate the distance between two geo-points, we use the Haversine formula (Figs. 3, 4 and 5).

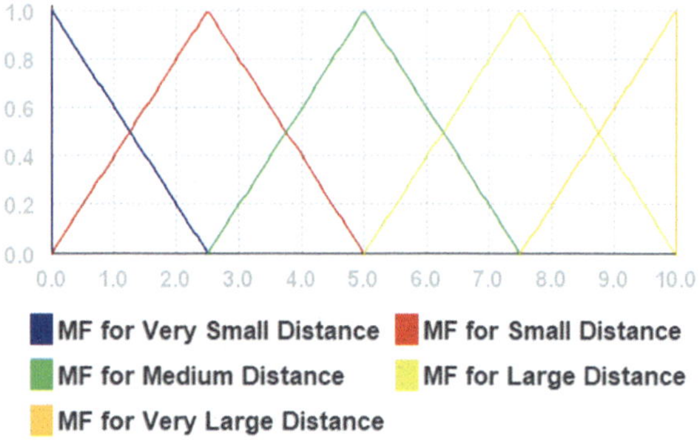

Fig. 3 The membership functions for distances

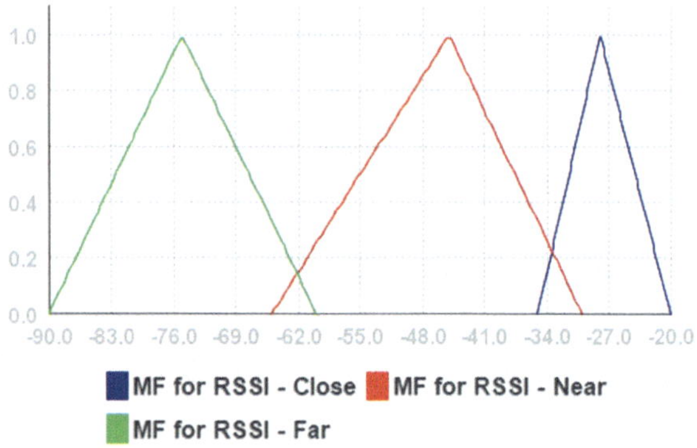

Fig. 4 The membership functions for RSSI

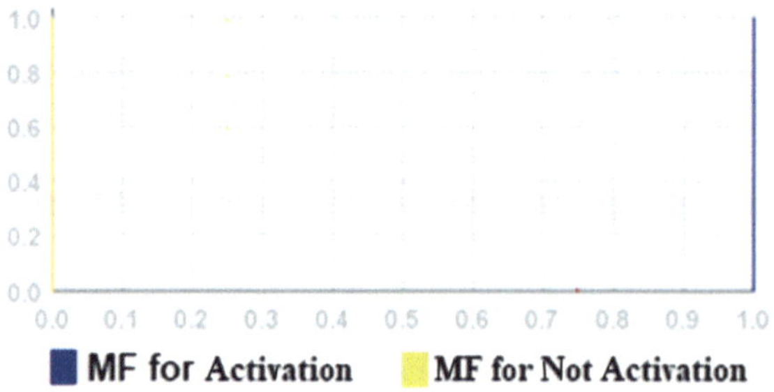

Fig. 5 The membership functions for activation IPS

$$d = 2rarcsin\left(\sqrt{\sin^2\left(\frac{\varphi_2 - \varphi_1}{2}\right) - \cos(\varphi_1)\cos(\varphi_2)\sin^2\left(\frac{\lambda_2 - \lambda_1}{2}\right)}\right)$$

The Fuzzifier is a triangular membership function. The inference engine is Mamdani max-min [25].

The defuzzifier is height method.

Fuzzy rules:

IF Distance value is Very Small **AND** RSSI is Close **THEN** Activation
IF Distance value is Very Small **AND** RSSI is Near **THEN** Activation
IF Distance value is Very Small **AND** RSSI is Very Far **THEN** Not Activation
IF Distance value is Small **AND** RSSI is Close **THEN** Activation
IF Distance value is Small **AND** RSSI is Near **THEN** Activation
IF Distance value is Small **AND** RSSI is Far **THEN** Not Activation
IF Distance value is Medium **AND** RSSI is Close **THEN** Activation
IF Distance value is Medium **AND** RSSI is Near **THEN** Not Activation
IF Distance value is Medium **AND** RSSI is Far **THEN** Not Activation
IF Distance value is Large **AND** RSSI is Close **THEN** Not Activation
IF Distance value is Large **AND** RSSI is Near **THEN** Not Activation
IF Distance value is Large **AND** RSSI is Far **THEN** Not Activation
IF Distance value is Very Large **AND** RSSI is Close **THEN** Not Activation
IF Distance value is Very Large **AND** RSSI is Near **THEN** Not Activation
IF Distance value is Very Large **AND** RSSI is Very Far **THEN** Not Activation.

4 Setup and the Progress of the Experiment

The experiment was made in our location at the Santaka Valley building, Barsausko 59, Kaunas, Lithuania (GPS coordinates to check the building on the map are: 54.899791, 23.961520) (see Fig. 6).

The 3 beacons are installed at 3 entrances to the buildings. The Geo-coordinates were given them. The configuration parameters of the BLE beacon:

- Transmit power (Tx): 4 dBm
- Advertising Interval: 200 ms.

The application was installed on a smartphone that launches GPS, if the system decided to launch the IPS then the application go into the IPS **mode** (Table 1).

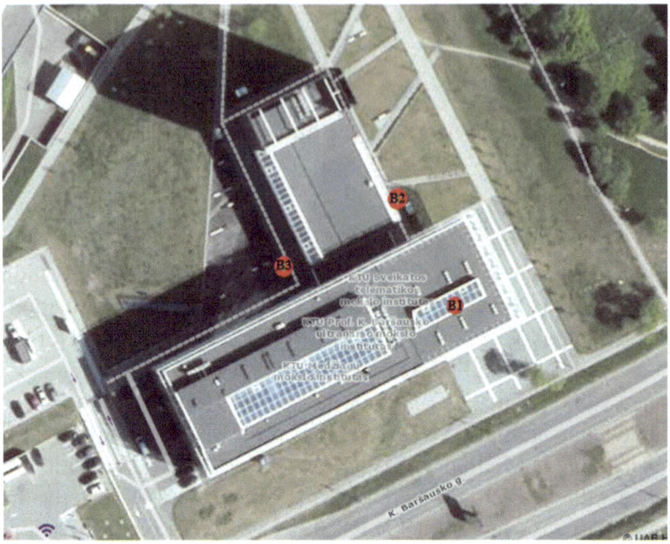

Fig. 6 The visual result of fingerprinting localization algorithms

Table 1 The coordinates of pre-planned location points

Fingerprint points	World Geodetic System (WGS) coordinates
B1	54.899759, 23.962091
B2	54.899963, 23.961646
B3	54.899799, 23.961337

5 Results of the Experiment

During the experiment, the subject passes by the test building with the running application.

The proposed method based on the signal strength and the distance showed a better result than the Proximity algorithm based only on the signal strength. Figure 7 shows the Error rate in percent.

As it can be seen from Fig. 7, the proposed fuzzy method showed the decision error, an error rate was 12% whereas the Proximity algorithm showed the worst result and the error rate was 23%.

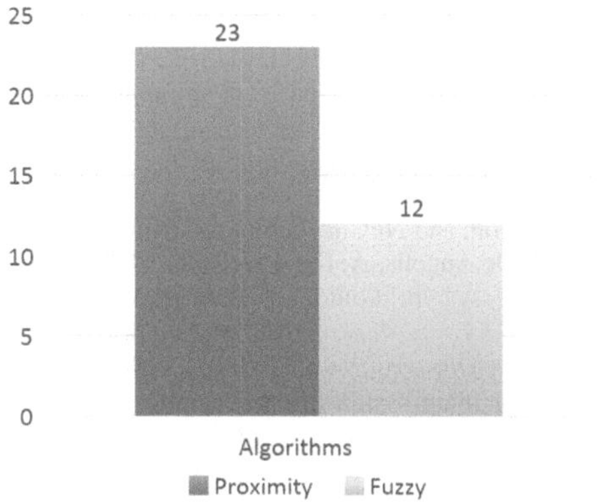

Fig. 7 Comparison of error rate of proximity and proposed algorithms

6 Conclusion

Overall, these results can be evaluated as positive. The number of beacons was established from 4 to 6 depending on the quantity of building entrances. It can be stated that:

- The signal is practically damped by the wall. Hence it is necessary to install at least one beacon from the outside that will avoid the problem with signal attenuation.
- According to results, the proposed fuzzy method showed relatively high result, the error rate was 12%.
- Using additional data from sensors on the smartphone (compass, etc.) can reduce the error rate.

References

1. A. Dey, J. Hightower, E. de Lara, N. Davies, Location-based services. Pervasive Comput. **9**, 11–12 (2010). https://doi.org/10.1109/mprv.2010.10
2. M. Yassin, E. Rachid, A survey of positioning techniques and location based services in wireless networks, in *2015 IEEE International Conference on Signal Processing, Informatics, Communication and Energy Systems (SPICES)* (Kozhikode, 2015), pp. 1–5. https://doi.org/10.1109/spices.2015.7091420
3. R.F. Brena, J.P. García-Vázquez, C.E. Galván-Tejada, D. Muñoz-Rodriguez, C. Vargas-Rosales, J. Fangmeyer, Evolution of indoor positioning technologies: a survey. J. Sens. **2017**, 21 pp. Article ID 2630413 (2017). https://doi.org/10.1155/2017/2630413
4. T. Ruiz-López, J.L. Garrido, K. Benghazi, L. Chung, A survey on indoor positioning systems: foreseeing a quality design, in *Distributed Computing and Artificial Intelligence. Advances in Intelligent and Soft Computing*, vol. 79, ed. by F. de Leon, A.P. de Carvalho, S. Rodríguez-González, J.F. De Paz Santana, J.M.C. Rodríguez (Springer, Berlin, Heidelberg, 2010)
5. K. Ogawa, E. Verbree, S. Zlatanova, N. Kohtake, Y. Ohkami, Toward seamless indoor-outdoor applications: developing stakeholder-oriented location-based services. Geo-spat. Inf. Sci. **14**, 109 (2011). https://doi.org/10.1007/s11806-011-0469-0

6. G. Retscher, A. Kealy, Ubiquitous positioning technologies for modern intelligent navigation systems. J. Navig. **59**(1), 91–103 (2006)
7. W.F. Hassen, F. Najjar, A positioning technology switch algorithm for ubiquitous pedestrian navigation systems, in *Proceedings—IEEE/IFIP 13th International Conference on Embedded and Ubiquitous Computing, EUC* (2015), pp. 124–131. https://doi.org/10.1109/euc.2015.26
8. R. Cheng, W. Hong, J. Wang, K.W. Lin, Seamless guidance system combining GPS, BLE beacon, and NFC technologies. Mob. Inf. Sys. (2016). https://doi.org/10.1155/2016/5032365
9. G. De Angelis, A. De Angelis, G. De Angelis, A. De Angelis, V. Pasku, A. Moschitta, P. Carbone, A hybrid outdoor/indoor positioning system for IoT applications, in *1st IEEE International Symposium on Systems Engineering, ISSE 2015—Proceedings*, pp. 1–6, https://doi.org/10.1109/syseng.2015.7302503. V. Pasku, A. Moschitta, P. Carbone, A hybrid outdoor/indoor positioning system for IoT applications, in *1st IEEE International Symposium on Systems Engineering, ISSE 2015—Proceedings*, pp. 1–6. https://doi.org/10.1109/syseng.2015.7302503
10. L. He, H. Lu, 3D indoor layered localization of multi-information fusion via intelligent terminal, in *China Satellite Navigation Conference (CSNC) 2017 Proceedings: Volume II. CSNC 2017*, vol 438, ed. by J. Sun, J. Liu, Y. Yang, S. Fan, W. Yu. Lecture Notes in Electrical Engineering (Springer, Singapore, 2017)
11. M. Jia, Y. Yang, L. Kuang, W. Xu, T. Chu, H. Song, An indoor and outdoor seamless positioning system based on android platform, in *Proceedings—15th IEEE International Conference on Trust, Security and Privacy in Computing and Communications, 10th IEEE International Conference on Big Data Science and Engineering and 14th IEEE International Symposium on Parallel and Distributed Processing with Applications, IEEE TrustCom/BigDataSE/ISPA 2016*, pp. 1114–1120. https://doi.org/10.1109/trustcom.2016.0183
12. T. Kulshrestha, D. Saxena, R. Niyogi, V. Raychoudhury, M. Misra, SmartITS: Smartphone-based identification and tracking using seamless indoor-outdoor localization. J. Netw. Comput. Appl. **98**, 97–113 (2017). https://doi.org/10.1016/j.jnca.2017.09.003
13. S. Li, Z. Qin, H. Song, C. Si, B. Sun, X. Yang, R. Zhang, A lightweight and aggregated system for indoor/outdoor detection using smart devices. Future Gener. Comput. Syst. (2017). https://doi.org/10.1016/j.future.2017.05.028
14. X. Li, J. Wang, T. Li, Seamless positioning and navigation by using geo-referenced images and multi-sensor data. Sensors (Basel, Switzerland) **13**(7), 9047–9069 (2013). https://doi.org/10.3390/s130709047
15. X. Li, D. Wei, Q. Lai, Y. Xu, H. Yuan, Smartphone-based integrated PDR/GPS/Bluetooth pedestrian location. Adv. Space Res. **59**(3), 877–887 (2017). https://doi.org/10.1016/j.asr.2016.09.010
16. X. Liu, Z. Wu, X. Lin, WLAN/MARG/GPS integrated positioning system based on a self-adaptive weighted algorithm. Qinghua Daxue Xuebao/J. Tsinghua Univ. **53**(7), 955–960 (2013)
17. H.S. Maghdid, Hybridisation of GNSS with other wireless/sensors technologies onboard smartphones to offer seamless outdoors-indoors positioning for LBS applications. Ph.D.Thesis, University of Buckingham, UK, 2015
18. B. Molina, E. Olivares, C.E. Palau, M. Esteve, A multimodal fingerprint-based indoor positioning system for airports. IEEE Access (2018). https://doi.org/10.1109/ACCESS.2018.2798918
19. G. Moon, S. Chun, M. Hur, G.-I. Jee, A T-DMB navigation system for seamless positioning in both indoor and outdoor environments. EURASIP J. Adv. Sig. Proc. **2014**, 99 (2014)
20. H. Takatsuka, S. Tokunaga, S. Saiki, S. Matsumoto, M. Nakamura, Integrating heterogeneous locating services for efficient development of location-based services, in *17th International Conference on Information Integration and Web-Based Applications and Services, iiWAS 2015—Proceedings* (2015). https://doi.org/10.1145/2837185.2837229
21. L. Wang, L. Feng, M. Zawodniok, Ubiquitous tracking using motion and location sensor with application to smartphone, in *2017 IEEE International Conference on Smart Computing, SMARTCOMP* (2017). https://doi.org/10.1109/smartcomp.2017.7947056
22. Q. Zhu, Y. Li, Q. Xiong, S. Zlatanova, Y. Ding, Y. Zhang, Y. Zhou, Indoor multi-dimensional location GML and its application for ubiquitous indoor location services. ISPRS Int. J. Geo-Inf. (2016), 5(12). https://doi.org/10.3390/ijgi5120220

23. H. Zou, H. Jiang, Y. Luo, J. Zhu, X. Lu, L. Xie, Bluedetect: an ibeacon-enabled scheme for accurate and energy-efficient indoor-outdoor detection and seamless location-based service. Sensors (Switzerland), **16**(2). https://doi.org/10.3390/s16020268
24. A. Küpper, *Location-Based Services: Fundamentals and Operation* (Wiley, 2005), pp. 365
25. R. Rahim, Comparative analysis of membership function on Mamdani fuzzy inference system for decision making. J. Phys. Conf. Ser. **930**(1), 012–029 (2017)

Neural Networks

From Traditional Neural Networks to Deep Learning: Towards Mathematical Foundations of Empirical Successes

Vladik Kreinovich

Abstract How do we make computers think? To make machines that fly, it is reasonable to look at the creatures that know how to fly: the birds. To make computers think, it is reasonable to analyze how we think—this is the main origin of neural networks. At first, one of the main motivations was speed—since even with slow biological neurons, we often process information fast. The need for speed motivated traditional 3-layer neural networks. At present, computer speed is rarely a problem, but accuracy is—this motivated deep learning. In this paper, we concentrate on the need to provide mathematical foundations for the empirical success of deep learning.

1 Traditional Neural Networks: A Brief History

Why traditional neural networks: (sanitized) history. How do we make computers think?

- To make machines that fly, it is reasonable to look at the creatures that know how to fly: the birds.
- To make computers think, it is reasonable to analyze how we humans think.

On the biological level, our brain processes information via special cells called *neurons*.

Somewhat surprisingly, in the brain, signals are electric—just as in the computer. The main difference is that in a neural network, signals are sequences of identical pulses. The intensity of a signal is described by the frequency of pulses.

A neuron has many inputs (up to 10^4). All the inputs $x_1, \ldots, x_n$ are combined, with some loss, into a frequency

V. Kreinovich (✉)
Department of Computer Science, University of Texas at El Paso,
500 W. University, El Paso, TX 79968, USA
e-mail: vladik@utep.edu

S. N. Shahbazova et al. (eds.), *Recent Developments and the New Direction in Soft-Computing Foundations and Applications*, Studies in Fuzziness and Soft Computing 393, https://doi.org/10.1007/978-3-030-47124-8_31

$$\sum_{i=1}^{n} w_i \cdot x_i.$$

Low inputs do not active the neuron at all, high inputs lead to largest activation. The output signal is a non-linear function

$$y = s_0\left(\sum_{i=1}^{n} w_i \cdot x_i - w_0\right).$$

In biological neurons,

$$s_0(x) = \frac{1}{1 + \exp(-x)}.$$

Traditional neural networks emulate such biological neurons; see, e.g., [2].

Why traditional neural networks: real history. At first, researchers ignored non-linearity and only used linear neurons. They got good results and made many promises.

The euphoria ended in the late 1960s when MIT's Marvin Minsky and Seymour Papert published a book [10]. Their main result was that a composition of linear functions is linear (I am not kidding). This ended the hopes of original schemes.

For some time, neural networks became a bad word. Then, smart researchers came us with a genius idea: let's make neurons non-linear. This revived the field.

Traditional neural networks: main motivation. One of the main motivations for neural networks was that computers were slow. Although human neurons are much slower than CPU, the human processing was often faster. So, the main motivation was to make data processing faster. The idea was that since we are the result of billion years of ever improving evolution, our biological mechanics should be optimal (or close to optimal).

To make processing faster, we need to have many fast processing units working in parallel. The fewer layers, the smaller overall processing time.

In nature, there are many fast linear processes—e.g., combining electric signals. As a result, linear processing (L) is faster than non-linear one.

For non-linear processing, the more inputs, the longer it takes. So, the fastest non-linear processing (NL) units process just one input. It turns out that two layers are not enough to approximate any function.

Why one or two layers are not enough. With one linear (L) layer, we only get linear functions. With one nonlinear (NL) layer, we only get functions of one variable.

With L $\to$ NL layers, we get

$$s_0\left(\sum_{i=1}^{n} w_i \cdot x_i - w_0\right).$$

For these functions, the level sets $f(x_1, \ldots, x_n) = \text{const}$ are planes

$$\sum_{i=1}^{n} w_i \cdot x_i = c.$$

Thus, they cannot approximate, e.g., $f(x_1, x_2) = x_1 \cdot x_2$ for which the level set is a hyperbola.

For NL → L layers, we get

$$f(x_1, \ldots, x_n) = \sum_{i=1}^{n} s_i(x_i).$$

For all these functions,

$$d \stackrel{\text{def}}{=} \frac{\partial^2 f}{\partial x_1 \partial x_2} = 0,$$

so we also cannot approximate $f(x_1, x_2) = x_1 \cdot x_2$ for which $d = 1 \neq 0$.

Why three layers are sufficient: Newton's prism and Fourier transform. In principle, we can have two 3-layer configurations: L → NL → L and NL → L → NL. Since L is faster than NL, the fastest is L → NL → L:

$$y = \sum_{k=1}^{K} W_k \cdot s_k \left(\sum_{i=1}^{n} w_{ki} \cdot x_i - w_{k0} \right) - W_0.$$

Newton showed that a prism decomposes while light (or any light) into elementary colors. In precise terms, elementary colors are sinusoids

$$A \cdot \sin(w \cdot t) + B \cdot \cos(w \cdot t).$$

Thus, every function can be approximated, with any accuracy, as a linear combination of sinusoids:

$$f(x_1) \approx \sum_{k} (A_k \cdot \sin(w_k \cdot x_1) + B_k \cdot \cos(w_k \cdot x_1)).$$

This result was theoretically proven later by Fourier.

For $f(x_1, x_2)$, we get a similar expression for each x_2, with $A_k(x_2)$ and $B_k(x_2)$. We can similarly represent $A_k(x_2)$ and $B_k(x_2)$, thus getting products of sines, and it is known that, e.g.:

$$\cos(a) \cdot \cos(b) = \frac{1}{2} \cdot (\cos(a + b) + \cos(a - b)).$$

Thus, we get an approximation of the desired form with $s_k = \sin$ or $s_k = \cos$:

$$y = \sum_{k=1}^{K} W_k \cdot s_k \left(\sum_{i=1}^{n} w_{ki} \cdot x_i - w_{k0} \right).$$

Which activation functions $s_k(z)$ should we choose. A general 3-layer NN has the form:

$$y = \sum_{k=1}^{K} W_k \cdot s_k \left(\sum_{i=1}^{n} w_{ki} \cdot x_i - w_{k0} \right) - W_0.$$

Biological neurons use

$$s_0(z) = \frac{1}{1 + \exp(-z)},$$

but shall we simulate it?

Simulations are not always efficient. E.g., airplanes have wings like birds but they do not flap them.

Let us analyze this problem theoretically. There is always some noise c in the communication channel. So, we can consider either the original signals x_i or denoised ones $x_i - c$. The results should not change if we perform a full or partial denoising $z \to z' = z - c$.

Denoising means replacing $y = s_0(z)$ with $y' = s_0(z - c)$. So, $s_0(z)$ should not change under shift $z \to z - c$.

Of course, $s_0(z)$ cannot remain the same: if $s_0(z) = s_0(z - c)$ for all c, then $s_0(z) =$ const. The idea is that once we re-scale x, we should get the same formula after we apply a "natural" y-re-scaling T_c: $s_0(x - c) = T_c(s_0(x))$. But which re-scalings are natural?

Which transformations are natural? Linear re-scalings are natural: they corresponding to changing units and starting points (like C to F).

An inverse T_c^{-1} to a natural re-scaling T_c should also be natural. A composition $y \to T_c(T_{c'}(y))$ of two natural re-scalings T_c and $T_{c'}$ should also be natural.

In mathematical terms, natural re-scalings form a *group*.

For practical purposes, we should only consider re-scaling determined by finitely many parameters. So, we look for a finite-parametric group containing all linear transformations.

A somewhat unexpected approach. N. Wiener, in his famous book *Cybernetics* [13], noticed that when we approach an object, we have distinct phases:

- first, we see a blob (the image is invariant under all transformations);
- then, we start distinguishing angles from smooth but not sizes (projective transformations);
- after that, we detect parallel lines (affine transformations);
- then, we detect relative sizes (similarities);
- finally, we see the exact shapes and sizes.

Are there other transformation groups?

Wiener argued that if there were other groups, after billions years of evolutions, we would use them. So he conjectured that there are no other groups.

Wiener was right. Wiener's conjecture was indeed proven in the 1960s. In 1-D case, this means that all our transformations are fractionally linear:

$$s_0(z - c) = \frac{A(c) \cdot s_0(z) + B(c)}{C(c) \cdot s_0(z) + D(c)};$$

see Appendix for technical details.

For $c = 0$, we get $A(0) = D(0) = 1$, $B(0) = C(0) = 0$. Differentiating the above equation by c and taking $c = 0$, we get a differential equation for $s_0(z)$:

$$-\frac{ds_0}{dz} = (A'(0) \cdot s_0(z) + B'(0)) - s_0(z) \cdot (C'(0) \cdot s_0(z) + D'(0)).$$

So,

$$\frac{ds_0}{C'(0) \cdot s_0^2 + (A'(0) - C'(0)) \cdot s_0 + B'(0)} = -dz.$$

Integrating, we indeed get

$$s_0(z) = \frac{1}{1 + \exp(-z)}.$$

(after an appropriate linear re-scaling of z and $s_0(z)$); see [9, 11, 12].

How to train traditional neural networks: main idea. *Reminder*: a 3-layer neural network has the form:

$$y = \sum_{k=1}^{K} W_k \cdot s_0\left(\sum_{i=1}^{n} w_{ki} \cdot x_i - w_{k0}\right) - W_0.$$

We need to find the weights that best described observations $\left(x_1^{(p)}, \ldots, x_n^{(p)}, y^{(p)}\right)$, $1 \le p \le P$.

We find the weights that minimize the mean square approximation error

$$E \stackrel{\text{def}}{=} \sum_{p=1}^{P} \left(y^{(p)} - y_{NN}^{(p)}\right)^2,$$

where

$$y_{NN}^{(p)} = \sum_{k=1}^{K} W_k \cdot s_0\left(\sum_{i=1}^{n} w_{ki} \cdot x_i^{(p)} - w_{k0}\right) - W_0.$$

The simplest minimization algorithm is gradient descent:

$$w_i \to w_i - \lambda \cdot \frac{\partial E}{\partial w_i}.$$

Towards faster differentiation. To achieve high accuracy, we need many neurons. Thus, we need to find many weights. To apply gradient descent, we need to compute all partial derivatives $\frac{\partial E}{\partial w_i}$. Differentiating a function f is easy:

- the expression f is a sequence of elementary steps,
- so we take into account that $(f \pm g)' = f' \pm g'$, $(f \cdot g)' = f' \cdot g + f \cdot g'$, $(f(g))' = f'(g) \cdot g'$, etc.

For a function that takes T steps to compute, computing f' thus takes $c_0 \cdot T$ steps, with $c_0 \le 3$.

However, for a function of n variables, we need to compute n derivatives. This would take time $n \cdot c_0 \cdot T \gg T$: this is too long.

Faster differentiation: backpropagation. Idea:

- instead of starting from the variables,
- start from the last step, and compute

$$\frac{\partial E}{\partial v}$$

  for all intermediate results v.

For example, if the very last step is $E = a \cdot b$, then

$$\frac{\partial E}{\partial a} = b$$

and

$$\frac{\partial E}{\partial b} = a.$$

At each step y, if we know

$$\frac{\partial E}{\partial v}$$

and $v = a \cdot b$, then

$$\frac{\partial E}{\partial a} = \frac{\partial E}{\partial v} \cdot b$$

and

$$\frac{\partial E}{\partial b} = \frac{\partial E}{\partial v} \cdot a.$$

At the end, we get all n derivatives

$$\frac{\partial E}{\partial w_i}$$

in time

$$c_0 \cdot T \ll c_0 \cdot T \cdot n.$$

This is known as *backpropagation*; see, e.g., [2].

2 Beyond Traditional Neural Networks (NN)

Need for deep learning. Nowadays, computer speed is no longer a big problem. What *is* a problem is *accuracy:* even after thousands of iterations, the NNs do not learn well. So, instead of computation speed, we would like to maximize learning accuracy.

We can still consider L and NL elements. For the same number of variables w_i, we want to get more accurate approximations. For given number of variables, and given accuracy, we get N possible combinations. If all combinations correspond to different functions, we can implement N functions. However, if some combinations lead to the same function, we implement fewer different functions.

For a traditional NN with K neurons, each of $K!$ permutations of neurons retains the resulting function; see, e.g., [6]. Thus, instead of N functions, we only implement

$$\frac{N}{K!} \ll N \text{ functions.}$$

Thus, to increase accuracy, we need to minimize the number K of neurons in each layer.

To get a good accuracy, we need many parameters, thus many neurons. Since each layer is small, we thus need many layers. This is the *main idea* behind *deep learning* [1, 5].

3 Empirical Formulas Behind Deep Learning Successes and How They Can Be Justified

Formulation of the problem. While the general idea of deep learning is natural, many specific formulas that lead to deep learning successes are purely empirical, and need to be explained. In this section, we list such formulas, and briefly mention how the corresponding formulas can be explained.

Rectified linear neurons. Instead of complex nonlinear neurons used in traditional neural networks, deep networks utilize *rectified linear neurons* for which the activation function $s_0(z)$ has the form $s_0(z) = \max(0, z)$.

Our explanation [3] is that:

- this activation function is invariant under re-scaling (changing of the measuring unit) $z \to \lambda \cdot x$;
- moreover, it is, in effect, the only activation function which is thus invariant, and
- it is, in effect, the only activation which is optimal with respect to any scale-invariant optimality criterion.

Combining several results. To speed up the training, the current deep learning algorithms use dropout techniques:

- they train several sub-networks on different portions of data, and then
- "average" the results.

A natural idea is to use arithmetic mean for this "averaging", but empirically, geometric mean works much better. In [4], we provide a theoretical explanation for the empirical efficiency of selecting geometric mean as the "averaging" in dropout training: namely, it turns out that

- this choice is scale-invariant—and,
- in effect, it is the only scale-invariant choice.

Softmax. In deep learning, instead of selecting an alternative for which the objective function $f(x)$ is the largest possible, we use so-called *softmax*—i.e., select each alternative x with probability proportional to $\exp(\alpha \cdot f(x))$, for some $\alpha > 0$.

In general, we could select any increasing function $F(z)$ and select probabilities proportional to $F(f(x))$, so why exponential function is the most successful? It turns out [8, 11] that:

- when we use this function, the resulting probabilities do not change if we simply shift all the values $f(x)$, i.e., change them to $f(x) + c$ for some c—which does not change the original optimization problem;
- moreover, exponential functions are the only ones which lead to such shift-invariant selection, and,
- in effect, the exponential functions are only ones which can be optimal under a shift-invariant optimality criterion.

Replacing Least Squares with a more complex expression. In deep learning, instead of the Least Squares, we often optimize the Kullback-Leibler (KL) divergence

$$-\sum_{k=1}^{K}\left[y^{(k)} \cdot \log\left(y_{\mathrm{NN}}^{(k)}\right) + \left(1 - y^{(k)}\right) \cdot \log\left(1 - y_{\mathrm{NN}}^{(k)}\right)\right].$$

A possible explanation for the empirical success of this formula is given in [7].

Acknowledgements This work was supported in part by the US National Science Foundation grant HRD-1242122.

Appendix: Why Fractional Linear

Every transformation is a composition of infinitesimal ones $x \to x + \varepsilon \cdot f(x)$, for infinitely small ε. So, it's enough to consider infinitesimal transformations.

The class of the corresponding functions $f(x)$ is known as a *Lie algebra* A of the corresponding transformation group. Infinitesimal linear transformations correspond to $f(x) = a + b \cdot x$, so all linear functions are in A. In particular, $1 \in A$ and $x \in A$.

For any λ, the product $\varepsilon \cdot \lambda$ is also infinitesimal, so we get

$$x \to x + (\varepsilon \cdot \lambda) \cdot f(x) = x \to x + \varepsilon \cdot (\lambda \cdot f(x)).$$

So, if $f(x) \in A$, then $\lambda \cdot f(x) \in A$.

If we first apply $f(x)$, then $g(x)$, we get

$$x \to (x + \varepsilon \cdot f(x)) + \varepsilon \cdot g(x + \varepsilon \cdot f(x)) = x + \varepsilon \cdot (f(x) + g(x)) + o(\varepsilon).$$

Thus, if $f(x) \in A$ and $g(x) \in A$, then $f(x) + g(x) \in A$. So, A is a linear space.

In general, for the composition, we get

$$x \to (x + \varepsilon_1 \cdot f(x)) + \varepsilon_2 \cdot g(x_1 + \varepsilon_1 \cdot f(x)) =$$

$$x + \varepsilon_1 \cdot f(x) + \varepsilon_2 \cdot g(x) + \varepsilon_1 \cdot \varepsilon_2 \cdot g'(x) \cdot f(x) + \text{quadratic terms}.$$

If we then apply the inverses to $x \to x + \varepsilon_1 \cdot f(x)$ and $x \to x + \varepsilon_2 \cdot g(x)$, the linear terms disappear, we get: $x \to x + \varepsilon_1 \cdot \varepsilon_2 \cdot \{f, g\}(x)$, where

$$\{f, g\} \stackrel{\text{def}}{=} f'(x) \cdot g(x) - f(x) \cdot g'(x).$$

Thus, if $f(x) \in A$ and $g(x) \in A$, then $\{f, g\}(x) \in A$. The expression $\{f, g\}$ is known as the *Poisson bracket*.

Let's expand any function $f(x)$ in Taylor series:

$$f(x) = a_0 + a_1 \cdot x + \cdots$$

If k is the first non-zero term in this expansion, we get

$$f(x) = a_k \cdot x^k + a_{k+1} \cdot x^{k+1} + a_{k+2} \cdot x^{k+2} + \cdots$$

For every λ, the algebra A also contains

$$\lambda^{-k} \cdot f(\lambda \cdot x) = a_k \cdot x^k + \lambda \cdot a_{k+1} \cdot x^{k+1} + \lambda^2 \cdot a_{k+2} \cdot x^{k+2} + \cdots$$

In the limit $\lambda \to 0$, we get $a_k \cdot x^k \in A$, hence $x^k \in A$.

Thus,

$$f(x) - a_k \cdot x^k = a_{k+1} \cdot x^{k+1} + \cdots \in A.$$

We can similarly conclude that A contains all the terms x^n for which $a_n \neq 0$ in the original Taylor expansion.

Since $g(x) = 1 \in A$, for each $f \in A$, we have

$$\{f, 1\} = f'(x) \cdot 1 + f(x) \cdot q' = f'(x) \in A.$$

Thus, for each k, if $x^k \in A$, we have $(x^k)' = k \cdot x^{k-1} \in A$ hence $x^{k-1} \in A$, etc. So, if $x^k \in A$, all smaller power are in A too. In particular, this means that if $x^k \in A$ for some $k \geq 3$, then we have $x^3 \in A$ and $x^2 \in A$; thus:

$$\{x^3, x^2\} = (x^3)' \cdot x^2 - x^3 \cdot (x^2)' = 3 \cdot x^2 \cdot x^2 - x^3 \cdot 2 \cdot x = x^4 \in A.$$

In general, once $x^k \in A$ for $k \geq 3$, we get

$$\{x^k, x^2\} = (x^k)' \cdot x^2 - x^k \cdot (x^2)' = k \cdot x^{k-1} \cdot x^2 - x^k \cdot 2 \cdot x = (k-2) \cdot x^{k+1} \in A,$$

hence $x^{k+1} \in A$. So, by induction, $x^k \in A$ for all k. Thus, A is infinite-dimensional—which contradicts to our assumption that A is finite-dimensional.

So, we cannot have Taylor terms of power $k \geq 3$; therefore we have:

$$x \to x + \varepsilon \cdot (a_0 + a_1 \cdot x + a_2 \cdot x^2).$$

This corresponds to an infinitesimal fractional-linear transformation

$$x \to \frac{\varepsilon \cdot A + (1 + \varepsilon \cdot B) \cdot x}{1 + \varepsilon \cdot D \cdot x} = (\varepsilon \cdot A + (1 + \varepsilon \cdot B) \cdot x) \cdot (1 - \varepsilon \cdot D \cdot x) + o(\varepsilon) =$$

$$x + \varepsilon \cdot (A + (B - D) \cdot x - D \cdot x^2).$$

So, to match, we need

$$A = a_0, \quad D = -a_2, \text{ and } B = a_1 - a_2.$$

We concluded that every infinitesimal transformation is fractionally linear. Every transformation is a composition of infinitesimal ones. Composition of fractional-linear transformations is fractional linear. Thus, all transformations are fractional linear.

References

1. C. Baral, O. Fuentes, V. Kreinovich, Why deep neural networks: a possible theoretical explanation, in *Constraint Programming and Decision Making: Theory and Applications*, ed. by M. Ceberio, et al. (Springer, 2018), pp. 1–6
2. C.M. Bishop, *Pattern Recognition and Machine Learning* (Springer, New York, 2006)
3. O. Fuentes, J. Parra, E. Anthony, V. Kreinovich, Why rectified linear neurons are efficient: a possible theoretical explanations, in *Beyond Traditional Probabilistic Data Processing Techniques: Interval, Fuzzy, etc. Methods and Their Applications*, ed. by O. Kosheleva, S. Shary, G. Xiang, R. Zapatrin (Springer, Cham, Switzerland, 2018) (to appear)
4. A. Gholamy, J. Parra, V. Kreinovich, O. Fuentes, E. Anthony, How to best apply neural networks in geosciences: towards optimal 'Averaging' in dropout training, in *Smart Unconventional Modelling, Simulation and Optimization for Geosciences and Petroleum Engineering*, ed. by J. Watada, S.C. Tan, P. Vasant, E. Padmanabhan, L.C. Jain (Springer, 2017) (to appear)
5. I. Goodfellow, Y. Bengio, A. Courville, *Deep Leaning* (MIT Press, Cambridge, Massachusetts, 2016)
6. P.C. Kainen, V. Kurkova, V. Kreinovich, O. Sirisaengtaksin, Uniqueness of network parameterization and faster learning. Neural Parallel Sci. Comput. **2**, 459–466 (1994)
7. O. Kosheleva, V. Kreinovich, Why deep learning methods use KL divergence instead of least squares: a possible pedagogical explanation. Math. Struct. Model. **46**, 102–106 (2018)
8. V. Kreinovich, Group-theoretic approach to intractable problems. Lecture Notes in Computer Science, vol. 417 (Springer, Berlin, 1990), pp. 112–121
9. V. Kreinovich, C. Quintana, Neural networks: what non-linearity to choose?, in *Proceedings of the 4th University of New Brunswick Artificial Intelligence Workshop* (New Brunswick Canada, Fredericton, 1991), pp. 627–637
10. M. Minsky, S. Papert, *Perceptions* (MIT Press, Cambridge, Massachusetts, 1969)
11. H.T. Nguyen, V. Kreinovich, *Applications of Continuous Mathematics to Computer Science* (Kluwer, Dordrecht, Netherlands, 1997)
12. O. Sirisaengtaksin, V. Kreinovich, H.T. Nguyen, Sigmoid neurons are the safest against additive errors, in *Proceedings of the First International Conference on Neural, Parallel, and Scientific Computations*, Atlanta, GA, 28–31 May 1995, vol. 1, pp. 419–423
13. N. Wiener, *Cybernetics: Or Control and Communication in the Animal and the Machine* (MIT Press, Cambridge, Massachisetts, 1948)

Dimensionality Reduction: From Fuzzy Partitions to F-Transforms

Irina Perfilieva

Abstract The appropriate representation of high-dimensional data is the main focus of machine learning, pattern recognition and computer vision. With the same motivation, the F-transform uses fuzzy partitions in order to establish a compressed representation of data. Two distinguished properties of the F-transform: the best approximation in a local sense and dimensionality reduction contributed to the fact that the F-transform has many successful applications. We show that the technique of F-transform fully agrees with the technique of dimensionality reduction, based on Laplacian eigenmaps. To justify this claim, we characterize the processed by the F-transform data in terms of the adjacency graph that reflects their (data) intrinsic geometry. An application to the problem of image restoration is given.

1 Introduction

Modeling and processing large data bases (texts, images, video signals, cash flows, etc.) motivate applying machine learning theory and algorithms. "Big data" mining is based on discovering structured knowledge from spatio-temporally correlated data. The first step is a certain granulation of data, which means developing a low dimensional representation of data that arises from sampling a complex high dimensional data. The generic problem of dimensionality reduction is to find a set of points $\mathbf{y}_1, \ldots, \mathbf{y}_k$ in the space $\mathbb{R}^m$ such that $\mathbf{y}_i$ "represents" a point $\mathbf{x}_i$ from the given dataset $\mathbf{x}_1, \ldots, \mathbf{x}_k$ that belongs to the space $\mathbb{R}^l$ with the substantially larger dimension so that $m \ll l$.

Let us give some remarks regarding the history of the problem of dimensionality reduction, see [1]. Classical approaches include principal components analysis (PCA) and multidimensional scaling. Various methods that generate nonlinear maps have also been considered. Most of them, such as self-organizing maps and other neural network-based approaches (e.g., [3]), set up a nonlinear optimization problem whose

I. Perfilieva (✉)
IRAFM, University of Ostrava, 30. dubna 22, 701 03 Ostrava, Czech Republic
e-mail: irina.perfilieva@osu.cz

S. N. Shahbazova et al. (eds.), *Recent Developments and the New Direction in Soft-Computing Foundations and Applications*, Studies in Fuzziness and Soft Computing 393, https://doi.org/10.1007/978-3-030-47124-8_32

solution is typically obtained by gradient descent that is guaranteed only to produce a local optimum; global optima are difficult to attain by efficient means. To our knowledge, the approach of generalizing the PCA through kernel-based techniques in [12] does not have this shortcoming. However, most of methods in [12] do not explicitly consider the structure of the manifold (space $\mathbb{R}^l$) on which the data may possibly lie.

In [1], an approach that builds a graph incorporating neighborhood information of the data set is proposed. Using the notion of the Laplacian of the graph, a low-dimensional representation of the data set that optimally preserves local neighborhood information is computed. The representation map generated by the proposed algorithm may be viewed as a discrete approximation to a continuous map that naturally arises from the geometry of the manifold. The most important feature of the solution consists in reflecting the intrinsic geometric structure of the manifold. The latter is approximated by the adjacency graph computed from the data points.

In fuzzy literature, the dimensionality reduction was hidden under the notions and techniques of granulation, clustering and fuzzy partition, [13]. The results were used in the form of collections of fuzzy sets and after that in fuzzy rules databases. Actually, the main advantage of modeling with fuzzy IF-THEN rules is in transforming a problem from an initial complex high dimensional data space to the low dimensional space of fuzzy sets that are new atomic units.

However, despite of this obvious similarity, the dimensionality reduction in the sense of machine learning is different. The difference is in the way of representation. Instead of center-shape (clustering) or membership function (fuzzy sets) representation, low-dimensional images are characterized in terms of features (eigenvectors). Therefore, a cluster (granule) is characterized as a collection of common features that are extracted from an initial data embedded into a particular manifold.

In this respect, the only fuzzy technique which is similar to the machine-learning-based dimensionality reduction is the F-(fuzzy) transform. The distinguished property of the F-transform is that it uses a fuzzy partition as a space characterization of data where the notion of closeness is determined by membership functions of partition units—basic functions. The extracted features are the so called (functionally expressed) F-transform components that are weighed projections on elementary functions (polynomials) where weights are fuzzy partition units. Below, we show how this is connected with eigenvectors that arise in the transformation of high to low-dimensional representation.

In the theoretical part of this contribution (Sect. 3), we show that the technique of F-transforms ("F" stands for fuzzy) fully agrees with the technique in [1]. The theory of fuzzy (F)-transforms provides a (dimensionally) reduced and robust representation of original data. The main characteristics with respect to input data: size reduction, noise removal, invariance to geometrical transformations, knowledge transfer from conventional mathematics, fast computation. The F-transform has been applied to: image processing, computer vision, on-line pattern recognition in big data bases, time series analysis and forecasting, mathematical finance, numerical methods for differential equations, deep learning neural networks.

To justify the main claim, we characterize the processed by the F-transform data in terms of the adjacency graph that reflects their intrinsic geometry. In the application part (Sect. 4), we consider the problem of image restoration and explain how it can be properly formulated and solved in the language of F-transform and using the proposed technique of dimensionality reduction.

2 Preliminaries

In this part, we remind basic facts regarding two main techniques used in this contribution: dimensionality reduction and F-transform.

2.1 Laplacian Eigenmaps for Dimensionality Reduction

We shortly remind useful details of the dimensionality reduction technique, as they were proposed in [1]. Following [1], we give an algorithm that constructs representatives $\mathbf{y}_i$'s from the space $\mathbb{R}^m$ for the set of points $\mathbf{x}_1, \ldots, \mathbf{x}_k$ from the space $\mathbb{R}^l$ where $m \ll l$. The result is in the form of embedding map and is provided by computing the eigenvectors of the graph Laplacian.

We start with the construction of a weighted graph $G = (V, E)$ with k nodes, one for each point $\mathbf{x}_i$, where the edges from E connect neighboring points. We assume that the graph is connected and that the resulting embedding $(\mathbf{y}_1, \ldots, \mathbf{y}_k)^T$ maps vertices of G to the real line $\mathbb{R}$ in such a way that connected points stay as close as possible.

Step 1 (constructing the adjacency graph). We put an edge between nodes i and j, if $\mathbf{x}_i$ and $\mathbf{x}_j$ are "close".

Step 2 (choosing the weights). In [1], two variants for weighting the edges were considered:

(a) *Heat kernel*. If vertices i and j are connected, then

$$W_{ij} = \exp(-\frac{\|\mathbf{x}_i - \mathbf{x}_j\|^2}{t}),$$

otherwise $W_{ij} = 0$. Parameter $t \in \mathbb{R}$ should be specified beforehand.

(b) *Simple-minded assignment*. $W_{ij} = 1$, if vertices i and j are connected by an edge and $W_{ij} = 0$, otherwise.

Step 3 (eigenmaps). Compute eigenvalues and eigenvectors for the generalized eigenvector problem

$$L\mathbf{f} = \lambda D\mathbf{f}, \tag{1}$$

where D is the diagonal weight matrix such that

$$D_{ii} = \sum_{j=1}^{k} W_{ij},$$

and $L = D - W$ is the Laplacian matrix. Below, we will show that matrix L is symmetric and positive semidefinite.

Let $\mathbf{f}_0, \ldots, \mathbf{f}_{k-1}$ be the solutions of Eq. (1), ordered according to their eigenvalues. If we leave out the eigenvector $\mathbf{f}_0$ corresponding to the eigenvalue 0 and use the next m eigenvectors, then the embedding in m-dimensional Euclidean space is given by

$$\mathbf{x}_i \to (\mathbf{f}_1(i), \ldots, \mathbf{f}_m(i).$$

Let us stress that the proposed solution reflects the intrinsic geometric structure of the manifold. It is reported in [1] that the above given Laplacian eigenmap algorithm is relative insensitive to outliers and noise. This is due to locality-preserving character of the adjacency graph. Moreover, by trying to preserve local information in the embedding, the algorithm implicitly emphasizes the natural clusters in the data.

In order to justify the proposed algorithm, we show that matrix L is symmetric and positive semidefinite. The first property is obvious, and to prove the second one, we choose an arbitrary vector $\mathbf{y} = (y_1, \ldots, y_k)$ and show that

$$\mathbf{y}^T L\mathbf{y} = \frac{1}{2} \sum_{i,j=1}^{k} (y_i - y_j)^2 W_{ij}. \tag{2}$$

Indeed,

$$\begin{aligned}\sum_{i,j=1}^{k} (y_i - y_j)^2 W_{ij} &= \sum_{i,j=1}^{k} (y_i^2 + y_j^2 - 2y_i y_j)^2 W_{ij} \\ &= \sum_{i=1}^{k} y_i^2 D_{ii} + \sum_{j=1}^{k} y_j^2 D_{jj} - 2 \sum_{i,j=1}^{k} y_i y_j^2 W_{ij} = 2\mathbf{y}^T L\mathbf{y}.\end{aligned}$$

It is known that vector $\mathbf{y}$ that minimizes $\mathbf{y}^T L\mathbf{y}$ is given by the minimum eigenvalue solution to the generalized eigenvalue problem (1). By (2), this minimum eigenvalue solution minimizes the following objective function

$$\sum_{i,j=1}^{k} (y_i - y_j)^2 W_{ij}. \tag{3}$$

Moreover, if the graph G is connected, then $\mathbf{y} = \mathbf{1}$ is the only eigenvector for $\lambda = 0$.

In the following sections, we show that the F-transform technique can be reinterpreted within the framework of dimensionality-reduction based on intrinsic geometry of data.

2.2 F-Transforms

The short characterization of the technique of F-transform (originally, *fuzzy transform*, [6]) is that it is an integral transform that uses a *fuzzy partition* of a universe of discourse (usually, a bounded interval of reals $[a, b]$). The F-transform has two phases: direct and inverse. The direct F-transform (FzT) is applied to functions from $L^2([a, b])$ and maps them linearly onto (finite) sequences of numeric/functional components. Each component is a weighted orthogonal projection of a given function on a certain linear subspace of $L^2([a, b])$. Dimensionality reduction by the F-transform is thought as a projection of an original data onto the set of the F-transform components, see Sect. 3 for the details.

Below, we recall some definitions from [7].

2.2.1 Fuzzy Partition

Let $[a, b]$ be an interval on the real line $\mathbb{R}$. Fuzzy sets on $[a, b]$ are identified by their membership functions; i.e., they are mappings from $[a, b]$ into $[0, 1]$.

Definition 1 Let $[a, b]$ be an interval on $\mathbb{R}$, $n \geq 2$, and let $x_0, x_1, \ldots, x_n, x_{n+1}$ be nodes such that $a = x_0 \leq x_1 < \cdots < x_n \leq x_{n+1} = b$. We say that fuzzy sets $A_1, \ldots, A_n : [a, b] \to [0, 1]$, which are identified with their membership functions, constitute a *fuzzy partition* of $[a, b]$ if for $k = 1, \ldots, n$, if they fulfill the following conditions:

1. (locality)—$A_k(x) = 0$ if $x \in [a, x_{k-1}] \cup [x_{k+1}, b]$,
2. (continuity)—$A_k(x)$ is continuous;
3. (covering)—$A_k(x) > 0$ if $x \in (x_{k-1}, x_{k+1})$.

The membership functions $A_1, \ldots, A_n$ are called basic functions.

We say that the fuzzy partition $A_1, \ldots, A_n$, $n \geq 2$, is *h-uniform* if nodes $x_0, \ldots, x_{n+1}$ are *h-equidistant*; i.e., for all $k = 1, \ldots, n+1$, $x_k = x_{k-1} + h$, where $h = (b - a)/(n + 1)$ and the following three additional properties are fulfilled:

4. for all $k = 1, \ldots, n$, $A_k(x)$ strictly increases on $[x_{k-1}, x_k]$ and strictly decreases on $[x_k, x_{k+1}]$,
5. for all $k = 1, \ldots, n$, and for all $x \in [0, h]$, $A_k(x_k - x) = A_k(x_k + x)$,
6. for all $k = 2, \ldots, n$, and for all $x \in [x_{k-1}, x_{k+1}]$, $A_k(x) = A_{k-1}(x - h)$.

It can be easily shown that for an h-uniform fuzzy partition $A_1, \ldots, A_n$, of $[a, b]$, there exists a continuous and even function $A_0 : [-1, 1] \to [0, 1]$ such that it vanishes on boundaries and for all $k = 1, \ldots, n$,

$$A_k(x) = A_0\left(\frac{x - x_k}{h}\right), \quad x \in [x_{k-1}, x_{k+1}]. \tag{4}$$

We call A_0 a *generating function* of an h-uniform fuzzy partition.

2.2.2 Hilbert Space with Weighted Inner Product

Let us fix $[a, b]$ and its fuzzy partition $A_1, \ldots, A_n$, $n \geq 2$. Let k be a fixed integer from $\{1, \ldots, n\}$. $L^2([a, b])$ denotes a set of square-integrable functions on $[a, b]$. Let us denote

$$s_k = \int\limits_{x_{k-1}}^{x_{k+1}} A_k(x)dx,$$

and consider $\frac{1}{s_k}\int_{x_{k-1}}^{x_{k+1}} f(x)g(x)A_k(x)dx$ as a Lebesgue integral $\int_{x_{k-1}}^{x_{k+1}} f(x)g(x)d\mu_k$, where $d\mu = A_k(x)dx/s_k$ and the measure μ_k on $[x_{k-1}, x_{k+1}]$ is defined as follows:

$$\mu_k(E) = \frac{\int_E A_k(x)dx}{\int_{x_{k-1}}^{x_{k+1}} A_k(x)dx}.$$

Let $L^2(A_k)$ be a set of square-integrable functions on $[x_{k-1}, x_{k+1}]$ with the weighted inner product

$$\langle f, g\rangle_k = \int\limits_{x_{k-1}}^{x_{k+1}} f(x)g(x)A_k(x)dx. \tag{5}$$

The functions $f, g \in L^2(A_k)$ are *orthogonal* in $L^2(A_k)$ if $\langle f, g\rangle_k = 0$. The function $f \in L^2(A_k)$ is orthogonal to a subspace B of $L^2(A_k)$ if $\langle f, g\rangle_k = 0$ for all $g \in B$.

Let us denote by $L_2^m(A_k)$ a linear subspace of $L^2(A_k)$ with the basis given by orthogonal functions $P_k^0, P_k^1, P_k^2 \ldots, P_k^m$ where $P_k^0(x) = 1,\ x \in [x_{k-1}, x_{k+1}]$.

The following lemma gives analytic representation of the orthogonal projection on the subspace $L_2^m(A_k)$.

Lemma 1 *([7]) Let function F_k^m be the orthogonal projection of $f \in L^2(A_k)$ on $L_2^m(A_k)$. Then,*

$$F_k^m = c_{k,0}P_k^0 + c_{k,1}P_k^1 + \cdots + c_{k,m}P_k^m, \tag{6}$$

where for all $i = 0, 1, \ldots, m$,

$$c_{k,i} = \frac{\langle f, P_k^i\rangle_k}{\langle P_k^i, P_k^i\rangle_k} = \frac{\int_{x_{k-1}}^{x_{k+1}} f(x)P_k^i(x)A_k(x)dx}{\int_{x_{k-1}}^{x_{k+1}} P_k^i(x)P_k^i(x)A_k(x)dx}. \tag{7}$$

The n-tuple $(F_1^m, \ldots, F_n^m)$ is an F^m-transform of f with respect to $A_1, \ldots, A_n$, or formally,

$$F^m[f] = (F_1^m, \ldots, F_n^m).$$

F_k^m is called the kth F^m-transform component of f.

In particular, let us consider the case where the basis of $L_2^m(A_k)$ is given by orthogonal polynomials $P_k^0, P_k^1, P_k^2 \ldots, P_k^m$ and P_k^0 is a constant function with the value 1. Then, the F^0-transform of f or simply, the F-transform of f with respect to the partition $A_1, \ldots, A_n$ is given by the n-tuple $(c_{1,0}, \ldots, c_{n,0})$ of constant functions (0-degree polynomials) where for $k = 1, \ldots, n$,

$$c_{k,0} = \frac{\langle f, 1\rangle_k}{\langle 1, 1\rangle_k} = \frac{\int_{x_{k-1}}^{x_{k+1}} f(x)A_k(x)dx}{\int_{x_{k-1}}^{x_{k+1}} A_k(x)dx}. \tag{8}$$

The F^1-transform of f with respect to $A_1, \ldots, A_n$ is given by the n-tuple $(c_{1,0} + c_{1,1}(x - x_1), \ldots, c_{n,0} + c_{n,1}(x - x_n))$ of linear functions (1-degree polynomials). The latter are fully represented by their 2D coefficients $((c_{1,0}, c_{1,1}), \ldots, (c_{n,0}, c_{n,1}))$, which in addition to (8), have the following particular representation:

$$c_{k,1} = \frac{\langle f, x - x_k\rangle_k}{\langle (x - x_k), (x - x_k)\rangle_k} = \frac{\int_{x_{k-1}}^{x_{k+1}} f(x)(x - x_k)A_k(x)dx}{\int_{x_{k-1}}^{x_{k+1}} (x - x_k)^2 A_k(x)dx}. \tag{9}$$

The *inverse* F^m-transform of function f with respect to the partition $A_1, \ldots, A_n$ is a function represented by the following *inversion formula*:

$$f_{F,n}^m(x) = \sum_{k=1}^{n} F_k^m A_k(x). \tag{10}$$

The following results demonstrate approximation properties of the direct and inverse F^m-transforms.

Lemma 2 *([7]) Let $m \geq 0$, and let functions F_k^m and F_k^{m+1} be the k-th F^m- and F^{m+1}-transform components of f, respectively. Then,*

$$\|f - F_k^{m+1}\| \leq \|f - F_k^m\|.$$

Theorem 1 *([6, 7]) Let $A_1, \ldots, A_n$, $n \geq 2$, be an h-uniform fuzzy partition of $[a, b]$, let functions f and A_k, $k = 1, \ldots, n$, be four times continuously differentiable on $[a, b]$, and let $F^1[f] = (c_{1,0} + c_{1,1}(x - x_1), \ldots, c_{n,0} + c_{n,1}(x - x_n))$ be the F^1-transform of f with respect to $A_1, \ldots, A_n$. Then, for every $k = 1, \ldots, n$, the following estimation holds true:*

$$c_{k,0} = f(x_k) + O(h^2),$$
$$c_{k,1} = f'(x_k) + O(h^2)$$

Theorem 2 *([6]) Let f be a continuous function on [a,b]. For every $\varepsilon > 0$, there exist an integer n_ε and the related fuzzy partition $A_1, \ldots, A_{n_\varepsilon}$ of $[a, b]$ such that for all $x \in [a, b]$,*

$$|f(x) - f_{F,n_\varepsilon}(x)| < \varepsilon,$$

where f_{F,n_ε} is the inverse F-transform of f with respect to $A_1, \ldots, A_{n_\varepsilon}$.

Theorem 3 *([7]) Let* $A_1, \ldots, A_n$, $n \geq 2$, *be an* h*-uniform fuzzy partition of* $[a, b]$ *that fulfills the Ruspini condition on* $[a+h, b-h]$. *Let functions* f *and* A_k, $k = 1, \ldots, n$, *be four times continuously differentiable on* $[a, b]$, *and let* $f^m_{F,n}$ *be inverse* F^m*-transform of* f *where* $m \geq 1$. *Then,*

$$\int_{a+h}^{b-h} |f^m_{F,n}(x) - f(x)|dx \leq O(h^2).$$

The discrete F-transforms were introduced in [6] and then further elaborated in a number of papers, see e.g., [4].

3 Dimensionality Reduction by the F-Transform

In this section, we will connect the machine-learning-based dimensionality reduction with the F-(fuzzy) transform. In more detail, we will show that the F^0- and the F^1-transform are solutions of Eq. (1), provided that the Laplacian matrix L is properly constructed. This fact confirms that these F-transforms provide with a low-dimensional representation of a given dataset.

At first, we specify a dataset that will be characterized by their F-transform-based low-dimensional representation in the form of embedding maps and finally, by components. This dataset will be connected with a discrete representation of a function, say f on some domain (subset of the Euclidean $\mathbb{R}^l$). This function can be a signal, time series, image, etc. For simplicity, we assume that the domain is an interval $[a, b]$ of the real line, and the function f is given on a discrete set, say P of points where $P \subseteq [a, b]$. Then, we assume that $[a, b]$ is partitioned into the collection of fuzzy sets $A_1, \ldots, A_n$ (not necessarily uniform), as it is described in Definition 1. Moreover, we assume that for every k, $1 \leq k \leq n$, there is one point $x_k \in P$ (we call it *node*) such that $A_k(x_k) = 1$ and $A_j(x_k) = 0$, $j \neq k$. Further on, we distinguish between points (denoted by using letter p) and nodes (denoted by using letter x).

The chosen partition determines a geometry of the set P, given by the following directed weighted graph $D = (V(D), A(D)$. Each vertex from $V(D)$ corresponds to one point in P, and if a point $p_i \in P$ is *covered by basic function* A_k, i.e. $A_k(p_i) > 0$, then the corresponding to it vertex i is connected by the directed edge (k, i) with the vertex k, corresponding to the node x_k. Let us emphasize that there is no connection between "pure" points, if there is no node among them. In Fig. 1, we illustrate the connection "from node-to-points, covered by basic function A_k".

The weight of the directed edge (k, i) is equal to $A_k(p_i)$. It is easy to see that graph D is covered by n weakly connected components $D_1, \ldots, D_n$, each has a "star" shape. In more detail, $D_k = (V(D_k), A(D_k))$ is a subgraph of D associated with a "source" vertex k, such that the *indegree* of k is zero, i.e. $deg^-(k) = 0$. Then,

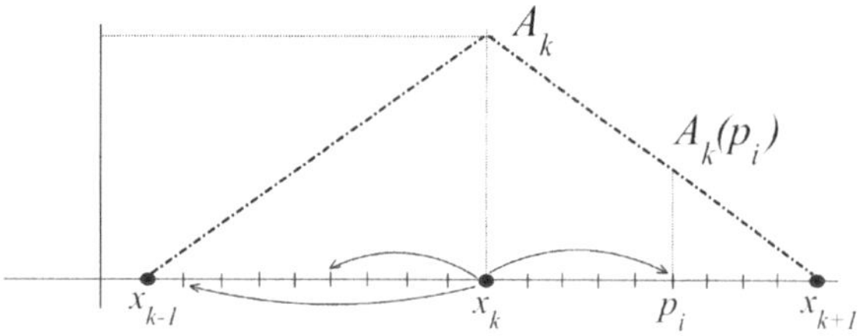

Fig. 1 Illustration of points $p_i \in P$ and the node x_k that are covered by basic function A_k. Connection from "node to point" is shown by a directed arrow

the vertex set $V(D_k)$ contains vertex k and moreover, vertex i belongs to $V(D_k)$, if $(k, i) \in A(D_k)$. Therefore, the set of directed edges $A(D_k)$ consists of the edges (k, i) where $i \in V(D_k)$ and $i \neq k$. We will continue with each connected component D_k separately, and construct the low-dimensional representation of the corresponding to it dataset.

Let a finite set of points $\{p_1, \ldots, p_{l_k}\}$ be covered by A_k (a fuzzy partition unit). Let us remark that the node x_k is among these points. Let $f_i = f(p_i)$, $1 \leq i \leq l_k$. The set X_k of data points allocated for the low-dimensional representation is $\{(p_1, f_1), \ldots, (p_{l_k}, f_{l_k})\}$. Let D_k be the corresponding directed subgraph of D and $G_k = (V(G_k), E(G_k)$ be the corresponding ordinary weighted graph derived from D_k after removing edge directions. Therefore, $V(G_k) = V(D_k)$, and if $(k, i) \in A(D_k)$, then $\{k, i\} \in E(G_k)$. In the corresponding to G_k weight matrix W, each edge, connecting i and k, is represented by the two symmetrical elements $W_{ik} = W_{ki} = A_k(p_i)$. Therefore, W is a symmetrical $l_k \times l_k$ matrix. Then, we proceed with *Step 3* of the algorithm described in Sect. 2.

The minimum eigenvalue solution to the generalized eigenvalue problem (1) is a constant vector that corresponds to the zero eigenvalue. This solution minimizes the objective function $\mathbf{y}^T L\mathbf{y}$ (see explanations in Sect. 2.1). To specify a solution, we put an additional constraint to this minimization problem and consider

$$\mathbf{f}_0 = \operatorname*{argmin}_{\langle \mathbf{f}, \mathbf{1} \rangle_k = (D\mathbf{y})_k} \mathbf{y}^T L\mathbf{y}, \tag{11}$$

where $\mathbf{f} = (f_1, \ldots, f_{l_k})$, $\langle \cdot, \cdot \rangle_k$ is the weighted inner product in $l^2(A_k)$ (discrete case of (5)) and $(\cdot)_k$ is the k-th vector component. Then, $\mathbf{f}_0$ is a constant vector, whose components are equal to each other and are equal to

$$(\mathbf{f}_0)_1 = \ldots (\mathbf{f}_0)_{l_k} = \frac{\sum_{j=1}^{l_k} A_k(p_j) f_j}{\sum_{j=1}^{l_k} A_k(p_j)}.$$

This solution agrees with the discrete version of the k-th F^0-transform component in (8). Moreover, it can be interpreted as a weighted projection of $\mathbf{f}$ on the constant vector $\mathbf{1}$, see illustration in Fig. 2.

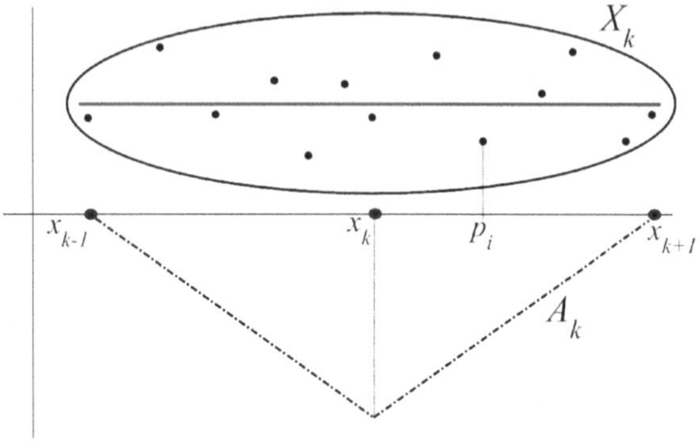

Fig. 2 Eigenvector (in red) of the k-th F^0-transform components as a low dimensional representation of data X_k

To obtain a non-constant vector solution to the minimization of $\mathbf{y}^T L\mathbf{y}$, let us put the constraint of orthogonality and consider

$$\mathbf{y}_1 = \operatorname*{argmin}_{\langle \mathbf{y},\mathbf{1}\rangle_k=0} \mathbf{y}^T L\mathbf{y}. \tag{12}$$

Then, the k-th F^1-transform component $\mathbf{f}_1$ has the following representation

$$\mathbf{f}_1 = \mathbf{f}_0 + \frac{\langle \mathbf{y}_1, \mathbf{f}\rangle_k}{\langle \mathbf{y}_1, \mathbf{y}_1\rangle_k}\mathbf{y}_1.$$

This representation agrees with (6) and (7).

To conclude, the vectorial F^0- and F^1-transform components, respectively, $\mathbf{f}_0$ and $\mathbf{f}_1$, provide with a low-dimensional representation of the considered data set X_k. In particular, every (x_i, f_i) from X_k is represented by $(\mathbf{f}_{0,i}, \mathbf{f}_{1,i})$.

4 Application to Image Restoration

Below, we discuss one typical problem of image processing—restoration and explain how it can be properly formulated and solved in the language of F-transform and using the proposed technique of dimensionality reduction.

Image restoration (inpainting) is a process of filling in damaged regions of images using information from surrounding areas. Formally, image I should be reconstructed from its damaged version f. This problem is connected with the two well developed problems of image processing: upscaling and denoising.

From the technical point of view, restoration is a type of (extra-) interpolation. However, there are many other approaches where restoration is considered as a boundary valued problem, based on nonlinear partial differential equations [2], as a particular inverse problem [11], or as a modification of the variational image denoising or segmentation. The latter model (called ROF—short of Rudin-Osher-Fatemi) focuses on the minimization of the functional

$$\int_{\Omega} |\nabla I| + \lambda \| f - GI \|^2_{L^2(\Omega)}, \tag{13}$$

where Ω is a domain of f, ∇ stands for the gradient, and G—a damaging operator. In (13), the first summand (regularization term) is a *total variation* (TV) of I, and the second one is the fidelity term.

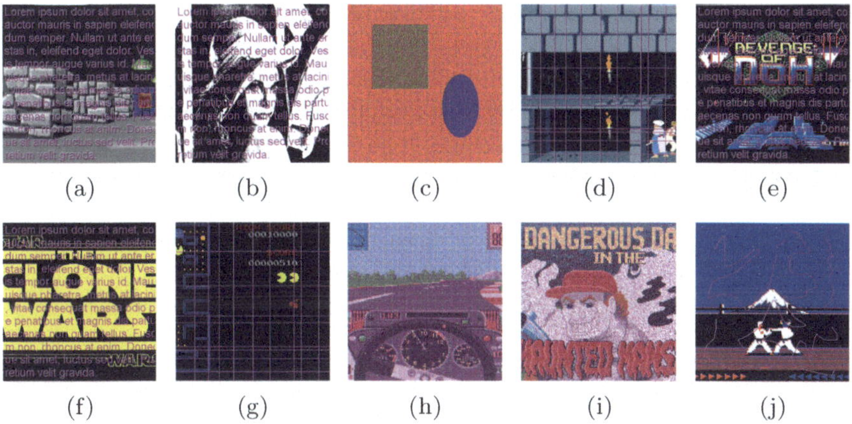

Fig. 3 Images with various damages: “text over image”, “blurring”,“noise”, etc.

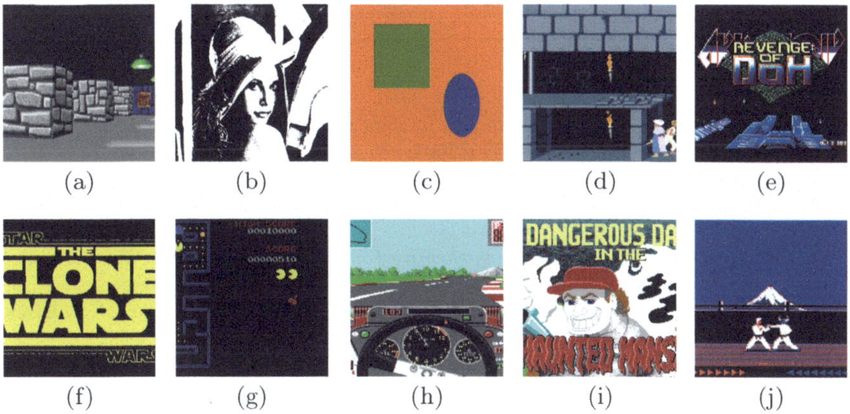

Fig. 4 Reconstruction on the basis of the proposed ROF-type model with the low-dimensional searching space

In [10], we proposed to use the modified ROF-type model, specified for a space with a fuzzy partition. Our proposal relates to both regularization and fidelity terms. Besides modification of the model, we introduced a searching space $\mathcal{I}$ where unknown images I are replaced by their low-dimensional representations. By this, we significantly reduced the complexity of the searching algorithm. Below, we show some results of our approach to the image restoration (Figs. 3 and 4).

5 Conclusions

In this contribution, we showed that the technique of F-transforms fully agrees with the technique of dimensionality reduction, based on Laplacian eigenmaps. To justify this claim, we characterized the processed by the F-transform data in terms of the adjacency graph that reflects their intrinsic geometry. Moreover, we showed that the F- and the F^1-transform of a function f are solutions of the generalized eigenvector problem where the Laplacian matrix L is properly constructed. This fact confirms that the F-transforms provide with a low-dimensional representation of a given data set.

Finally, we discussed the problem of image restoration and explained how it can be formulated and solved in the language of F-transform and using the proposed technique of dimensionality reduction.

Acknowledgements This work was supported by the project LQ1602 IT4Innovations excellence in science. The additional support was also provided by the Czech Science Foundation (GAČR) through the project of No.18-06915S. The SW and the implementation of the technique of F-transforms to image restoration was performed by Dr. Pavel Vlašánek.

References

1. M. Belkin, P. Niyogi, Laplacian eigenmaps for dimensionality reduction and data representation. Neural Comput. **15**(6), 1373–1396 (2003)
2. M. Bertalmio, G. Sapiro, V. Caselles, C. Ballester, Image inpainting, in *Proceedings of 27th Annual Conference on Computer Graphics and Interactive Techniques* (ACM Press/Addison-Wesley Publishing Co., 2000), pp. 417–424
3. S. Haykin, *Neural Networks: A Comprehensive Foundation, Upper Saddle River* (Prentice Hall, NJ, 1999)
4. V. Novák, I. Perfilieva, A. Dvořák, *Insight into Fuzzy Modeling* (Wiley, Hoboken, New Jersey, 2016)
5. I. Perfilieva, Fuzzy transform: application to the Reef growth problem, in *Fuzzy Logic in Geology*, ed. by R.V. Demicco, G.J. Klir (Academic Press, Amsterdam, 2003), pp. 275–300
6. I. Perfilieva, Fuzzy transform: theory and application. Fuzzy Sets Syst. **157**, 993–1023 (2006)
7. I. Perfilieva, M. Danková, B. Bede, Towards a higher degree F-transform. Fuzzy Sets Syst. **180**, 3–19 (2011)
8. I. Perfilieva, P. Števuliáková, R. Valášek, F-transform-based shooting method for nonlinear boundary value problems. Soft Comput. **21**, 3493–3502 (2017)

9. I. Perfilieva, P. Števuliáková, R. Valášek, F-transform for numerical solution of two-point boundary value problem. Iran. J. Fuzzy Syst. **14**(6), 1–13 (2017)
10. I. Perfilieva, P. Vlašánek, Total variation with nonlocal FT-Laplacian for patch-based inpainting. Soft Computing **23**, 1833–1841(2019) https://doi.org/10.1007/s00500-018-3589-8
11. L. Rudin, S. Osher, E. Fatemi, Non linear total variation based noise removal algorithms. Physica **60**, 259–268 (1992)
12. B. Scholkopf, A. Smola, K.-R. Mulller, Nonlinear component analysis as a Kernel eigenvalue problem. Neural Comput. **10**, 1299–1319 (1998)
13. L.A. Zadeh, Toward a theory of fuzzy information granulation and its centrality in human reasoning and fuzzy logic. Fuzzy Sets Syst. **90**, 111–127 (1997)

Development of Methodological and Functional Principles of the Intelligent Demographic Forecasting System

Masuma Mammadova and Zarifa Jabrayilova

Abstract The article proposes a strategy of population forecasting using the fuzzy time series model. It proposes a technique based on fuzzy time series for the prediction of the number of population (total population, able-bodied population, economically active population, population of different age groups, deaths and births, etc.). Based on the technique, the predictive values are estimated for the total population in Azerbaijan and the results are compared to those from other forecast models. An alternative forecasting system based on the proposed approach is supposed to be developed, and an intellectual demographic prediction system is developed. Operating principle of the system is presented, its functional scheme is developed, and the working principle of each block is described. Knowledge base of the system is based on production rules and supports prospective decisions on demographic policy by referring to the analysis of the forecast results of various demographic indicators.

Keywords Population growth · Social-demographic factors · Fuzzy time series · Intelligent demographic forecasting system

1 Introduction

The development of information technologies and software resources has opened up new opportunities for modeling demographic processes and handling forecasting problems. Researches carried out over the recent years prove that the application of traditional analyzing methods and modeling of population growth process on the basis of processing numeric/quantitative data don't produce the desired results and even involve considerable risks and errors. One of the main causes for this undesirable circumstance stems from the fact that a great many forecasting models are not

M. Mammadova · Z. Jabrayilova (✉)
Institute of Information Technology, National Academy of Science of Azerbaijan, Baku, Azerbaijan
e-mail: djabrailova_z@mail.ru

M. Mammadova
e-mail: mmg51@mail.ru

S. N. Shahbazova et al. (eds.), *Recent Developments and the New Direction in Soft-Computing Foundations and Applications*, Studies in Fuzziness and Soft Computing 393, https://doi.org/10.1007/978-3-030-47124-8_33

sufficiently efficacious owing to the incompatibility of highly accurate quantitative methods of classical mathematical apparatus with the great complexity of population growth process [1–3]. The other cause, in our opinion, is that these methods aimed at the mathematical analysis of accurately determined systems are not capable to encompass certain characteristics of the research sphere.

Thus, population is a large dynamic (economic, social, ecological) system irrespective of a specific territory and definite group. A distinguishing feature of this system consists of its functioning under indefinite, uncertain conditions due to a multitude of causes. These uncertainties are, first of all, associated with: (a) the impossibility of identifying all the factors that determine the development dynamics of human population; (b) variability and inconstancy of boundaries of many indicators used in demographic analysis and considerable variations in the values of some indicators; (c) the lack of comprehensive prior information pertaining to demographic processes associated with the data source problems and impossibility of recording all the demographic events. To this list of causes can be added data incompleteness, uncertainties involved while collecting some indicators from various sources: results of population census and demographic researches, current registration of population movement etc. Discrepancies between official and unofficial data, expert evaluations can cause further difficulties [4–7].

Population growth is a multi-factored and time-dependent process. But it is not possible to consciously influence this process by varying certain parameters and observing changes in others. Various uncontrollable factors (wars, inter-ethnic conflicts, natural disasters, ecological factors etc.) significantly affect population growth. If systematic statistical material with regard to the concerned problem is lacking, empirical data becomes the sole information source.

Thus, keeping in mind that the demographic data are incomplete and accuracy of some or all the available data are questionable for any of several reasons, a demographic analysis based on this incomplete, inaccurate information bears a special significance. The vagueness, inaccuracy, incompleteness, fuzziness of the data on the demographic events and processes and the prevailing evaluation methods excluding this data characteristics necessitate taking a new approach to the analysis and evaluation of demographic situations, particularly, population growth forecasting.

Within the context of the above-mentioned arguments, exploring the possibilities of the application of the fuzzy sets theory or the apparatus of sets, going by the name of fuzzy logic, to modeling demographic processes bears a special interest.

2 Applications of the Fuzzy Sets Theory in Demographic Forecasting

The advent of fuzzy logic made it possible to tackle a great many problems with fuzzy input data [8–10]. One of them was a forecasting problem. Many of the structural elements of the latter (input data and interdependence between its components, interval evaluation of indicators and their interdependence, expert evaluations and judgments etc.) are either of a fuzzy nature or, by being in fuzzy relationships, condition the fuzzy description of the problem.

The application of fuzzy logic to the handling of forecasting problems was undertaken by the researches in which the mathematical models of fuzzy time series were described in a fuzzy form for handling the problem with fuzzy input data [11–13]. This approach was developed later by other scientists dealing with the solution of analogous problems [4–7, 14]. To tackle the task, the authors proposed a model of fuzzy time series and tried to reduce the average forecasting error by making adequate alterations in the model.

The above-mentioned features of population, functioning under indefinite, uncertain circumstances, condition the fuzziness of input data or "loads" the task onto fuzzy environment. Therefore, from both theoretical and practical standpoints, handling the concerned problem based on fuzzy time series would be more expedient.

Thus, the major purpose of the proposed approach is methodological: (1) putting forth an evaluation method based on fuzzy time series for estimating model parameters; (2) testing the extent to which the model is adequate to reflect the real process, that is to say, computing the method error; (3) conducting the comparative analysis of computation results; (4) revealing the practical and theoretical importance of the model.

3 A Brief Information on Fuzzy Time Series

Time series represents a consecutive series of observation that is conducted by equal time intervals and lies at the root of exploring real processes in economics, meteorology and natural sciences etc.

The analysis of time series of observation consists of the followings: (1) constructing the mathematical model of time series of observation of real processes; (2) model identification or selection of quantitative evaluation/estimation method for assessing model parameters in order to test the extent to which the model is adequate to reflect the real process; (3) the conversion of identification model into time series through the statistical evaluation of model parameters.

Formally, time series can be defined as a discrete function x(t) whose argument and function values are dependent on discrete time moments as well as argument values, function values at different time intervals.

It is assumed, the time interval $0 \leq t \leq T$ of process $x(t)$ is observed, that is to say, the parameter t varies along the time interval [0, T] (set R) or assumes any integer belonging to this interval. For every fixed time moment $t = s$, the value of function, beginning from this moment, is generally determined by the values of function arguments at all the time moments ranging from $t = 0$ *to* $t = s - 1$, and value of function at all the time moments ranging from $t = 0$ *to* $t = s - 2$.

Fuzzy time series. Let us assume that $U = \{u_1, u_2, \ldots, u_n\}$ is a universal time set. The fuzzy set A of universal set U is defined as follows:

A = {($\mu_A(u_1)/u_1$), ($\mu_A(u_2)/u_2$), …, ($\mu_A(u_n)/u_n$)} or A = {($\mu_A(u_i)/u_i$)}, $i = \overline{1, n}$, $u_i \in U$, $\mu_A(u_i) \in [0, 1]$.

Where $\mu_A(u_i)$—membership function, $\mu_A(u_i)$: $U \Rightarrow [0, 1]$, is a degree of belonging of u_i to the set A, "/" is a division sign.

Let us assume that $Y(t)$ (t = …, 0, 1, 2 …), which is a subset of set R of real numbers, is simultaneously a universal set on which is defined a fuzzy set $\mu_i(t)$, (t = 1, 2, …), that is to say, the membership function is time-dependent. Let us define a set F(t) arranged out of $\{\mu_i(t), t = 1, 2, \ldots\}$. More precisely, $F(t)$ is a set of fuzzy sets $F(t) = \{\mu_i(t), t = 1, 2, \ldots\}$. Then $F(t)$ is a fuzzy time series defined on a universal set Y(t) (t = 1, 2, 3 ….). It is evident, if $F(t)$ is accepted as a linguistic variable, the fuzzy sets$\{\mu_i(t)$, t = 1, 2, …}. Out of which we arranged $F(t)$ will assume the possible corresponding values of $F(t)$. Besides, as is evident, F(t) is time-dependent, which means, the function $F(t)$ will assume different values at different time moments.

4 Fuzzy Time Series in Demographic Foresting

Problem statement. The intensive changes in demographic processes that are caused by the influence of the social-demographic factors, have rendered the determination of perspective variation in total population one of the most important tasks to be tackled for demographic forecasting. To solve task of forecasting total population, we have introduced a model of fuzzy time series in this article. More precisely, the problem is described as follows: (a) for a given time interval, data pertaining to the total population in Azerbaijan or to be more clear, the dynamics and respective variation of total population are available. The point is to find the anticipated total population based on the variations of the previous years.

Problem solution. In accordance with the description of the problem, the following forecasting methodology is proposed:

1. Finding appropriate variations of population dynamics in the country for a certain period (e.g. 1980–2001);
2. Definition of universal set U containing the interval between the least and greatest variations in total population.
3. Division of the universal set U into equal-length intervals containing variation values corresponding to different population growth rates and the arithmetic mean of each interval is found by the following formula:

$$u_m^j = \frac{\max(u_j) + \min(u_j)}{2},$$

   where $\max(u_j)$ and $\min(u_j)$ are the largest and lowest values of interval j respectively.
4. The qualitative description of variation values of total population as a linguistic variable, that's to say, determining the respective values of linguistic variable or the set of fuzzy sets $F(t)$.

5. Fuzzification the input data or the conversion of numerical values into fuzzy values, by the following formula:

$$\phi_{A_j}(u_j) = \frac{1}{1 + [C \cdot (U - u_m^j)]^2}$$

where U-variations, u_m^j—is the middle point of the corresponding interval; C is a constant. C is chosen in such a way that it ensures the conversion of definite quantitative values into fuzzy values or their belonging to the interval [0, 1]: $A_j = (\phi_{A_j}(u_j)/u_j \; u_j \in U, \phi_{A_j}(u_j) \in [0, 1]$—is a fuzzy set (in our case $C = 0.0001$).
   This operation enables us to reflect the corresponding numerical/qualitative values of qualitative representations of population growth rates in the value of membership function.
6. Selection of parameter $W > 1$, corresponding to the time period prior to the concerned year, calculation of fuzzy relationships matrix R(t) and forecasting of population growth in the next year.
   According to the method, the relationship matrix R(t) is calculated at the next step:

$$R(t)[i, j] = O^{w}(t)[i, j] \cap K(t)[j],$$

or

$$R(t) = O^{W}(t) \otimes K(t) = \begin{vmatrix} R_{11} & R_{12} & \ldots & R_{1j} \\ R_{21} & R_{22} & \ldots & R_{2j} \\ \ldots & \ldots & \ldots & \ldots \\ R_{i1} & R_{i2} & \ldots & R_{ij} \end{vmatrix}$$

   Here, $O^W(t)$—operation matrix, $K(t)$—is an criterion matrix, ⊗—is an operation *min* (∩). Later there is defined the forecasted value $F(t)$ for the t year in a fuzzy form as follows:
   $F(t) = [Max(R_{11}, R_{21} \ldots, R_{i1}) \; Max(R_{12}, R_{22}, \ldots, R_{i2}) \ldots Max(R_{1j}, R_{2j}, R_{ij})]$.
7. Defuzzification the obtained results or conversion of fuzzy values into qualitative values. The application of the proposed technique in forecasting demographic indicators (total population, able-bodied population, economically active population, births, deaths, different age groups) of the population in Azerbaijan is described in [4–7].

5 Intelligent Demographic Forecasting System

Based on the proposed technique, the estimates of population numbers for 2002–2016 are calculated based on the total number of population in Azerbaijan for 1981–2001 and compared with the results obtained from other forecasting models (SPSS software referencing to the World Bank's PROST model and statistical forecasting model). The results are given in Table 1.

The relative error of the submitted method is calculated by the following formula:

$$\delta(t) = \frac{\left|N(t)_{act.} - N(t)_{forec.}\right|}{N(t)_{forec.}} \cdot 100\%$$

$N(t)_{act.}$—is the number of actual population for year t; $N(t)_{forec.}$—forecasted number of population for year t, $2002 \leq t \leq 2016$. The relative errors of the forecast results show that the proposed method can be used as an alternative approach to predicting demographic parameters (7–8% of estimation errors are acceptable).

Relevant algorithm and software tool were developed on the basis of the proposed technique, and an intelligent demographic forecasting system (IDFS) was built. The system was implemented in the Delphi 7 programming system. The system provides predictive estimates for various demographic indicators for any year or for coming

Table 1 Dynamics of population (thousand people) in Azerbaijan for 2002–2017 with the use of different forecasting models and relative errors (source for actual population number [15])

Years	PROST	SPSS model	FSM	Actual	Error %		
					PROST	SPSS	FTS
2002	8200.189	8141.41	8155.3	8191.4	0.11	0.61	0.44
2003	8277.965	8205.278	8233.8	8269.2	0.11	0.77	0.43
2004	8357.393	8277.211	8315.2	8349.1	0.10	0.86	0.41
2005	8438.496	8356.911	8398	8447.4	1.20	1.08	0.70
2006	8521.262	8450.611	8481.9	8553.1	1.37	1.2	0.83
2007	8606.523	8540.611	8566.7	8666.1	1.69	1.45	1.15
2008	8694.348	8646.278	8652.3	8779.9	0.96	1.52	1.45
2009	8784.848	8758.811	8738.5	8897.0	1.26	1.55	1.78
2010	8878.074	8876.011	8825.3	8997.6	1.33	1.35	1.91
2011	8974.047	8989.394	8912.5	9111.1	1.50	1.36	2.18
2012	9070.694	9099.561	9000	9235.1	1.78	1.47	2.54
2013	9168.131	9209.906	9087.8	9356.5	2.01	1.57	2.81
2014	9266.445	9322.628	9175.9	9477.1	2.22	1.63	3.18
2015	9365.678	9437.628	9264.2	9593.0	2.37	1.62	3.43
2016		9544.478	9352.6	9705.6		1.66	3.64

years. For example, Fig. 1 and Fig. 2 depict software windows that show the total population for 2018, the estimated values for years of 2018–2034 respectively.

The IDFS functional scheme includes the following blocks (Fig. 3).

Interface Block enables communication between the system and the user. The user selects the system operating mode via the interface block. The system can operate in three modes: knowledge base building mode, forecasting mode of forecasting

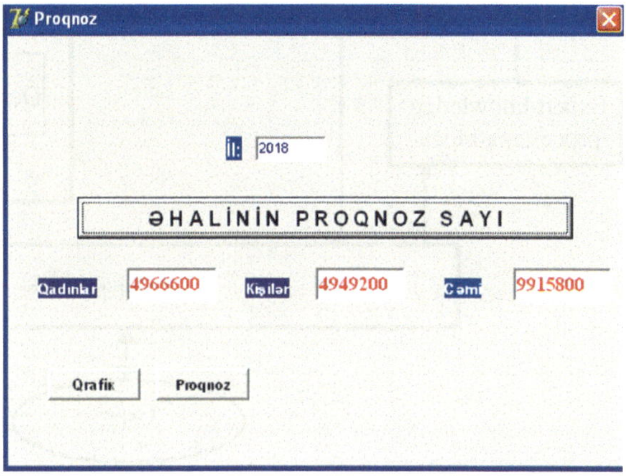

Fig. 1 Estimations for 2018

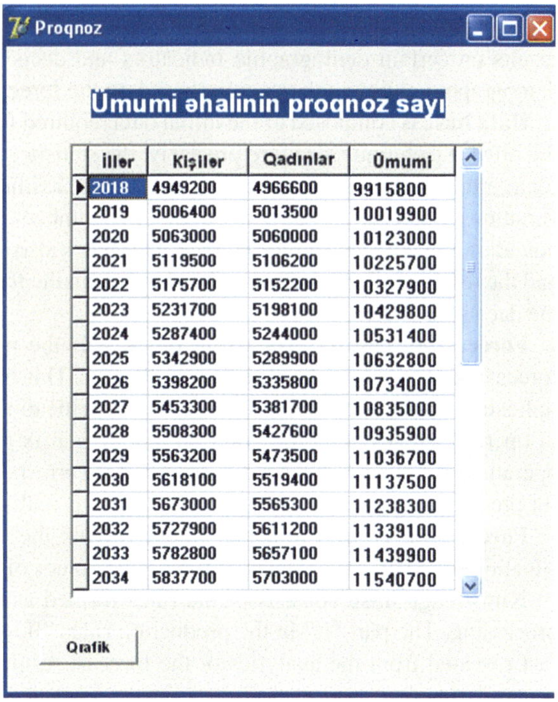
Proqnoz

Ümumi əhalinin proqnoz sayı

illər	Kişilər	Qadınlar	Ümumi
2018	4949200	4966600	9915800
2019	5006400	5013500	10019900
2020	5063000	5060000	10123000
2021	5119500	5106200	10225700
2022	5175700	5152200	10327900
2023	5231700	5198100	10429800
2024	5287400	5244000	10531400
2025	5342900	5289900	10632800
2026	5398200	5335800	10734000
2027	5453300	5381700	10835000
2028	5508300	5427600	10935900
2029	5563200	5473500	11036700
2030	5618100	5519400	11137500
2031	5673000	5565300	11238300
2032	5727900	5611200	11339100
2033	5782800	5657100	11439900
2034	5837700	5703000	11540700

Qrafik

Fig. 2 Table of estimated population number

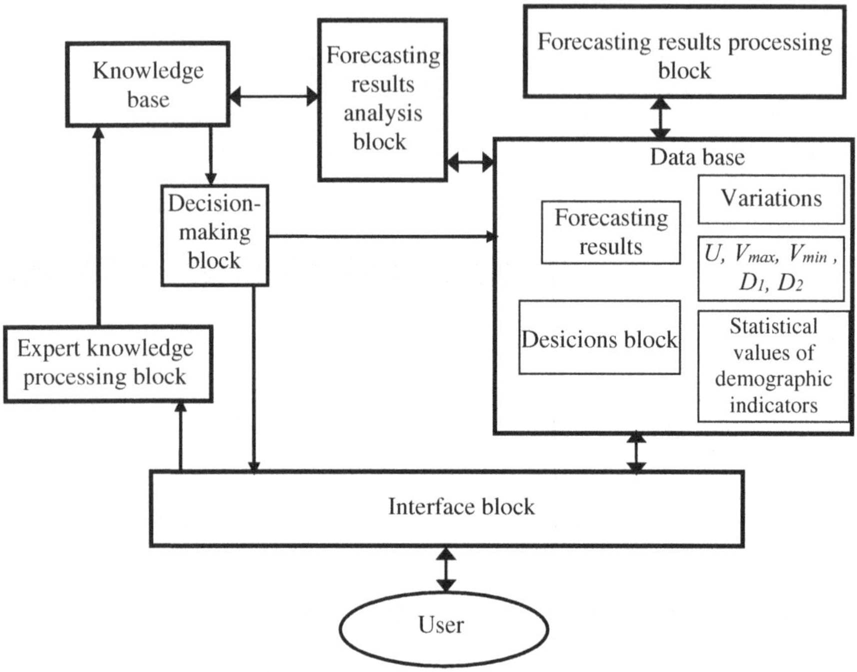

Fig. 3 IDPS functional scheme

results on certain demographic indicators and decision-making mode that shapes demographic policy of the country based on the forecasting results analysis.

Data base is composed of the initial data required for forecasting results based on the offered technique, or more precisely, the statistical values for the studied period for a certain demographic indicator, variations, maximum and minimum values of the variations, universal multiplicity U, and the numbers D_1, D_2 selected to smooth the boundaries of universal multiplicity U. This block also includes the forecasting results and the result of enabled rule in accordance with the forecasting results analysis, i.e., the decisions made.

Forecasting results processing block gets the results based on the proposed forecasting method for demographic indicators. This block calculates the fuzzy multiplication matrix of the forecasted year by fuzzification of variations and calculation of operation matrix and the criterion unique matrix of the chosen basis. Here, the operation of obtaining forecasting results for corresponding demographic indicator for predicted year based on its defuzzification is realized.

Forecasting results analysis block reveals the facts about the demographic situation based on the analysis of forecasted values of certain indicators.

Knowledge base consists of the rules formed as a result of expert knowledge processing. The part "if" in the production rules "If, then" corresponds to a certain fact revealed from the analytics of the forecast results, and the part "then" can be any decision or measure to be taken in accordance with the appropriate demographic policy.

If the condition of the rule coincides with any fact received from the analytics block, then the rule is enabled, the second part "then" enters the **decision-making block** and presented to the user as a valid decision and stored in the DB.

6 Conclusion

The methodology proposed in this article enables us to forecast demographic processes on the basis of fuzzy time series. A peculiar trait of the methodology consists of its capability to forecast the required indicator by utilizing incomplete, fuzzy input data.

The described approach, by entering the dynamics of total population until some previous year into an experimental base, helps make forecast calculations for any distant perspective. This, in its turn, allows us to take into account the trend of previous population growth rates and as a result achieve more accurate forecasts.

Such forecasting results can be used for decision support in the following cases:

- Determining the demand for food, energy, housing, social-household, health, education, transport and other services (population forecast by general and individual age groups);
- Developing pensions, social and health insurance programs (population forecast by age-gender and family structure, including the proportion of able-bodied and disabled people);
- Developing national and regional development programs, drafts on individual areas (forecasting the total number of population, predictions of individual socio-demographic groups and reproduction of population);
- Developing population growth policy and/or human development program;
- Determining the economic growth rate (number of (employed) population, its education, age and family structure).

Thus, the perfection of the intelligent predictive decision support system for the demographic policy involves expert-demographers and analysts in the knowledge base formation.

References

1. A.E. Ganiev, Demographic forecasting methods, Sociol. Sci. (46). NovaInfo.ru (2016), http://novainfo.ru/article/6270/pdf
2. V. Zaqores, I. Zaqores, Demographic policy: methods and tools, application experience and efficiency analysis. J. Int. Law Int Rel. (2) (2014), www.evolutio.info/content/view/2257/235
3. I.I. Eliseeva, M.A. Klupt, Demography and population statistics, 3rd edn. Trans. and add., Textbook (Moscow-Yurayt, 2016), 405 pp

4. M.H. Mammadova, Z.G. Jabrayilova, M.I. Manafli, Intelligent demographic forecasting system (express information). Inf. Technol. 55 pp. Publishing house, Baku (2009)
5. M.H. Mammadova, Z.G. Jabrayilova, Application of fuzzy time series for population forecasting, in *Proceedings on Labor and Social Problems CSRE*, vol. 1, Baku (2002), pp. 41–63
6. M.H. Mammadova, Z.G. Jabrayilova, Application of fuzzy logic in the demographic forecast. Inf. Technol. (3), 45–53 (2004)
7. M.H. Mammadova, Z.G. Jabrayilova, Fuzzy logic in forecasting demographic aspects of the labor market. Artif. Intell. (3), 450–460 (2005)
8. L.A. Zadeh, The concept of a linguistic variable and its application to approximate reasoning. Inf. Sci. **8**(4), 301–357 (1975)
9. A.M. Abbasov, M.H. Mammadova, V.A. Gasimov, Fuzzy relational model for knowledge processing and decision making. Adv. Math. NY **1**, 191–223 (2002)
10. M.H. Mammadova, Decision-making based on knowledge bases with fuzzy relational structure, Baku, Elm, 1997, 256 pp
11. Q. Song, B.S. Chissom, Forecasting enrollments with fuzzy time series—part II. Fuzzy Sets Syst. **62**(1), 1–8 (1994)
12. Q. Song, B.S. Chissom, Fuzzy time series and its models. Fuzzy Sets Syst. **54**, 269–277 (1993)
13. S.M. Chen, Forecasting enrollments based on fuzzy time series. Fuzzy Sets Syst. **81**(3), 311–319 (1996)
14. M.Z. Ahmadov, Forecasting based on fuzzy time series, Dissertation work, Baku, 2015, 136 pp
15. Demographic Indicators of Azerbaijan, Baku, 2017, 472 p. (St41), www.stat.gov.az/source/demoqraphy/

Correlation Between Low Level Properties of Image and Semantic Using Algorithm BP Neural Network

Guschin Roman, Rozaliev Vladimir, Orlova Yulia, and Zaboleeva-Zotova Alla

Abstract The paper proposes a method to improve the modeling of subjectivity and human understanding based on the annotation of emotional semantics of images using the theory of fuzzy sets. Adaptive acceleration and reverse propagation of the neural network (BP). The annotation method was tested by analyzing images from the solar image database. In addition, the accuracy of the search is based on 85% of our method. This study is the basis for a more accurate semantic analysis and verification.

Keywords Human emotions · Image identification · Neural networks · Adaboost algorithm · Information system

1 Introduction

Emotional and semantic analysis of images is an important step in the study of semantic image at a high level, pattern recognition and computer vision. When solving problems such as image classification, face detection, outdoor advertising monitoring and military intelligence, it is necessary to analyze the emotional behavior of a person, get semantic features from images, and then calculate the degree of similarity of the object.

The final objective of emotional semantic image analysis is to allow computers to define human emotional response to an image. Emotional semantic image's features

G. Roman · R. Vladimir (✉) · O. Yulia · Z.-Z. Alla
SAS Department, Volgograd State Technical University, Volgograd, Russia
e-mail: vladimir.rozaliev@gmail.com

G. Roman
e-mail: romanshadow3@gmail.com

O. Yulia
e-mail: yulia.orlova@gmail.com

Z.-Z. Alla
e-mail: zabzot@gmail.com

S. N. Shahbazova et al. (eds.), *Recent Developments and the New Direction in Soft-Computing Foundations and Applications*, Studies in Fuzziness and Soft Computing 393, https://doi.org/10.1007/978-3-030-47124-8_34

are extracted based on low-level visual features. The image semantics can be divided into behavioral and emotional one which is the top-level semantics.

Low-level properties of images, such as color, texture, shapes and contours is the first extracted from the related processing technologies; further attempt be made to find a correlation between low-level properties and emotional semantics at a high level [1]. This automatic method is one of acquiring the high-level semantics. Semantic of emotional annotation is an advanced process in the field of digital image analysis. Scene images—this is the main data type generated from multimedia sources. This method provides a clear image search and is an effective tool for the introduction of multimedia information retrieval systems. The study of their semantic annotation is the foundation for the creation of the emotional semantic search for other image types and has a strong theoretical and practical significant consequently.

The launch of the study on computer emotional calculations comes under 1980-th years. Currently, a reasonable approach to analyzing emotions is a popular topic of research, but development difficult in the computing fields. As of today, there have been many studies to examine the relations between the visual image elements and emotions. Mao, Ding and Mauting defined a mathematical model by analyzing image's features and came up with an analytical image method of fluctuation to calculate harmonious feelings for images [2]. Their deduction revealed that images in compliance with the fluctuation law provoke harmonious sentiments in humans. Chen and Wang draw on semantic quantization and factor analysis to create an emotional room based on dimensional analysis in the field of psychology [3]. Colombo, Bimbo, and Pala identified several basically used words, such as: warm, cool and natural, to describe the emotional images' semantics and to set an emotional room [4].

Baek, Hwang, Chung and Kim defined 52 of images templates and 55 emotional factors corresponding to patterns have been used questionnaires and have been measured the relations between low-level visual features and high-level emotions [5]. The accuracy factor based on their method has reached approximately 70%. Nevertheless, most previous elaborations have been dealt to mapping/depicting/comparing interrelations between the affective expression and the low-level visual properties of images. Research on the probabilistic degree of association between the affective terms and the semantics involved in the images are rare. However, the two problems are unsolved:

- The receipt and analysis of emotion semantic data in scene images.
- The semantic equivocality and particularly of scene images have been realized.

Particularly and equivocality play a key role in the human consciousness of the image. To identify effective method of subjectivity and ambiguity modeling over the human image comprehension will greatly enhance the search efficiency and result in human-centered image retrieval [6].

This paper is organized along the following lines. The second section presents the theory, which used in emotional semantic analysis of scene image. The third one highlights the experimental procedure of the current study. The final section sums up the main results of this paper.

2 Theories

2.1 Fuzzy Set Theory

Fuzzy set theory is an important theory in the field of artificial intelligence. It was proposed by Zadeh in 1965. He first used the term "fuzzy set" to classify fuzzy objects and used the concept of "membership degrees" to accurately describe the relationship between an element and its fuzzy set. The development of fuzzy set theory marks the birth of fuzzy mathematics. Although membership degrees are the foundation of fuzzy sets, there is no one recognized standard for their determination, which is a psychological process. Currently, fuzzy theories mainly cover fuzzy set theory, fuzzy logic, fuzzy inference, and fuzzy control. The related theories and technology have matured and been widely applied.

The present paper takes fuzzy theory as automated describes for images. The fuzzy theory was used to represent the levels of human emotions when comprehending the scene images. Thus, the feeling of "relaxation" will be provoked by when subject watch the image or background of image with nature.

The degree of this feelings type will be reviewed in this study.

In order to define human emotions excited, by images, this research demands on the definitions set out below:

- Emotional variables: Variables comprise a five-dimensional vector: (x, E(x), U, G, T), with x—as the name of the variables, E(x) representing the emotional value set of x, U—as the domain (the extraction space of image features in this work), G being the grammar laws for the generation of the emotional value of E(x), and T which serves as the semantic rules for the calculate of the degree of emotional membership [7].
- Basic emotional value set: this refers to array of emotional value, which can't be divided semantically.
- Extended emotional value set: this value of basic emotional with numeric degree description.

Follow the font type sizes specified in Table 1. The font type sizes are given in points, same as in the MS Word font size points. Times New Roman is the preferred font.

2.2 Principal Component Analysis—PCA

Follow the font type size PCA had first been proposed by Karl and Pearson in 1901 and used as a tool for data analysis and the development of a mathematical model [8]. This is tool for a multicriteria set statistical analysis to transfer through more than one index on the linear combination of multiple independent indices, which contain a large amount of information contained in the original indexes [7].

Table 1 Enhanced motional values

Base emotional value	Semantic feature value	Degree to enhance value		
		Very	Neutrally	Slightly
Natural	0.92	0.875	0.230	0.071
Romantic	0.31	0.094	0.837	0.68
Soft	0.44	0.213	0.978	0.56
Relaxed	0.87	0.782	0.350	0.67
Bright	0.88	0.646	0.563	0.18
Regenerating	0.79	0.605	0.648	0.21
Changeful	0.26	0.075	0.780	0.74

PCA algorithm is based on the following principles. Assume, the samples matrix might be represented as

$$X = \{x_{ij}\}_{n \times N}, \tag{1}$$

With n—as signs dimension, and N—as quantity of samples.

(1) The sample data have to be standardized following the:

$$x'_{ij} = \frac{(x_{ij} - \overline{x_i})}{\sigma_i}, \tag{2}$$

with $i = 1, 2, \ldots, n$, x_i as the average value of the features of the i—and σ_i being the standard error of the i-th feature.

$$r_{ij} = \frac{1}{n} \sum_{k=1}^{n} \frac{(x_{ki} - x_i)(x_{kj} - x_j)}{\sigma_i \sigma_j} \tag{3}$$

(3) On the basis of the feature equation $|R - \lambda I| = 0$, the characteristic root λ_i and the eigenvector $\alpha(i = 1, 2, \ldots, N)$ of R are determined, and the characteristic $\lambda_1 < \lambda_2 < \cdots < \lambda_N$.

(4) Contribution ratio $e_{i\psi}$ and accumulated contribution ratio E_m of each basic component is calculated as follows:

$$e_i = \frac{\lambda_i}{\sum_{k=1}^{N} \lambda'_k} \tag{4}$$

$$e_m = \frac{\sum_{k=1}^{M} \lambda_m}{\sum_{k=1}^{N} \lambda_k}, k = 1, 2, \ldots, N \tag{5}$$

The contributions rate of the first base component e1 represents the share of variance component to total variance.

(5) The base parts are calculated based on:

$$F_i = \alpha_{1i}X_1 + \alpha_{2i}X_2 + \cdots + \alpha_{Ni}X_N, i = 1, 2, \ldots, N \quad (6)$$

3 The Approach of the Analysis of Emotional Semantic and Automatic Annotation

After the image automatically analyzed and semantically annotated it is first necessary to extract the lower-level visual features. For the images, colors are the important features that may define the emotional semantics. In this connection, this study used a segmentation method to extract the visual features of the scene images. Then there should be created an emotional model and has to be executed a semantic mapping to implement a semantic mapping from low-level color features to emotional semantics of a higher level. Further, semantic analysis and automatic annotation can be completed.

3.1 Visual Color Reproduction

(1) Picking up the colour-space and quantization: Since the colour space of hue, saturation, and value (HSV) can mirror in satisfactory manner the human perception of colours, HSV was used as a working space. One 60-dimensional vector of colour features was established to create the domain of emotional variables. Since human visual system is more sensitive to colour than to saturation and value, we quantitated HSV space, in accordance with the methods used in previous studies [7, 8].

    Such quantitative methods have different advantages, including a better understanding of the human visual system, reduce the redundancy of colour and are to be applied to the grey surfaces simplifying the calculation.

(2) *Visual Feature. Colour extracting*: The release strategy was taken for to extract local features of images [3, 7]. The commonly used method is to divide the images into m * n number of blocks, while different blocks have different weights. The central area of an image plays a crucial role in the semantic comprehension of a man and therefore has the greatest weight. In the case research, we sized the image into 4 * 4 segments, as shown in Fig. 1.

The weight of each unit depends on the characteristics of the image to use. As the central segment or user-assigned one has usually a large weight that may better reflect information about the location of the image thus the weight of the central blocks of the image has been raised. The blocks weight was distributed as follows.

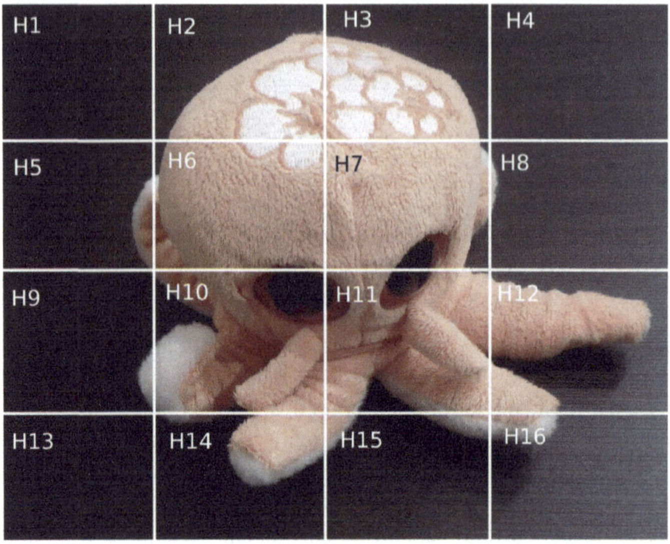

Fig. 1 Segment layout of image

As shown in Fig. 1, the area with the red lines was 1/4 of the total area of the image. Thus, his weight has been raised to 1/3. Therefore, weight H6, H7, H10 and H11 were 1/12 each, and the weight of each of them surrounding the 12 segments was 1/18.

3.2 Emotional Model Building Up

The making up of emotional model is critical for emotional semantic scene image analysis and the definition of a core set of emotional value E(x) is a critical component in building up this model. Normally, the creation of emotional patterns consists of three stages. Firstly, emotional adjectives are being collected to determine the emotional values. Secondly, the experiment on semantic quantization is being carried out and an emotional database is being created, based on the evaluation of images of individual subjects of the shooting. Thirdly, an analysis of the data is being implemented to establish the emotional room.

(1) *The determination of a set of emotional values*: Seven adjectives—terms have been carefully selected to set E(X) = (natural, romantic, soft, relaxed, bright, regenerating, changeful). An expanded set of emotional values have been subsequently constructed as very, neutrally and slightly. For example, the enhanced emotional value of a set of basic emotional value of "soft" very soft, soft, slightly soft.: The grammar law of G emotional variables was thereafter formulated in the following way: emotionally expressive formula :: == enhanced emotional

value—base emotional value Enhanced emotional value :: == listed variable & base emotional value Affiliation variables :: == very—neutrally—slightly base emotional value :: == natural—romantic—soft—relaxed—energetic—regenerating— changeful.

Building on our experiment, three extended emotional values (very neutral, and slightly) have been quantized as follows:

$$V_e(\mathrm{x}) = \{\mathrm{T}_e^2(x)|x \in U\}, \tag{7}$$

$$N_e(x) = \{\sin T_e(x) \times \pi|x \in U\}, \tag{8}$$

$$H_e(x) = \{1 - T_e(x)|x \in U\}, \tag{9}$$

with e—taken as base emotional value set, $\mathrm{T_e(x)}$—the degree of fuzzy affiliation, derived from neural network training BP, x constitutes a base emotional value, $\mathrm{V_e(x)}$, $\mathrm{N_e(x)}$ and $\mathrm{H_e(x)}$ are affiliation degree of very, neutrally and slightly x consequently.

(2) *Emotional database building up*: The SUN data base is a free image database for researchers in the field of computer-understand manner. In this study it was used images from the database for the experiments [9]. A total of 100 databases SUN typical image was chosen. These images provided different colors, cyberspace layouts and content. The users included 10 members; their ages ranged from 20 to 26. Emotional database had been produced on the base of user ratings.
(3) *Emotional room fostering and semantic mapping*: For a set of samples where {(V1, y1), (V2, y2), ..., (Vn, yn)}—the extracted 60-element vector color features Vi ∈ U, (i = 1, 2, ..., 60) and yi(i = 1, 2, ..., n)—membership class of base emotional value included in an enhanced emotional value, must mapping be set Te : V → y, e ∈ E(x).

Considering that neural BP network [10] has a simple, structure, high speed instruction and a strong learning ability, which uses the processing of fuzzy sets and that the basic idea Adaboost algorithm is to combine the results of several weak predicates for effective prediction, Adaboost algorithm was combined with BP neural network. The flowchart and learning audio BP neural network are shown in Figs. 2 and 3, respectively.

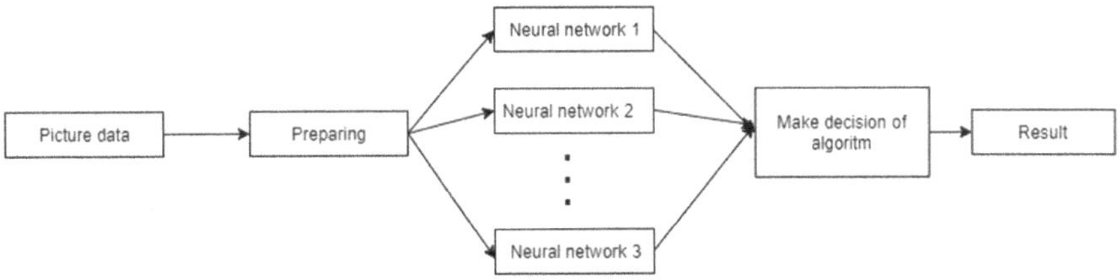

Fig. 2 Flowchart neural network

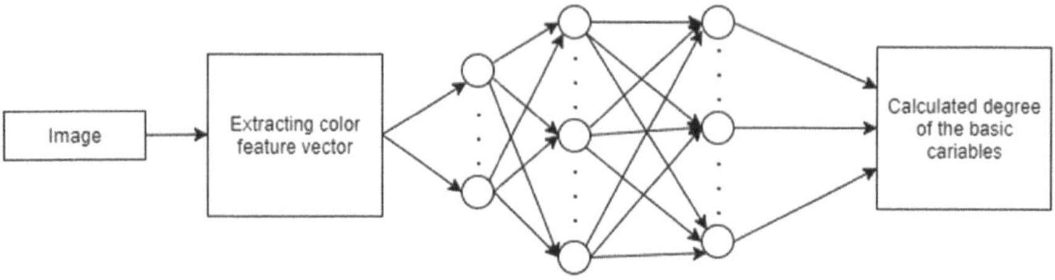

Fig. 3 Learning process neural network

BP neural network is a neural network with feedback, which consists of an input layer, hidden layer and output layer. In this study, as a network entry isolated 60-dimensional visual color low-level particularities were used. The number of elements in the hidden layer was determined to be 20. As a function of the activation of hidden layers Gaussian function had been used:

$$\Phi(x) = e^{\frac{-(x-\mu)^2}{\sigma^2}} \tag{10}$$

The lead predicate has been set based on the Adaboost algorithm [7]. The procedures were as follows:

(1) Choice of data and initialization of the neural network: from the sample space m the sample data group were randomly selected. The distribution of weights of the training data was initialized according to the neural network structure have been determined in accordance with the size of the input and output sample data. Weight value and threshold of neural network RBF have been initialized.
(2) The sum of the predictive errors of the predicted sequence was obtained based on the following equation:

$$e_t = \sum_i D_i(i), i = 1, 2, \ldots, m(g(t) \neq y), \tag{11}$$

with g(t) as the predicted result, and y—as he expected one.
(3) Computation of the sequence weight: according to the of the predicted sequence g(t), the weight of sequence was computed by reference to the following equation:

$$a_t = \frac{1}{2} \ln \frac{1 - e_t}{e_t}, \tag{12}$$

(4) The weight adjusting of the checked data: vary from the weight of the predicted sequence, the weight of the next round of prepared samples was corrected referring to the following equation:

$$D_{i+1}(i) = \frac{D_t(i)}{B_t} \times \exp^{-a_y y_i g_t(x_i)}, i = 1, 2, \ldots, m, \tag{13}$$

where Bt—being the normalization factor, which allows the sum of the distributional weights to be equal to 1 since the weight proportions rest the same.

(5) Strong prediction function: the weight a_t of weak predicates were normalized after T rounds of training follows:

$$a_t = \frac{a_t}{\sum_{t=1}^{T} a_t} \tag{14}$$

The results of predict prediction function was calculated using the following equation:

$$y(x) = a_t h(x), \tag{15}$$

with h(x) being the predictive value.

At the stage of learning, 250 images with styles were chosen from the 500 scene images which used as a set for training. After training, the degree of membership of seven emotional variables has been identified. For example, or the base emotional term "natural" since the membership degree to be 0.81, the degree of natural feeling inspired by input image being equal to 0.83. In accordance with the above procedures, the image features have been presented in the form of a 7-dimensional vector; that is, where the degree of fuzzy membership of sets basic emotional variables. Corresponding base emotional value of a plurality scene of images in space was (natural, changeful, romantic, regenerating, soft, relaxed, and bright).

For example, if the semantic scene image feature vector after training was F = 0.92, 0.33, 0.44, 0.88, 0.87, 0.79, 0.26, its enhanced emotional values are given in Table 1, using the formula (4).

4 Upshots

PCA was used to collect the original emotional semantic data to annotate artificially the scene images. The collected raw data are enriched in Table 2.

Table 3 shows that the total amount of the dispersion deposits of the first three principle components of the system reached 75%. The rate of introduction of the dispersion of the three main components into more than 85% of the samples reached about 81 in common. This revelation shows that the subjects had good attitude towards the experiments, and that data collected was reliable. The first three main components may well describe the emotional semantic information associated with the images. The open-environment-based behavioural experiment was held successfully, the emotional semantic data obtained can be called effective.

Table 2 Subset of initial data of testing

Subject	Img1	Img2	Img3	Img4	Img5	Img6	Img7
1	0	4	2	5	6	3	4
2	0	2	6	0	5	4	0
3	2	3	1	5	5	5	4
4	4	5	4	4	3	2	4
5	3	3	5	5	6	3	0
6	3	3	1	4	6	6	0
7	0	3	6	0	5	3	0
8	1	4	2	4	6	3	6
9	5	3	2	4	4	3	3
10	2	4	1	6	3	4	1

Table 3 The analysis rules of emotional semantic data for scene images

The key component	Defining value	Percentage difference	Accumulated contribution ratio
1	11.6990	47.2836	47.2836
2	4.1186	19.3736	66.6572
3	3.8496	15.4956	82.1528
4	1.9562	7.8459	89.9986
5	1.4307	5.8841	95.8827
6	0.6734	2.6791	98.5618
7	0.1655	0.9396	99.5014
8	0.0614	0.4986	100.0000

5 Insights

A serious field of artificial intelligence issue is discovering how machines can replicate the human perception and expression to fulfil the human-machine harmony. The current study had resorted to theory of fuzzy set to find the relations between human comprehension and scene images [11]. The model was created based on Adaboost algorithm and BP neural network for the implementation of the automated description of emotional semantics in scene images. The findings revealed that the combination of algorithm and neural network has more advantages when they are applied to solve problems of human subjectivity and ambiguity of emotional subjects. The study not only suggests one approach to solving the problem of human subjectivity and ambiguities, but also provides new ideas for comprehension another kind of types of emotional semantics on images.

However, this study has several limitations. Firstly, all test users were of approximately the same ones; so the errors can be made out in our insights. The number increase of test users could help improve research results and make it more accurate and objective. Secondly, the model of artificial neural network requires more data. Clearly, if the experiment is to use a large set of training data, the result would be more correct. On the other hand, the algorithm in this situation would take excessive computing time and human resources. Thirdly, it is necessary to standardize the classification of the semantics of the image of emotions, since scene images have a rich semantic content.

Acknowledgements This work was partially supported by RFBR (grants 17-29-07021, 18-07-00220, 19-47-343001, 19-47-343002, 19-47-340003, 19-47-340009, 19-47-340013, 19-37-90060).

References

1. J. Lv, J. Xiang, J.J. Chen, Research of image affection based on feature extraction technology of ROI. Comput. Eng. Des. **31**(3), 660–662 (2010)
2. X. Mao, Y.K. Ding, Y. Moutian, Analysis of affective characteristics and evaluation on harmonious feeling of image. Acta Electron. Sin. **29**(12A), 1923–1927 (2001)
3. S.F. Wang, E.H. Chen, S.H. Wang, An image retrieval based on emotion model. J. Circuits Syst. **8**(6), 48–52 (2003)
4. C. Colombo, A. del Bimbo, P. Pala, Semantics in visual information retrieval. IEEE Multimed. **6**(3), 38–53 (1999)
5. S. Baek, M. Hwang, P. Kim, Kansei factor space classified by information for Kansei image modeling. Appl. Math. Comput. **205**(2), 874–882 (2008)
6. Kansei sessions, in *Proceedings of the IEEE International Conference on Systems Man and Cybernetics*, Tokyo, Japan, 1999
7. J. Cao, L. Chen, Fuzzy emotional semantic analysis and automated annotation of scene images. Comput. Intell. Neurosci. **2015**, 115–125 (2015)
8. W. Han, G. Li, The applied study of principal component analysis in evaluation of science and technology competence. Appl. Stat. Manag. **25**(5), 512–517 (2006)
9. L.S. Thomas, *The Analytic Hierarchy Process*, 2nd edn. (RWS Publications, Pittsburgh, PA, USA, 1996)
10. X. Sun, Y.D. Xie, D.C. Ren, Study on image registration technique based on wavelet transform and sub-graph. Comput. Eng. Des. **31**(21), 46534654 (2010)
11. V. Rozaliev, Y. Orlova, R. Guschin, V. Verishev, General approach to the synthesis of emotional semantic information from the video, in *Creativity in Intelligent Technologies and Data Science. Second Conference*. Communications in Computer and Information Science, vol. 754 (Springer International Publishing, 2017), pp. 201–214

Granular Computing and Relations

Piecewise Multi-linear Model Based Lookup Table Controller for Nonlinear Systems with Input Constraints

Tadanari Taniguchi and Michio Sugeno

Abstract This paper proposes a piecewise multi-linear model based lookup table controller for nonlinear systems with input constraints. We construct a piecewise multi-linear model of a nonlinear system. The model is a nonlinear approximation and it can be derived from fuzzy if-then rules with singleton consequents. The model can be expressed as a lookup table. The dynamics is described by multi-linear interpolation of the lookup table elements. We design a piecewise multi-linear controller for the piecewise model via feedback linearization. The controller can be also represented as a lookup table. We propose the lookup table controller for nonlinear systems with input constraints and apply the controllers to TORA (Translational Oscillator with Rotating Actuator) system. Examples are shown to confirm the feasibility of our proposals by computer simulations.

1 Introduction

A lookup table (LUT) is an array of data that maps input values to output values. Because LUT can reduce computational load and time, it is widely used in various fields. LUT is also widely used in control engineering [1, 2], especially industrial fields [3, 4]. However it is very difficult to analyze the stability of control systems with LUT controllers. We design an LUT controller using the following steps:

1. Constructing the piecewise multi-linear (PML) model of nonlinear systems
2. Designing the PML stabilizing controller
3. Transforming from the controller to the LUT controller.

T. Taniguchi (✉)
IT Education Center, Tokai University, Hiratsuka 2591292, Japan
e-mail: taniguchi@tokai-u.jp

M. Sugeno
Tokyo Institute of Technology, Yokohama 2268503, Japan
e-mail: michio.sugeno@gmail.com

S. N. Shahbazova et al. (eds.), *Recent Developments and the New Direction in Soft-Computing Foundations and Applications*, Studies in Fuzziness and Soft Computing 393, https://doi.org/10.1007/978-3-030-47124-8_35

The PML model is a nonlinear approximation. The model is built on hyper cubes partitioned in state space and is found to be multi-linear [5], so the model has simple nonlinearity. The PML model is derived from fuzzy if-then rules with singleton consequents. It has a general approximation capability for nonlinear systems. It is a piecewise nonlinear model and second simplest after the piecewise linear (PL) model. It is continuous and fully parametric.

The stabilizing conditions are represented by bilinear matrix inequalities (BMIs) [6], therefore, it takes long computing time to obtain a stabilizing controller. To overcome these difficulties, we derived the stabilizing conditions [7–9] based on feedback linearization, where [7, 9] applied input-output linearization and [8] applied full-state linearization. In this method, only partial knowledge of vertices in piecewise regions is necessary, not overall knowledge of an objective plant. The control system is applicable to a wider class of nonlinear systems than conventional feedback linearization. The PML model can be also expressed as an LUT representation. The internal model dynamics is described by multi-linear interpolation of the LUT values.

In some cases the LUT controller values might be huge. As a solution for this problem, we propose the stabilizing LUT controller for nonlinear systems with input constraints in this paper.

2 Canonical Forms of Piecewise Multi-Linear Models

In this section, we introduce PML models suggested in [5]. We deal with the two-dimensional case without loss of generality. Define vector $d(\sigma, \tau)$ and rectangle $R_{\sigma\tau}$ in two-dimensional space as $d(\sigma, \tau) \equiv (d_1(\sigma), d_2(\tau))^T$, $R_{\sigma\tau} \equiv [d_1(\sigma), d_1(\sigma + 1)] \times [d_2(\tau), d_2(\tau + 1)]$. σ and τ are integers: $-\infty < \sigma, \tau < \infty$ where $d_1(\sigma) < d_1(\sigma + 1), d_2(\tau) < d_2(\tau + 1)$ and $d(0, 0) \equiv (d_1(0), d_2(0))^T$. Superscript ^{T} denotes a *transpose* operation.

We consider a two-dimensional nonlinear control system.

$$\begin{cases} \dot{x} = f(x) + g(x)u(x), \\ y = h(x). \end{cases} \tag{1}$$

For $x \in R_{\sigma\tau}$, the PML model (2) is constructed from the nonlinear system (1).

$$\begin{cases} \dot{x} = f_p(x) + g_p(x)u(x), \\ y = h_p(x), \end{cases} \tag{2}$$

where

$$
\begin{cases}
f_p(x) = \sum_{i=\sigma}^{\sigma+1} \sum_{j=\tau}^{\tau+1} \omega_1^i(x_1)\omega_2^j(x_2) f(i,j), \\
g_p(x) = \sum_{i=\sigma}^{\sigma+1} \sum_{j=\tau}^{\tau+1} \omega_1^i(x_1)\omega_2^j(x_2) g(i,j), \\
h_p(x) = \sum_{i=\sigma}^{\sigma+1} \sum_{j=\tau}^{\tau+1} \omega_1^i(x_1)\omega_2^j(x_2) h(i,j), \\
x = \sum_{i=\sigma}^{\sigma+1} \sum_{j=\tau}^{\tau+1} \omega_1^i(x_1)\omega_2^j(x_2) d(i,j),
\end{cases}
\tag{3}
$$

$f(i,j), g(i,j), h(i,j)$ and $d(i,j)$ are vertices of the nonlinear system (1).

$$
\begin{cases}
\omega_1^\sigma(x_1) = \frac{(d_1(\sigma+1) - x_1)}{(d_1(\sigma+1) - d_1(\sigma))}, & \omega_1^{\sigma+1}(x_1) = \frac{(x_1 - d_1(\sigma))}{(d_1(\sigma+1) - d_1(\sigma))}, \\
\omega_2^\tau(x_2) = \frac{(d_2(\tau+1) - x_2)}{(d_2(\tau+1) - d_2(\tau))}, & \omega_2^{\tau+1}(x_2) = \frac{(x_2 - d_2(\tau))}{(d_2(\tau+1) - d_2(\tau))}
\end{cases}
$$

$\omega_1^i(x_1), \omega_2^j(x_2) \in [0, \ 1]$ are the membership functions of the PML model. In the above, we assume $f(0,0) = 0$ and $d(0,0) = 0$ to guarantee $\dot{x} = 0$ for $x = 0$.

The modeling procedure in region $R_{\sigma\tau}$ is as follows:

1. Assign vertices $d(i,j)$ for $x_1 = d_1(\sigma), d_1(\sigma+1), x_2 = d_2(\tau), d_2(\tau+1)$ of state vector x, then partition state space into piecewise regions.
2. Compute vertices $f(i,j), g(i,j)$ and $h(i,j)$ in Eq. (3) by substituting values of $x_1 = d_1(\sigma),\ d_1(\sigma+1)$ and $x_2 = d_2(\tau),\ d_2(\tau+1)$ into original nonlinear functions $f(x), g(x)$ and $h(x)$ in the system (1).

The overall PML model is obtained automatically when all vertices are assigned. Note that $f(x), g(x)$ and $h(x)$ in the PML model coincide with those in the original system at vertices of all regions.

3 TORA System

The TORA (Translational Oscillator with Rotating Actuator) system [10], shown in Fig. 1 has a cart of mass M connected to a wall with a linear spring (constant k). The cart can, show in 1, oscillate without friction in the horizontal plane. A rotating mass m in the cart is actuated by a motor. The mass is eccentric with a radius of eccentricity e and can be imagined to be a point mass mounted on a massless rotor.

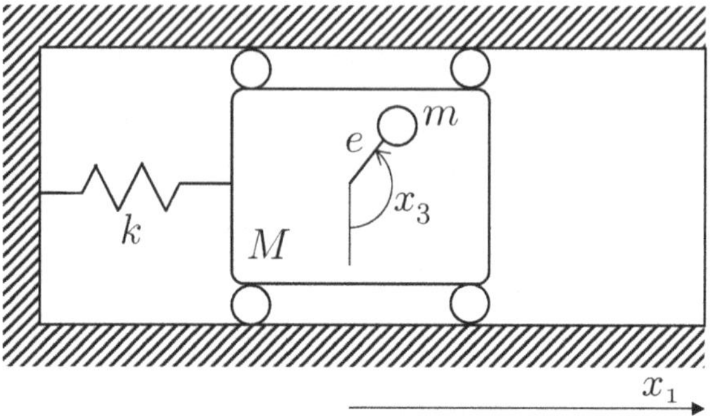

Fig. 1 Kinematic model of TORA system

The rotating motion of the mass m controls the oscillation of the cart. The motor torque is the control variable.

The coordinate transformed dynamics [10] of TORA system is obtained as

$$\begin{cases} \dot{x} = f + gu = \begin{pmatrix} x_2 \\ -x_1 + \varepsilon \sin x_3 \\ x_4 \\ 0 \end{pmatrix} + \begin{pmatrix} 0 \\ 0 \\ 0 \\ 1 \end{pmatrix} u \\ y = h = x_1, \end{cases} \tag{4}$$

where $x \in R^4$, $y \in R$, and the parameter ε depends on the eccentricity.

4 Controller Design of TORA System via Exact Feedback Linearization

We show the controller of TORA system (4) via the exact feedback linearization [11]. The time derivatives of the output y are calculated until the input u appears. Then the controller is obtained as

$$u = \frac{-x_1 + \varepsilon \sin x_3 + \varepsilon x_4^2 \sin x_3}{\varepsilon \cos x_3} + \frac{1}{\varepsilon \cos x_3} \mu, \tag{5}$$

where μ is the linear controller for the linearized system.

$$\begin{cases} \dot{\xi} = A\xi + B\mu \\ y = C\xi, \end{cases}$$

where $\xi = (h, L_f h, L_f^2 h, L_f^3 h)^T$,

$$A = \begin{pmatrix} 0\,1\,0\,0 \\ 0\,0\,1\,0 \\ 0\,0\,0\,1 \\ 0\,0\,0\,0 \end{pmatrix}, B = \begin{pmatrix} 0 \\ 0 \\ 0 \\ 1 \end{pmatrix}, C = \begin{pmatrix} 1 \\ 0 \\ 0 \\ 0 \end{pmatrix}^T. \tag{6}$$

However the controller (5) is only well defined at $-\pi/2 < x_3 = \theta < \pi/2$ because the denominator of the controller is $\varepsilon \cos x_3$.

5 PML Model

We construct the PML model of TORA system (4). The state variable x is divided by $m_1 \times m_2 \times m_3 \times m_4$ vertices,

$$\begin{aligned} x_1 &\in \{d_1(1), \ldots, d_1(m_1)\}, & x_2 &\in \{d_2(1), \ldots, d_2(m_2)\}, \\ x_3 &\in \{d_3(1), \ldots, d_3(m_3)\}, & x_4 &\in \{d_4(1), \ldots, d_4(m_4)\}. \end{aligned}$$

The PML model is expressed as

$$\begin{cases} \dot{x} = f_p + g_p u \\ y = h_p = x_1, \end{cases} \tag{7}$$

where $x \in R_{\rho\sigma\tau\upsilon}$,

$$\begin{aligned} f_p &= \sum_{i=\rho}^{\rho+1} \sum_{j=\sigma}^{\sigma+1} \sum_{k=\tau}^{\tau+1} \sum_{\ell=\upsilon}^{\upsilon+1} \omega_1^i(x_1)\omega_2^j(x_2)\omega_3^k(x_3)\omega_4^\ell(x_4) \\ &\quad \times \left(d_2(j)\ -d_1(i) + \varepsilon \sin d_3(k)\ d_4(\ell)\ 0\right)^T, \\ g_p &= \left(0\,0\,0\,1\right)^T, \\ \omega_1^\rho(x_1) &= (d_1(\rho+1) - x_1)/(d_1(\rho+1) - d_1(\rho)), \\ \omega_1^{\rho+1}(x_1) &= (x_1 - d_1(\rho))/(d_1(\rho+1) - d_1(\rho)). \end{aligned}$$

The model is found to be fully parametric and the internal model dynamics is described by multi-linear interpolation of the vertices: $d_1(i)$, $d_2(j)$, $d_3(k)$ and $d_4(\ell)$. The PML model can be represented by a lookup table (LUT).

Note that trigonometric functions of TORA system (4) are smooth functions and are of class C^∞. The PML models are not of class C^∞. In TORA system control, we have to calculate the fourth derivatives of the output y. Therefore the derivative PML models lose some dynamics. However the PML model based control for TORA system can be applied to a wider region than the conventional one.

Note that there are some modeling errors because the PML model is a nonlinear approximation. In proposed method the vertices $d_i(j)$ of an arbitrary number can be set on arbitrary position of the state space. Therefore it is easily possible to adjust the approximated error.

6 LUT Controller for TORA System via Exact Feedback Linearization

We define the output as $y = x_1$ in the same manner as the previous section, the time derivative of y is calculated as

$$\dot{y} = L_{f_p} h_p = x_2$$

The time derivative of y doesn't contain the control inputs u. We calculate the time derivative of $\dot{y}$. We get

$$\ddot{y} = \sum_{i=\rho}^{\rho+1} \sum_{k=\tau}^{\tau+1} \omega_1^i(x_1)\omega_3^k(x_3)\left(-d_1(i) + \varepsilon \sin d_3(k)\right),$$

where $x \in R_{\rho\sigma\tau\upsilon}$. The time derivative of $\dot{y}$ also doesn't contain the control inputs u. We continue to calculate the time derivative of $\ddot{y}$. We get

$$y^{(3)} = \sum_{j=\sigma}^{\sigma+1} \sum_{\ell=\upsilon}^{\upsilon+1} \omega_2^j(x_2)\omega_4^\ell(x_4)\left(-d_2(j) + \varepsilon \frac{\sin d_3(\tau+1) - \sin d_3(\tau)}{d_3(\tau+1) - d_3(\tau)} d_4(\ell)\right)$$

We continue to calculate the time derivative of $y^{(3)}$. We obtain

$$y^{(4)} = \sum_{i=\rho}^{\rho+1} \sum_{k=\tau}^{\tau+1} \omega_1^i(x_1)\omega_3^k(x_3)\left(d_1(i) - \varepsilon \sin d_3(k)\right) + \varepsilon \frac{\sin d_3(\tau+1) - \sin d_3(\tau)}{d_3(\tau+1) - d_3(\tau)} u$$

Finally the stabilizing controller of (7) is designed as

$$u = \frac{\sum_{i=\rho}^{\rho+1} \sum_{k=\tau}^{\tau+1} \omega_1^i(x_1)\omega_3^k(x_3)\left(-d_1(i) + \varepsilon \sin d_3(k)\right)}{\varepsilon \frac{\sin d_3(\tau+1) - \sin d_3(\tau)}{d_3(\tau+1) - d_3(\tau)}} + \frac{d_3(\tau+1) - d_3(\tau)}{\varepsilon(\sin d_3(\tau+1) - \sin d_3(\tau))} \mu \tag{8}$$

where $\mu = -F\zeta$ is the linear controller of the linear system (9).

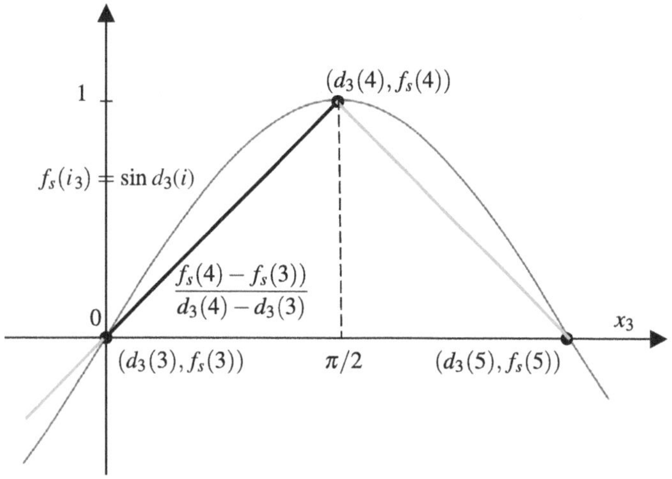

Fig. 2 PML modeling

$$\begin{cases} \dot{\zeta} = A\zeta + B\mu, \\ y = C\zeta, \end{cases} \tag{9}$$

where $\zeta = (h_p, L_{f_p}h_p, L_{f_p}^2 h_p, L_{f_p}^3 h_p)^T$, the matrix A and the vectors B and C are the same as (6).

If $f_s(i) \neq f_s(i+1)$ and $d_3(i) \neq d_3(i+1)$, $i = 1, \ldots, m$, there exists the controller (8) of TORA system (7) since $\det(L_{g_p} L_{f_p}^3 h_p) \neq 0$. Figure 2 shows the PML modeling. Thus we have to construct the PML model of TORA system such that $f_s(i) \neq f_s(i+1)$ and $d_3(i) \neq d_3(i+1)$, where $i = 1, \ldots, m$. Note that the PML model based controller (8) can be applied to a wider region than the conventional feedback linearized controller.

The PML controller (8) is represented by convex combinations of 2^4 vertices $(u(i, j, k, \ell))$ in the case of TORA system ($x \in R^4$). The PML controller can be expressed as an LUT controller using 2^4 vertices (see Tables 1 and 2). The internal dynamics is described by multi-linear interpolation of the vertices.

7 LUT Controller for TORA System with Input Constraints

It is necessary to consider a controller with input constraints when the input value of the LUT controller (8) are too large. We propose an LUT controller with input constraints.

Table 1 LUT controller for TORA system without input constraints

$u(x)$		$x_4 = -10$					
		$x_2 = 0$			$x_2 = 2.5$		
		x_1			x_1		
		-2.5	0	2.5	-2.5	0	2.5
x_3	$-\pi$	106.6	67.86	29.11	131.7	92.98	54.24
	$-7\pi/8$	108.0	62.29	16.59	137.6	91.93	46.22
	$-3\pi/4$	120.9	52.46	-15.94	165.2	96.81	28.41
	$-5\pi/8$	205.3	10.55	-184.2	331.6	136.9	-57.92
	$-\pi/2$	-64.89	129.9	324.7	-191.2	3.572	198.4
	$-3\pi/8$	19.58	87.98	156.4	-24.77	43.63	112.0
	$-\pi/4$	32.45	78.15	123.9	2.810	48.51	94.21
	$-\pi/8$	33.84	72.58	111.3	8.710	47.45	86.20
	0	29.11	67.86	106.6	3.988	42.73	81.48
	$\pi/8$	16.59	62.29	108.0	-13.05	32.65	78.35
	$2\pi/8$	-15.94	52.46	120.9	-60.30	8.101	76.50
	$3\pi/8$	-184.2	10.55	205.3	-310.5	-115.8	79.02
	$\pi/2$	324.7	129.9	-64.89	**451.0**	256.2	61.42
	$5\pi/8$	156.4	87.98	19.58	200.7	132.3	63.94
	$3\pi/4$	123.9	78.15	32.45	153.5	107.8	62.08
	$7\pi/8$	111.3	72.58	33.84	136.4	97.70	58.96
	π	106.6	67.86	29.11	131.7	92.98	54.24

$u(x)$		$x_4 = 0$					
		$x_2 = 0$			$x_2 = 2.5$		
		x_1			x_1		
		-2.5	0	2.5	-2.5	0	2.5
x_3	$-\pi$	38.74	0	-38.74	63.87	25.13	-13.62
	$-7\pi/8$	40.13	-5.569	-51.27	69.77	24.07	-21.63
	$-3\pi/4$	53.00	-15.40	-83.80	97.35	28.95	-39.44
	$-5\pi/8$	137.5	-57.31	-252.1	263.8	69.01	-125.80
	$-\pi/2$	-132.8	62.03	256.8	-259.1	-64.29	130.50
	$-3\pi/8$	-48.27	20.12	88.52	-92.63	-24.23	44.17
	$-\pi/4$	-35.41	10.29	55.99	-65.05	-19.35	26.36
	$-\pi/8$	-34.02	4.722	43.47	-59.15	-20.40	18.34
	0	-38.74	0	38.74	-63.87	-25.13	13.62
	$\pi/8$	-51.27	-5.569	40.13	-80.91	-35.21	10.50
	$2\pi/8$	-83.80	-15.40	53.00	-128.2	-59.76	8.641
	$3\pi/8$	-252.1	-57.31	137.5	-378.4	-183.6	11.16
	$\pi/2$	256.8	62.03	-132.8	383.1	188.3	-6.44
	$5\pi/8$	88.52	20.12	-48.27	132.9	64.48	-3.919
	$3\pi/4$	55.99	10.29	-35.41	85.63	39.93	-5.774
	$7\pi/8$	43.47	4.722	-34.02	68.59	29.85	-8.897
	π	38.74	0	-38.74	63.87	25.13	-13.62

$u(x)$		$x_4 = 10$					
		$x_2 = 0$			$x_2 = 2.5$		
		x_1			x_1		
		-2.5	0	2.5	-2.5	0	2.5
x_3	$-\pi$	-29.11	-67.86	-106.6	-3.988	-42.73	-81.48
	$-7\pi/8$	-27.73	-73.43	-119.1	1.912	-43.79	-89.49
	$-3\pi/4$	-14.86	-83.26	-151.7	29.49	-38.90	-107.3
	$-5\pi/8$	69.62	-125.2	-319.9	195.9	1.1490	-193.6
	$-\pi/2$	-2.600	-5.83	188.9	-326.9	-132.1	62.64
	$-3\pi/8$	-116.1	-47.73	20.66	-160.5	-92.09	-23.69
	$-\pi/4$	-103.3	-57.57	-11.86	-132.9	-87.20	-41.50
	$-\pi/8$	-101.9	-63.14	-24.39	-127.0	-88.26	-49.52
	0	-106.6	-67.86	-29.11	-131.7	-92.98	-54.24
	$\pi/8$	-119.1	-73.43	-27.73	-148.8	-103.1	-57.36
	$2\pi/8$	-151.7	-83.26	-14.86	-196.0	-127.6	-59.22
	$3\pi/8$	-319.9	-125.2	69.62	**-446.3**	-251.5	-56.70
	$\pi/2$	188.9	-5.83	-2.600	315.3	120.5	-74.30
	$5\pi/8$	20.66	-47.73	-116.1	65.02	-3.379	-71.78
	$3\pi/4$	-11.86	-57.57	-103.3	17.77	-27.93	-73.63
	$7\pi/8$	-24.39	-63.14	-101.9	0.7332	-38.01	-76.76
	π	-29.11	-67.86	-106.6	-3.988	-42.73	-81.48

Table 2 LUT controller for TORA system with input constraints

$u(x)$		$x_4 = -10$					
		$x_2 = 0$			$x_2 = 2.5$		
		x_1			x_1		
		-2.5	0	2.5	-2.5	0	2.5
x_3	$-\pi$	38.84	28.83	18.81	37.76	27.74	17.72
	$-7\pi/8$	39.35	27.53	15.71	38.07	26.25	14.43
	$-3\pi/4$	42.93	25.24	7.557	41.01	23.32	5.639
	$-5\pi/8$	65.86	15.49	-34.87	60.39	10.03	-40.33
	$-\pi/2$	-7.105	43.26	93.62	-1.642	48.72	**99.09**
	$-3\pi/8$	15.82	33.51	51.19	17.74	35.43	53.11
	$-\pi/4$	19.4	31.22	43.04	20.69	32.50	44.32
	$-\pi/8$	19.91	29.93	39.94	20.99	31.01	41.03
	0	18.81	28.83	38.84	19.89	29.91	39.93
	$\pi/8$	15.71	27.53	39.35	17.0	28.81	40.63
	$2\pi/8$	7.557	25.24	42.93	9.475	27.16	44.85
	$3\pi/8$	-34.87	15.49	65.86	-29.41	20.95	71.32
	$\pi/2$	93.62	43.26	-7.105	88.16	37.80	-12.57
	$5\pi/8$	51.19	33.51	15.82	49.28	31.59	13.90
	$3\pi/4$	43.04	31.22	19.40	41.76	29.94	18.12
	$7\pi/8$	39.94	29.93	19.91	38.86	28.84	18.82
	π	38.84	28.83	18.81	37.76	27.74	17.72

$u(x)$		$x_4 = 0$					
		$x_2 = 0$			$x_2 = 2.5$		
		x_1			x_1		
		-2.5	0	2.5	-2.5	0	2.5
x_3	$-\pi$	10.02	0	-10.02	8.931	-1.080	-8.10
	$-7\pi/8$	10.52	-1.296	-13.11	9.239	-2.570	-14.39
	$-3\pi/4$	14.10	-3.584	-21.27	12.18	-5.500	-23.19
	$-5\pi/8$	37.03	-13.33	-63.70	31.57	-18.80	-69.16
	$-\pi/2$	-35.93	14.43	64.80	-30.47	19.90	70.26
	$-3\pi/8$	-13.00	4.682	22.37	-8.08	6.600	24.29
	$-\pi/4$	-9.423	2.395	14.21	-8.141	3.670	15.49
	$-\pi/8$	-8.919	1.099	11.12	-7.833	2.180	12.20
	0	-10.02	0	10.02	-8.931	1.080	11.10
	$\pi/8$	-13.11	-1.296	10.52	-8.83	-0.012	11.80
	$2\pi/8$	-21.27	-3.584	14.10	-19.35	-1.660	16.02
	$3\pi/8$	-63.70	-13.33	37.03	-58.24	-7.870	42.49
	$\pi/2$	64.80	14.43	-35.93	59.33	8.970	-41.39
	$5\pi/8$	22.37	4.682	-13.00	20.45	2.760	-14.92
	$3\pi/4$	14.21	2.395	-9.423	12.93	1.110	-10.70
	$7\pi/8$	11.12	1.099	-8.919	10.03	0.010	-10.01
	π	10.02	0	-10.02	8.931	-1.080	-8.10

$u(x)$		$x_4 = 10$					
		$x_2 = 0$			$x_2 = 2.5$		
		x_1			x_1		
		-2.5	0	2.5	-2.5	0	2.5
x_3	$-\pi$	-18.81	-28.83	-38.84	-19.89	-29.91	-39.93
	$-7\pi/8$	-18.31	-30.12	-41.94	-19.59	-31.40	-43.22
	$-3\pi/4$	-14.72	-32.41	-50.10	-16.64	-34.33	-52.01
	$-5\pi/8$	8.203	-42.16	-92.52	2.741	-47.62	**-97.99**
	$-\pi/2$	-64.76	-14.39	35.97	-59.29	-8.931	41.43
	$-3\pi/8$	-41.83	-24.14	-6.458	-39.91	-22.23	-4.540
	$-\pi/4$	-38.25	-26.43	-14.61	-36.97	-25.15	-13.33
	$-\pi/8$	-37.75	-27.73	-17.71	-36.66	-26.64	-16.62
	0	-38.84	-28.83	-18.81	-37.76	-27.74	-17.72
	$\pi/8$	-41.94	-30.12	-18.31	-40.66	-28.84	-17.02
	$2\pi/8$	-50.10	-32.41	-14.72	-48.18	-30.49	-12.81
	$3\pi/8$	-92.52	-42.16	8.203	-87.06	-36.70	13.67
	$\pi/2$	35.97	-14.39	-64.76	30.51	-19.86	-70.22
	$5\pi/8$	-6.458	-24.14	-41.83	-8.377	-26.06	-43.75
	$3\pi/4$	-14.61	-26.43	-38.25	-15.90	-27.71	-39.53
	$7\pi/8$	-17.71	-27.73	-37.75	-18.80	-28.81	-38.83
	π	-18.81	-28.83	-38.84	-19.89	-29.91	-39.93

To design the LUT controller

$$u = \sum_{i=\rho}^{\rho+1} \sum_{j=\sigma}^{\sigma+1} \sum_{k=\tau}^{\tau+1} \sum_{\ell=\upsilon}^{\upsilon+1} \omega_1^i(x_1)\omega_2^j(x_2)\omega_3^k(x_3)\omega_4^\ell(x_4)\,(\alpha(i,k) - \beta F \zeta(i,j,k,\ell)) \quad (10)$$

for nonlinear systems with input constraints, we calculate a feedback gain F of the linearized system (9) that satisfies the following conditions:

$$\begin{aligned} &|u(i,j,k,\ell)| < \delta,\ R_e\lambda(A - BF) < 0, \\ &i = 1,\dots,m_1,\ j = 1,\dots,m_2,\ k = 1,\dots,m_3,\ \ell = 1,\dots,m_4, \end{aligned}$$

where δ is the value of the input constraints,

$$\alpha(i,k) = \frac{-d_1(i) + \varepsilon \sin d_3(k)}{\varepsilon \frac{\sin d_3(\tau+1) - \sin d_3(\tau)}{d_3(\tau+1) - d_3(\tau)}},\ \beta = \frac{d_3(\tau+1) - d_3(\tau)}{\varepsilon(\sin d_3(\tau+1) - \sin d_3(\tau))},$$

$$\zeta(i,j,k,\ell) = \begin{pmatrix} d_1(i) \\ d_2(j) \\ -d_1(i) + \varepsilon \sin d_3(k) \\ -d_2(j) + \varepsilon \frac{\sin d_3(\tau+1) - \sin d_3(\tau)}{d_3(\tau+1) - d_3(\tau)} d_4(\ell) \end{pmatrix}.$$

8 Simulation Results

We apply the PML model based controllers (8) and (10) to TORA system in a computer simulation. In the simulations, the state variables x_1, x_2, x_3, x_4 of TORA system are divided by the following vertices.

$$\begin{aligned} &x_1 \in \{-2.5,\ 0,\ 2.5\},\ x_2 \in \{-2.5,\ 0,\ 2.5\}, \\ &x_3 \in \{-\pi, -7\pi/8, \dots, \pi\},\ x_4 \in \{-10,\ 0,\ 10\} \end{aligned}$$

and the parameter ε is 0.5.

8.1 LUT Controller Without Input Constraints

We consider the feedback gain $F = (4.472,\ 11.68,\ 13.02,\ 6.786)$ such that the linearized control system (9) is stable. The initial condition is $x(0) = (1.5,\ 0,\ 0,\ 0)^T$. Due to lack of space, Table 1 shows a part of the LUT controller (8). In the table, the bold numbers mean the minimum and maximum values of the LUT controller. Figure 3 shows the control responses of TORA system and the control input of the

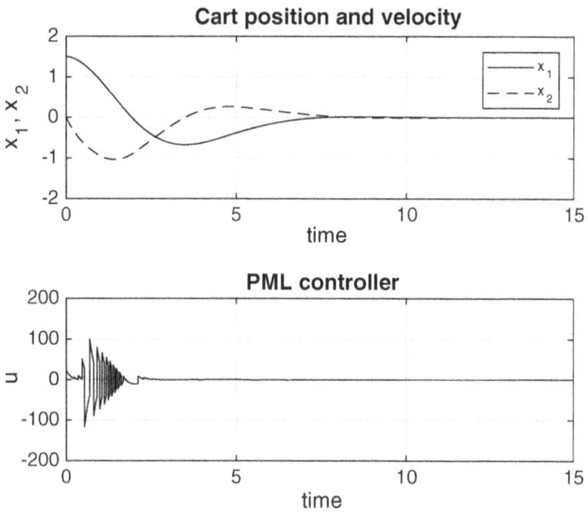

Fig. 3 Control and input responses without input constraints

controller. We apply the LUT controller (10) to TORA system in the next subsection since the control input is too large in Fig. 3.

8.2 LUT Controller with Input Constraints

We design the LUT controller (10) with input constraints ($\delta = 100$) for TORA system. The feedback gain is calculated as $F = (0.8452,\ 2.671,\ 3.798,\ 2.883)$. Table 2

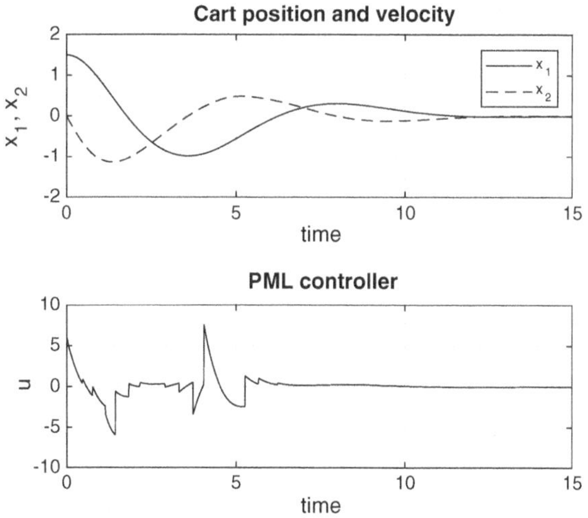

Fig. 4 Control and input responses with input constraints

shows that a part of the LUT controller. In the table, the bold numbers mean the minimum and maximum values of the LUT controller. Figure 4 shows the control responses of TORA system and the control input of the controller. This result shows the feasibility of our proposed controller by the computer simulation.

9 Conclusion

This paper has proposed a piecewise multi-linear model based lookup table controller for nonlinear systems with input constraints. We have constructed a piecewise multi-linear model of a nonlinear system. The model is a nonlinear approximation and it can be derived from fuzzy if-then rules with singleton consequents. The model can be expressed as a lookup table. The dynamics is described by multi-linear interpolation of the lookup table elements. We have designed a piecewise multi-linear controller for the piecewise model via feedback linearization. The controller can be also represented as a lookup table. We have proposed the lookup table controller for nonlinear systems with input constraints and have applied the proposed method to TORA (Translational Oscillator with Rotating Actuator) system. Examples have been shown to confirm the feasibility of our proposals by computer simulations.

In future work, we will apply the proposed methods to real systems and will design an LUT controller for nonlinear systems using PML models.

References

1. K.H. Ang, G. Chong, Y. Li, PID control system analysis, design, and technology (2005)
2. Y.F. Li, C.C. Lau, Development of fuzzy algorithms for servo systems. IEEE Control Syst. Mag. **9**(3), 65–72 (1989)
3. K. Tan, S. Islam, Optimum control strategies in energy conversion of pmsg wind turbine system without mechanical sensors. IEEE Trans. Energy Convers. **19**(2), 392–399 (2004)
4. Q. Wang, L. Chang, An intelligent maximum power extraction algorithm for inverter-based variable speed wind turbine systems. IEEE Trans. Power Electron. **19**(5), 1242–1249 (2004)
5. M. Sugeno, On stability of fuzzy systems expressed by fuzzy rules with singleton consequents. IEEE Trans. Fuzzy Syst. **7**(2), 201–224 (1999)
6. K.C. Goh, M.G. Safonov, G.P. Papavassilopoulos, A global optimization approach for the BMI problem, in *Proceedings of the 33rd IEEE CDC* (1994), pp. 2009–2014
7. T. Taniguchi, M. Sugeno, Piecewise bilinear system control based on full-state feedback linearization. SCIS ISIS **2010**, 1591–1596 (2010)
8. T. Taniguchi, M. Sugeno, Stabilization of nonlinear systems with piecewise bilinear models derived from fuzzy if-then rules with singletons. FUZZ-IEEE **2010**, 2926–2931 (2010)
9. T. Taniguchi, M. Sugeno, Design of LUT-controllers for nonlinear systems with PB models based on I/O linearization. FUZZ-IEEE **2012**, 997–1022 (2012)
10. R. Sepulchre, M. Jankovic, P. Kokotovic, *Constructive Nonlinear Control* (Springer, Berlin, 1997)
11. A. Isidori, The matching of a prescribed linear input-output behavior in a nonlinear system. IEEE Trans. Autom. Control **30**(3), 258–265 (1985)

Difference of Spatiotemporal Human Brain Activities on Recalling Some Fruit Names

Koji Sugimoto, Takahiro Yamanoi, Hisashi Toyoshima, Mika Otuki, Shin-ichi Ohnishi, and Toshimasa Yamazaki

Abstract The authors have measured electroencephalograms (EEGs) from subjects observing images of fruit and recalling them silently. The equivalent current dipole source localization (ECDL) method has been applied to those event related potentials (ERPs): averaged EEGs. ECDs were localized to the primary visual area V1 around 100 ms, to the ventral pathway (TE) around 270 ms, to the parahippocampal gyrus (paraHip) around 380 ms. Then ECDs were localized to the Broca's area around 450 ms, to the fusiform gyrus (FuG) around 600 ms, and again to the Broca's area around 760 ms. Process of search and preservation in the memory has been done from the result of some ECDs to the paraHip. From the results of the present experiment, we supposed that long (not normal) visual shape stimulus processed on the Wernicke's area after through the angular gyrus (AnG), but that round shape (normal) visual stimulus processed on the Wernicke without passing through the AnG.

Keywords Equivalent current dipole source localization · Recalling · Fruit names · Spatiotemporal brain activities · Angular gyrus

---

K. Sugimoto · T. Yamanoi (✉) · S. Ohnishi
Graduate School of Engineering, Hokkai-Gakuen University, Sapporo, Hokkaido 064-0926, Japan
e-mail: yamanoi@hgu.jp

K. Sugimoto
e-mail: sugimoto@sapporoseishu.ed.jp

S. Ohnishi
e-mail: ohnishi@eli.hokkai-s-u.ac.jp

H. Toyoshima
Japan Technical Software, Sapporo, Hokkaido 064-0926, Japan
e-mail: toyoshima@jtsnet.co.jp

T. Yamanoi · M. Otuki
Faculty of Health Sciences, Hokkaido University, Sapporo, Hokkaido 060-0808, Japan
e-mail: lasteroideb612@pop.med.hokudai.ac.jp

T. Yamazaki
Department of Bioscience and Bioinformatics, Kyushu Institute of Technology, Iizuka, Fukuoka 680-4, Japan
e-mail: t-ymzk@bio.kyutech.ac.jp

S. N. Shahbazova et al. (eds.), *Recent Developments and the New Direction in Soft-Computing Foundations and Applications*, Studies in Fuzziness and Soft Computing 393, https://doi.org/10.1007/978-3-030-47124-8_36

1 Introduction

By presenting Kanji word and Hiragana word to the subjects, some of the present authors have measured electroencephalograms (EEGs) under those stimuli [1] and both data were summed and averaged according to the type of the stimuli and the subjects in order to get event related potentials (ERPs). Each peak of ERPs were detected and analyzed by the equivalent current dipole source localization (ECDL) [2] at that latency by use of two dipole model. In both cases of Kanji and Hiragana recognition, they localized equivalent current dipole (ECD) nodes from early components of ERPs to the V1, V2 and the inferior temporal gyrus (ITG), after then ECDs are localized to the Wernicke's area and the Broca's area [1]. These results agree with the results on MEG, PET or fMRI.

On the other hand, from the clinical lesion studies, it is known that the lesions causing disabilities of naming and comprehension of objects are dissociate with the target categories, e.g. artificial or biological things. These symptoms are called category-specific disorders [3].

Using the same methodology as the preceding research [4–10], some of the present authors had cleared human brain activities during language recognition and during image recognition.

In the present study, we had measured electroence- phalograms (EEGs), in order to investigate the brain activity during watching monochrome line drawings of fruit and recalling a name of presented fruit. And both data were summed and averaged according to the type of the stimuli in order to get event related potentials (ERPs). Each peak of ERPs were detected and analyzed by the equivalent current dipole source localization (ECDL) method [2].

2 EEG Measurement Experiments

One subject Y.Y is a 22 year-old female and has normal visual acuity. She is a right handed. Subjects K.S, H.T and Y.K are 22 year-old males and also have normal visual acuity. They are right handed except Y.K. The subjects put on 19 active electrodes and watched the 21 inch CRT 30 cm in front of them. Their heads were fixed on the table on a chin rest.

Each image was displayed on the CRT. Stimuli are simple monochrome image (line drawings) of fruit. Presented images were strawberry, persimmon, cherry, etc. (Fig. 1). First, a fixation point was presented, then a stimulus was presented, both of them were during 3000 ms (Fig. 2). EEGs were measured on the multi-purpose portable bio-amplifier recording device (Polymate AP1524, TEAC) by way of the electrodes and the frequency band is between 1.0 and 2000 Hz. Outputs were transmitted to a recording PC. Each position of electrode was measured in the three dimensional coordinate on each experiment by the subjects. These electrode positions were used in applying ECDL method.

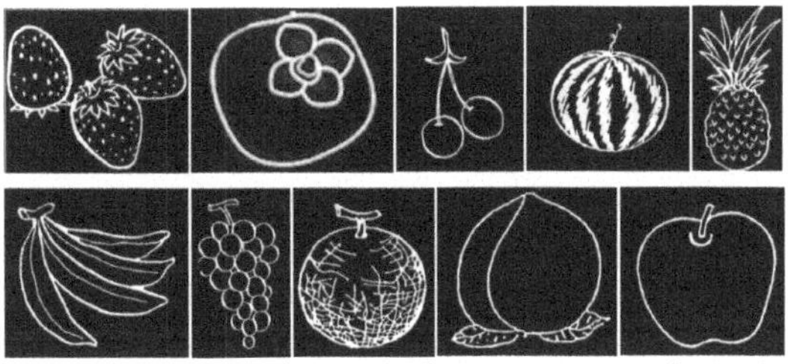

Fig. 1 Presented images of fruit

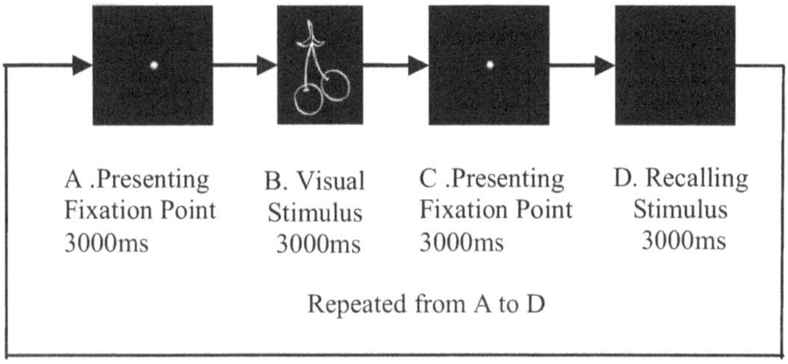

Fig. 2 Time chart of the present experiment

We have measured EEGs on each visual stimulus. So as to effectively execute the ECDL method, both data were summed and averaged according to the type of parts to get event-related potentials (ERPs).

Then the ECDL method was applied to each ERP. Because of the number of the recording electrodes was 19, theoretically, at most three ECDs could be estimated by use of the PC-based ECDL analysis software "SynaCenterPro [2]" (NEC Corporation). The goodness of fit (GOF) of ECDL was more than 99%. Estimated EEGs were super imposed on MRIs for a subject by using three dimensional electrodes positions measured in each experiment.

Analyzed EEGs were images of strawberry, persimmon, cherry, watermelon, banana, and apple.

3 Results of ECDL Analysis

After the latency around 400 ms, the ECDs were localized to the right parahippocampal gyrus (R ParaHip) (Fig. 3), the right fusiform gyrus (FuG), the Broca area, the right ParaHip, the Broca's area, and the right fusiform gyrus (FuG).

Table 1 Physical information of subjects

Subject	Dominant hand	Age	Sex
Y.Y.	Right	22	Female
K.S.	Right	22	Male
H.T.	Right	22	Male
Y.K.	Left	22	Male

Above mentioned spatiotemporal pathway accords with so called the ventral pathway which is said to be related with the primitive process of visual recognition. These areas are also related to the integrated process of visual recognition of picture and the recalling of word. Especially, the angular gyrus (AnG) is said to integrate information of some modalities, so there might have recalled a word already at this stage. On the right inferior temporal white matter, a process progresses from recognition of a picture to recalling of a word. The ECDs were localized on the AnG in case of some fruits, however, no ECDs were localized in some cases.

These processes are done in series or in parallel. Table 1 is physical information of four subjects. The estimated relationship between position of ECDs and latencies are resumed in Tables 2 and 3, and some estimated spatiotemporal pathways are shown in Fig. 3.

According to Tables 2 and 3, we found spatiotemporal pathway of the human brain activities on recalling a fruit name as follows.

(pathway1) V1 → R TE → R ParaHip → R FuG→R ParaHip
→ Broca
(pathway2) Broca → L Insula → R ParaHip → Broca
→ (R FuG) → (R AnG) → R FuG → Wernicke
→ Broca

In this study, we have further estimated the human brain activities during the subject watching a line drawing of watermelon, cherry, apple, and strawberry. Some of the present authors have applied the same methodology to EEGs on watching a line drawing of banana and persimmon.

4 Conclusion

We have estimated the human brain activities during the subject watching a line drawing of fruit and recalling its name silently. ECDs were localized to the word generation area and the image recognition area.

We have detected a pathway regarding with the recalling fruit names. By use of ECDL method, ECDs were localized to the right angular gyrus (R AnG), the Broca area and the Wenicke area. These areas are related to the integrated process of visual recognition of picture and the recalling of word. Some of these areas are also related to the image recognition and word generation.

Table 2 Relationship between localized ECDs and its latencies I

Subject	Y.Y.	K.S.	K.S.	K.S.
Image	Cherry	Cherry	Watermelon	Apple
V1	88	119	84	114
R TE	276	277	248	330
R ParaHip	350	334	311	353
R FuG	361	377	337	363
R ParaHip	375 380	380	386	387
Broca	451	387	439	457
L Insula	466	468	500	468
R ParaHip	485 487	430 470	504	535
Broca	540	477 530		575
R FuG	606	585		602 630
Broca	645	601		
R AnG	652 655	683	648	
R FuG	678	754		
Wernicke	729	764	759	778
Broca	760(R)	828	784	792

According to the brain activities of subjects, we can confirm that most of activity such as Broca area and the Wernicke area considered to be a word generation area, these are concentrated on the left cerebral hemisphere. Therefore, the word generation area of subjects is predominated in the left hemisphere. On the other hand, the activity on the right Broca area is seen at latency 760 ms for the subject Y.Y, also. From this, the activity on the right hemisphere is seen occasionally.

Also, Hippocampus, ParaHip, and Fusiform gyrus, known as a part about memory, it is said that language memory in the left hemisphere of the brain mainly, control non-language memory in the right hemisphere mainly. Because the activity with the right ParaHip was estimated over multiple times in this study, some kind of processing in the non-language memory is performed in recalling fruit names, and it is thought that a search, rearranging, maintenance of the memory is always performed.

Comparing with the precedent study [10], for an input pathway, both in the precedent study and this study, the brain activity is found in ParaHip and FuG gyrus which are concerned with memory processing. In the output pathway, both in the precedent studies and the study, brain activities are found in the left insula. The insula is said to be related to taste of the food, and it is thought that both cases are related to conscious greed of the taste in fruit shown as stimulus.

Table 3 Relationship between localized ECDs and its latencies II

Subject	Y.K.	H.T.	Y.K.	H.T.
Image	Strawberry	Strawberry	Persimmon	Banana
V1	129	122	87	146
R TE	269	235	316	291
R ParaHip	300	298	384	387
R FuG	318	302	401	391
Broca	399	337		
R ParaHip	456	342	416	401
Broca			512	436
L Insula	473	355	522	442
R ParaHip	493	374	548	508 534
Broca	508			569
R FuG	545	427		584
Broca	585			
R AnG	699	583		626
Broca		592		
R FuG	734	614		655
Wernicke	787	655	680(R)	662(R)
Broca	813	672	711	

On the other hand, in the case of "cherry" and "banana", the activity in the AnG, Broca area, and Wernicke area are found. The AnG is said to be concerned with a language. The ECD estimated with angular gyrus was estimated afterwards in Wernicke area. It is thought that integration of input information was carried out in the area. However, in the case of "persimmon", ECD was not localized to the AnG. We thought that the some factor which is common to "cherry", "watermelon", "banana", and "strawberry" relates here, but "persimmon" and "apple" are different from these. However, it fact. The factor may be the difference between "round fruit" and "unusual shape fruit". Although, watermelon has a round shape itself, but the difference from other round shape fruits is stripes. One of the present author Ms. Mika Otsuki had tried an experimental research on patients with the name recalling disabilities [10]. She asked to recall the name of monochrome image of fruits and tetrapod animals. From her research results on fruits, right answered group of fruits agree with a group which has activities on the AnG in the present paper. If the decision is depend only on the visual information, discrimination of fruits would be done near the AnG (Fig. 3).

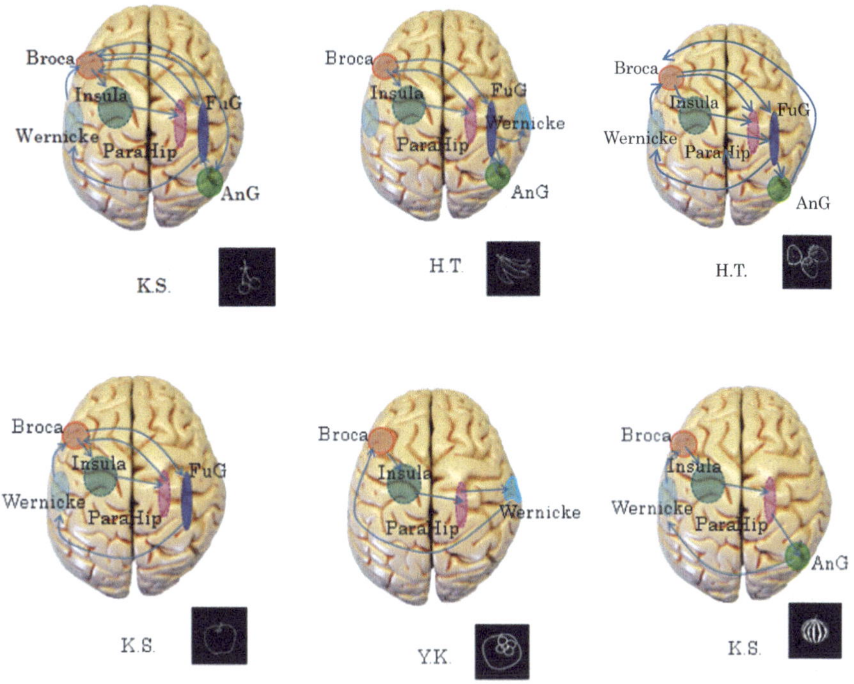

Fig. 3 Some estimated spatiotemporal pathways by presented image of fruit (Pathway2)

Fig. 4 Results of name recalling task by Otuki [11]

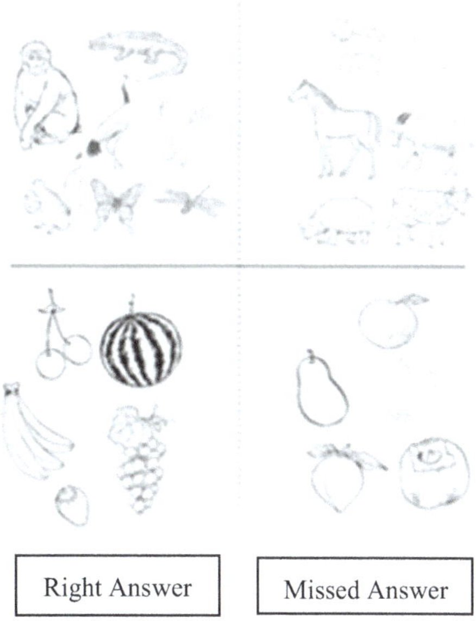

Acknowledgements This research was supported by the project of the High-tech Research Center of Hokkai-Gakuen University with the grant-in-aide from the Japanese Ministry of Education, Culture, Sports, Science and Technology ended in March 2013. And the work was partly supported by JSPS KAKENHI Grant Number 16H02852.

References

1. T. Yamanoi, T. Yamazaki, J.-L. Vercher, E. Sanchez, M. Sugeno, Dominance of recognition of words presented on right or left eye-comparison of Kanji and Hiragana, in *Modern Information Processing, From Theory to Applications* (Elsevier Science B.V., Oxford), pp. 407–416 (2006)
2. T. Yamazaki, K. Kamijo, T. Kiyuma, Y. Takaki, Y. Kuroiwa, A. Ochi, H. Otsubo, PC-based multiple equivalent current dipole source localization system and its applications. Res. Adv. Biomed. Eng. **2**, 97–109 (2001)
3. A. Martin, L.G. Ungerleider, J.V. Haxby, Category specificity and the brain: the sensory/motor model of semantic representations of objects, in *Higher Cognitive Functions, in the New Cognitive Neurosciences*, ed. by M.S. Gazzaniga (MIT Press, Cambridge, MA, 2000), pp. 1023–36
4. T. Yamanoi, H. Toyoshima, H. Ichihashi, Spatiotemporal brain activities in recalling sentences by loci mnemonic system, in *2007 IEEE International Conference on Systems, Man and Cybernetics* (2007), pp. 1878–1883
5. T. Yamanoi, A. Moritaka, F. Kakumori, H. Takayanagi, T. Yamazaki, M. Sugeno, H. Nonaka, Micro-robot control by use of EEG from frontal lobe at silent reading, in *18th International Congress on Brain Electromagnetic Topography*, PS509 (2009), p. 84
6. M. Sugeno, T. Yamanoi, Spatiotemporal analysis of brain activity during understanding honorific expressions. J. Adv. Comput. Intell. Intell. Inf. **15**(9), 1211–1220 (2011)
7. H. Toyoshima, T. Yamanoi, T. Yamazaki, S. Ohnishi, Spatiotemporal brain activity during hiragana word recognition task. J. Adv. Comput. Intell. Intell. Inf. **15**(3), 357–361 (2011)
8. I. Hayashi, H. Toyoshima, T. Yamanoi, A measure of localization of brain activity for the motion aperture problem using electroencephalogram, in *Developing and Applying Biologically-Inspired Vision System: Interdisciplinary Concept*, Chapter 9 (2012), pp. 208–223
9. T. Yamanoi, H. Toyoshima, H. Takayanagi, T. Yamazaki, S. Ohnishi, M. Sugeno, Discrimination of electroencephalograms on recognizing and recalling playing cards—a magic without trick (in Japanese). J. Jpn. Soc. Fuzzy Theory Intell. Inf. **28**(3), 639–646 (2016)
10. K. Sugimoto, T. Yamanoi, H. Toyoshima, M. Otsuki, S. Ohnishi, T. Yamazaki, Spatiotemporal human brain activities on recalling 4-legged mammal and fruit names, in *Joint 17th World Congress of International Fuzzy Systems Association and 9th International Conference on Soft Computing and Intelligent Systems*, No. 140 (2017)
11. M. Otsuki, Neuropsychology of language (in Japanese). Neuropsychology **32**, 104–119 (2016)

A Granular Computing Based Approach for Improving the Consistency of Intuitionistic Reciprocal Preference Relations

F. J. Cabrerizo, I. J. Pérez, J. A. Morente-Molinera, S. Alonso, and E. Herrera-Viedma

Abstract Pairwise comparison between alternatives is the preference elicitation method most used in decision making. To model it, a mathematical model based on the concept of preference relation has been proposed. When preference relations are used, consistency, which is commonly related to the concept of transitivity, is a fundamental concern needing attention, especially when many alternatives are involved in the decision-making problem. Because inconsistent preferences could lead to not logical solutions, the study of consistency is essential in decision-making. Therefore, we develop in this study a new approach based on granular computing that improves the consistency when intuitionistic reciprocal preference relations are used to represent the preferences. This approach uses information granularity to develop the concept of granular intuitionistic reciprocal preference relations in which each entry is formed as an information granule instead of an exact numeric value. As well, it uses the multiplicative transitivity property to model the consistency associated with the intuitionistic reciprocal preference relations. Finally, an experimental example is shown to test and illustrate the performance of this approach.

F. J. Cabrerizo (✉) · E. Herrera-Viedma
Department of Computer Science and Artificial Intelligence, University of Granada, 18071 Granada, Spain
e-mail: cabrerizo@decsai.ugr.es

E. Herrera-Viedma
e-mail: viedma@decsai.ugr.es

I. J. Pérez
Department of Computer Science and Engineering, University of Cádiz, 11519 Puerto Real, Cádiz, Spain
e-mail: ignaciojavier.perez@uca.es

J. A. Morente-Molinera
Department of Engineering, International University of La Rioja (UNIR), 26006 La Rioja, Logroño, Spain
e-mail: juan.morente@unir.net

S. Alonso
Department of Software Engineering, University of Granada, 18071 Granada, Spain
e-mail: zerjioi@ugr.es

S. N. Shahbazova et al. (eds.), *Recent Developments and the New Direction in Soft-Computing Foundations and Applications*, Studies in Fuzziness and Soft Computing 393, https://doi.org/10.1007/978-3-030-47124-8_37

1 Introduction

In decision making scenarios, a decision maker (or a group of them) has to provide his/her preferences about a possible set of alternatives to solve a given problem [1]. The objective is to rank the alternatives from best to worst to solve the problem under study.

An essential issue requiring attention in that kind of scenarios is that of the preference elicitation method used by the decision makers to express their preferences. Among them, preference orderings, utility values or pairwise comparisons between alternatives are generally used. Millet studied different preference elicitation methods in [2] and he concluded that pairwise comparison methods are more precise that non-pairwise comparison ones.

Given two alternatives, x_i and x_j, of a collection of them, $X = \{x_1, \ldots, x_n\}, n \geq 2$, either the decision maker chooses one alternative to the other one (for instance, the alternative x_i over the alternative x_j) or he/she is indifferent between them. Of course, there exists other potential action, that of the decision maker is not able to compare them.

Based on the concept of a preference relation, two mathematical frameworks have been developed to model the above preference states. The first one defines a preference relation for each one of the three preference states, which is commonly known as a preference structure on the collection of alternatives [3]. In the second one, a single preference relation integrates the three preference states [4]. Following on from this, we can adopt two representations in each case: (i) [0, 1]-valued (fuzzy) preference relations, or (ii) binary (crisp) preference relations.

In fuzzy set theory, the first representation has been frequently used for representing intensities of preferences [5]. In this representation, the preferences provided by a decision maker on the collection of the alternatives, X, are represented via a preference relation, $P \subset X \times X$, depicted by a mapping $\mu_P : X \times X \rightarrow [0, 1]$. Here, $\mu_P(x_i, x_j) = p_{ij}$ indicates the preference degree that the alternative x_i has over the alternative x_j, being $0 \leq p_{ij} \leq 1$ and $p_{ij} + p_{ji} = 1, \forall i, j$. These kind of preference relations are usually referred to as just reciprocal preference relations, which are a particular case of (weakly) complete fuzzy preference relations [6], that is, fuzzy preference relations that satisfy $p_{ij} + p_{ji} \geq 1, \forall i, j$.

The reciprocal preference relations are not the only ones used to capture the uncertainty of the information. Xu gave a thorough survey of preference relations [7], being the more used in the literature the following ones: linguistic preference relations, reciprocal preference relations, multiplicative preference relations, and intuitionistic reciprocal preference relations.

The last ones are based on Atanassov's intuitionistic fuzzy sets [8], which are based on the use of a membership degree, a non-membership degree and a hesitation index to represent the decision makers' preferences. In fact, in real world decision making scenarios, a decision maker may present some hesitation and, as a consequence, he/she may be unable to communicate with precision his/her preferences for all or some of the alternatives [9]. In such a situation, intuitionistic fuzzy values are appropriate to represent the decision makers' preferences [10].

An issue to address in decision making problems with intuitionistic reciprocal preference relations is the consistency associated with the opinions expressed by the decision makers, a problem thoroughly studied in the case of reciprocal preference relations [11, 12]. The concept of transitivity has been widely used to model consistency of reciprocal preference relations, in the sense that if a decision maker prefers the alternative x_i to the alternative x_j ($p_{ij} \geq 0.5$) and this decision maker also prefers the alternative x_j and x_k ($p_{jk} \geq 0.5$), then this decision maker should prefer the alternative x_i to the alternative x_k ($p_{ik} \geq 0.5$). This notion of transitivity is known as weak stochastic transitivity [5], but it is not the only one and different stronger conditions have been proposed [12]. Among them, the multiplicative transitivity property has been characterized to be the most suitable one for modeling cardinal consistency of reciprocal preference relations [11]. Because a reciprocal preference relation is a particular case of an intuitionistic reciprocal preference relation, the same previous claim may be applied to it. Based on this, the multiplicative transitivity property for intuitionistic reciprocal preference relations was formalized in [13], and a method to quantify the level of consistency associated with this kind of preference relations was also presented.

Given the importance of obtaining consistent preferences in real world decision making problems, this study aims to develop a method to improve the consistency associated with an intuitionistic reciprocal preference relation. The proposed method presents the following characteristics. On the one hand, the multiplicative transitivity property is used to model the consistency associated with an intuitionistic reciprocal preference relation. On the other hand, it is assumed that the improvement of the consistency implies the necessity for some degree of flexibility in the decision maker, in the sense that he/she has to allow some modifications in his/her first preferences, and here information granularity becomes of importance [14, 15]. Given that intuitionistic reciprocal preference relations are considered, the required flexibility is brought by allowing granular entries instead of single numeric values. This means the entries of an intuitionistic reciprocal preference relation are information granules [16] rather than exact numeric values. The proposed concept of granular intuitionistic reciprocal preference relation is utilized to maximize a performance index that expresses the level of consistency associated with the intuitionistic reciprocal preference relation.

The rest of this study is set out as follows. Section 2 presents the preliminaries required throughout the rest of this study, that is, the definition of an intuitionistic fuzzy set, the definition of an intuitionistic reciprocal preference relation, and the consistency concept associated with an intuitionistic reciprocal preference relation. Section 3 describes our method in detail. In Sect. 4, an experimental example is given to illustrate and test its performance. Finally, Sect. 5 concludes this study and points out some future studies.

2 Preliminaries

In this section, both an intuitionistic fuzzy set and an intuitionistic reciprocal preference relation are defined. In addition, we present the consistency concept associated with an intuitionistic reciprocal preference relation and a method to obtain the consistency level associated with it.

A. *Intuitionistic Fuzzy Sets*

Atanassov introduced the concept of an intuitionistic fuzzy set in [8]:

Definition 2.1 An intuitionistic fuzzy set A over a universe of discourse X is represented as $A = \mu_A, \nu_A$ where $\mu_A : X \to [0, 1]$, $\nu_A : X \to [0, 1]$, and $0 \leq \mu_A(x) + \nu_A(x) \leq 1$, $\forall x \in X$. For each $x \in X$, the numbers $\mu_A(x)$ and $\nu_A(x)$ are known as the degree of membership and the degree of non-membership of x to A, respectively.

When $\mu_A(x) = 1 - \nu_A(x), \forall x \in X$, an intuitionistic fuzzy set becomes a fuzzy set. Nevertheless, if there is at least a value $x \in X$ such that $\mu_A(x) < 1 - \nu_A(x)$, an extra parameter, which is referred to as hesitancy degree, is required when working with intuitionistic fuzzy sets. It is defined as $\tau_A(x) = 1 - \mu_A(x) - \nu_A(x)$ and represents the amount of lacking information in determining the membership degree of x to A. When $\tau_A(x) = 0$, the reciprocal relationship between the non-membership degree and the membership degree makes the former one unnecessary in the formulation as it may be derived from the latter.

B. *Intuitionistic Reciprocal Preference Relations*

The concept of an intuitionistic reciprocal preference relation was defined by Xu as a generalization of the concept of a reciprocal preference relation [10]:

Definition 2.2 An intuitionistic reciprocal preference relation R on a finite set of alternatives X is characterized by a membership function $\mu_R : X \times X \to [0, 1]$ and a non-membership function $\nu_R : X \times X \to [0, 1]$, such that $0 \leq \mu_R(x_i, x_j) + \nu_R(x_i, x_j) \leq 1$, $\forall (x_i, x_j) \in X \times X$. The value $\mu_R(x_i, x_j) = \mu_{ij}$ represents the certainty degree up to which x_i is preferred to x_j, while the value $\nu_R(x_i, x_j) = \nu_{ij}$ represents the certainty degree up to which x_i is non-preferred to x_j. The following conditions are also imposed:

$$\mu_{ii} = \nu_{ii} = 0.5 \forall i \in \{1, \ldots, n\} \tag{1}$$

$$\mu_{ji} = \nu_{ij} \forall i, j \in \{1, \ldots, n\} \tag{2}$$

C. *Consistency of Intuitionistic Reciprocal Preference Relations*

Rationality is associated with consistency [17], which is related to the transitivity property [12]. To model transitivity in reciprocal preference relations many properties have been proposed, but many of them have been proved to be inappropriate [11]. In a study developed by Chiclana et al., it was concluded that the multiplicative transitivity property, which was proposed by Tanino when $p_{ij} > 0, \forall i, j$ [5], is the most appropriate property to model the cardinal consistency associated with reciprocal preference relations [11].

Definition 2.3 A reciprocal preference relation $P = \left(p_{ij}\right)$ on a finite set of alternatives X is multiplicative transitive if and only if:

$$p_{ij} \cdot p_{jk} \cdot p_{ki} = p_{ik} \cdot p_{kj} \cdot p_{ji} \forall i, k, j \in \{1, \ldots, n\} \tag{3}$$

The multiplicative transitivity property for reciprocal preference relations was formally generalized to intuitionistic reciprocal preference relations in [13]. To do so, the Zadeh's extension principle [18] was applied to the case when the preference values are fuzzy sets instead of crisp values in [0, 1]. Then, the representation theorem [19] was applied to the related fuzzy sets obtained so that the related multiplicative transitivity property of fuzzy preference values was obtained. This result was next applied to the case when the fuzzy sets are interval-valued fuzzy sets. This process allowed to properly extend the multiplicative transitivity property from reciprocal preference relations to interval-valued reciprocal preference relations. Finally, the mathematical isomorphism between intuitionistic fuzzy sets and interval-valued fuzzy sets [9] was used to obtain the correct formulation of the multiplicative transitivity property for intuitionistic reciprocal preference relations. According to this process, the multiplicative transitivity property of intuitionistic reciprocal preference relations is defined as follows [13]:

Definition 2.4 An intuitionistic reciprocal preference relation $R = \left(r_{ij}\right) = \left(\mu_{ij}, \nu_{ij}\right)$ is multiplicative transitive if and only if:

$$\forall i, j, k : \begin{cases} \mu_{ij} \cdot \mu_{jk} \cdot \mu_{ki} = \mu_{ik} \cdot \mu_{kj} \cdot \mu_{ji} \\ \left(1 - \nu_{ij}\right) \cdot \left(1 - \nu_{jk}\right) \cdot (1 - \nu_{ki}) = \\ = (1 - \nu_{ik}) \cdot \left(1 - \nu_{kj}\right) \cdot \left(1 - \nu_{ji}\right) \end{cases} \tag{4}$$

The intuitionistic preference value between a pair of alternatives $\left(x_i, x_j\right)$ with $(i < j)$ can be estimated by (4) using another different intermediate alternative x_k $(k \neq i, j)$ as follows [13]:

$$mr_{ij}^{k-} = \frac{\mu_{ik} \cdot \mu_{kj} \cdot \mu_{ji}}{\mu_{jk} \cdot \mu_{ki}} \tag{5}$$

$$mr_{ij}^{k+} = 1 - \frac{(1-\nu_{ik})\cdot\left(1-\nu_{kj}\right)\cdot\left(1-\nu_{ji}\right)}{\left(1-\nu_{jk}\right)\cdot(1-\nu_{ki})} \tag{6}$$

providing that the denominators are not zero.

The partially multiplicative transitivity based estimated intuitionistic preference value of the pair of alternatives $\left(x_i, x_j\right)$ obtained using the intermediate alternative x_k is called $mr_{ij}^k = mr_{ij}^{k-}, mr_{ij}^{k+}$.

When two of the three sub-indexes are equal, both equations in (4) are always true. Let $k = i$, if $\mu_{ij} \neq 0$, then $mr_{ij}^{i-} = \mu_{ij}$, while if $\nu_{ij} \neq 1$, then $mr_{ij}^{i+} = \nu_{ij}$. Because $\mu_{ji} = \nu_{ij}$, then we have that: $\mu_{ji} \neq 0$ if and only if $\nu_{ij} \neq 0$, and $\nu_{ji} \neq 1$ if and only if $\mu_{ij} \neq 1$. Therefore, if $k = i$ and $\left(r_{ij}, r_{ji}\right) \neq (1, 0, 0, 1)$ we have $r_{ij} = m_{ij}^i$. In similar manner, if $k = i$ and $\left(r_{ij}, r_{ji}\right) \neq (0, 1, 1, 0)$ we have $r_{ij} = m_{ij}^j$. In summary, although it is possible to obtain the multiplicative transitivity based estimated intuitionistic preference value of the pair of alternatives $\left(x_i, x_j\right)$ when $k \in \{i, j\}$ and $\left(r_{ij}, r_{ji}\right) \neq \{(1, 0, 0, 1), (0, 1, 1, 0)\}$, it is also true that there is no indirect estimation process as described above. Finally, when $i = j$ we have by definition that $r_{ii} = 0.5, 0.5$ and we would have $mr_{ii}^k = r_{ii}$ whenever $r_{ik} \notin (0, 1, 1, 0)$.

The average of all possible partially multiplicative transitivity based estimated values of the pair of alternatives $\left(x_i, x_j\right)$ can be interpreted as their global multiplicative transitivity based estimated value:

$$mr_{ij}^{-} = \frac{\sum_{k\in R_{ij}^{01}} mr_{ij}^{k-}}{\#R_{ij}^{01}};\ mr_{ij}^{+} = \frac{\sum_{k\in R_{ij}^{01}} mr_{ij}^{k+}}{\#R_{ij}^{01}} \tag{7}$$

where we have $R_{ij}^{01} = \left\{k \neq i, j | \left(r_{ik}, r_{kj}\right) \notin R^{01}\right\}$, $\#R_{ij}^{01}$ is the cardinality of R_{ij}^{01}, and $R^{01} = \{(1, 0, 0, 1), (0, 1, 1, 0)\}$.

Thus, given an intuitionistic reciprocal preference relation, the following multiplicative transitivity based intuitionistic reciprocal preference relation, $MR = \left(mr_{ij}^{-}, mr_{ij}^{+}\right)_{n\times n}$, may be constructed [13]:

$$mr_{ij}^{-} = \begin{cases} \frac{\sum_{k\in R_{ij}^{01}} mr_{ij}^{k-}}{\#R_{ij}^{01}}, i < j \\ 0.5, i = j \\ \frac{\sum_{k\in R_{ji}^{01}} mr_{ji}^{k+}}{\#R_{ji}^{01}}, i > j \end{cases} \tag{8}$$

$$mr_{ij}^{+} = \begin{cases} \frac{\sum_{k\in R_{ij}^{01}} mr_{ij}^{k+}}{\#R_{ij}^{01}}, i < j \\ 0.5, i = j \\ \frac{\sum_{k\in R_{ji}^{01}} mr_{ji}^{k-}}{\#R_{ji}^{01}}, i > j \end{cases} \tag{9}$$

If an intuitionistic reciprocal preference relation R is multiplicative transitive, then $R = MR$. Indeed, if R is multiplicative transitive both equations in (4) are true

$\forall i, j, k$. Particularly, we have:

$$\mu_{ij} = \frac{\mu_{ik} \cdot \mu_{kj} \cdot \mu_{ji}}{\mu_{jk} \cdot \mu_{ki}} \tag{10}$$

$$\nu_{ij} = 1 - \frac{(1 - \nu_{ik}) \cdot \left(1 - \nu_{kj}\right) \cdot \left(1 - \nu_{ji}\right)}{\left(1 - \nu_{jk}\right) \cdot (1 - \nu_{ki})} \tag{11}$$

whenever $k \in R_{ij}^{01}$. Consequently, $mr_{ij}^{k-} = \mu_{ij}$ and $mr_{ij}^{k+} = \nu_{ij}$, $\forall i, j, k \in R_{ij}^{01}$. It proves that $r_{ij} = mr_{ij}$, $\forall i, j$. An intuitionistic reciprocal preference relation R is referred to as multiplicative consistent when $R = MR$ [13].

Then, a methodology to measure the consistency level of an intuitionistic reciprocal preference relation was proposed in [13]. It is based on the similarity with the related multiplicative transitivity based intuitionistic reciprocal preference relation. In this approach, the similarity between the values r_{ij} and mr_{ij} is used to measure the consistency level of an intuitionistic reciprocal preference relation at its three different levels: pair of alternatives, alternatives, and relation.

(1) Consistency level of pair of alternatives x_i and x_j, cl_{ij}:

$$cl_{ij} = 1 - d\left(r_{ij}, mr_{ij}\right) \forall i, j \tag{12}$$

Here, a metric d, which describes the distance between intuitionistic fuzzy sets, is used to define a similarity function between intuitionistic fuzzy sets.

(2) Consistency level of alternative x_i, cl_i:

$$cl_i = \frac{\sum_{j=1; i \neq j}^{n} cl_{ij}}{n - 1} \tag{13}$$

(3) Consistency level of an intuitionistic reciprocal preference relation, cl:

$$cl = \frac{\sum_{i=1}^{n} cl_i}{n} \tag{14}$$

3 A Method for Improving Consistency of Intuitionistic Reciprocal Preference Relations

Pairwise comparison has the advantage that the decision makers focus only on two alternatives at a time. It facilitates to the decision makers the expression of their preferences. Nevertheless, it produces more information than is really required and limits the global perception of the decision makers on the alternatives. Therefore, situations in which the decision makers express inconsistent preferences are very frequent [11, 20].

In decision making situations, illogical solutions could be led by inconsistent opinions provided by the decision makers. For this reason, it is of utmost importance the study of methods for measuring and improving the level of consistency associated with the opinions given by the decision makers involved in the problem under consideration [12, 21]. In such a way, a method to improve the level of consistency associated with an intuitionistic reciprocal preference relation becomes desirable.

The level of consistency associated with an intuitionistic reciprocal preference relation may be improved by allowing some flexibility, which is reached by handling pairwise comparison as information granules [22] rather than exact numeric values, in the first preferences communicated by the decision makers [14, 23]. In this scenario, an intuitionistic reciprocal preference relation R is denoted as $\boldsymbol{G}(R)$ to highlight that granular intuitionistic reciprocal preference relations are considered. The objective is to take advantage of the use of information granularity [24] to improve the consistency level associated with an intuitionistic reciprocal preference relation. In short, the flexibility is delivered by the information granularity, which arises here as a valuable resource both conceptual and computational that is used to maximize the consistency level.

This optimization (maximization) task is solved by applying a technique of global optimization. In particular, the particle swarm optimization method, proposed by Kennedy and Eberhart [25], is employed. We use this technique because it has a limited computing overhead and a fast convergence speed [26, 27]. Regardless, we may apply other techniques of global optimization to solve this task [28–30].

The origin of the particle swarm optimization method is encountered in the social interactions with biological swarms. In this probabilistic search optimization technique, some particle containing a collection of parameters in the searching range are organized in a random manner to form the swarm. The position associated with each particle corresponds to a collection of parameter values to be determined. The particle's position is used to calculate a fitness value representing the quality of the particle to solve the problem under study. On the one hand, a particle is able to compare its current position with the best one by means of the value returned by the fitness function in order to save the parameter values associated with its best-known position. On the other hand, a particle is also able to obtain the best-known position of the complete swarm. After some iterations, the particles' position converges to the best-known position. In what follows, we describe in detail the relevant aspects of the particle swarm optimization method.

A. *Representation of the Particle*

In order to define the representation of the particle, the first task concerns the determination of the granular formalism $\boldsymbol{G}(.)$ being used as entries of the granular intuitionistic reciprocal preference relations. In this study, intervals are used and, thus, $\boldsymbol{G}(R) = \boldsymbol{P}(R)$, being the length of the intervals used as a synonym of the level of granularity. Here, $\boldsymbol{P}(.)$ denotes a family of intervals.

Then, a particle is modeled by a vector whose entries are appropriately placed in the unit interval, which corresponds to the search space. Considering a collection of n alternatives, a particle consists of $n \cdot (n-1)$ entries.

Let us suppose a decision maker who expresses an intuitionistic reciprocal preference relation. Considering an entry $r_{ij} = \left(\mu_{ij}, \nu_{ij}\right)$, $\mu_{ij}, \nu_{ij} \in [0, 1]$, and a level of granularity α, the interval of possible values of the entry $r_{ij} = \left(\mu_{ij}, \nu_{ij}\right)$ is computed as follows:

$$[d, e] = \left[\max\left(0, \mu_{ij} - \alpha/2\right), \min\left(1, \mu_{ij} + \alpha/2\right)\right] \tag{15}$$

$$[f, g] = \left[\max\left(0, \nu_{ij} - \alpha/2\right), \min\left(1, \nu_{ij} + \alpha/2\right)\right] \tag{16}$$

For the purpose of illustration, let us take an example in which $r_{ij} = (0.5, 0.4)$, $\alpha = 0.2$, and the related values of the particle x and y are 0.3 and 0.5, respectively. The corresponding intervals obtained using (15) and (16) are $[d, e] = [0.4, 0.6]$ and $[f, g] = [0.3, 0.5]$, respectively. Then, applying the expressions $\mu_{ij} = d + (e - d) \cdot x$ and $\nu_{ij} = f + (g - f) \cdot y$, we get $\mu_{ij} = 0.46$ and $\nu_{ij} = 0.40$, respectively.

B. *Fitness Function*

The fitness function evaluates the quality of a particle to solve the problem. In this study, we aim to maximize the consistency level of an intuitionistic reciprocal preference relation. Then, given a fixed level of granularity α, this task is carried out for all the potential interval-valued intuitionistic reciprocal preference relations. Hence, the fitness function f is defined as:

$$f = cl \tag{17}$$

C. *The Algorithm*

In the existing literature, we can find many variants of the particle swarm optimization method [31]. In this study, the generic form of the particle swarm optimization method is applied [25], in which the particles are controlled as follows:

$$\boldsymbol{v}(t+1) = \omega \times \boldsymbol{v}(t) + c_1 \times \boldsymbol{a} \cdot \left(\boldsymbol{z}_p(t) - \boldsymbol{x}(t)\right) + c_2 \times \boldsymbol{b} \cdot \left(\boldsymbol{z}_g(t) - \boldsymbol{x}(t)\right) \tag{18}$$

$$\boldsymbol{x}(t+1) = \boldsymbol{x}(t) + \boldsymbol{v}(t+1) \tag{19}$$

The particle is represented as $\boldsymbol{x}$ and its velocity as $\boldsymbol{v}$; "t" represents the index of the generation (iteration); c_1 and c_2 represent the individual factor and the societal factor (positive constants), respectively; $\cdot$ is a vector multiplication performed coordinate-wise; and $\boldsymbol{a}$ and $\boldsymbol{b}$ are vectors of random values computed from the uniform distribution in the unit interval. The best position achieved by the particle is represented and recorded as z_p while the best position overall and developed up to now across the swarm is represented as z_g. The inertia weight, ω, scales the actual velocity: a small inertia weight facilitates a local exploration while a large inertia weight facilitates a global one.

4 Experimental Example

This section aims to illustrate our method and to test its performance. To do so, let us suppose the following intuitionistic reciprocal preference relation expressed by a decision maker:

$$R = \begin{pmatrix} 0.50, 0.50 & 0.50, 0.30 & 0.50, 0.40 & 0.30, 0.50 \\ 0.30, 0.50 & 0.50, 0.50 & 0.50, 0.40 & 0.30, 0.40 \\ 0.40, 0.50 & 0.40, 0.50 & 0.50, 0.50 & 0.50, 0.40 \\ 0.50, 0.30 & 0.40, 0.30 & 0.40, 0.50 & 0.50, 0.50 \end{pmatrix}$$

According to the methodology shown in Sect. 2-C, the consistency level associated with this intuitionistic reciprocal preference relation is equal to 0.72. Note that, in this case, the entries of this intuitionistic reciprocal preference relation are plain numbers ($\alpha = 0$).

Before applying the proposed methodology, the particle swarm optimization method was set up with the following values: the swarm was composed of 50 particles, 100 iterations (generations) were run, ω was set to 0.2, and c_1 and c_2 were set as 2. We chose these values as they are usually used in the existing literature. In addition, the Hamming distance [32] was used as metric d in (12).

Figure 1 depicts the performance of the proposed methodology in relation to the values reported by the fitness function according to different values of the granularity level α. In particular, the values 0.2, 0.4, 0.6, and 0.8 are used. As it can be observed, the higher the granularity level, the higher the improvement in the consistency level associated with the intuitionistic reciprocal preference relation. It is due to the higher the granularity level, the higher the flexibility level introduced and, as a result, the likelihood of improving the consistency level is higher. In any case, the proposed methodology improves the consistency level associated with the original intuitionistic reciprocal preference relation, which is equal to 0.72.

On the other hand, the higher the granularity level α, the more different the values associated with the entries of the intuitionistic reciprocal preference relation in comparison with the first ones. This should be considered because if the values

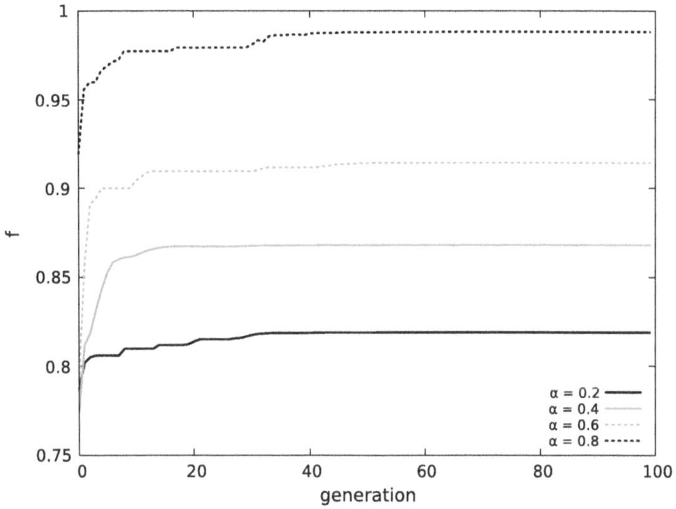

Fig. 1 Progression of f in successive generations for different α values

are very different in comparison with the original ones, the decision maker might reject them. To clarify this issue, we show the new intuitionistic reciprocal preference relations obtained by the proposed methodology in relation to the granularity levels 0.2 and 0.8, respectively (the are reported with two decimal points):

$$R_{0.2} = \begin{pmatrix} 0.50, 0.50 & 0.42, 0.24 & 0.43, 0.33 & 0.23, 0.43 \\ 0.24, 0.42 & 0.50, 0.50 & 0.40, 0.33 & 0.20, 0.32 \\ 0.33, 0.43 & 0.33, 0.40 & 0.50, 0.50 & 0.40, 0.32 \\ 0.43, 0.23 & 0.32, 0.20 & 0.32, 0.40 & 0.50, 0.50 \end{pmatrix}$$

$$R_{0.8} = \begin{pmatrix} 0.50, 0.50 & 0.15, 0.22 & 0.13, 0.21 & 0.20, 0.21 \\ 0.22, 0.15 & 0.50, 0.50 & 0.12, 0.20 & 0.10, 0.15 \\ 0.21, 0.13 & 0.20, 0.12 & 0.50, 0.50 & 0.11, 0.21 \\ 0.21, 0.20 & 0.15, 0.10 & 0.21, 0.11 & 0.50, 0.50 \end{pmatrix}$$

5 Conclusions

In this study, we have proposed a granular computing-based approach for improving the consistency of intuitionistic reciprocal preference relations. This method characterizes the consistency by means of the multiplicative transitivity property and uses the information granularity as an important asset to maximize the level of consistency associated with an intuitionistic reciprocal preference relation. Furthermore, the particle swarm optimization method has been applied to solve this maximization task.

The consistency itself implicates two matters: (i) when a decision maker is individually considered consistent, and (ii) when a group of them is considered consistent, which is usually referred to as consensus [33–35]. Hence, we propose as a future study the use of the consensus as a complementary parameter of the optimization criterion.

Acknowledgements The authors would like to thank FEDER financial support from the Project TIN2016-75850-R.

References

1. Y. Zhang, Z.S. Xu, H. Liao, A consensus process for group decision making with probabilistic linguistic preference relations. Inf. Sci. **414**, 260–275 (2017)
2. I. Millet, The effectiveness of alternative preference elicitation methods in the analytic hierarchy process. J. Multi-Criteria Decis. Anal. **6**(1), 41–51 (1997)
3. P. Perny, M. Roubens, Fuzzy preference modelling, in *Fuzzy Sets in Decision Analysis, Operations Research and Statistics*, vol. I, ed. by R. Slowinski (Kluwer Academic Publishers, 1998), pp. 3–30
4. J.C. Bezdek, B. Spillman, R. Spillman, A fuzzy relation space for group decision theory. Fuzzy Sets Syst. **1**(4), 255–268 (1978)
5. T. Tanino, Fuzzy preference orderings in group decision making. Fuzzy Sets Syst. **12**(2), 117–131 (1984)
6. J. Fodor, M. Roubens, *Fuzzy Preference Modelling and Multicriteria Decision Support* (Kluwer Academic Publishers, 1994)
7. Z.S. Xu, A survey of preference relations. Int. J. Gen Syst **27**(36), 179–203 (2007)
8. K.T. Atanassov, Intuitionistic fuzzy sets. Fuzzy Sets Syst. **20**(1), 87–96 (1986)
9. G. Deschrijver, E.E. Kerre, On the relationship between some extensions of fuzzy set theory. Fuzzy Sets Syst. **133**(2), 227–235 (2003)
10. Z.S. Xu, Intuitionistic preference relations and their application in group decision making. Inf. Sci. **177**(11), 2363–2379 (2007)
11. F. Chiclana, E. Herrera-Viedma, S. Alonso, F. Herrera, Cardinal consistency of reciprocal preference relations: a characterization of multiplicative transitivity. IEEE Trans. Fuzzy Syst. **17**(1), 14–23 (2009)
12. E. Herrera-Viedma, F. Herrera, F. Chiclana, M. Luque, Some issues on consistency of fuzzy preference relations. Eur. J. Oper. Res. **154**(1), 98–109 (2004)
13. J. Wu, F. Chiclana, Multiplicative consistency of intuitionistic reciprocal preference relations and its application to missing values estimation and consensus building. Knowledge-Based Syst. **71**, 187–200 (2014)
14. F.J. Cabrerizo, R. Ureña, W. Pedrycz, E. Herrera-Viedma, Building consensus in group decision making with an allocation of information granularity. Fuzzy Sets Syst. **255**, 115–127 (2014)
15. W. Pedrycz, From logic descriptors to granular logic descriptors: a study in allocation of information granularity. J. Ambient Intell. Humaniz. Comput. **4**(4), 411–419 (2013)
16. E. Kim, W. Pedrycz, Information granulation as a basis of fuzzy modeling. J. Intell. Fuzzy Syst. **18**(2), 123–148 (2007)
17. V. Cutello, J. Montero, Fuzzy rationality measures. Fuzzy Sets Syst. **62**(1), 39–54 (1994)
18. L.A. Zadeh, The concept of a linguistic variable and its application to approximate reasoning-I. Inf. Sci. **8**(3), 199–249 (1975)

19. L.A. Zadeh, Fuzzy sets. Inf. Control **8**(3), 338–353 (1965)
20. Y. Xu, R. Patnayakuni, H. Wang, The ordinal consistency of a fuzzy preference relation. Inf. Sci. **224**, 152–164 (2013)
21. F.Y. Meng, Q.X. An, C.Q. Tan, X.H. Chen, An approach for group decision making with interval fuzzy preference relations based on additive consistency and consensus analysis. IEEE Trans. Syst. Man Cybern.-Syst. **47**(8), 2069–2082 (2017)
22. X. Zhu, W. Pedrycz, Z. Li, Granular data description: designing ellipsoidal information granules. IEEE Trans. Cybern. **47**(12), 4475–4484 (2017)
23. W. Pedrycz, M. Song, Analytic hierarchy process (AHP) in group decision making and its optimization with an allocation of information granularity. IEEE Trans. Fuzzy Syst. **19**(3), 439–527 (2011)
24. W. Lu, L. Zhang, W. Pedrycz, J. Yang, X. Liu, The granular extension of Sugeno-type fuzzy models based on optimal allocation of information granularity and its application to forecasting of time series. Appl. Soft Comput. **42**, 38–52 (2016)
25. J. Kennedy, R.C. Eberhart, Particle swarm optimization, in *Proceedings of IEEE International Conference on Neural Networks*, vol. 4 (IEEE Press, NJ, 1995), pp. 1942–1948
26. M.R. Bonyadi, Z. Michalewicz, Stability analysis of the particle swarm optimization without stagnation assumption. IEEE Trans. Evol. Comput. **20**(5), 814–819 (2016)
27. Y. Li, X. Bai, L. Jiao, Y. Xue, Partitioned-cooperative quantum-behaved particle swarm optimization based on multilevel thresholding applied to medical image segmentation. Appl. Soft Comput. **56**, 345–356 (2017)
28. M. Akbari, H. Rashidi, S.H. Alizadeh, An enhanced genetic algorithm with new operators for task scheduling in heterogeneous computing systems. Eng. Appl. Artif. Intel. **61**, 35–46 (2017)
29. D. Bucur, G. Iacca, M. Gaudesi, G. Squillero, A. Tonda, Optimizing groups of colluding strong attackers in mobile urban communication networks with evolutionary algorithms. Appl. Soft Comput. **40**, 416–426 (2016)
30. Y. Lin, Z. Bian, X. Liu, Developing a dynamic neighborhood structure for an adaptive hybrid simulated annealing – tabu search algorithm to solve the symmetrical traveling salesman problem. Appl. Soft Comput. **49**, 937–952 (2016)
31. Y. del Valle, G.K. Venayagamoorthy, S. Mohagheghi, J.-C. Hernandez, R.G. Harley, Particle swarm optimization: Basic concepts, variants and applications in power systems. IEEE Trans. Evol. Comput. **12**(2), 171–195 (2008)
32. Y. Yang, F. Chiclana, Consistency of 2D and 3D distances of intuitionistic fuzzy sets. Expert Syst. Appl. **39**(10), 8665–8670 (2012)
33. E. Herrera-Viedma, F.J. Cabrerizo, J. Kacprzyk, W. Pedrycz, A review of soft consensus models in a fuzzy environment. Inform. Fusion **17**, 4–13 (2014)
34. Y. Dong, Z.P. Fan, S. Yu, Consensus building in a local context for the AHP-GDM with the individual numerical scale and prioritization method. IEEE Trans. Fuzzy Syst. **23**(2), 354–368 (2015)
35. F.J. Cabrerizo, J.A. Morente-Molinera, W. Pedrycz, A. Taghavi, E. Herrera-Viedma, Granulating linguistic information in decision making under consensus and consistency. Expert Syst. Appl. **99**, 83–93 (2018)

Method for Proactive Quality Control of Aviation Security of Airports Using Inverse Interval Weighted Average

Dan Tavrov, Olena Temnikova, and Volodymyr Temnikov

Abstract Level of aviation security (AS) of an airport depends on the state of officers of the airport security service (SS) and engineering systems and devices used by them during passengers and hand luggage inspection, baggage inspection, passenger pass control, etc. In practice, it is important to determine, which factors influence the overall level of quality of the AS the most. Due to the subjective nature of the notion of quality, the AS needs to be modeled using linguistic concepts. In this paper, we propose to model the AS using a specific instance of a perceptual computer. Using this model, it is possible to perform proactive quality control of AS by setting the desired target level of the overall quality and relative weights of each input factor, and then obtaining the values for each factor. This approach can be considered as inverse to the linguistic weighted average used in perceptual computers to obtain the output. In the paper, we introduce the method for proactive quality control of airport AS in a simpler case when the linguistic concepts in the perceptual computer are modeled using intervals, and propose a method for solving the inverse interval weighted average problem.

D. Tavrov (✉) · O. Temnikova
Applied Mathematics Department, Igor Sikorsky Kyiv Polytechnic Institute, Kiev, Ukraine
e-mail: dtavrov@kse.org.ua

O. Temnikova
e-mail: temnikova_elena@ukr.net

D. Tavrov
Kyiv School of Economics, Kyiv, Ukraine

V. Temnikov
Institute of Information and Diagnostic Systems, National Aviation University, Kiev, Ukraine
e-mail: temnikov_v@ukr.net

S. N. Shahbazova et al. (eds.), *Recent Developments and the New Direction in Soft-Computing Foundations and Applications*, Studies in Fuzziness and Soft Computing 393, https://doi.org/10.1007/978-3-030-47124-8_38

1 Introduction

According to various analytical materials, aviation security (AS) in many countries is unsatisfactory [1, 2], which is manifested, among other things, in failures to prevent terror attacks. According to [3–5], providing AS nowadays requires that a collection of measures be applied. Such measures include organizational, technological measures, and measures aimed at reducing the human factor.

A key role in providing AS of airports is played by the officers of the airport security service (SS), including their actions in protecting passengers, airport workers, aircrafts, baggage, and information from the acts of unlawful interference. In its own turn, efficiency of the SS depends on the quality of organization of their work, equipment, and, to a great extent, functional state of its officers [6]. In practice, it is important to determine, which factors influence the overall quality of the AS the most, and therefore to (proactively) adjust their quality in order to reach the desired output level of overall AS quality.

International standard [7] defines quality as the behavior, attitudes, activities, and processes that deliver value through fulfilling the needs and expectations of customers and other relevant interested parties. This definition presumes subjective nature of the concept of quality, which is why it is natural to assess the level of quality using words rather than (arbitrary) numbers. As a result, it is necessary to create a model of the overall AS that is based on processing linguistic information instead of numbers, and then use this model in the backward direction to determine the input factors whose quality needs to be adjusted to achieve the overall desired quality of AS.

In this paper, we propose to use perceptual computer (Per-C) as such a model of AS, which is described in Sect. 2. Due to complicated nature of inference mechanisms in this model, we limit ourselves to discussing only the simplest case, in which all the words are modeled using intervals. Then, it is possible to obtain the intervals for quality levels of input factors by solving the inverse interval weighted average problem that we introduce in Sect. 3. We conclude the paper with a simple numerical example given in Sect. 4 that illustrates the application of the proposed approach.

2 Perceptual Computer Based Model of Aviation Security

A. *General Architecture of a Perceptual Computer*

Perceptual computing is a realization of computing with words (CWW) [8], in which objects of computations are words and propositions expressed in a natural language. The usage of a Per-C is natural because CWW is used for making subjective judgments [9], which were identified above as the main kinds of judgments involved in assessing the quality of AS.

Every Per-C consists of three components [10]:

(1) *Encoder*: This component takes words as inputs and transforms them into their type-2 fuzzy set (T2FS) representation. A collection of words used in Per-C, along with associated T2FS representations, is called a codebook.
(2) *CWW Engine*: This component processes T2FSs and outputs one or more other T2FSs. One of the CWW engines widely used in practice is the linguistic weighted average (LWA) [11, 12].
(3) *Decoder*: This component transforms the output of the CWW engine into the output of the Per-C, which can be one of the following: a word from the codebook most similar to the CWW engine output, a rank of several alternatives, or a class.

B. *Perceptual Computer for Determining the Quality of Aviation Security of Airports*

The decision making process in the Per-C in the context of AS is presented in Fig. 1. This process is hierarchical and distributed, in that it is made [10] based on aggregating independently made expert evaluations of input factors of AS (depicted as rectangles in Fig. 1).

Nodes in the graph in Fig. 1 represent various factors influencing the quality of the AS of the airport. Factors are grouped by levels of aggregation, whose number in general can be relatively high (up to 6). In Fig. 1, three levels of aggregation are shown.

The output node S represents the overall quality of the AS of the airport. The input nodes A–K stand for factors that characterize functional state of SS officers [13, 14] and technical conditions of equipment used by SS (security systems and equipment used for passengers and hand luggage inspection, baggage inspection, passenger pass control, etc.).

Intermediate levels of the Per-C consist of nodes L–P that represent the quality of preflight inspection of passengers and hand luggage, quality of passenger pass control, quality of preflight baggage inspection, quality of organization and provision of intrasite mode, and quality of the organization and provision of the access regime of the employees of the aviation enterprise. These nodes are grouped in the aggregate nodes Q (quality of preflight control of passengers and luggage) and R (quality of organization and security of controlled areas and aircraft), which can be viewed as the two main (independent) areas of airport security.

Nodes in Fig. 1 are connected with arcs that are weighted by words assigned by experts.

The model presented in Fig. 1 should be taken with a grain of salt, as it is possible to include or exclude security factors on each level depending on the real circumstances. In this paper, we present the generic version that can be applied to a wide selection of real cases.

In this paper, in order to simplify the calculations, we discuss the simplest scenario when the target value of the overall quality of AS, levels of quality of each factor, and weights of connecting arcs are intervals.

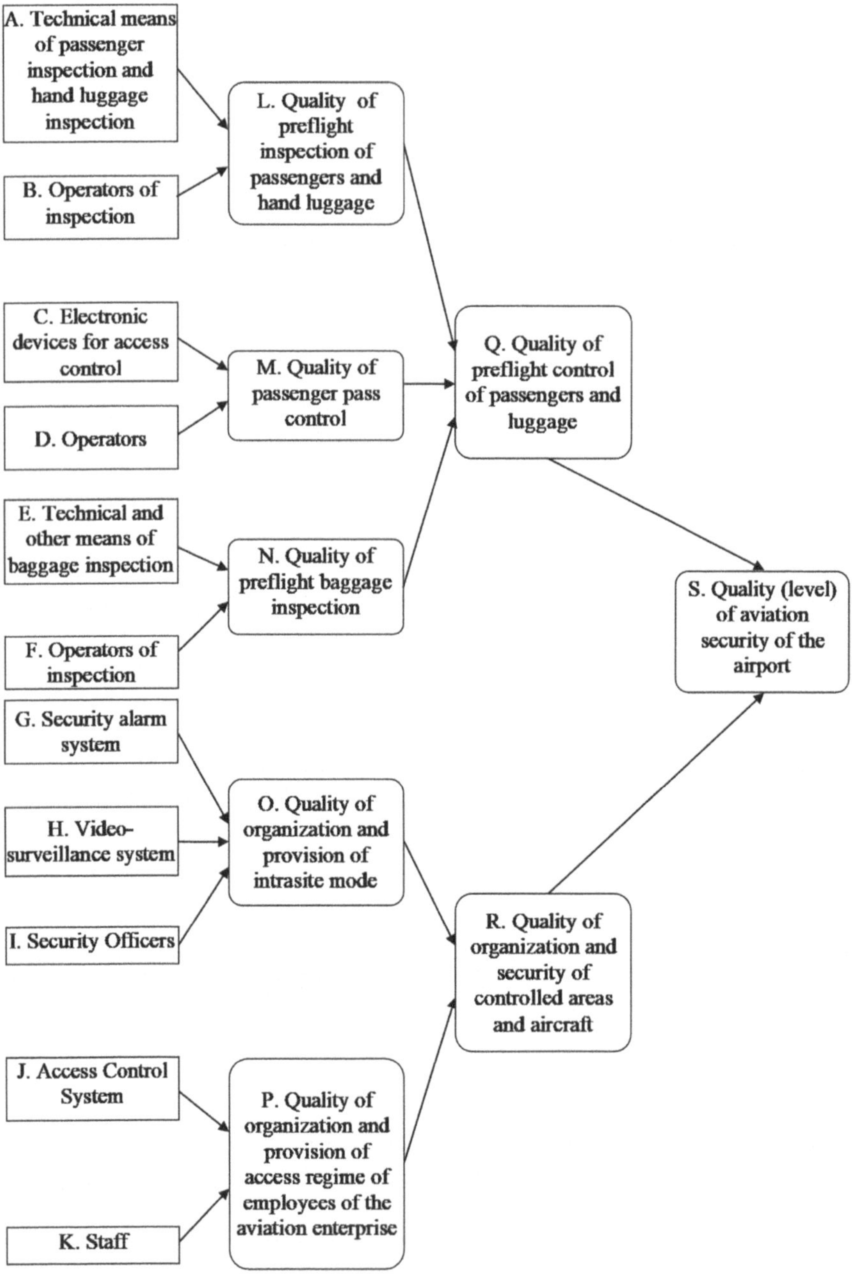

Fig. 1 Decision making process in the Per-C for AS of airports

3 Proactive Quality Control Using Inverse Interval Weighted Average

A. *Interval Weighted Average*

Let us be given a collection of intervals $x_i = [a_i, b_i] \equiv X_i$, $a_i, b_i \in [0, 1]$, and a collection of weights $w_i = [c_i, d_i] \equiv W_i$, $c_i, d_i \in [0, 1]$, $i = 1, \ldots, n$. Then, the *interval weighted average* (IWA) [10] is defined as the interval

$$y_{IWA} \equiv [y_l, y_r] = \frac{\sum_{i=1}^{n} X_i W_i}{\sum_{i=1}^{n} W_i}. \tag{1}$$

From the computational point of view, left and right ends of the interval (1) can be obtained using the following formulas:

$$y_l = \frac{\sum_{i=1}^{L} a_i d_i + \sum_{i=L+1}^{n} a_i c_i}{\sum_{i=1}^{L} d_i + \sum_{i=L+1}^{n} c_i},\ y_r = \frac{\sum_{i=1}^{R} b_i c_i + \sum_{i=R+1}^{n} b_i d_i}{\sum_{i=1}^{R} c_i + \sum_{i=R+1}^{n} d_i}, \tag{2}$$

where $L \in [1, n-1]$ and $R \in [1, n-1]$ are *switch points* that can be found, e.g., using the enhanced Karnik-Mendel algorithms (EKM) described in [15]. Before calculating y_l (y_r) using (2), all a_i (b_i) must be sorted in ascending order first.

B. *Inverse Interval Weighted Average*

Let us discuss the problem of determining such intervals X_i, $i = 1, \ldots, n$, that their substitution in (2) yields some prespecified output interval $[\hat{y}_l, \hat{y}_r]$ given some weights $\hat{W}_i = \left[\hat{c}_i, \hat{d}_i\right]$, $i = 1, \ldots, n$. Let $\hat{L}$ and $\hat{R}$ be some switch points. Then, we can rewrite (2) as follows:

$$\sum_{i=1}^{\hat{L}} a_i d_i + \sum_{i=\hat{L}+1}^{n} a_i c_i = w_{\hat{L}},\quad \sum_{i=1}^{\hat{R}} b_i c_i + \sum_{i=\hat{R}+1}^{n} b_i d_i = w_{\hat{R}}, \tag{3}$$

where $w_{\hat{L}} = \hat{y}_l \cdot \left(\sum_{i=1}^{\hat{L}} d_i + \sum_{i=\hat{L}+1}^{n} c_i\right)$, $w_{\hat{R}} = \hat{y}_r \cdot \left(\sum_{i=1}^{\hat{R}} c_i + \sum_{i=\hat{R}+1}^{n} d_i\right)$.

Obviously, for any given choice of $\left[\hat{c}_i, \hat{d}_i\right]$, $[\hat{y}_l, \hat{y}_r]$, $\hat{L}$, and $\hat{R}$, there is an infinite choice of $[a_i, b_i]$ that satisfy (3). Therefore, in practice, it is necessary to define some function of a_i and b_i, whose point of minimum can be considered as the feasible choice of intervals, given the constraint (3). In this paper, in the context of the problem discussed above, we propose to use the following function:

$$f(\mathbf{a}, \mathbf{b}) = \sum_{i=1}^{n} C_i \left(\beta_i (b_i - a_i)^2 + (1 - \beta_i)\left(\frac{a_i + b_i}{2}\right)^2\right), \tag{4}$$

where $\mathbf{a} = (a_1, \ldots, a_n)$, $\mathbf{b} = (b_1, \ldots, b_n)$, $C_i \in [0, 1]$ are the relative costs associated with each security factor, $\beta_i \in [0, 1]$ are the coefficients that show the preferences towards interval length (which can be thought of as the certainty of our choice) as opposed to the average target value that is represented by the interval. In other words, the higher the β_i, the more we are interested in obtaining narrow interval estimates of the target intervals for factor i, even if the center of this interval will be high. Relative costs C_i represent how expensive it is to achieve the target value of factor i.

For the point of minimum of (4) to be considered a valid collection of intervals for our problem, we need to impose the following additional inequality constraints:

$$\begin{aligned} &a_i \geq 0, \quad b_i \leq 1, \quad a_i \leq b_i, \\ &a_1 \leq a_2 \leq \ldots \leq a_n, \quad b_1 \leq b_2 \leq \ldots \leq b_n. \end{aligned} \tag{5}$$

Then, the problem to be solved can be stated as follows: minimize (4) subject to inequality constraints (5) and equality constraints (3) for *every possible combination* of $\hat{L}$ and $\hat{R}$, and select the intervals $\left[a_i^*, b_i^*\right]$ that yields the minimum value of (4) among all choices.

A word of caution is in order. In (5), it is required that a_i and b_i be ordered in a nondecreasing sequence, because otherwise application of EKM algorithms or similar algorithms will yield different results, as the first step of such algorithms would be to sort a_i and b_i in ascending order. However, introduction of this constraint in (5) necessarily restricts the possible solution set. To overcome this obstacle, the problem needs to be solved *for every possible order of* $\left[\hat{c}_i, \hat{d}_i\right]$, and then the collection of intervals $\left[a_i^*, b_i^*\right]$ that yields the minimum value of (4) among all choices must be selected.

C. *Method for Proactive Quality Control Using Inverse Interval Weighted Average*

The main idea of proactive quality control is to determine the factors that have the most significant impact on the overall quality of AS, and then apply appropriate measures to increase their quality to the levels that are given by solving the inverse interval weighted average problem.

The method consists of the following stages:

(1) *Expert Evaluation*: Define with the help of the experts the target value $\left[\hat{y}_l, \hat{y}_r\right]$ of the overall quality of AS, relative costs C_i and coefficients β_i from (4) for each factor i, $i = 1, \ldots, n$.
(2) *Factor Determination*: Solve the inverse interval weighted average problem for all the nodes in the architecture of the Per-C that models the AS for a given airport, and obtain the intervals $[a_i, b_i]$ for the levels of quality of each security factor i, $i = 1, \ldots, n$, that must be attained to achieve the overall quality of AS equal to $\left[\hat{y}_l, \hat{y}_r\right]$.
(3) *Quality Adjustment*: Increase the quality of those security factors whose current assessed level of quality q_i lies outside $[a_i, b_i]$. Detailed discussion of how this can be done lies beyond the scope of this paper.

4 Numerical Example

In order to illustrate the application of the proposed method, we will discuss a numerical example involving model numbers given in Tables 1 and 2.

Suppose that the target output level of the overall quality of AS is $[\hat{y}_l, \hat{y}_r] = [0.7; 1.0]$. Then, to obtain levels of qualities for nodes Q and R from Fig. 1, we need to solve two optimization problems:

(1) The one when $a_Q \leq a_R, b_Q \leq b_R$.
(2) The one when $a_R \leq a_Q, b_R \leq b_Q$.

Since in this case the number of intervals being averaged is 2, switch points can be only $L = R = 1$.

The objective function for both problems is

$$f(a_Q, a_R, b_Q, b_R) = 0.85\left(0.8(b_Q - a_Q)^2 + 0.2\left(\frac{a_Q + b_Q}{2}\right)^2\right) + \\ + 0.85\left(0.7(b_R - a_R)^2 + 0.3\left(\frac{a_R + b_R}{2}\right)^2\right).$$

The inequality constraints for both problems are

Table 1 Arc weights

Arc label	Weight
AL	[0.75; 0.90]
BL	[0.65; 0.80]
CM	[0.55; 0.70]
DM	[0.30; 0.50]
EN	[0.70; 0.90]
FN	[0.55; 0.80]
GO	[0.50; 0.70]
HO	[0.75; 0.90]
IO	[0.80; 0.95]
JP	[0.75; 0.95]
KP	[0.65; 0.85]
LQ	[0.70; 0.90]
MQ	[0.40; 0.50]
NQ	[0.60; 0.80]
OR	[0.70; 0.90]
PR	[0.60; 0.80]
QS	[0.50; 0.70]
RS	[0.60; 0.80]

Table 2 Parameters of the inverse interval weighted average problem for the example

Node i label	Cost C_i	Coefficient β_i
A. Technical means of passenger inspection and hand luggage inspection	0.85	0.90
B. Operators of inspection	0.40	0.70
C. Electronic devices for access control	0.55	0.80
D. Operators	0.40	0.60
E. Technical and other means of baggage inspection	0.85	0.80
F. Operators of inspection	0.40	0.60
G. Security alarm system	0.55	0.60
H. Video surveillance system	0.55	0.60
I. Security officers	0.40	0.50
J. Access control system	0.55	0.70
K. Staff	0.40	0.60
L. Quality of preflight inspection of passengers and hand luggage	0.85	0.80
M. Quality of passenger pass control	0.55	0.40
N. Quality of preflight baggage inspection	0.55	0.60
O. Quality of organization and provision of intrasite mode	0.55	0.70
P. Quality of the organization and provision of the access regime of the employees of the aviation enterprise	0.55	0.60
Q. Quality of preflight control of passengers and luggage	0.85	0.80
R. Quality of organization and security of controlled areas and aircraft	0.85	0.70

$$
\begin{aligned}
&a_Q \geq 0, \quad a_R \geq 0, \quad b_Q \leq 1, \quad b_R \leq 1, \\
&a_Q \leq b_Q, \quad a_R \leq b_R.
\end{aligned}
$$

Additional inequality constraints for problem 1 are $a_Q \leq a_R, b_Q \leq b_R$, additional inequality constraints for problem 2 are $a_R \leq a_Q, b_R \leq b_Q$.

Equality constraints for problem 1 are

$$
\begin{aligned}
0.80a_Q + 0.85a_R &= 0.70 \cdot (0.80 + 0.85), \\
0.85b_Q + 0.70b_R &= 1.00 \cdot (0.85 + 0.70).
\end{aligned}
$$

Equality constraints for problem 2 are

$$
\begin{aligned}
0.70a_R + 0.85a_Q &= 0.70 \cdot (0.70 + 0.85), \\
0.85b_R + 0.80b_Q &= 1.00 \cdot (0.85 + 0.80).
\end{aligned}
$$

Solutions to both problems were obtained using the `fmincon` routine in MathWorks® MATLAB. For problem 1, optimal intervals are $[a_Q, b_Q] = [0.7; 1.0]$,

Table 3 Solutions to the inverse interval weighted average problem for the example

Node i label	Resulting interval
A. Technical means of passenger inspection and hand luggage inspection	[0.982; 1.000]
B. Operators of inspection	[0.903; 1.000]
C. Electronic devices for access control	[0.684; 1.000]
D. Operators	[0.413; 1.000]
E. Technical and other means of baggage inspection	[0.890; 1.000]
F. Operators of inspection	[0.738; 1.000]
G. Security alarm system	[0.662 1.000]
H. Video surveillance system	[0.636; 1.000]
I. Security officers	[0.447; 1.000]
J. Access control system	[0.815; 1.000]
K. Staff	[0.729; 1.000]
L. Quality of preflight inspection of passengers and hand luggage	[0.941; 1.000]
M. Quality of passenger pass control	[0.555; 1.000]
N. Quality of preflight baggage inspection	[0.809; 1.000]
O. Quality of organization and provision of intrasite mode	[0.560; 1.000]
P. Quality of the organization and provision of the access regime of the employees of the aviation enterprise	[0.770; 1.000]
Q. Quality of preflight control of passengers and luggage	[0.790; 1.000]
R. Quality of organization and security of controlled areas and aircraft	[0.644; 1.000]

$[a_R, b_R] = [0.7; 1.0]$, and the value of the objective function is 0.422. For problem 2, optimal intervals are $\left[a_Q, b_Q\right] = [0.790; 1.000]$, $[a_R, b_R] = [0.644; 1.000]$, and the value of the objective function is 0.404, which is lower than above, so these intervals are retained and used as target values for inverse interval weighted average problems for calculating intervals for nodes L–N, O–P, and so on until all the intervals are calculated. The final results are given in Table 3.

As we can see, significant influence on the overall quality of the AS of the airport in this example is exercised by the quality of preflight inspection of passengers and hand luggage (left end of the interval is 0.941) and the quality of preflight baggage inspection (left end is 0.809). Such high requirements can be explained by the costs necessary to increase the functional state of inspectors (B), as well as costs and levels of confidence that experts have in the quality of technical equipment used for inspection (A, E).

These observations do not mean that it is not necessary to pay attention to other factors influencing the overall quality level of the AS of the airport, because high level of airport security can be achieved only by carrying out all the necessary measures in all directions at a specified level (which is lower for some factors and higher for others).

5 Conclusions

In the paper, we introduced a method for proactive quality control of aviation security of airports that is based on the perceptual computer model of the aviation security. Unlike in the regular application of Per-C when the output is determined by applying CWW engine to its inputs, we set the target output value of the overall quality of AS, and attempt to determine the levels of quality of input factors that must be attained to achieve this output.

In order to simplify the calculations, we assumed that the words in the Per-C are modeled using intervals instead of type-2 fuzzy sets. This enabled us to state the inverse interval weighted average problem and propose a method of solving it.

Obvious direction for improving the method is to replace intervals with proper T2FS models of the words in the Per-C, which entails formulating and solving the inverse linguistic weighted average problem.

References

1. *Safety Report* (International Civil Aviation Organization, Published in Montréal, Canada, 2017)
2. *Safety Review* (State Aviation Administration of Ukraine, 2016 (in Ukrainian))
3. *Doc 9859: Safety Management Manual,* 3rd edn. (International Civil Aviation Organization, 2013)
4. *Doc 8973: Aviation Security Manual*, 9th edn. (International Civil Aviation Organization, 2014)
5. *Annex 17: Security: Safeguarding International Civil Aviation Against Acts of Unlawful Interference* (International Civil Aviation Organization, 2017)
6. V.M. Mihaylenko, V.A. Temnikov, A.V. Temnikov, Principles of constructing support systems for operational decision-making in the air traffic control tower. Sci. Educ. New Dime. Nat. Tech. Sci. **15**(140), 33–35 (2017)
7. *ISO/FDIS 9000:2015*, Quality Management Systems, Fundamentals and Vocabulary, 2015
8. L.A. Zadeh, From computing with numbers to computing with words—from manipulation of measurements to manipulation of perceptions. IEEE Trans. Circuits Syst. I: Fundam. Theory Appl. **46**(1), 105–119 (1999)
9. J.M. Mendel, The perceptual computer: an architecture for computing with words, in *Proceedings of Modeling with Words Workshop in the Proceedings of FUZZ-IEEE* (2001), pp. 35–38
10. J.M. Mendel, D. Wu, *Perceptual Computing. Aiding People in Making Subjective Judgments* (Wiley, Hoboken, New Jersey, 2010)
11. D. Wu, J.M. Mendel, Aggregation using the linguistic weighted average and interval type-2 fuzzy sets. IEEE Trans. Fuzzy Syst. **15**(6), 1145–1161 (2007)
12. D. Wu, J.M. Mendel, Corrections to 'Aggregation using the linguistic weighted average and interval type-2 fuzzy sets'. IEEE Trans. Fuzzy Syst. **16**(6), 1664–1666 (2008)
13. V.A. Temnikov, Principles of automatic intra-shift control of operators' access to information systems resources, Bull. Volodymyr Dahl East Ukrainian Nat. Univ. **8**(179), 184–190 (2012) (in Russian)
14. R.M. Bayevsky, A.P. Berseneva, *Assessment of the Adaptive Capabilities of the Body and the Risk of Developing Diseases* (Medicine, Moscow, 1997) (in Russian)
15. D. Wu, J.M. Mendel, Enhanced Karnik-Mendel algorithms. IEEE Trans. Fuzzy Syst. **17**(4), 923–934 (2009)

Towards Automatic Textual Summarization of Movies

Chang Liu, Mark Last, and Armin Shmilovici

Abstract With the rapidly increasing number of online video resources, the ability of automatically understanding those videos becomes more and more important, since it is almost impossible for people to watch all of the videos and provide textual descriptions. The duration of online videos varies in a extremely wide range, from several seconds to more than 5 h. In this paper, we focus on long videos, especially on full-length movies, and propose the first pipeline for automatically generating textual summaries of such movies. The proposed system takes an entire movie as input (including subtitles), splits it into scenes, generates a one-sentence description for each scene and summarizes those descriptions and subtitles into a final summary. In our initial experiment on a popular cinema movie (*Forrest Gump*), we utilize several existing algorithms and software tools for implementing the different components of our system. Most importantly, we use the S2VT (Sequence to Sequence—Video to Text) algorithm for scene description generation and MUSEEC (MUltilingual SEntence Extraction and Compression) for extractive text summarization. We present preliminary results from our prototype experimental framework. An evaluation of the resulting textual summaries for a movie made of 156 scenes demonstrates the feasibility of the approach—the summary contains the descriptions of three out of the four most important scenes/storylines in the movie. Although the summaries are far from satisfactory, we argue that the current results can be used to prove the merit of our approach.

C. Liu (✉) · M. Last · A. Shmilovici
Department of Software and Information Systems Engineering,
Ben-Gurion University of the Negev, Beer Sheva, Israel
e-mail: liuc@post.bgu.ac.il

M. Last
e-mail: mlast@bgu.ac.il

A. Shmilovici
e-mail: armin@bgu.ac.il

S. N. Shahbazova et al. (eds.), *Recent Developments and the New Direction in Soft-Computing Foundations and Applications*, Studies in Fuzziness and Soft Computing 393, https://doi.org/10.1007/978-3-030-47124-8_39

1 Introduction

With the rapid development of Internet and mobile devices, making videos becomes more and more popular and convenient and the number of uploaded videos on websites such as YouTube has been increasing at astonishing speed. As a typical challenge in the Artificial Intelligent (AI) domain, video understanding, especially video description generation has attracted much research interest over the past two decades. Generating descriptions of either images or videos is challenging since it requires not only understanding of the visual elements but also the ability to concisely express the interaction between the visual elements as an interesting "story". Recently, with the increasing attempts of combining visual and textual information, and the development of deep neural networks, the task of description generation enjoys a great number of opportunities. Most of the current approaches towards describing videos are aimed at short video clips collected from YouTube with the length varying from 10 s to 1 min [1–4]. Other movie description works focus on segmented movie clips [5–7]. Longer videos with dozens of minutes duration or even full length cinema movies are much more difficult to describe, because the visual elements and actions have much more complex relationships, which tell a more complex story. As currently almost no research is geared towards long videos, we focus our efforts on summarizing long videos, especially full-length cinema movies, in natural language sentences. In this work, we first split the input movies into scenes based on frame change and take use of S2VT (Sequence to Sequence—Video to Text) [4], a state-of-the-art video description generation algorithm to generate descriptions for each scene. Then we apply MUSEEC (MUltilingual SEntence Extraction and Compression) [8], a tool implementing several single-document summarization algorithms, to summarize the scene descriptions and obtain a textual summary of the movie.

The main contributions of this paper are:

a. The first attempt to design a pipeline for automatic generation of full-length cinema movie summaries in textual format.
b. Proposing an experimental framework for evaluating textual summaries of full-length movies.

The rest of the paper is organized as follows: Sect. 2 describes the related work; Sect. 3 presents the software architecture; Sect. 4 presents the experiment which are evaluated in Sect. 5, and Sect. 6 concludes with a discussion.

2 Related Work

2.1 Video Description Generation

The Video Description Generation task produces natural language sentences as output for a given input video. There are many efforts towards describing videos in natural language, such as describing GIF [9] (a sequence of very few frames presenting a very simple action), real life videos with a variety of actions [1, 3, 4], long duration surveillance videos [10], and video clips from commercial movies [5–7]. The work of Venugopalan et. al [4], known as S2VT, has been regarded as the state-of-the-art method for video description generation. It has been evaluated on several different datasets such as Microsoft Video Description Corpus (MSVD) [11], MPII-MD [7] and M-VAD [12] and achieved acceptable results. More recently, the Hierarchical & Multimodal Video Caption (HMVC) model [13] outperformed the S2VT model on the same three datasets, and achieved competitive results on another new dataset for 2016 MSR Video to Text challenge (MSR-VTT) [14].

The current video description methods mostly work on video or movie clips, and generate very simple sentence descriptions such as "*A man is swimming in a pool*". There are works of generating multi-sentence descriptions for daily life videos with minutes length [15] or long duration surveillance videos [10], but it is still not the main stream in the video description research community. This is the main gap between the existing research and the goal of our work, which is generating textual story-like summaries for entire movies.

2.2 *Video Summarization*

Video summarization aimed at generating short version of input video in order to include the key events, specifically, this task selects a subset of the original video's frames which represent its most important content. In general, the current techniques for automatic video summarization fall into two categories based on the learning strategies: unsupervised [16, 17] and supervised [18, 19]. Most of the current works are applied to videos with length of one to five minutes, and the following two benchmark datasets are popular among researchers: SumMe [20] and TVSum [21]. SumMe consists of 25 user videos of various interesting events such as holiday or sports, with their human created summaries and TVSum consists of 50 videos from YouTube, with their crowdsourcing annotated shot-level importance scores.

As a branch of video summarization, the task of movie summarization focuses on cinema movies and generates a short version of a given movie, which is composed of several movie clips with important content extracted from the original movie [22, 23]. It can help with the applications such as movie retrieval or movie preview. This task has not attracted as much research interest as video summarization because of its high complexity.

2.3 *Text Summarization*

Text summarization is a task of generating summaries of textual documents, which has gained significant improvements [24] during its almost sixty-year history [25].

Classified by the source of words in the summaries, there are two sub-tasks of text summarization: extractive and abstractive summarization. Extractive summarization selects words and sentences only from the input documents [8, 26], whereas abstractive summarization utilizes external vocabulary to summarize ideas or concepts taken from the source documents [27, 28].

A very recent work that implemented several extractive summarization techniques is the MUSEEC (MUltilingual SEntence Extraction and Compression) summarization tool [8]. This is also the tool we used in our prototype system for summarizing scene descriptions.

3 Architecture

The initial design of the proposed system pipeline is shown in Fig. 1.

Our system accepts an entire full-length movie as input and splits it first into scenes first, which is done by detecting the difference between two subsequent frames and cut the movie when the difference exceeds a pre-defined threshold. Then, those scenes are sent to a scene description generation model, where for each scene, a one-sentence description of its main event is generated. An example of such a description is the that of the first scene in the movie "*Forrest Gump*", see Fig. 2, which is "*A man is talking*". Since movies may contain plenty of different characters, it is necessary to identify the characters in the descriptions, e.g. "*Forrest Gump is talking*" for the first scene. This could be regarded as a classification problem addressed by training a classifier to identify the faces in each scene. In our initial experiment, it is done manually.

Movie subtitles are informative. The main content of the subtitles is the dialogs among movie characters within the movie, e.g., "*Hello*" in the subtitles for the scene in Fig. 2. Dialogs provide plenty of extra information about the movies and their stories.

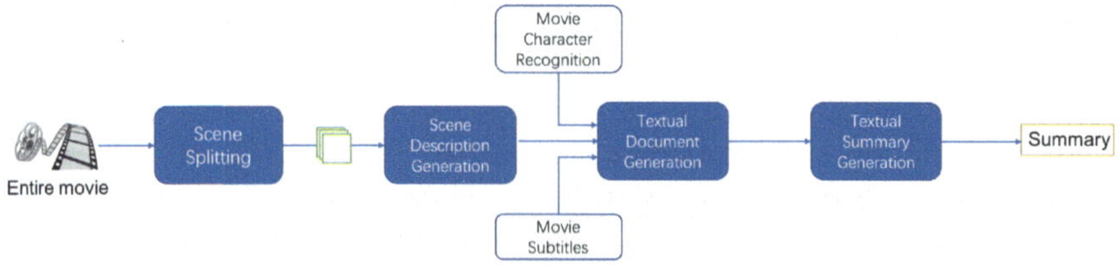

Fig. 1 Pipeline of the textual movie summarization prototype system

Fig. 2 Example of the first scene (after opening) in "*Forrest Gump*" (part) with subtitles

Besides, they include notifications of some background noise such as noise caused by a bus, noise made by people applause, etc, e.g. "*(BUS APPROACHING)*" in Fig. 2, which is also additional information that is hard to find if we only watch the movie, and that information sometimes can significantly improve the understanding of the movie stories. In this example, this simple line in the subtitles gives the audiences an understanding of where this scene is taking place, even without watching the movie itself. Therefore we import the subtitles text file into our system by matching it with the scene descriptions and try to enrich the scene description with the subtitles information.

Together with the subtitles, we store the scene descriptions and all additional information into plain text document as the MUSEEC text summarization tool can only accept plain text as input. The final step is summary generation and it is currently done by applying a text summarization algorithm to the plain text document containing the scene descriptions and the subtitle texts. The best results obtained so far by applying extractive text summarization methods to such a document will be presented and discussed in Sect. 5.

4 Experiments

As a very initial experiment, we run and test our system on one single full-length movie, *Forrest Gump*. To evaluate the results, we obtained the ground truth textual summary of *Forrest Gump* (with 675 words in total) from the CMU Movie Summary Corpus [29], which consists of 42,306 plot summaries of the related movies extracted from Wikipedia.

We use a public scene detection software, PySceneDetect,[1] for scene detection. The total length of the movie *Forrest Gump* is 2 h and 22 min and we obtained 900 scenes after scene splitting. Then we filtered out the scenes shorter than 3 s and the opening and ending scenes (related with the movie credits and casts). 533 scenes remain after filtering. In this initial experiment, we used only the first 156 scenes (out of the 533 scenes), corresponding to the first 40 min and 40 s part of the movie (after the movie credits).The ground truth summary of the movie was also cut with respect to the movie duration, with 153 words left.

We use S2VT (Sequence to Sequence—Video to Text) [4] for scene description generation. This algorithm accepts a set of video frames as input and extracts the RGB and optical flow features by the 16-layer VGG model [30] from the raw frames. A Long Short Term Memory (LSTM) network is then used to generate a one-sentence description of the original video based on its frame features. In our experiment, we used ffmpeg[2] to sample every 24th frame (approximately one frame per second) of the selected scenes and finally obtained 2208 frames in total. By feeding these frames into S2VT algorithm, we obtained a list of one-sentence descriptions of the

[1] Available online: https://pyscenedetect.readthedocs.io/en/latest/.

[2] Available online: https://www.ffmpeg.org/.

156 scenes. Then we manually replaced the terms "*man*", "*woman*" and "*person*" in the scene descriptions with the names of recognizable movie characters within the corresponding scene.

We combined the one-sentence descriptions movie's subtitles file obtained from the Internet: We manually aligned the subtitles of dialogs in each scene with the respective scene description. For each scene, the plain text begins with the scene description and it is followed by the dialogs happened in that scene. Those descriptions are then summarized by MUSEEC [8], a text summarization tool that integrates various text summarization algorithms and techniques. We used both MUSE (MUltilingual Sentence Extractor) and POLY (POLYnomial summarization with POLYtopes) algorithms from MUSSEC, with several objective functions such as Maximal weighted term sum of TF-IDF (Term Frequency—Inverse Document Frequency) and McDonald. The generated summaries are post-processed by anaphora resolution (AR). We limited the number of sentences in the output summary two six since this number produced the most meaningful results.

For comparison, we used another video description generation model which claimed to be the improved version of S2VT, know as Language-Fusion [31], and used these descriptions for the above experiments as well. In addition, we summarized only the descriptions from both description generation algorithms, without subtitles, to explore the effect of using subtitles.

We present and discuss below only the experiments, which produced meaningful outputs.

5 Results and Discussion

5.1 Experimental Results

In this section we present two output summaries. They are both generated from the S2VT generated descriptions, with and without subtitles. They both used POLY algorithm, with TF-IDF objective function and anaphora resolution. An overview comparison is shown in Fig. 3. Obviously, there is still a significant gap between the machine generated summaries and the human-written summary, yet, some important information is already successfully recognized and included in the generated summaries, such as "*JENNY was like an angel*" and "*And Jenny was thrown out of school*". The later information does not even appear in the gold standard summary but it is indeed an important sub story within the first 40 min of the movie.

It is not hard to tell from Fig. 3 that the summary with subtitles is more informative than that without subtitles, since only one sentence out of six is meaningless and others can easily bring the plot back to the readers.

Table 1 shows the type (description or subtitle) of each sentence in the summary and their origin (which scene it belongs to). Half of the summary sentences are selected from the first scene with the scene description and two sentences from the

As he waits at a bus stop, Forrest Gump starts recounting his life story to nearby strangers. His story starts with the leg braces he had to wear as a child, which resulted in him being bullied by children. He lives with his mother , who tells him that "stupid is as stupid does". Forrest teaches one of their guests, a young Elvis Presley, a hip-swinging dance. At school, Forrest meets Jenny , with whom he immediately falls in love, and they become best friends. Forrest discovers that he can run very fast which, despite his below-average intelligence, earns him a scholarship to the University of Alabama from Bear Bryant. While in college, he witnesses George Wallace's Stand in the Schoolhouse Door, is named an All-American football player, and meets President John F. Kennedy. After graduating, Forrest enlists in the United States Army, where he becomes friends with Benjamin Buford "Bubba" Blue.

Forrest Gump talks.
The doctor talks.
Mrs. Gump shows a bag.
The principle cuts a piece of a pair.
A man walks in a sidewalk.
Forrest Gump kicks a soccer.

Forrest Gump talks.
(BUS APPROACHING) Hello.
Those must be comfortable shoes.
But our didn't mind.
JENNY was like an angel.
And Jenny was thrown out of school.

Fig. 3 Comparison between the gold standard summary (left) and our two generated summaries. The top-right summary is the one without subtitles and the bottom-right summary is the one with subtitles

Table 1 Detail information of summary with subtitles

#	Description	Type	Origin
1	Forrest Gump talks	Description	Scene 1
2	(BUS APPROACHING) Hello	Subtitle	Scene 1
3	Those must be comfortable shoe	Subtitle	Scene 1
4	But our didn't mind	Subtitle	Scene 23
5	JENNY was like an angel	Subtitle	Scene 44
6	And Jenny was thrown out of school	Subtitle	Scene 131

subtitles of the scene. The scene description is accurate because the main event of this scene is Forrest Gump talking (Fig. 4). This implies that the summarization system considered the first scene as a very important scene from this part of the movie, and this is reasonable because the first scene of this movie introduces the whole story via Forrest Gump's narrative. From the phrase "*(BUS APPROACHING)*" it could be inferred that the first scene is taken place at a bus stop, which is true in the movie.

On the other hand, the 5th sentence "JENNY was like an angel." is selected from the 44th scene, where Forrest Gump first met his beloved woman Jenny, and this sentence is originally spoken by Forrest Gump as "*She was like an angel*", as shown in Fig. 5. Here the anaphora resolution replaced the "she" with "JENNY", which significantly improved the informativeness of this sentence and also this is the most important sentence to express Forrest Gump's feeling towards Jenny. Regarding the last sentence, it is selected from almost the end of this movie part, the 131th scene, and it plays an important role to depict the personality of Jenny in the movie by telling the truth that she was forced to quit school. Similarly, the anaphora resolution

Fig. 4 Detail preview the first scene (after opening) in "*Forrest Gump*" with subtitles and the corresponding summary of this scene

Fig. 5 Detail preview the 44th scene in "*Forrest Gump*" with subtitles and the corresponding summary of this scene

Fig. 6 Detail preview the 131th scene in "*Forrest Gump*" with subtitles and the corresponding summary of this scene

replaced "*she*" in the original sentence with "*Jenny*" and improved the sentence's comprehensiveness (Fig. 6).

The 4th sentence in this summary is an obvious mistake made by the algorithm because (a) The sentence it self is grammatically incorrect; (b) It does not include any meaningful information from the movie. If we watch carefully the 23rd scene, it tells about Forrest Gump's home together with the scenes before and after it. This part of the movie is less important compared with the other selected scenes. We assume that this error is possibly caused by the tendency of the POLY text summarization algorithm to choose short sentences.

5.2 *Discussion and Future Work*

We believe that it is still too early to perform a quantitative evaluation of results. The summary covered 4 scenes in total and 3 of them are considered to be the important scenes from the movie by a human judgment, so at this point we have a precision of 75%. However, the number of scenes covered by the gold standard is about 72 (we have 153 scenes from this chosen period of movie), which means our recall is far from being acceptable.

The state-of-the-art video description generation algorithm we used made significant number of mistakes when we applied it to our movie. Since it is trained on videos of at most 1 min, we believe it will improve the overall performance if we re-split all scenes longer than 1.5 min into shorter clips. Detecting the key scenes instead of using all scenes from the movie may also help to generate a better summary. On the other hand, we already discovered that subtitles can make a positive effect on the generated summaries, but still we need a better method to merge the subtitles with the descriptions instead of simply putting them together as a sequence of sentences. With these enhancements, the text summarization algorithm is expected to generate a more informative movie summary. Moreover, it is also necessary to run the system on the whole movie instead of part of it and on multiple movies rather than one, so that we can have more meaningful and statistically significant evaluation results.

6 Conclusions

In this work, we proposed our prototype system for automatically generating textual summaries of full-length movies. The system includes components of scene splitting, scene description generation, textual document generation and extractive text summarization. The initial results of our experimental framework show several directions for future improvement. Describing a short video clip is a challenging task, summarizing the entire movie is even more challenging. There are several potential future areas for this research: (1) Movie character recognition in each scene; (2) Speaker identification in each dialogue; (3) Improved description generation for each movie scene; (4) Training the summarization algorithms on a collection of movies and their human-generated summaries. Specifically, scene description generation could also help the visually impaired people to "watch" the movies online by listening to the dynamically generated scene descriptions.

References

1. T.-H. Chen, K.-H. Zeng, W.-T. Hsu, M. Sun, Video captioning via sentence augmentation and spatio-temporal attention (Springer International Publishing, Cham, 2017), pp. 269–286. https://doi.org/10.1007/978-3-319-54407-6_18
2. G.A. Sigurdsson, G. Varol, X. Wang, I. Laptev, A. Farhadi, A. Gupta, Hollywood in homes: crowdsourcing data collection for activity understanding (2016). arXiv:1604.01753
3. G.A. Sigurdsson, S. Divvala, A. Farhadi, A. Gupta, Asynchronous temporal fields for action recognition (2016). arXiv.org
4. S. Venugopalan, M. Rohrbach, J. Donahue, R. Mooney, T. Darrell, K. Saenko, Sequence to sequence—video to text, in *2015 IEEE International Conference on Computer Vision (ICCV* (IEEE, 2015), pp. 4534–4542
5. A. Rohrbach, A. Torabi, M. Rohrbach, N. Tandon, C. Pal, H. Larochelle, A. Courville, B. Schiele, Movie description. Int. J. Comput. Vis. **123**(1), 94–120 (2017)

6. D. Kaufman, G. Levi, T. Hassner, L. Wolf, Temporal tessellation for video annotation and summarization (2016). arXiv.org
7. A. Torabi, C.J. Pal, H. Larochelle, A.C. Courville, Using descriptive video services to create a large data source for video annotation research. *CoRR*, vol. cs.CV (2015)
8. M. Litvak, N. Vanetik, M. Last, E. Churkin, Museec: a multilingual text summarization tool, in *Proceedings of ACL-2016 System Demonstrations* (2016), pp. 73–78
9. Y. Li, Y. Song, L. Cao, J. Tetreault, L. Goldberg, A. Jaimes, J. Luo, TGIF: a new dataset and benchmark on animated GIF description (2016). arXiv.org
10. S. Sah, S. Kulhare, A. Gray, S. Venugopalan, E. Prud'Hommeaux, R. Ptucha, Semantic text summarization of long videos, in *2017 IEEE Winter Conference on Applications of Computer Vision (WACV)* (IEEE, 2017), pp. 989–997
11. D.L. Chen, W.B. Dolan, Collecting highly parallel data for paraphrase evaluation, in *Proceedings of the 49th Annual Meeting of the Association for Computational Linguistics: Human Language Technologies*, vol. 1, ser. HLT '11 (Association for Computational Linguistics, Stroudsburg, PA, USA, 2011), pp. 190–200. http://dl.acm.org/citation.cfm?id=2002472.2002497
12. A. Rohrbach, M. Rohrbach, N. Tandon, B. Schiele, A dataset for movie description, in *The IEEE Conference on Computer Vision and Pattern Recognition (CVPR)* (2015)
13. A.-A. Liu, N. Xu, Y. Wong, J. Li, Y.-T. Su, M. Kankanhalli, Hierarchical & multimodal video captioning: discovering and transferring multimodal knowledge for vision to language, in *Computer Vision and Image Understanding*, vol. 163, no. Supplement C, pp. 113 – 125, (2017), language in Vision. http://www.sciencedirect.com/science/article/pii/S1077314217300735
14. J. Xu, T. Mei, T. Yao, Y. Rui, Msr-vtt: a large video description dataset for bridging video and language, in *2016 IEEE Conference on Computer Vision and Pattern Recognition (CVPR)* (2016), pp. 5288–5296
15. H. Yu, J. Wang, Z. Huang, Y. Yang, and W. Xu, Video paragraph captioning using hierarchical recurrent neural networks, in *The IEEE Conference on Computer Vision and Pattern Recognition (CVPR)* (2016)
16. A. Khosla, R. Hamid, C.-J. Lin, N. Sundaresan, Large-scale video summarization using web-image priors, in *2013 IEEE Conference on Computer Vision and Pattern Recognition (CVPR)* (IEEE, 2013), pp. 2698–2705
17. D. Potapov, M. Douze, Z. Harchaoui, C. Schmid, *Category-Specific Video Summarization* (Cham: Springer International Publishing, 2014), pp. 540–555. https://doi.org/10.1007/978-3-319-10599-4_35
18. M. Gygli, H. Grabner, L. Van Gool, Video summarization by learning submodular mixtures of objectives, in *The IEEE Conference on Computer Vision and Pattern Recognition (CVPR)* (2015)
19. K. Zhang, W.-L. Chao, F. Sha, K. Grauman, Summary transfer: exemplar-based subset selection for video summarization, in *The IEEE Conference on Computer Vision and Pattern Recognition (CVPR)* (2016)
20. M. Gygli, H. Grabner, H. Riemenschneider, L. Van Gool, *Creating Summaries from User Videos* (Cham: Springer International Publishing, 2014) pp. 505–520. https://doi.org/10.1007/978-3-319-10584-0_33
21. Y. Song, J. Vallmitjana, A. Stent, A. Jaimes, Tvsum: summarizing web videos using titles, in *The IEEE Conference on Computer Vision and Pattern Recognition (CVPR)* (2015)
22. X. Li, T. Utsuro, H. Uehara, Movie Summarization Based on Alignment of Plot and Shots, in *2017 IEEE 31st International Conference on Advanced Information Networking and Applications (AINA)* (IEEE, 2017), pp. 189–196
23. Q.D. Tran, D. Hwang, O.-J. Lee, J.E. Jung, Exploiting character networks for movie summarization. Multimed. Tools Appl. **76**(8), 10,357–10,369 (2016)
24. K. Spärck Jones, Automatic summarising: a review and discussion of the state of the art, University of Cambridge, Computer Laboratory, Tech. Rep. UCAM-CL-TR-679 (2007). http://www.cl.cam.ac.uk/techreports/UCAM-CL-TR-679.pdf
25. H.P. Luhn, The automatic creation of literature abstracts. IBM J. Res. Dev. **2**(2), 159–165 (1958). Apr

26. Y. Ko, J. Seo, An effective sentence-extraction technique using contextual information and statistical approaches for text summarization. Pattern Recognit. Lett. **29**(9), 1366–1371 (2008). http://www.sciencedirect.com/science/article/pii/S0167865508000676
27. J.S. Kallimani, K. Srinivasa, et al., Information extraction by an abstractive text summarization for an Indian regional language, in *2011 7th International Conference on Natural Language Processing and Knowledge Engineering (NLP-KE)* (IEEE, 2011), pp. 319–322
28. A. Khan, N. Salim, Y.J. Kumar, A framework for multi-document abstractive summarization based on semantic role labelling. Appl. Soft Comput. **30**, 737–747 (2015)
29. D. Bamman, B. O'Connor, N.A. Smith, Learning latent personas of film characters, in *Proceedings of the Annual Meeting of the Association for Computational Linguistics (ACL)* (2014), p. 352
30. K. Simonyan, A. Zisserman, Very deep convolutional networks for large-scale image recognition, *CoRR*, abs/1409.1556, (2014). http://arxiv.org/abs/1409.1556
31. S. Venugopalan, L.A. Hendricks, R. Mooney, K. Saenko, Improving lstm-based video description with linguistic knowledge mined from text, in *Conference on Empirical Methods in Natural Language Processing (EMNLP)* (2016)

Fuzzy Control Applications

Zadehian Paradigms Shaping 21st Century Artificial Intelligence

B. E. Bărbat

Abstract Starting from the premise that Zadeh's research heritage is irreducible to his 20th Century work, the paper aims to show that his *Generalized Theory of Uncertainty* is even more influential now, for 21st Century service-oriented engineering, than his papers on fuzzy sets were for the product-based industrial era. To mirror the whole architectonics of Zadeh's work, the paper highlights the lasting puissance and evolution of 20th Century Zadehian paradigms. On this groundwork, two paradigmatic breakthroughs follow: (a) moving from 'information is statistical in nature' to 'information is a generalized constraint'; (b) setting as target 'achievement of NL-capability'. Next, two cardinal upshots: reshaping the relation between numbers and words and scaling down the importance of algorithmic paradigms. Both are needed to meet the challenge of modern artificial intelligence: interacting with living systems; emphasis is on model tractability (for efficiency) and on model interpretability (for user acceptance).

Keywords Artificial intelligence (AI) · General theory of uncertainty (GTU) · Computing with words (CWW) · Natural language (NL)-capability · Large scale living systems (LS2)

1 Introduction. Zadeh Alive in the 21st Century

"Because of its unorthodoxy, it has been and will continue to be controversial for some time. Eventually, though, the theory of fuzzy sets is likely to be recognized as a natural development in the evolution of scientific thinking. In retrospect, the scepticism about its usefulness will be viewed as a manifestation of the human attachment to tradition and resistance to innovation" [26], quoted in [20]. Zadeh was right about the generations of accomplished scientists (Kálmán's reaction was notorious). On the contrary, for generations born after WWII "The Fuzzy Revolution: Goodbye to

B. E. Bărbat (✉)
Freelance researcher, Sibiu, Romania
e-mail: bbarbat@gmail.com

S. N. Shahbazova et al. (eds.), *Recent Developments and the New Direction in Soft-Computing Foundations and Applications*, Studies in Fuzziness and Soft Computing 393, https://doi.org/10.1007/978-3-030-47124-8_40

the Aristotelian Weltanschauung" [20] was the very first technological/scientific revolution they ever faced. Moreover, for mathematically or technologically-oriented students Zadeh became their spiritual guide: "The impact of his seminal paper is enormous with 95,000 citations in Google Scholar. [...]. Without his paper, my professional career would have been completely different" [11]). The pathfinder was relentless: "Zadeh continued to propose new ideas for the advances of the fuzzy systems research—fuzzy decision making (*Management Science*, 1970), possibility theory (*Fuzzy Sets and Systems*, 1978), fuzzy quantifiers (*Computers and Mathematics with Applications*, 1983), computing with words (*IEEE Trans. on Fuzzy Systems*, 1996), and fuzzy information granulation (*Fuzzy Sets and Systems*, 1997)" [11].

However, this impressive list is far too incomplete, excluding from the research heritage, some of Zadeh's brainchildren, not less brilliant than the others. In this light, the paper aims to show that his *Generalized Theory of Uncertainty* (GTU) [34] is even more influential now for service-oriented engineering than his papers on fuzzy sets were for the product-based industrial era.

Thus, to mirror the whole architectonics of Zadeh's work, after clarifying the *terminology* (Sect. 2), very recent references highlight the lasting importance of Zadehian *industrial era paradigms in modern settings* (Sect. 3). On this groundwork, the next two sections outline the "*breakthrough: information is comprehensible*", (Sect. 4) and its cardinal upshots: *reshaping the relation between numbers and words* and *scaling down the importance of algorithmic paradigms* (Sect. 5). This armamentarium is (badly) needed to meet *the challenge: interacting with living systems* (Sect. 6). (Details and examples about applying Zadehian paradigms in agent-oriented software engineering will be given in a future paper.) *Remarks* (Sect. 7): It is expecting too much from 'Kelvin-Number-Oriented' scientists to accept a paradigm reverse to the reigning one.

Other abbreviations: *CI*: Computational Intelligence; *PT*: Probability Theory; *GC*: Generalized Constraint; *HMI*: Human-Machine Interface; *wiki*: https://en.wikipedia.org/wiki.

2 Terminology. A Carnap-Like Mini-Glossary

The mini-glossary is 'Carnap-like' for five reasons: it is inspired by Carnap's "Introduction to Semantics" in both spirit (within the semiotic triad, syntax has no anymore the lion's share) and letter (emphasis is on contextual usage); it is *sine qua non* in any position paper expressing opinions outside the reigning paradigm; it reflects an explicit application-oriented stance (Zadeh described himself as a "mathematically oriented, electrical engineer" [wiki]; it reflects also an implicit Zadehian mindset (reconsidering the Western cultural tradition about numbers prevailing over words); it is useful for young researchers, with less mathematical background.

A. *Explaining the Title*

Artificial Intelligence. The paper's "stance is that artificial intelligence is reflected by *word*-oriented, Turing-*test*-like, anthropocentric interfaces (aimed at managing situations, as in 'The Imitation Game'), rather than by *number*-oriented, Turing-*machine*-like algorithmic software (aimed at solving problems, in line with the Church-Turing thesis)" [3].

21st Century Artificial Intelligence. The qualifier suggests "that novel approaches to uncertainty and time challenge last century software engineering paths [since] human beings and robots are distinct cognitive species" [4].

Paradigm has the widespread meaning given by Kuhn. However, "Kuhn's paradigms are in fact Fleck's thought styles. Since Kuhn had used his notion of a paradigm extremely loosely and received various, incisive criticisms […], he later substituted for it the notion of a 'disciplinary matrix' […]. The disciplinary matrix in a particular scientific area is the set of shared basic rules, methods, and beliefs which underlie theory formation and knowledge acquisition in this area" [20].

Zadehian Paradigms are results, and consequences of his creative thoughts considered by scientific authorities [2, 11, 12, 18, 21] as having major impact and—in line with this paper—his cardinal 21st century work outlined in Sect. 5.

B. *"Inadequacy of the Conventional Mathematics"* [25]

A congratulations letter to Zadeh [20] shows, in his own words, why biological systems (*'the challenge'*, Sect. 6) cannot be modelled using conventional concepts (*'moving away'*, Sect. 5), forty years before redefining information (*'the breakthrough'*, Sect. 4): "In fact, there is a fairly wide gap between what might be regarded as 'animate' system theorists and 'inanimate' system theorists at the present time, and it is not at all certain that this gap will be narrowed, much less closed, in the near future. There are some who feel this gap reflects the fundamental inadequacy of the conventional mathematics—the mathematics of precisely-defined points, functions, sets, probability measures, etc.—for coping with the analysis of biological systems, and that to deal effectively with such systems, which are generally orders of magnitudes more complex than man-made systems, we need a radically different kind of mathematics, the mathematics of fuzzy or cloudy quantities which are not described in terms of probability distributions" [25], quoted in [20].

Probability has its usual meaning as "the measure of the likelihood that an event will occur" [wiki]. However, in both its interpretations the number expressing it is of little benefit for human reasoning: "relative frequency 'in the long run' of outcomes [or] includes expert knowledge as well as experimental data to produce probabilities" [wiki]. Initially, Computational intelligence had as "Field of Interest […] the theory, design, application, and development of biologically and linguistically motivated computational paradigms. [I]f we approach CI from the point of view of components […], maybe we miss in the case of CI the field of Probabilistic Reasoning, a topic that had an important role in SC. This absence is also supported by the first definition, since Probabilistic Reasoning is neither biologically nor linguistically motivated" [13].

Algorithm. The conceptual confusion about defining the procedure usually called algorithm (the strange journey from 'Turing machine' to 'metaheuristic(s)' seems yet unfinished) is irrelevant for robotics and even for (bodiless) AI when human-machine interaction is low. The problem is only the syntagma 'algorithmic software', when computing with words.

Information. Although confusion is alike, its sequels are huge: applying successfully algorithms without terminological soundness is just uncomfortable but having incompatible ideas about the kernel concept of an entire technologic era reveals a deep scientific crisis. In short: GTU was not just badly needed, it was inexorable.

On the contrary, a theory able to spawn other theories after half a century (see below) is an 'Unfinished Symphony'.

3 Industrial Era Paradigms in Modern Settings

To increase relevance, subfields are chosen from [11] whereas references—for both theories and applications—are published in 2017 or 2018. Computing with Words (CWW), seen as direct ancestor of GTU, is interpreted in Sect. 4.

A. *Extending successful 20-th Century Fuzziness Theories*

Dubois and Prade extend their more than forty-year-old pivotal work on possibility theory and possibilistic logic with a "generalized possibilistic logic (GPL), a logic for epistemic reasoning based on possibility theory. [T]he logic generalizes possibilistic logic (PL) [...] GPL can encode any set of epistemic states. This feature makes GPL particularly suitable for reasoning about what an agent knows about the beliefs of another agent" [7]. In [12] "the automatic rule base synthesis for fuzzy control and decision-making systems" is modified. [21] "tries to recognize a few important examples of extracting value from social network data" [21]. The prolific and polyvalent Yager published (2018) research in the fields of Granular Computing [22], Information Fusion [23], and Internet of Things [24].

B. *Soft Computing Methods and Applications*

The examples stem from System Sciences where service-oriented engineering is essential for both efficiency and acceptance: "More complex systems arisen in biology, medicine, the humanities, management sciences, and similar fields often remained intractable to conventional analytical methods" [16]. "Despite the widespread use of techniques and tools for causal analysis, existing methodologies still fall short as they largely regard causal variables as independent elements, thereby failing to appreciate the significance of the interactions of causal variables. [W]e introduce a new approach for the causal complexity with the goal of making the solution set closer to deterministic by taking into consideration the underlying patterns embedded within a dataset: the interactions of causal variables" [17].

A comprehensive collection of relevant recent Soft Computing applications is given in [2].

4 Breakthrough: Information Is Comprehensible

The assertion about the paramount significance of GTU "is based on two cardinal paradigmatic breakthroughs: (a) moving from "information is statistical in nature" to "information is a generalized constraint"; (b) setting as target "achievement of NL-capability"" [3].

A. *GTU Claim: Information Is a Generalized Constraint*

All quotes in Sects. 4 A and B are from [34].

"There is a deep-seated tradition in science of dealing with uncertainty – whatever its form and nature – through the use of probability theory. Successes of this tradition are undeniable. But as we move further into the age of machine intelligence and automated decision-making, a basic limitation of probability theory becomes a serious problem. More specifically, in large measure, standard probability theory, call it PT, cannot deal with information described in natural language; that is, to put it simply, PT does not have NL-capability [...] The generalized theory of uncertainty (GTU) differs from other theories in three important respects. First, the thesis that information is statistical in nature is replaced by a much more general thesis that information is a generalized constraint [28] with statistical uncertainty being a special, albeit important case. Equating information to a generalized constraint is the fundamental thesis of GTU. In symbolic form, the thesis may be expressed as

$$\mathrm{I(X)} = \mathrm{GC(X)}$$

where X is a variable taking values in U; I(X) is information about X; and GC(X) is a generalized constraint on X.

Second, bivalence is abandoned throughout GTU, and the foundation of GTU is shifted from bivalent logic to fuzzy logic [29]. As a consequence, in GTU everything is or is allowed to be a matter of degree or, equivalently, fuzzy. Concomitantly, all variables are, or are allowed to be granular, with a granule being a clump of values defined by a generalized constraint [...].

And third, one of the principal objectives of GTU is achievement of NL-capability" (see next subsection).

"The center piece of GTU is the concept of a generalized constraint—a concept drawn from fuzzy logic. The principal distinguishing features of fuzzy logic are (a) graduation and (b) granulation. More specifically, in fuzzy logic everything is, or is allowed to be, graduated, that is, be a matter of degree or, more or less equivalently, fuzzy. Furthermore, in fuzzy logic all variables are allowed to be granulated, with a granule being a clump of values drawn together by indistinguishability, similarity, proximity or functionality [...] Graduation and granulation underline the concept of a linguistic variable". (This crucial concept was research object both before and after publishing GTU [6, 18, 19, 27, 30].)

"More fundamentally, graduation and granulation have a position of centrality in human cognition. This is one of the basic reasons why fuzzy logic may be viewed in a model of human reasoning". (Zadeh's research was transdisciplinary.)

"*A key idea—the meaning postulate.* [It] assents that the meaning of a proposition may be represented as a generalized constraint. It is this postulate that makes the concept of a generalized constraint the centerpiece of GTU".

The "Concluding remark" is telling: "The theory of uncertainty which is outlined in this paper may be viewed as a radical step toward abandonment of bivalence and shifting the foundation of the theory from bivalent logic to fuzzy logic. Though only a step, it is a step which has wide-ranging ramifications. Standing out in importance is achievement of NL-capability. This capability opens the door to extensive enlargement of the role of natural languages in themes of uncertainty, decision analysis, economics and other fields in which human perceptions play an important role".

B. *GTU Target: Achieving Natural-Language capability*

"Why is NL-capability an important capability? Principally because much of human knowledge and real-world information is expressed in natural language. Basically, a natural language is a system for describing perceptions. Perceptions are intrinsically imprecise, reflecting the bounded ability of human sensory organs, and ultimately the brain, to resolve detail and store information. Imprecision of perception is passed on to natural languages. It is this imprecision that severely limits the ability of PT to deal with information described in natural language. NL-capability of GTU is the focus of attention in the present paper [...] NL-Computation is the core of PNL (precisiated natural language) [...], a fuzzy-logic-based system for computation and deduction with information described in natural language". In line with the paper's aim, Zadeh's question is slightly changed: Why should be NL-capability the paramount capability of post-industrial software engineering? From an end-user acceptance stance, because of the (goal-oriented) implication chain:

modern AI → high HMI → human-like language → (model) interpretability → NL-capability → GTU-based approach.

C. *What Computing with Words Means to Me*

"Computing with words (CWW) means different things to different people. This article is the start of a position paper, [...] that answers the question 'What does CWW mean to me?'" [14]. Below, '*Me*' is Zadeh: "Humans have many remarkable capabilities. [...] First, the capability to converse, communicate, reason and make rational decisions in an environment of imprecision, uncertainty, incompleteness of information and partiality of truth. And second, the capability to perform a wide variety of physical and mental tasks without any measurements and any computations. In large measure, Computing with Words is inspired by these remarkable capabilities. Achievement of human level machine intelligence has long been one of the principal objectives of AI. Progress toward achievement of this objective has been and continues to be slow. Anybody who had the experience of struggling with a dumb automated customer service system will readily agree. In [33] I argued that one of the principal reasons for slowness of progress has been AI's failure to address the problem of reasoning and decision-making with perception-based information. A key idea that was suggested in my paper was that of dealing not with perceptions per se but with their descriptions in a natural language [...] There are many misconceptions about what Computing with Words (CW or CWW) is and what it has to offer. A common misconception is that CW and natural language processing are closely related. In reality, the opposite is the case. More importantly, what is widely unrecognized at this juncture is that moving from computation with numbers to computation with words has the potential for evolving into a basic paradigm shift […] which would open the door to a wide-ranging enlargement of the role of natural languages in scientific theories" [14].

After eight years, Zadeh's argument appears stronger than ever; it is detailed below, grounded on six phrases quoted above: 'much of human knowledge and real-world information is expressed in natural language'; 'natural language is a system for describing perceptions'; 'perceptions are intrinsically imprecise'; 'imprecision of perception is passed on to natural languages'; 'standing out in importance is achievement of NL-capability'; 'PT does not have NL-capability'.

5 Demythifying the Number, Shaking the Algorithm

From this paper's stance, CWW has a threefold meaning: *reshaping the relation between numbers and words*, in line with human nature (Sect. A, confined to AI), *scaling down the importance of algorithmic paradigms* (Sect. B, referring to Computer Science), and conjecturing about the 'basic paradigm shift': accepting *rationales against precision* (Sect. C, extended to modern science as a whole).

A. *In the Beginning Was the Word (or was it the Number?)*

Here, any reference to cultural traditions is confined to the historical sequence of (three) disciplines having the *Word* as basic concept, ordered by decreasing generality. In Western tradition, *logic* (Aristotelian system based on Chryssippean bivalence) was first, *linguistics* followed after two millennia, and *semiotics* in the nineteenth century. In [1] Al-George outlines for the Indian culture a (chronologically reverse) "phylogenetic perspective: semiology – linguistics – logic" [1].

"There are several echelons or stages in quitting the Western tradition when developing decision-oriented software […] and embracing the Eastern approach to logic in tackling real-world situations (epitomised by Lao Tzu and promoted vigorously by Zadeh)" [8] (where, in a subchapter entitled 'Back to Lao Tzu: Synergy, Symbols, Semiotics', among the 'fuzziness quotes' from Lao Tzu, the most non-deterministic-oriented one asserts 'preterminologic future contingency': "Those who have knowledge, don't predict. Those who predict, don't have knowledge" [8] (applied in software in [9]).

B. *Moving Away from Bayes, Kelvin, and Chrysippus*

"For over 40 years, determinism, and bivalence of Chrysippean logic were the pillars of Computer Science; likewise, algorithms were the backbone of computer programs, complying with their etymon: *pro-gramma* = what is written in advance.

They sufficed for both FORTRAN-like number crunching and COBOL-like business data processing. When early real-time applications (firstly, operating systems) needed less autistic programs, algorithms tried to adapt and bizarre terms, such as ‘unsolicited input’, were coined to fit the incipient non-determinism” [5].

In the part of [3] below, all quotes stem from [34]: “the fundamental role of GTU is illustrated summarizing – in order of increasing paradigmatic gap – the pillars of algorithmic/mechanistic stance (still ruling in system modelling and design) it shakes, *because* of their lacking or reduced NL-capability (in brackets are hints to causes that impair NL-capability, as perceived from the other side of the gap), [...]:

- *Probability*. [see Sect. 4 A] (Two decades after Dubois, Prade, and Smets won the challenge on undeniable mathematical grounds, defending possibility theory in GTU was helpful, underlining that probability is inapplicable in real-world word-oriented decision making. Indeed, ‘0’ and ‘1’ are used at most as placeholders of truth values, not as real numbers. As regards model theory, after Carnap reinstated in mathematical logic the right of meaning—through semantics—no logician felt the need of a universe of ‘probable worlds’.)
- *Precision*. [see Sect. 4 B]
- *Bivalence*. [see Sect. 4 A] “[T]he conceptual structure of bivalent logic [...] is much too limited to allow a full development of the concept of precisiation”. (The attempt to reconcile the Boolean infrastructure of IT with fuzzy sets theory yielded frustrating side effects: ‘Kelvin-number-oriented’ scientists moulded complex theories where instead of dealing with two integers, programmers have to consider the continuum of reals. Conversely, there is not even a ‘third IF value’ expressing uncertainty as generalized constraint according to the meaning postulate.)” [3].

C. *Zadeh Beyond Zadeh: Rationales Against Precision*

In [31, 32], where Zadeh launched and refined CWW, he “goes beyond ‘tolerance for imprecision’. In fact, he adds—albeit implicitly—to his ‘Rationale 2 for granulation: precision is costly’, a ‘Rationale 3: precision is unnatural’ (from a bounded rationality stance, in the very meaning of Simon)” [3]. In [8], within a post-industrial setting, from a definite (service-oriented) software engineering stance, two working assumptions (*Wa*) made the claim explicit: “*Wa5*: Precision is useless [...]. *Wa6*: Precision could be harmful when decision is urgent (“Just-in-Time” decision making), [...]. For the same reason, this assumption is called “*Rationale 4*” [8]. Indeed, neither nature nor humans have enough reasons to waste precious time for optimizing (for both, suboptimal solutions suffice). Nevertheless, the crucial incompatibility is between *precision* and *living systems*.

6 The Challenge: Interacting with Living Systems

All beings (B) are interactants (I). $\forall x\ (B(x) \rightarrow I(x))$. Living systems (LS2) are *structurally* complex as (large scale) living systems and *cognitively* complex as *interactants* (to be graspable to entities they interact with); they are outlined in Subsection A.

Some B are modelers (M). $\exists x\ (B(x) \rightarrow M(x))$. They must make the model efficient. Model *tractability* is outlined in Subsection B.

Some B are decision makers (D). $\exists x\ (B(x) \rightarrow D(x))$. Hence, they must understand the (model of the) real-world part they interact with. *Interpretability* is outlined in Subsection C.

A. *Modelling Living Systems Needs GTU-Based Approach*

Living systems from cell, to human, to society, to Gaia are too intricate to be regarded otherwise than 'large scale' (LS2). Moreover, they are *given as such* (modelers do not have the 'demiurgic privilege' of robot creators to *assign* features). Their general characteristics are summarized after [4]:

- "*LS2 are complex*. Their complexity has two dissimilar components. Cognitive complexity depends on model *interpretability*, while structural complexity depends on model *relevance*. […]
- *LS2 are nondeterministic and open*. Taken together, the concepts express user perception that uncertainty is the epistemic facet of nondeterminism, while any kind of 'Closed World Assumption' is out of question. […]
- *LS2 evolve in irreversible time*. 'Irreversible' has here the old, simple Augustinian meaning: even for God '*Undo*' is impossible. Real-world events are unique and cannot unhappen. 'Irreversible' time is called *Bergsonian* (after Wiener). Hence, LS2 must be modelled as *processes*, requiring Bergsonian time – at most simulated through closed Newtonian time. […] In short: LS2 models cannot be precise, algorithmic, apodictic, atemporal or sequential" [4].

Indeed, the goal-oriented implication chain, adapted for LS2, and reshaped as cause-oriented, becomes:

LS2 are given → structural/cognitive complexity → simple/fuzzy model → NL-capability → tractability/interpretability.

B. *LS2 Model Tractability Requires Computing With Words*

CWW in its semiotic-oriented meaning (Sect. 5 A, B) "was tried in *Cybernetic Modelling*, a new modelling sub-species aimed at enhancing LS2 model *efficiency* (software engineering stance) and user *acceptance* (service-oriented engineering stance). Starting from the premises that model tractability is crucial for both efficiency and acceptance, and that NL-capability is sufficient condition for tractability, the method avoids anything with low or without NL-capability (e.g., probabilities, precision, bivalence, difficult numerical mathematics). Intractable differential equations are evaded via discrete-time modelling" [4]. Its first application in [9] shows "a model for homeostasis with hysteretic delay [...] based on an (over)simplified Lotka-Volterra model for predators-prey species" [4]. Discrete time was—in the context of [9]—"the only way to avoid intractable mathematics (mainly partially suitable differential equations) in modelling. [...] The time granule [...] has semantic value, [...] is variable within a very large range (e.g., prey population lifespan), [is] appropriate for ['What-if scenarios] (simple movies are a sequence of pictures), [is] simulating (future or past) discrete events [...] that occur in [Bergsonian time], and [...] is Kripke-compatible with irreversible time (the time granule Δt belongs undisputable to [Newtonian time] since it is reversible" [9].

C. *End-User Acceptance Requires Model Interpretability*

Recent research quoted in [4] shows that there is an overt mistrust of computers as advisors to human decision makers. Such scepticism is paradoxical: humans consider computers reliable when decision is (a *synchronous*) part of *problem-solving* mechanisms (as in automatic control) but consider them far too unconvincing when decision is (an *asynchronous*) part of *situation-managing* processes (as in risk management). This paradox suggests profound epistemic limits, since it is strongly related to NL-capability, showing that for decision makers trust is *doxastic*, while distrust is *credal*. "Real-world decision makers do not trust help the computers could give them [...], while, [...] they entrust computers to make decisions in [...] vital areas [...]. An outstanding example is the field of Automatic Control [...] where billions of 'imprecise *IF*s' are executed every second. [Computers] cannot advise because they are not *living* systems; they [have no] human-like sense of time" [4].

To impair redundancy with [4] and another recent paper, below is just a minor example of boosting interpretability by replacing (some) numbers with words: The 'golden ratio' was employed in [8, 9] for reasons reaching from applying non-numeric mathematics to its relevance as lasting, ubiquitous, and viable meme: "not even passionate mathematicians would prefer the exact trigonometric formulas for the golden ratio [...] to the rational appearance as *relevant geometric ratio* or as quotient of successive Fibonacci numbers" [8]. The Fibonacci path has threefold relevance: the 'golden ratio' can be calculated in mind; it inspires modelling in various areas, from ecosystem stability to analysis of financial markets (Fibonacci retracement); it suggests a temporal dimension (key attribute of both services and living systems).

7 Remarks. 21st Century AI Logo: 'Zadeh Inside'

Drawing conclusions would be too arrogant for a position paper defending a paradigm, not just diverging with the reigning one, but pursuing a reverse paradigm shift. Thus, the remarks are rather 'Beliefs, Desires, Intentions'—in line with the oldest agent paradigm conveying Zadehian ideas.

When 'exact sciences' are king, suggesting that *precision* is a particular case of *imprecision* is more than legitimizing fuzziness: it means thinking the other way around—in short, scientific heresy. Perhaps it is expecting too much from a single generation of scientists, after assimilating, extending, and valorising, in impressing work, fuzzy sets (and logic), possibility theory, CWW, and so on for over half a century.

Yet, uninhibited young researchers have good reasons to suspect number-crunching IT and to strive for more meaning through CWW: indeed, the 'Big Data Conundrum' suggests a systemic weakness of the 'Data Science' paradigm. There is already a risky positive feedback loop: data accumulates exponentially, while extracting the knowledge it hides seems doomed to leg behind, despite the host of new sophisticated applications, revealing the gap between *problem* (real world requirements) and *solution* (AI offer). (The gap deepens since, instead of offering more *interpretable* models, IT firms offer more *accurate/precise* ones. No solution arises, at a medium-range horizon, within the algorithmic paradigm.) Even more: they can exploit the huge openings of CWW, via 'precisiated body language' (some gestures express perceptions better than words). First step? Promote the concept of non-algorithmic software in the free market economy of competing memes.

Since there is no 'methodology for shifting paradigms', accepting GTU with its new Weltanschauung, giving up Western cultural tradition, requires more time, mainly for 'Kelvin-Number-Oriented' scientists.

Fortunately, young researchers can explore already very promising 21CZ (trans)disciplines moulded by 'Zadeh-Word-Oriented' computer scientists. The following quote stems from a 'brain-father' (of the science of anticipation; some of his seminal work was referred to in [4]): "human beings [...], constitute themselves through their activity, testimony to their abilities. Alas, they prove theorems, but not in a mechanical [...] manner. Moreover, they are not subject to the infinite loop of the halting problem: that is, can a computer recognize when the programs task is finished (or will it continue to process indefinitely)? The human being [...] would halt. In other words, it understands whatever is performed and stops, either when it cannot achieve what it wants or after achieving it. [O]ne can infer that, [...], human beings are not reducible to algorithmic machines" [15].

Likewise, the paper—*In memoriam*: Lotfi A. Zadeh—stops here, without knowing whether it achieved its aim.

Acknowledgements I am indebted to all BISCers I met on this very fine forum, and to Emilian Guţuleac for talks about the unsure role of 'Big Data' in (micro)electronic research.

References

1. S. Al-George, *Language and Thought in Indian Culture. An Introduction to Indian Semiology* [Limbă şi gândire în cultura indiană – Introducere în semiologia indiană] (ESE, Romanian, Bucharest, 1976)
2. V.E. Balas, L.C. Jain, M.M. Balas (eds.), *Soft Computing Applications: Proceedings of the 7th International Workshop Soft Computing Applications (sofa 2016)*, vol. 1 (Springer, 2017)
3. B.E. Bărbat, *Letter* (submitted in support of the nomination of Prof. Lotfi A. Zadeh for the 2014 ACM A. M. Turing Award), 2013, unpublished

4. B.E. Bărbat, F.G. Filip, Uncertainty and time in modern artificial intelligence, in *Proceedings of 9th International Conf. of Microelectronics and Computer Science* (Chişinău, Moldova, 2017), pp. 217–220
5. B.E. Bărbat, S.C. Negulescu, Symbolic inferences in multi-agent systems. *Int. J. Comput. Commun. Control* **1**, Suppl. Issue Proc. ICCCC, I, ed. by Dzitac et al. (Agora, Oradea, 2006), pp. 68–73
6. B. Blair, Interview with Lotfi Zadeh, Creator of Fuzzy Logic by Azerbaijada Int. **2**(4) (1994). http://www.azer.com/aiweb/categories/magazine/24_folder/24_articles/24_
7. D. Dubois, H. Prade, S. Schockaert, Generalized possibilistic logic: foundations and applications to qualitative reasoning about uncertainty. *Artif. Intell.* **252**, 139–174 (2017)
8. R.D. Fabian, *Bounded Rationality in Agent Orientation—"Just-in-Time" Visual Pattern Recognition*. http://scribe.ulbsibiu.ro/dspace/bitstream/123456789/846/1/Ralf_Fabian_Phd_Thesis.pdf (2011)
9. R.D. Fabian, *Validating Non-Algorithmic Software in Service-Oriented Engineering Research*. http://digital-library.ulbsibiu.ro/dspace/handle/123456789/824 (2013)
10. M.M. Gupta, R.K. Ragade, R.R. Yager (eds.), *Advances in fuzzy set theory and applications* (North-Holland, Amsterdam, 1979), pp. 3 ± 18
11. H. Ishibuchi, Father of fuzzy logic [Editor's Remarks].*IEEE Comput. Intell. Mag.* **13**(1), 2 (2018)
12. Y.P. Kondratenko, O.V. Kozlov, O.V. Korobko, Two modifications of the automatic rule base synthesis for fuzzy control and decision-making systems, in *Information Processing and Management of Uncertainty in Knowledge-Based Systems. Theory and Foundations. IPMU 2018. Communications in Computer and Information Science*, vol. 854, ed. by J. Medina et al. (Springer, Cham, 2018)
13. L. Magdalena, What is soft computing? Revisiting possible answers. *Int. J. Comput. Intell. Syst.* **3**(2), 148–159 (2010)
14. J.M. Mendel et al., What computing with words means to me [discussion forum]. *IEEE Comput. Intell. Mag.* **5**(1), 20–26 (2010)
15. M. Nadin, *'In folly ripe. In reason rotten' Putting machine theology to rest* (Institute for Research in Anticipatory Systems, University of Texas at Dallas, Richardson, TX, USA, 2017)
16. G. Pasi, R. Kruse, M. Viviani, Introduction to the Minitrack on soft computing: methods and applications, in *Proceedings of the 51st Hawaii International Conference on System Sciences* (2018)
17. S.K. Poon, S. Henry, N. Gorji, A network-based deterministic model for causal complexity, in *Proceedings of the 51st Hawaii Intl Conference on System Sciences* (2018)
18. R. Seising, What is soft computing?—Bridging gaps for 21st century science! *Int. J. Comput. Intell. Syst.* **3**(2), 160–175 (2010)
19. R. Seising, V. Sanz (eds.), *Soft Computational in Humanity and Social Sciences, STUDFUZZ 273* (Springer, Berlin, Heidelberg, 2012), pp. 3–36
20. K. Sadegh-Zadeh, The fuzzy revolution: goodbye to the Aristotelian Weltanschauung. *Artif. Intell. Med.* **21**(1), 1–25 (2001)
21. S.N. Shahbazova, S. Shahbazzade, Big data analytics and fuzzy technology: extracting information from social data, in *Recent Developments and the New Direction in Soft-Computing Foundations and Applications* (Springer, Cham, 2018), pp. 3–13
22. R.R. Yager, Decision making under measure-based granular uncertainty. *Granular Comput.* 1–9 (2018)
23. R.R. Yager, Refined expected value decision rules.*Inf. Fusion* **42**, 174–178 (2018)
24. R.R. Yager, J. Pascual Espada (eds.), *New Advances in the Internet of Things* (Springer, 2018)
25. L.A. Zadeh, From circuit theory to systems theory.*Proc. IRE* **50**, 856 ± 65 (1962)
26. L.A. Zadeh, Fuzzy sets versus probability. *Proc. IEEE* **68**, 421 (1980)
27. L.A. Zadeh, Fuzzy sets and information granularity, in [10], pp. 3–18 (1979)
28. L.A. Zadeh, Outline of a computational approach to meaning and knowledge representation based on the concept of a generalized assignment statement, in *Proceedings of the International Seminar on Artificial Intelligence and Man–Machine Systems*, ed. by M. Thoma, A. Wyner (Springer, Heidelberg, 1986), pp. 198–211

29. L.A. Zadeh, The concept of a linguistic variable and its applications to approximate reasoning, Part I. *Inf. Sci.* **8**, 199–249 (1975)
30. L.A. Zadeh, The concept of a linguistic variable and its application to approximate reasoning, III. *Inform Sci.* **9**, 43 $\pm$ 80 (1976)
31. L.A. Zadeh, Fuzzy logic = computing with words. *IEEE Trans. Fuzzy Syst.* **4**(2), 103–111 (1996)
32. L.A. Zadeh, From computing with numbers to computing with words—from manipulation of measurements to manipulation of perceptions 1, in *Intelligent Systems and Soft Computing: Prospects, Tools and Applications* (Dec 2006), p. 3
33. L.A. Zadeh, A new direction in AI—toward a computational theory of perceptions. *AI Mag.* **22**(1), 73–84 (2001)
34. L.A. Zadeh, Generalized theory of uncertainty (GTU)—principal concepts and ideas. *Comput. Stat. Data Anal.* **51**(1), 15–46 (2006)

The Development of Interval Data Type for Analytical Information Processing

A. K. Pogodaev, Alexander Galkin, Pavel Saraev, and A. I. Miroshnikov

Abstract The development of the interval analysis theory and its applications in solving applied problems led to the need to collect, store and extract efficiently interval-significant data in relational databases. Modern relational database management systems (DBMS) either allow to work with built-in data types which however are not friendly for use or provide the ability to create your own user-defined data types in several ways. The article describes the creation of a user-defined data type iInterval, functions and aggregates to work with interval data in MS SQL Server using the Common Language Runtime (CLR).

Keywords Interval analysis · Database · User-defined data type · Interval data type

1 Introduction

The current state of information technology allows to accumulate large amounts of information in various spheres: technological, economic, financial, etc. An important problem is how to extract knowledge from available information [1]. The received knowledge can be used at decision making that lead to qualitative improvement of considered processes. The quality of the decisions largely depends on the correctness

This work is partially supported by Russian Foundation for Basic Research (RFBR) and Lipetsk regional administration Grant#16-47-480929-r_a.

A. K. Pogodaev · A. Galkin (✉) · P. Saraev · A. I. Miroshnikov
Faculty of Automation and Informatics, Lipetsk State Technical University, Lipetsk, Moskovskaya 30 398055, Russian Federation
e-mail: avgalkin82@mail.ru

A. K. Pogodaev
e-mail: pak@stu.lipetsk.ru

P. Saraev
e-mail: psaraev@yandex.ru

A. I. Miroshnikov
e-mail: a.i.miroshnikov@yandex.ru

S. N. Shahbazova et al. (eds.), *Recent Developments and the New Direction in Soft-Computing Foundations and Applications*, Studies in Fuzziness and Soft Computing 393, https://doi.org/10.1007/978-3-030-47124-8_41

of the knowledge extracted and the time of its origination [2]. Information, on the basis of which management decisions are made, can come from various sources. It can be obtained through measurements containing an error which eventually leads to its uncertainty and inconsistency [3]. There are different methods of processing and extracting knowledge which take into account the uncertainty of information. An interval representation of information can be one of these methods which puts the lower and upper limits of the interval in the value of the numerical parameter in which the value is reliably contained. Interval storage in database management systems can be implemented by organizing two fields corresponding to the minimum and maximum values (the lower and upper limits of the interval).

2 Theoretical Analysis

A. *Methods for implementing the interval data type*

At the moment various methods and algorithms for working with interval data have been developed as well as specialized software that implements them [4–6]. The use of computations over interval data makes it possible to obtain an estimate of all possible variants which helps in decision making that minimizes risks.

The following ways can be proceeded in case when it is necessary to take into account interval parameters:

- create the structure of tables in the database that specify two columns in the table containing the lower and upper boundaries of the interval. The disadvantage of this approach is the complexity of implementing operations on interval parameters. When performing queries the user must each time implement operations on the intervals. That can lead to errors because of the ambiguity in the definition of such operations [7–9];
- create interval-valued data types and processing operations which can greatly simplify the work of users programmers. In this case they don't have to think constantly about determining the operations of processing interval data. An additional point is that SQL queries will not differ from standard queries to relational databases that use scalar data types.

B. *Analysis of built-in data types*

In [10] it is stated that in Microsoft SQL Server database management system spatial intervals can be represented by GEOMETRY and GEOGRAPHY spatial data types. There are special methods to operate with these types.

The term geometric object (GEOMETRY) represents data in two-dimensional space in the form of points, lines and polygons using one of the existing flat spatial models.

A geometric object can be imagined as a data type with several subtypes. Subclasses are divided into two categories: basic geometric subclasses and subclasses of homogeneous collections. Basic geometric subclasses include Point, LineString and Polygon subclasses. Homogeneous collections include the MultiPoint, Multi-LineString and MultiPolygon subclasses. A list and brief description of the types being accessed is given in [11].

GEOMETRY and GEOGRAPHY data types are implemented by Microsoft SQL Server application developers using the Common Language Runtime (CLR) layer that main task is to load and manage.NET types. In addition the CLR is responsible for a number of issues such as memory management and security control.

But it should be pointed that when GEOMETRY and GEOGRAPHY types are used the data is displayed in hexadecimal format which is not suitable for instant processing. So it is necessary to use the expression parser which reduces the accuracy of the calculations.

The SQL Server Management Studio environment has the ability to display spatial data in a graphic format. This functionality can be demonstrated in the following example in Transact-SQL language:

```
DECLARE @polygon1 GEOMETRY = 'POLYGON((1 1, 1
5, 5 5, 5 1, 1 1))';
DECLARE @line GEOMETRY = 'LINESTRING (0 1, 7 5)';
SELECT @polygon1
UNION ALL
SELECT @line
```

After the query execution the rectangle specified by the @polygon1 variable is displayed as well as the line according to the variable @line, as shown in Fig. 1.

There is no special support for intervals of other types provided in Microsoft SQL Server. These intervals are often represented by two attributes that contain the lower and upper values using in queries predicates related to these intervals attributes.

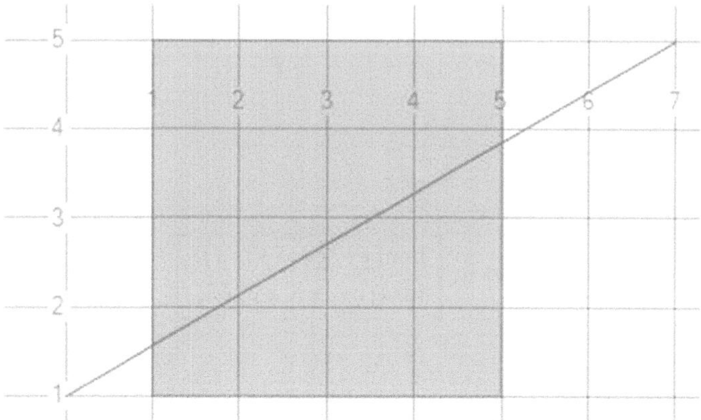

Fig. 1 Visual query result in SQL server management studio

The main problem of this method is the database rows access inauthenticity that contain fields associated with rounding errors when the information is presented in digital form. To eliminate this problem it is necessary to implement special algorithms for reliable processing of stored data.

There are different types of intervals that need to be represented in databases including temporary (for example, different terms), spatial (for example, road sections) and numerical (for example, temperature ranges) ones.

C. *Basic definitions of interval analysis*

The main term of interval analysis is the interval [4, 6]. An interval is any simply connected subset of the set of real numbers:

$$[x] = [\underline{x}, \bar{x}] = \{x | \underline{x} \le x \le \bar{x}\}. \tag{1}$$

The following basic operations are defined over the intervals:

- multiplication by scalar

$$a[x] = \begin{cases} [a\underline{x}, a\bar{x}], a \ge 0, \\ [a\bar{x}, a\underline{x}], a < 0, \end{cases}, \tag{2}$$

- addition of intervals

$$[x] + [y] = [\underline{x} + \underline{y}, \bar{x} + \bar{y}], \tag{3}$$

- subtraction of intervals

$$[x] - [y] = [\underline{x} - \bar{y}, \bar{x} - \underline{y}], \tag{4}$$

- multiplication of intervals

$$[x][y] = \begin{bmatrix} \min\{\underline{x}\underline{y}, \underline{x}\bar{y}, \bar{x}\underline{y}, \bar{x}\bar{y}\}, \\ \max\{\underline{x}\underline{y}, \underline{x}\bar{y}, \bar{x}\underline{y}, \bar{x}\bar{y}\} \end{bmatrix}, \tag{4}$$

- division of intervals

$$\frac{[x]}{[y]} = [x]\left[\frac{1}{\bar{y}}, \frac{1}{\underline{y}}\right], 0 \notin [\underline{y}, \bar{y}] \quad (5)$$

Such mathematical functions as root, exponential, trigonometric and others can also be used in SQL queries. If these functions are applied to interval arguments the result will also be the interval containing the possible values of the function. The problem of determining the interval expansion of functions in many cases is equivalent to the search for a range of values (search for optima) for a function. Therefore such functions must be overrided in the query language too. One way to implement it is to use natural or centered interval extensions.

The natural interval extension is defined as:

$$f_n([x]) = \begin{bmatrix} \min(f(x)|x \in [x]); \\ \max(f(x)|x \in [x]) \end{bmatrix}. \quad (6)$$

The centered interval expansion is defined as:

$$f_{mv}([x]) = f(mid[x]) + f_n^{'}([x])([x] - mid[x]), \quad (7)$$

where $mid[x] = \frac{\underline{x}+\bar{x}}{2}$ is the middle of interval.

The calculation of the natural and centered interval extensions is performed to obtain a more accurate result and to narrow the resulting interval. The maximum lower limit of interval extensions is selected as the lower limit of the final interval and the minimum upper limit is chosen as the upper limit of the result.

3 Methodology

A. *User-defined data types implementation ways in SQL Server*

The SQL Server includes the Database Engine, Analysis Services, Reporting Services, Integration Services Integration Services and the SQLXML extension. It allows to manage structured, semi-structured and nonstructured data. SQL Server provides a comprehensive, integrated system software that implements both an OLTP system and a data analysis system [11, 12].

There are three ways to create a user-defined data type using the CREATE TYPE statement in Microsoft SQL Server 2012 DBMS:

- a type alias creation based on one of the standard types provided by SQL Server;
- a user-defined table type creation;
- a user-defined data type creation using the CLR.

The first method is not suitable for the implementation of the user-defined "interval" type because it is an alias of only one standard type and the interval type must contain at least two fields.

The second method allows to create a type based on two fields but it returns a separate table that can not be used as a field when creating data tables. In this case it is necessary to create triggers on the INSERT, UPDATE and DELETE statements to maintain data consistency each time an interval field is created and when creating relationships between tables that use fields of this type.

The proposed implementation of the interval-significant data type for MS SQL Server 2012 is based on the third method by using of the.NET common language runtime. In this case the entire description of the data type is implemented in C# as a dynamic-link class library class that binds to SQL Server as an assembly. A user-defined type, user-defined functions, procedures and aggregates can be created based on classes, structures and methods.

The advantage of this approach is the ability to create Interval type table fields and work with them later using the usual call of DML call instructions in Transact-SQL.

B. *Base class of the user-defined type iInterval description*

The iInterval class includes the following main fields and their properties:

- private double down;
- private double up;
- private bool isNull.

An INullable interface is specified for the class. The use of the Format. Native argument determines the need for a serialization format. The IBinarySerialize interface provides a user-defined serialization of the type. The Read() and Write() methods must be defined to serialize the data in BinaryReader and BinaryWriter [12].

The constructor of the class takes the lower and upper bounds of the interval. The lower boundary is rounded to the minus infinity and the upper one to plus infinity:

```
public iInterval(double down, double up) {
      if (down >= up)
         this = Null;
      else {
         isNull = false;
         this.down = Math.Round(down, precision,
idpointRounding.ToEven);
         this.up = Math.Round(up, precision,
MidpointRounding.AwayFromZero);
      }
   }.
```

The INullable interface requires defining the IsNull property to support the ability to store null values: public bool IsNull. The static Null property is used to create an object that represents a null value: public static iInterval Null.

The UDT type implements the type redefinition in the string. The following implementation of this override is proposed:

```
public override string ToString() {
   if (this.isNull)
      return null;
   return String.Format("{0};{1}", down, up);
}.
```

The static method Parse() is used to reverse the conversion from to iInterval type string in SQL Server:

```
public static iInterval Parse(SqlString s) {
   if (s.IsNull)
      return iInterval.Null;
   try {
      string[] boundaries = s.Value.Split(';');
      return new
iInterval(Convert.ToDouble(boundaries[0]),
Convert.ToDouble(boundaries[1]));
   }
   catch (Exception ex) {
      throw new ArgumentException("Неверный
синтаксис аргумента.");
   }
}.
```

C. *User-defined interval functions and aggregates*

Not only the standard aggregates of the quantity count (COUNT) and the sum (SUM) of the values of all intervals were implemented but special functions that return the lower value of the interval, the upper value of the interval, the width, the median, the interval multiplied by the scalar and the product of intervals (Table 1).

Aggregates of the sum and product of intervals are realized the same way. The four methods described above should be prescribed for each unit: Init, Accumulate, Merge and Terminate.

The difference in intervals is not an aggregating function so it is necessary to use the above operations of addition and multiplication by minus one for its implementation.

Table 1 User-defined aggregates and functions description

Name	Input parameters	Return type
ICOUNT	iInterval	SqlInt32
IDOWN	iInterval	SqlDouble
IUP	iInterval	SqlDouble
IWIDTH	iInterval	SqlDouble
IMID	iInterval	SqlDouble
ISCALAR_MULT	iInterval, SqlDouble	iInterval

The mathematical function that take interval values creation can be represented by the example of an exponential function. First of all it is necessary to determine the functions of the natural interval expansion (NEXP) and the centered interval expansion (CEXP) in accordance with formulas (6) and (7). After that it is necessary to execute intersection of return values of various interval extensions in interval function EXP:

```
public static iInterval EXP(iInterval Value) {
    iInterval n_exp = new iInterval();
    iInterval c_exp = new iInterval();
    iInterval exp = new iInterval();
    n_exp = NEXP(Value); c_exp = CEXP(Value);
    exp.down = Math.Min(c_exp.down, n_exp.down );
    exp.up = Math.Max(c_exp. up, n_exp. up);
    return exp;
}.
```

4 Experiment

An Intervals table was created in SQL Server to demonstrate the interval data type operations (Tables 2, 3, 4 and 5):

Table 2 The interval-significant data type storing format

Id	Interval
1	1,1;2,1
2	1,2;2,2
3	1,3;2,3
4	1,4;2,4
5	1,5;2,5

Table 3 ICOUNT aggregation function result

Number of columns
5

Table 4 Execution result of the user-defined functions IDOWN and IUP and the aggregate functions IWIDTH and IMID

Id	Down	Up	Width	Medium
1	1,1	2,1	1	1,6
2	1,2	2,2	1	1,7
3	1,3	2,3	1	1,8
4	1,4	2,4	1	1,9
5	1,5	2,5	1	2

Table 5 the ISUM aggregate function result

Down	Up
6,5	11,5

```
CREATE TABLE Intervals (
    id INT NOT NULL PRIMARY KEY IDENTITY,
    interval iInterval NOT NULL
).
```

To insert walues in this table execute:

```
INSERT INTO Intervals (interval)
VALUES('1,1;2,1'), ('1,2;2,2'), ('1,3;2,3'), ('1,4;2,4'), ('1,5;2,5').

CREATE AGGREGATE ICOUNT (@value iInterval)
RETURNS INT
EXTERNAL NAME [INTERVALS].[iInterval.ICOUNT];
    -- output of the number of intervals
SELECT dbo.ICOUNT(interval)
FROM Intervals
    -- user-defined function that returns the lower bound of the interval creation
CREATE FUNCTION dbo.IDOWN(@value iInterval)
RETURNS FLOAT
AS EXTERNAL NAME
INTERVALS.[iInterval.IDOWN].[DOWN]
    -- user-defined function that returns the upper bound of the interval creation
CREATE FUNCTION dbo.IUP(@value iInterval)
RETURNS FLOAT
AS EXTERNAL NAME INTERVALS.[iInterval.IUP].[UP]
    -- user-defined function that returns the width of the interval creation
```

```
CREATE FUNCTION dbo.IWIDTH(@value iInterval)
RETURNS FLOAT
AS EXTERNAL NAME
INTERVALS.[iInterval.IWIDTH].[WIDTH]
    -- user-defined function that returns the median of the
interval creation
CREATE FUNCTION dbo.IMID(@value iInterval)
RETURNS FLOAT
AS EXTERNAL NAME
INTERVALS.[iInterval.IMID].[MID]
    -- user-defined functions execution
SELECT dbo.IDOWN(interval) AS down,
        dbo.IUP(interval) AS up,
        dbo.IWIDTH(interval) AS width,
        dbo.IMID(interval) AS medium
FROM Intervals;
    -- user-defined function that returns the result of
multiplying an interval by a number creation
CREATE FUNCTION dbo.ISCALAR_MULT(@value
iInterval, @a FLOAT)
RETURNS iInterval
AS EXTERNAL NAME
INTERVALS.[iInterval.ISCALAR_MULT].[SCALAR_MUL
T]
    -- all intervals of the interval column multiplication by 2
UPDATE Intervals
SET interval = dbo.ISCALAR_MULT(interval, 2.0);
    -- user-defined aggregate creation

CREATE AGGREGATE ISUM (@value iInterval)
RETURNS iInterval
EXTERNAL NAME INTERVALS.[iInterval.ISUM];
    -- the sum of intervals on an interval column scaling
SELECT dbo.IDOWN(dbo.ISUM(interval))
FROM Intervals
```

To implement interval function expansion the classes presented in Table 6 were implemented.

5 Conclusion

The introduced interval data type and operations over it allow query developers not to think about the nature of the data stored in the database when processing them. This avoids errors in the operations implementation on interval data because there

Table 6 Functions with interval extension

Class	Description
IPOW(iInteral Value, SqlDouble a)	a^x interval extension
IEXP(iInteral Value)	e^x interval extension
ILOG(iInteral Value, SqlDouble a)	$\log_a x$ interval extension
ISQ(iInteral Value)	x^2 interval extension
ISQRT(iInteral Value)	$\surd x$ interval extension
ISIN(iInteral Value)	$\sin(x)$ interval extension
ICOS(iInteral Value)	$\cos(x)$ interval extension
ITAN(iInteral Value)	$\tan(x)$ interval extension
IASIN(iInteral Value)	$a\sin(x)$ interval extension
IACOS (iInteral Value)	$a\cos(x)$ interval extension
IATAN(iInteral Value)	$a\tan(x)$ interval extension

is no need to create two separate fields of the same type to maintain constant data integrity in these fields and while implementing operations on the data in these fields. Developed functions of to the interval data type execution is carried out in the form of instructions of the Transact-SQL.

References

1. A.K. Pogodaev, P.V. Saraev, E.P. Tatarinov, Universal information and software for analytical data processing, in *Information Technologies of Modeling and Management*, vol. 4 (63) (Science Book, Russian Federaion, Voronez, 2010), pp. 543–550
2. A.V. Galkin, D.V. Djatchina, A.K. Pogodaev, The efficiency of information processing increasing in databases using materialized views, in *Control Systems and Information Technology*, vol. 3.1 (57) (Science Book, Russian Federaion, Voronez, 2014), pp. 136–140
3. P.V. Saraev, A.V. Galkin, A.I. Miroshnikov, A.A. Nikol'skaya, Object processing interval type in SQL SERVER DBMS, in *Big systems control* (Perm National Research Polytechnic University, Russian Federaion, Permian, 2017), pp. 397–404
4. B.S. Dobronec, *Interval mathematics* (Krasnoyarsk State University, Russian Federaion, Krasnoyarsk, 2004)
5. S.A. Kalmikov, U.A. Shokin, Z.H. Juldashev, *Methods of interval analysis* (Science, Russian Federaion, Novosibirsk, 1986)
6. S.P. Sharij, *Finite-interval analysis* (ICT SB RAS, Russian Federaion, 2016)
7. R. Vieira, *Beginning Microsoft SQL Server 2008 programming* (Wrox Press Ltd., Birmingham, UK, 2009)
8. J. Groff, P. Weinberg, A. Oppel, *SQL: The Complete Reference* (Mcgraw Hill Education, US, 2011)
9. B. Shvarc, P.N. Vainberg, V. Tkachenko, J. Zavodni, A. Lenc, D. Belling, *MySQL. Optimizing Performance* (Symbol Plus, St. Petersburg, Russian Federaion, 2010)
10. A.I. Miroshnikov, A.A. Nikolskaya, The research of the interval databases design and interval queries execution methods, in *Modern Informatization Problems in the Technological and*

Telecommunication Systems Analysis and Synthesis: Proceedings of the XXII-th International Open Science Conference (Science Book Publishing House, Yelm, WA, USA, 2017), pp. 273–278
11. D. Petkovic, *Microsoft SQL Server 2012 A Beginners Guide 5/E* (McGraw-Hill Education, NY, 2012)
12. A.K. Pogodaev, A.V. Galkin, A.I. Miroshnikov, Interval data types and operations development in MS SQL Server, in *Control Systems and Information Technology*, vol. 1 (67), (Science Book, Russian Federaion, Voronez, 2017), pp. 48–51

How to Monitor Possible Side Effects of Enhanced Oil Recovery Process

Jose Manuel Dominguez Esquivel, Solymar Ayala Cortez, Aaron Velasco, and Vladik Kreinovich

Abstract To extract all the oil from a well, petroleum engineers pump hot reactive chemicals into the well. These Enhanced Oil Recovery (EOR) processes need to be thoroughly monitored, since the injected fluids can seep out of the production oil wells and, if unchecked, eventually pollute sources of drinking water. There is a need to measure the corresponding effects. One way to measure these underground effects is by observing seismic waves resulting from hot fluids-induced fracturing. Seismic waves generated by this fracturing are, however, weak in comparison with the background noise. Thus, the accuracy with which we can locate the spreading liquids based on these weak signals is low. Hence, we get only an approximate understanding of how those liquid propagate in the reservoir. To get a more accurate picture of the propagation of these fluids, we propose to use active seismic analysis: namely, we propose to generate strong seismic waves and use a large-N array of sensors to observe their propagation.

J. M. Dominguez Esquivel
Mexican Petroleum Institute, Ejec Central Lázaro Cárdenas 152, Col. San Bartolo Atepehuacan Ciudad de M´exico, C.P07730 Mexico City, Mexico
e-mail: jmdoming@msn.com; jmdoming@imp.mx

S. Ayala Cortez · A. Velasco
Department of Geological Sciences, University of Texas at El Paso, 500 W. University, El Paso, TX 79968, USA
e-mail: sayalacortez@miners.utep.edu

A. Velasco
e-mail: aavelasco@utep.edu

V. Kreinovich (✉)
Department of Computer Science, University of Texas at El Paso, 500 W. University, El Paso, TX 79968, USA
e-mail: vladik@utep.edu

S. N. Shahbazova et al. (eds.), *Recent Developments and the New Direction in Soft-Computing Foundations and Applications*, Studies in Fuzziness and Soft Computing 393, https://doi.org/10.1007/978-3-030-47124-8_42

1 Introduction

What is enhanced oil recovery process. Traditional oil and gas industry mostly rely on locations where oil and gas are stored under high pressure. Because of this pressure, oil and gas flow out of the well on their own.

As the pressure decreases, production decreases accordingly. Hence, higher pressure pumping is needed to recover physical push to enhance mobility of oil and gas to the surface; this is performed by water, nitrogen, or CO_2 injection.

Alternatively, instead of pumping high-pressure fluids, we can pump chemicals that convert difficult-to-extract heavy carbohydrates into easier-to-extract lighter ones. This is known as *enhanced oil recovery process*; see, e.g., [1, 7, 8] and references therein.

The resulting chemical reaction must be as efficient as possible. It is known that the speed of chemical processes exponentially grows with temperature. Hence, to speed up the corresponding processes, chemicals at high temperatures—between 200 and 350° C—are injected into the well. This leads to a better extraction of oil from the production wells which are near the injection well.

Enhanced oil recovery process: successes and problems. The enhanced oil recovery process has enabled us to extract a large amount of remaining oil—up to 75% of the remaining oil [8].

However, the problem is that the chemically aggressive hot liquids seep out, and the corresponding chemicals can eventually pollute the sources of drinking water.

Need to monitor the enhanced oil recovery process. To avoid unnecessary pollution, it is important to monitor how the pumped liquids propagate at the corresponding depths. Also, we need to monitor the location of the injected liquids after the injection process is over.

How the enhanced oil recovery process is monitored now. When the liquid propagates, it fractures the minerals and thus, causes minor earthquakes. Just like major earthquakes, the location of these minor earthquakes can be detected by the seismic waves that they generate; see, e.g., [2]. This passive seismic approach is indeed used for the desired monitoring.

Limitations of passive seismic monitoring. In contrast to major earthquakes, disturbances caused the pumped liquid are small. As a result, the generated seismic waves are very weak (they are imperceptible to human senses), the signal-to-noise ratio is very low, and hence, the accuracy with which we can trace the spreading of the pumped liquid is very low—we only get a very crude approximate understanding of how and where the hot liquids propagate.

What we do in this paper. In this paper, we propose an alternative, *active* seismic technique, that enables us to provide a more accurate picture:

- of liquid propagation and
- of the resulting location of the liquids.

Future plans. At this stage, we only have a theoretical idea, an idea supported by computer simulations. However, we are already planning real-life tests on a location in Mexico.

Need to take into account uncertainty, in particular, fuzzy uncertainty. How is all this related to fuzzy and soft computing? The relation is straightforward: we do not know the exact characteristics describing the propagation of the corresponding seismic waves. Instead, we need to rely on expert understanding of this process—and this understanding is often described in terms of imprecise ("fuzzy") words from natural language. To describe this knowledge in precise terms, it is reasonable to use techniques specifically developed for processing such expert statements—namely, the technique of fuzzy logic; see, e.g., [4, 6, 9–14].

2 Our Main Idea and the Corresponding Physics

Main idea. The low accuracy of the existing techniques is caused by the fact that the micro-quakes generated by enhanced oil recovery process are very *weak*. Thus, to improve this accuracy, a natural idea is to generate *stronger* seismic waves and to see how these waves propagate—by measuring the signals detected by the seismic sensor located on the Earth's surface. Such techniques, when we actively generate seismic waves, is known as *active seismic analysis*.

To describe this idea in detail, we need to describe:

- what kind of seismic signals we can generate,
- how the generated signals propagate, and
- how we can determine the location of the liquid based on the measurement results.

Let us consider these topics one by one.

What seismic signals can be generated. To generate an active seismic signal, we have basically two main options:

- we can use all the available energy at once, thus producing an explosion, or
- we can spread this energy over time, thus generating a periodic seismic signal; this is done by using especially equipped truck called a *vibroseis*.

In this paper, we consider both options.

How seismic waves propagate: a brief reminder. When the medium is reasonably homogeneous, with some inhomogeneities whose size is much larger than the wavelength of the corresponding seismic wave, then the waves propagate geometrically, by following paths. Specifically, the path between points A and B followed by a wave is the path for which the propagation time is the smallest possible; see, e.g., [2–5]. This shortest-time idea leads to the known Snell's Law of propagation, according to which, when a wave crosses the border between the two layers with different wave propagation speeds v_1 and v_2, then the angles α_1 and α_2 between the paths in both areas and the direction orthogonal to the border between the layers are related by the following formula:

$$\frac{\sin(\alpha_1)}{v_1} = \frac{\sin(\alpha_2)}{v_2}.$$

In such homogeneous situations, waves behave as if they were particles.

The situation changes drastically if we have inhomogeneities whose size is smaller than the wavelength. In this case, in the analysis of the wave propagation, we can no longer view the wave as a single whole, we need to take into account that different parts of the wave encounter areas with different wave propagation velocity and thus, get reflected differently. As a result, instead of the wave simply changing its direction and continuing as a single ray, we get a *scattering* phenomenon, when the wave that was initially a single ray starts going in several different directions.

How pumped liquid affects the propagation of seismic signals. The liquid spreads via the cracks—both the existing cracks and the cracks it generates. So, its trajectories are composed of linear paths whose width is definitely much smaller than the wavelengths of the seismic waves. Thus, the pumped liquid produced scattering.

In relative terms, the amount of liquid is small in comparison with the amount of surrounding minerals. Thus, the angle of the resulting scattering is mostly also small.

What we know before we start the enhanced oil recovery process. Usually, for an oilfield or a gas field, we know the velocities at different locations and different depths. Indeed, this is one of the main techniques based on which we decide that there is oil and/or gas in a given location—by:

- analyzing the seismic data,
- extracting the velocities from this data, and
- looking for patterns of the corresponding 3-D velocities model that are typical for oil and gas fields.

In this case, in the pre-pumping stage, if we use an explosion at some location E to generate a pulse wave, and we use a 2-D grid of surface sensors to monitor the resulting waves, then for each sensor location S, we also observe a single pulse. The time delay of this pulse is affected by the velocities along the smallest-time path that connects the explosion location E and the sensor location S.

Comment. In some locations, we may have small inhomogeneities. In this case, at the corresponding sensor, instead of a single instantaneous pulse, we observe a longer signal, a signal that combines the original pulse and the signals scattered by this inhomogeneity.

How we can determine the location of the liquid based on the measurement results: general idea. On the surface, we have a 2-D array of sensors that detect the signals on all possible surface locations.

Based on the previously obtained description of seismic velocities v at different depths, for each sensor S, we know the 1-D path following which the seismic signal propagates to reach this sensor S.

When in some underground location, the liquid appears, this liquid scatters the original seismic wave. Due to this scattering, the duration of the observed seismic wave becomes longer than it was before we started injecting the liquid.

So:

- if for some sensor, after the injection of the liquid, the observed seismic signal becomes wider that before,
- this means that somewhere along the corresponding 1-D path, there was the injected liquid.

How we can determine the location of the liquid based on the measurement results: first approximate idea. We know that the liquid is somewhere along the path, but based simply on the fact that the signal has become wider, we do not know where exactly on this path is the location of the liquid.

One way to find this location is it take into account that since, in geological terms, the amount of injected liquid is reasonably small in comparison to the amount of the surrounding minerals, the scattering angle α is small. How does this affect the duration of the observed signal?

Let D be the original distance from the source of the seismic wave to the sensor. Let us assume that at distance d from the sensor, the path changes the angle by α. Now, the overall path consists of two segments:

- the first segment of length $D - d$, and
- the second segment of length d at the angle α with the first segment.

Because of the angle, the length of the second segment in the original direction is no longer d, but the hypothenuse of the triangle in which d is one of the sides, i.e., the length is

$$d' = \frac{d}{\cos(\alpha)}.$$

For small α, we have

$$\cos(\alpha) \approx 1 - \frac{\alpha^2}{2},$$

thus

$$d' = \frac{d}{\cos(\alpha)} \approx \frac{d}{1 - \frac{1}{2} \cdot \alpha^2} \approx d + \frac{1}{2} \cdot \alpha^2 \cdot d^2.$$

The increase in path is proportional to d^2, thus the increase Δt in the duration of the observed signal is also proportional to d^2: $\Delta t \approx c \cdot d^2$, for some constant c.

So, based on the increase Δt in the duration of the observed signal:

- we can not only find the 1-D path along which the liquid is located,
- we can also find the location of the liquid along this path:

Namely, the liquid is located at a distance d from the sensor, where

$$d \approx \sqrt{\frac{\Delta t}{c}}.$$

Case of the periodic active seismic signal. For the sinusoid (periodic) signal, scattering appears as smoothing of the signal, with amplitude decreasing as

$$\exp(-d^2 \cdot \alpha^2) \approx 1 - \alpha^2 \cdot d^2,$$

i.e., with a change in amplitude proportional to d^2.

In this case, we can similarly estimate d based on the observed decrease in the amplitude of the observed signal.

This location is still approximate. While the 1-D path can be determined reasonably accurately, the exact distance d on this path is determined only approximately – since:

- the constant c depends on the scattering angle, and
- this angle may be somewhat different for different locations of the liquid.

How can we get a more accurate location of the liquid?

How we can determine the location of the liquid based on the measurement results: second idea that leads to much more accurate location. The above analysis shows that:

- if we have only one source of active seismic signals,
- then we cannot find the distance between the liquid and the sensor very accurately.

Thus, to make a more accurate location, a natural idea is to use *two* different sources of the active seismic waves.

Based on each source, we find the 1-D paths that contain the desired liquid locations. We can find the actual location of each liquid mass as the intersection of the two corresponding 1-D paths.

As we have mentioned earlier, the paths are determined very accurately, as a result we can find the location of the liquid very accurately.

This way, we can determine the size of the liquid, not just its location. As we have mentioned earlier, the scattering effect occurs only when the size of the obstacle starts being commeasurable with the wavelength. The generated seismic wave is usually a combination of waves of several wavelengths. The corresponding frequencies range from 1 to 475 Hz, with:

- the smallest frequency 1 Hz corresponding to the longest wavelength, and
- the largest frequency of 475 Hz corresponding to the smallest wavelength.

Thus:

- on the shortest wavelengths, which are much smaller than the size of the liquid mass, we will not see any scattering,
- on the other hand, on the longest wavelengths, we will see an increase in the duration of the observed seismic signal—which is an indication of scattering.

Thus, by comparing the signals on different wavelengths, we can find the wavelength at which the scattering starts—and thus, find:

- not only the location of the liquid mass,
- but also its size.

Acknowledgements This work was supported in part by the US National Science Foundation grant HRD-1242122.

References

1. H. Aguilar-Cisneros, V. Uribe-Vargas, B. Carreón-Calderón, J.M. Domínguez-Esquivel, M. Ramirez-de-Santiago, Hydrogen solubility in heavy undefined petroleum fractions using group contributions methods. Oil Gas Sci. Technol.–Rev. IFP Energ. Nouv. **72**(2) (2017)
2. K. Aki, P.G. Rochards, *Quantitative Seismology* (University Science Books, Sausalito, California, 2009)
3. M.G. Averill, K.C. Miller, G.R. Keller, V. Kreinovich, R. Araiza, S.A. Starks, Using expert knowledge in solving the seismic inverse problem. Int. J. Approx. Reason. **45**(3), 564–587 (2007)
4. S. Ayala Cortez, A.A. Velasco, V. Kreinovich, Soft computing approach to detecting discontinuities: seismic analysis and beyond, in *Proceedings of the IEEE Conference on Systems, Man, and Cybernetics SMC2017* (Banff, Canada, October 5–8 2017)
5. S. Ayala Cortez, A.A. Velasco, V. Kreinovich, Need for a large-N array (and wavelets and differences) to determine the assumption-free 3-D Earth model. J. Uncertain Syst. (to appear)
6. R. Belohlavek, J.W. Dauben, G.J. Klir, *Fuzzy Logic and Mathematics: A Historical Perspective* (Oxford University Press, New York, 2017)
7. B. Carreón-Calderón, V. Uribe-Vargas, M. Ramirez-De-Santiago, E. Ramírez-Jaramillo, Thermodynamic characterization of heavy petroleum fluids using group contribution methods. Ind. Eng. Chem. Res. **53**(13), 5598–5607 (2014)
8. Enhanced Oil Recovery, US Department of Enegry website. https://energy.gov/fe/science-innovation/oil-gas-research/enhanced-oil-recovery
9. G. Klir, B. Yuan, *Fuzzy Sets and Fuzzy Logic* (Prentice Hall, Upper Saddle River, New Jersey, 1995)
10. J.M. Mendel, *Uncertain Rule-Based Fuzzy Systems: Introduction and New Directions* (Springer, Cham, Switzerland, 2017)
11. J.M. Mendel, D. Wu, *Perceptual Computing: Aiding People in Making Subjective Judgments* (IEEE Press and Wiley, New York, 2010)
12. H.T. Nguyen, E.A. Walker, *A First Course in Fuzzy Logic* (Chapman and Hall/CRC, Boca Raton, Florida, 2006)
13. V. Novák, I. Perfilieva, J. Močkoř, *Mathematical Principles of Fuzzy Logic* (Kluwer, Boston, Dordrecht, 1999)
14. L.A. Zadeh, Fuzzy sets. Inf. Control **8**, 338–353 (1965)

Customer and Product Clustering in Retail Business

Ondřej Sokol, Vladimír Holý, and Tomáš Cipra

Abstract The customer segmentation plays an important role in modern retail marketing strategies. It is a common practice to classify customers involved in a loyalty programme based on frequency and value of their shopping. In addition to this information, we also utilize the characteristics of bought products, specifically the product category. We use k-means to cluster the customers and determine the key factors in the clustering. We also explore the relationship between customer clustering and product clustering. We cluster products into broader categories based on shopping behavior of the customer segments. This approach significantly reduces the size of the product clustering problem and results in meaningful product supercategories. As an application, we analyze a Czech drugstore company.

1 Introduction

Retail chains have huge amount of market basket data available. The analysts have access to every single basket—a set of items that were bought together along with the revenue. The specific of the Czech retail market is a big popularity of loyalty programs—in the case of drugstore market data, over three quaters of revenue can be linked to a member of the loyalty program. One of the reasons of popularity of loyalty programs in the Czech retail market is the number of marketing promotions available only to members of the loyalty program. The Czech retail market, especially

O. Sokol (✉) · V. Holý
Department of Econometrics, University of Economics, Prague, Winston Churchill Square 1938/4, 130 67 Prague 3, Czech Republic
e-mail: ondrej.sokol@vse.cz

V. Holý
e-mail: vladimir.holy@vse.cz

T. Cipra
Department of Probability and Mathematical Statistics, Charles University, Sokolovská 49/83, 186 75 Prague 8, Czech Republic
e-mail: cipra@karlin.mff.cuni.cz

S. N. Shahbazova et al. (eds.), *Recent Developments and the New Direction in Soft-Computing Foundations and Applications*, Studies in Fuzziness and Soft Computing 393, https://doi.org/10.1007/978-3-030-47124-8_43

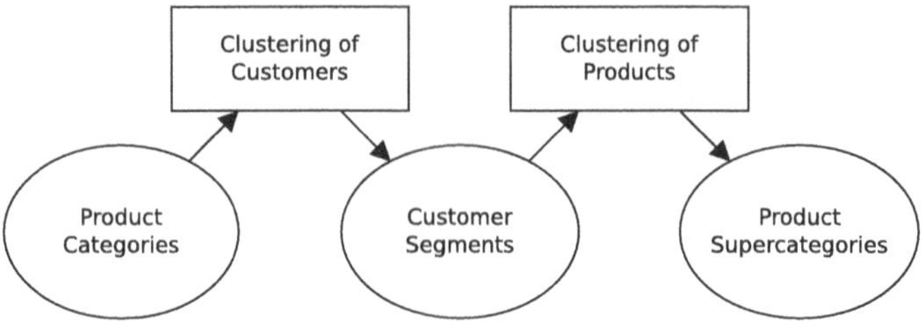

Fig. 1 Customer and product clustering process

drugstore, is well known for the huge amount of promotion. Some brands make over 90% of revenue in the sales promotion.

The customer segmentation focuses on getting knowledge about the structure of customers. With large amount of data, the importance of proper customer segmentation increases. The customer segmentation can be done from many perspectives. The most popular include RFM segmentation based on the recency, frequency and monetary aspects of customers, price sensitivity based on the price level of bought products and promotional segmentation based on the ratio of products bought during the promotion.

We focus on the basket segmentation of customers in which customers are segmented based on the categories of products they bought. There are two goals of this paper.

1. Firstly, we determine the customer segmentation based on the categories of products bought.
2. Secondly, we use the customer segmentation and estimate the supercategories of the products.

The clustering process is demonstrated in Fig. 1. Both of our goals can be utilized in sales promotions planning, especially in collaboration with e-mail and online marketing.

The rest of the paper is structured as follows. In Sect. 2, we briefly review some related literature about customer segmentation and product clustering. In Sect. 3, we describe the data available in retail business. In Sect. 4, we present the clustering of customers based on shopping behavior and demonstrate its potential in application to real drugstore's market basket data. In Sect. 5, we follow with the clustering of products based on customer shopping behavior. The paper concludes with a summary in Sect. 6.

2 Clustering in Retail Business

Jonker et al. [1] dealt with customer segmentation and its usability in marketing. He focused not only on the segmentation into homogenous groups of customers but on the determination of the optimal policy to each segment. In this paper, we try to follow this basic idea, so after this determine homogenous clusters of customers, we use it to determine groups of products using the found segments.

Another approach to determining customer segmentation was used by Lockshin et al. [2] who clustered the shoppers in the wine shops. The novel of his approach was in adding brand decision involvement to the clustering.

Customer segmentation based on self-organizing maps with a priori prioritization in direct marketing was proposed in Seret et al. [3]. Another two-stage approach for dynamically deriving behaviorally persistent segments was presented in Reutterer et al. [4]. In their article, the segments are used to subsequently target marketing selection using basket histories from loyalty-program members.

Ammar et al. [5] presented a meta-clustering approach which uses the results from clustering of customers as an input to cluster the products, and recursively uses clustering of products as input to the clustering customers. In this paper, we use a similar idea.

Desai and Talukdar [6] focused on the product groups and its relationship using customer segmentation based on the size of their baskets and price level of products they bought.

One of the most important area in marketing (especially in Czech market) is the promotional planning, this topic was studied by Hruschka et al. [7]. The authors focused on the cross-category sales promotion effects.

More general clustering of retail market baskets along with visualization can be found in Nikhil [8]. The recent survey of the heuristic and metaheuristic clustering methods was published in José García [9].

3 Retail Data

Nowadays, retail chains have a database with every receipt of every customer in the form of a list of products with prices. The products can be linked to the table with product's description which involves all information about the categorization of the product and its properties such as weight, brand or producer.

Similarly, a lot of data can be retrieved about customer if he is a member of the loyalty program. For example, his age, place of residence, children etc. Some data such as e-mail, phone number or web account on the retailer's website can be used to determine the involvement of the client to the retailer. This type of data is gathered from the customer directly and while some data is invalid (either obsolete or some customers give deliberately false data), the vast majority of data is valid.

Other approach to get data about the clients is to use their purchase history. By analysis of purchase history, the retailer can estimate the lifestyle of the customer, for example estimate if the customer owns a pet or if the customer has newly born children. In order to simplify the process, various customer segmentations are used based on the recency, frequency and monetary aspects, price level of bought products, the ratio of products bought during the promotion or, as we mentioned above, the category of products bought by the customer. We follow this idea and determine the customer segments based on the categories of bought products.

Similar approach can be used to clustering products based on the basket data. This includes the analysis of price sensitivity of the product, whether the product is often bought together with some other products, ratio of purchases during sales promotions or the analysis of which segment of customers buys a product or category over-proportionally—this is what we study in the second part of this paper.

We use a sample from real dataset of Czech drugstore retail chain. The dataset contains all receipts of the sample of the most valuable customers (based on RFM segmentation, see for example [10]) from the 4th quarter of 2017. The size of our dataset is over 600,000 receipts of over 9000 customers. The number of products is over 10,000 divided into 320 categories. The categories involve huge ones such as various cleaning products, tooth-brushes or hair dyes and colorants to small ones such as car accessories or paints and varnishes. In our dataset, we also have a category of e-shop delivery, which involves various pharmacy products. Those products can be ordered at the e-shop with the delivery to the drugstore.

4 Customer Clustering

We are looking for the most elementary archetypes of customers. Because of that, we expect the number of customer clusters to be lesser than ten. We also expect the vast portion of customers to be assigned to single, *general*, cluster. This cluster would describe the average customer, e.g. a customer who is a target of the current *general* marketing campaigns. Even though the other clusters are expected to be much smaller in numbers of customers, it is expected that the on-top revenue we could get by the targeted marketing should at least pay the additional costs.

We aggregated the revenue by the customer and the product category which is based mostly on product purpose and sometimes on the price level and was set by experts of the procurement department. We compute the customer's ratio of revenue in each category. The absolute values of revenue would skew the results as the customers have very different monthly revenues. However, this information is not lost as it is included in different segmentation such as previously mentioned RFM.

Using these data, we performed k-means to find clusters of customers. For k-means we use the implementation in R from package *stats* with a maximal number of iteration set to 300 and number of starting point set to 30. These parameters were chosen after numerous testing. With these parameters, in the vast majority of instances the found clusters are the global extremes we were looking for.

For the evaluation of clustering we use Davies-Bouldin index [11] and the ratio of between cluster variance and the total variance. We run the k-means for 2–70 cluster and based on these statistics (see Figs. 2 and 3) we have concluded that the optimal number of clusters should be around 7. This number is also convenient for the targeting of the mass marketing campaign.

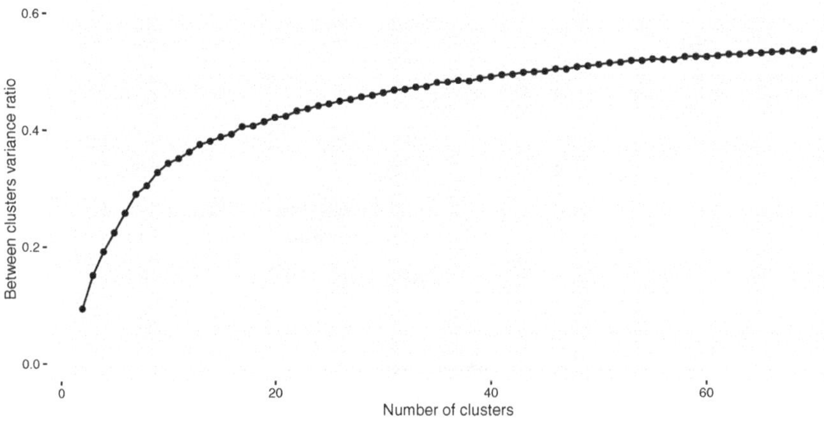

Fig. 2 Ratio of between clusters variance and total variance—customers clustering

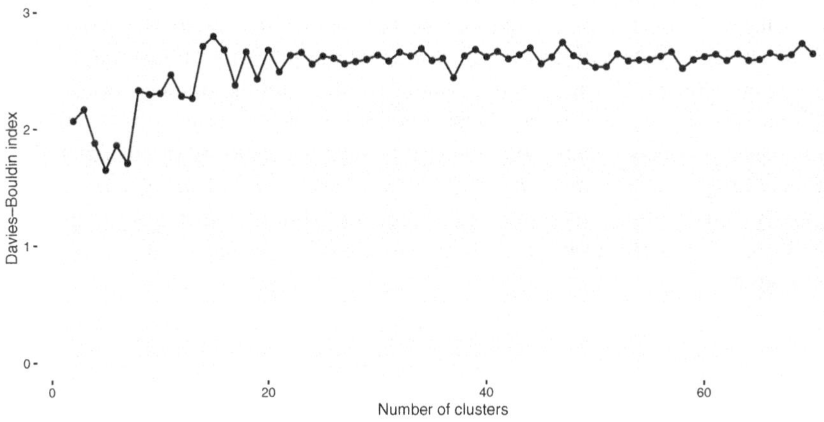

Fig. 3 Davies–Bouldin index—customers clustering

The resulting clustering revealed some interesting groups; however, as the dataset is confidential, we do not mention the statistics that could lead to the improper revealing of the data. The resulting groups with respective size are to be seen in Table 1, where for simplicity we renamed the clusters based on properties of its center point.

Cluster 1 and 4 consists of customers with over-proportional revenue of perfumes categories for men and women respectively. This is the example of clusters which are well-defined as it known that perfumes are bought usually by wealthy customers. The same can be said about clusters 3 and 5 which involves the products for children. Customers with a child are one of the most valuable in the drugstore market.

Table 1 Description and statistics of customer clustering

	Short description	Size
1	Cologne	6357
2	General	60461
3	Child diapers	5254
4	Perfume	10512
5	Child alimentation	2739
6	E-shop pharmacy	1760
7	Products for men	4463

The cluster 6 surprisingly consists of customers who use the drugstore as the collection point for the pharmacy without prescription which can be ordered online. This is one of the extra services that is done by drugstore chain we get the data from. This category is very small and therefore it was not expected that it could form the own cluster. However, the customers who use this service have different shopping behavior in contrasts to the other groups.

The cluster 7 consists of customers who buy mostly the products for men, this is a minority in Czech drugstore retail market as the majority of customers in the drugstore are women. Cluster 2 is the *general* cluster with the standard behavior.

Clusters are fairly small except the second cluster which includes 2/3 of all premium customers. This is expected behavior as the vast majority of customers buy traditional drugstore customer. Our goal in this phase is to find customers who differ greatly and to which the company should use different marketing approach. While those specialized groups are small, they share on the revenue of drugstore is inconsiderable.

5 Product Clustering

In this section, we focus on the second goal which consists in the estimation of the product supercategories. The second goal could be done straightforwardly. However, due to the size of database, it is computationally expensive (see for example [12]).

In the next step, we utilize the newly found clusters of customer and estimate the supercategories—clusters of categories. We based the clustering on the revenue ratios of the products in each of the 7 customer segments given by Table 1.

Again, we use the k-means method. Similarly to the previous section, we used the k-means implementation in R from package *stats* with a maximal number of iteration set to 300 and number of starting point set to 30. We run the clustering for 2 to 60 clusters and based on the ratio of between clusters variance and total variance (see Fig. 4) and Davies-Bouldin index (see Fig. 5) we decided to use 17 clusters.

The resulting 17 clusters along with size can be found in Table 2. For simplicity, we added the short description of the resulting clusters based on the clustered categories

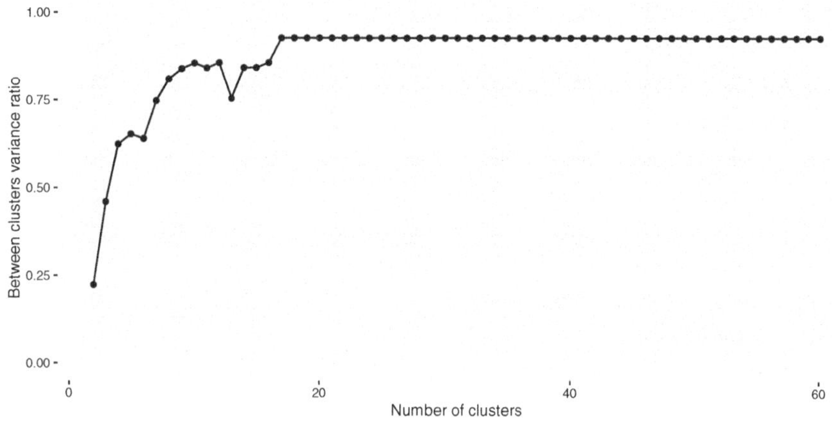

Fig. 4 Ratio of between clusters variance and total variance—category clustering

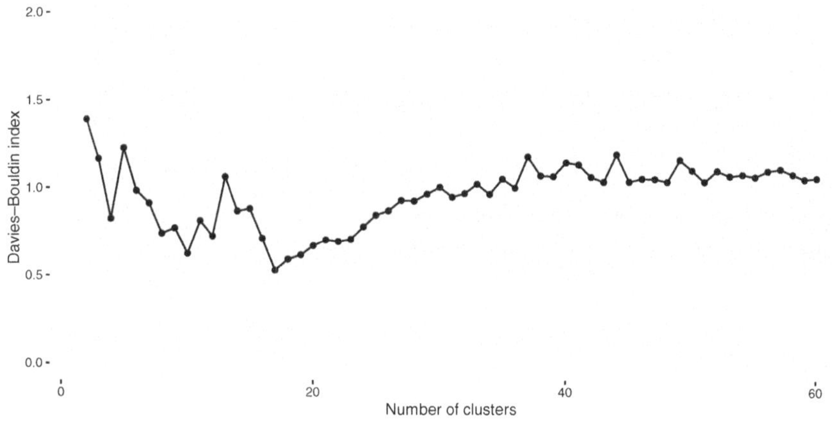

Fig. 5 Davies–Bouldin index—category clustering

with higher revenue. As it was described earlier, there is a total of 320 categories; however, the revenue is unevenly distributed so that top 60 categories do over 98% of revenue.

Similarly to the customer segmentation, the original categories are unevenly distributed to the found clusters. The reason behind it is, that small categories are very hard to cluster due to their fractional revenue. Many of them are clustered together into the clusters 1 and 5.

Many similar original categories were clustered together, for example sunbathing cream and oil and small skin care categories in cluster 16 or shower gel and shampoo categories in cluster 6. The child categories were divided into the two clusters as

Table 2 Description and statistics of category clustering

	Short characteristic	Categories
1	Condoms, various small cleaning categories	28
2	Child diapers	1
3	Cleaning products, feminine hygiene	31
4	Laundry products, make-up and hair colours	13
5	Small categories (products for cars, painting colours) and pet products	126
6	Men shaving products	1
7	Hygiene products and dental whitening	40
8	Cologne for men	1
9	Toys	3
10	Child alimentation	1
11	Cheaper beauty and body products	17
12	Premium Beauty, shower gel and shampoo	6
13	E-shop pharmacy	1
14	Dental products for children	3
15	Other child categories	6
16	Sunbathing and skin care	41
17	Perfumes for women	1

diapers and alimentation have different customers. This comes from the customer segmentation where we deducted the same thing.

6 Conclusion

We present the application of both customer and product clustering in the Czech drugstore market. The customers are clustered based on their shopping behavior in each of the product categories. Product categories are then clustered into supercategories based on the customer segmentation. Our approach revealed 7 segments of customers—1 cluster for general customers and 6 specialized clusters. This segmentation can be used in marketing promotions and bring additional profit for the drugstore companies.

Acknowledgements The work of Ondřej Sokol and Vladimír Holý was supported by the Internal Grant Agency of University of Economics, Prague under Grant F4/58/2017. The work of Tomáš Cipra was supported by the Czech Science Foundation under Grant P402/12/G097 (DYME).

References

1. J.J. Jonker, N. Piersma, D. Van den Poel, Joint optimization of customer segmentation and marketing policy to maximize long-term profitability. Expert Syst. Appl. **27**(2), 159–168, (2004), [Online]. http://linkinghub.elsevier.com/retrieve/pii/S095741740400003X
2. L.S. Lockshin, A.L. Spawton, G. Macintosh, Using product, brand and purchasing involvement for retail segmentation. J. Retail. Consum. Serv. **4**(3), 171–183, (1997), [Online]. http://linkinghub.elsevier.com/retrieve/pii/S0969698996000483
3. A. Seret, T. Verbraken, B. Baesens, A new knowledge-based constrained clustering approach: theory and application in direct marketing. Appl. Soft Comput. **24**, 316–327, (2014), [Online]. http://linkinghub.elsevier.com/retrieve/pii/S1568494614002658
4. T. Reutterer, A. Mild, M. Natter, A. Taudes, A dynamic segmentation approach for targeting and customizing direct marketing campaigns. J. Interact. Mark. **20**(3–4), 43–57, (2006), [Online]. http://linkinghub.elsevier.com/retrieve/pii/S1094996806700548
5. A. Ammar, Z. Elouedi, P. Lingras, Meta-clustering of possibilistically segmented retail datasets. Fuzzy Sets Sys. **286**, 173–196, (2016), [Online]. http://linkinghub.elsevier.com/retrieve/pii/S0165011415003619
6. K.K. Desai D. Talukdar, Relationship between product groups' price perceptions, shopper's basket size, and grocery store's overall store price image. Psychol. Mark. **20**(10), 903–933, (2003), [Online]. http://doi.wiley.com/10.1002/mar.10102
7. H. Hruschka, M. Lukanowicz, C. Buchta, Cross-category sales promotion effects. J. Retail. Consum. Serv. **6**(2), 99–105, (1999), [Online]. http://linkinghub.elsevier.com/retrieve/pii/S0969698998000265
8. L.C.J. Nikhil Pal, *Advanced Techniques in Knowledge Discovery and Data Mining* (*Advanced Information and Knowledge Processing*), 1st edn. (Springer, 2005)
9. A. Jos'e-Garc'a, W. G'omez-Flores, Automatic clustering using nature-inspired metaheuristics: a survey. Appl. Soft Comput. **41**, 192–213, (2016), [Online]. http://linkinghub.elsevier.com/retrieve/pii/S1568494615007772
10. A.X. Yang, How to develop new approaches to RFM segmentation. J. Target. Measur. Anal. Mark. **13**(1), 50–60, (2004), [Online]. https://doi.org/10.1057/palgrave.jt.5740131
11. D.L. Davies, D.W. Bouldin, A cluster separation measure. IEEE Trans. Pattern Anal. Mach. Intell. **PAMI-1**(2), 224–227, (1979), [Online]. http://ieeexplore.ieee.org/document/4766909/
12. V. Holý, O. Sokol, M. Černý, Clustering retail products based on customer behaviour. Appl. Soft Comput. (2017), [Online]. http://linkinghub.elsevier.com/retrieve/pii/S1568494617300728

Analysis of Word Embeddings Using Fuzzy Clustering

Shahin Atakishiyev and Marek Z. Reformat

Abstract In data dominated systems and applications, a concept of representing words in a numerical format has gained a lot of attention. There are a few approaches used to generate such a representation. An interesting issue that should be considered is the ability of such representations—called embeddings—to imitate human-based semantic similarity between words. In this study, we perform a fuzzy-based analysis of vector representations of words, i.e., word embeddings. We use two popular fuzzy clustering algorithms on count-based word embeddings, known as GloVe, of different dimensionality. Words from WordSim-353, called the gold standard, are represented as vectors and clustered. The results indicate that fuzzy clustering algorithms are very sensitive to high-dimensional data, and parameter tuning can dramatically change their performance. We show that by adjusting the value of the fuzzifier parameter, fuzzy clustering can be successfully applied to vectors of high—up to one hundred—dimensions. Additionally, we illustrate that fuzzy clustering allows to provide interesting results regarding membership of words to different clusters.

Keywords Fuzzy clustering · Fuzzy C-means · Fuzzy Gustafson-Kessel · Cluster validity · Word embeddings · Word vectors

1 Introduction

Word embeddings become an essential element of methods that focus on analysis and comparison of texts. One of the most popular embeddings is *GloVe* [1]. The embedding is obtained via analysis of word-word co-occurrences in a text corpus. A natural question is related to the ability of the embeddings to represent a human-based semantic similarity of words.

S. Atakishiyev (✉) · M. Z. Reformat
Department of Electrical and Computer Engineering, University of Alberta, Edmonton, Canada
e-mail: shahin.atakishiyev@ualberta.ca

M. Z. Reformat
e-mail: marek.reformat@ualberta.ca

S. N. Shahbazova et al. (eds.), *Recent Developments and the New Direction in Soft-Computing Foundations and Applications*, Studies in Fuzziness and Soft Computing 393, https://doi.org/10.1007/978-3-030-47124-8_44

Data clustering is the process of grouping objects in a way that similarity between data points that belong to the same group (cluster) becomes as high as possible, while similarity between points from different groups gets as small as possible. It is an important task in analysis processes and has been successfully applied to pattern recognition [2, 3], image segmentation [4], fault diagnosis and search engines [5]. Fuzzy clustering which allows data points to belong to several numbers of clusters with different membership grades has been proved to have useful applications in many areas. Specifically, fuzzy C-Means (FCM) clustering and its augmented version—fuzzy Gustafson-Kessel (FGK) clustering are the most popular fuzzy clustering techniques. High-dimensional spaces often have a devastating effect on data clustering in terms of performance and quality; this issue is regarded as the *curse of the dimensionality*.

Our study sheds some light on the comparative analysis of the above fuzzy clustering methods to observe if and how the results of fuzzy clustering change with the dimensionality of word embeddings. Additionally, we illustrate 'usefulness' of fuzzy clustering via analysis of degrees of belongings of words to different clusters.

The paper is structured as follows: Sect. 2 contains the description of Fuzzy C-Means (FCM) and Fuzzy Gustafson-Kessel (FGK) algorithms. In Sect. 3, we provide an overview of the validity indices applied for fuzzy clustering processes. The theoretical background behind a construction process of *GloVe* embeddings is covered in Sect. 4. Section 5 outlines the methodology, while Sect. 6 shows our experimental results. Finally, the obtained conclusion is presented in Sect. 7.

2 Fuzzy Clustering

Zadeh's fuzzy sets theory [6] has triggered a number of studies focused on the application of theoretical and empirical concepts of fuzzy logic to data clustering. In contrast to hard clustering techniques, where one point is assigned exactly to only one cluster, fuzzy clustering allows data points to pertain to several clusters with different grades of membership.

We have analyzed the behavior of FCM and FGK in clustering of vector representations of words in different dimensions. The details of these clustering methods have been described in the following sections.

2.1 *Fuzzy C-Means Clustering*

Fuzzy C-means algorithm was introduced by Bezdek [7] in 1981. It allows an observation to belong to multiple clusters with varying grades of membership. Having D as the number of data points, N as the number of clusters, m as the fuzzifier parameter, x_i as the i-th data point, c_i as the center of the j-th cluster, μ_{ij} as the membership degree of x_i for the j-th cluster, FCM aims to minimize

$$J = \sum_{i=1}^{D} \sum_{j=1}^{N} \mu_{ij}^{m} ||x_i - c_j||^2 \tag{1}$$

The FCM clustering proceeds in the following way:

1 Cluster membership values μ_{ij} and initial cluster centers are initialized randomly.
2 Cluster centers are computed according to the following formula:

$$c_j = \frac{\sum_{i=1}^{D} \mu_{ij}^{m} x_i}{\sum_{i=1}^{D} \mu_{ij}^{m}} \tag{2}$$

3 Membership grades μ_{ij} are updated in the following way:

$$\mu_{ij} = \frac{1}{\sum_{1}^{N} \left(\frac{||x_i - c_j||}{||x_i - c_k||} \right)^{\frac{2}{m-1}}}, u_{ij} \in [0, 1] and \sum_{i=1}^{c} u_{ij} = 1 \tag{3}$$

4 The objective function J is calculated
5 The steps 2, 3, 4 are repeated until the value of objective function become less than a specified threshold.

Fuzzy C-means (FCM) has many useful applications in medical image analysis, pattern recognition, and software quality prediction [7, 8], to name just a few. The most important factors affecting the performance of this algorithm is the fuzzifier parameter m, the size of and the dimensionality of data. The performance analysis of the algorithm for high dimensional clustering will be discussed in Sect. 7 in detail.

2.2 *Fuzzy Gustafson-Kessel Clustering*

Fuzzy Gustafson-Kessel (FGK) extends FCM by introducing an adaptive distance norm that allows the algorithm to identify clusters with different geometrical shapes [9]. The distance metric is defined in the following way:

$$D_{GK}^2 = (x_k - \vee_i)^T A_i (x_k - \vee_i) \tag{4}$$

where A_i itself is computed from fuzzy covariance matrix of each cluster:

$$A_i = (\rho_i |C_i)^{1/d} C_i^{-1}, \tag{5}$$

$$C_i = \frac{\sum_{k=1}^{N} \mu_{ik}^m (x_k - \vee_i)^T (x_k - \vee_i)}{\sum_{k=1}^{N} \mu_{ik}^m} \tag{6}$$

Here the parameter ρ_i is the constrained form of the determinant of A_i:

$$|A_i| = \rho_i, \rho_i > 0, \forall i \tag{7}$$

Enabling the matrix A_i to change with fixed determinant serves to optimize the shape of clusters by keeping the cluster's volume constant [9]. Gustafson-Kessel clustering minimizes the following criterion:

$$J = \sum_{i=1}^{c} \sum_{k=1}^{N} \mu_{ik}^m D_{GK}^2 = \sum_{i=1}^{c} \sum_{i=1}^{N} \mu_{ik}^m (x_k - \vee_i)^T A_i (x_k - \vee_i) \tag{8}$$

Like FCM, this optimization is also subject to the following constraints:

$$u_{ik} \in [0, 1], \forall i, k \; and \; \sum_{i=1}^{c} u_{ik} = 1, \forall k \tag{9}$$

We see that the computation of the FGK algorithm is more convoluted than FCM clustering.

3 Validity Indices

There are several validity indices to analyze the performance of the fuzzy clustering algorithms. One of them was proposed by Bezdek [7]. It is called a *fuzzy partition coefficient (FPC)*. This index is calculated as follows:

$$FPC = \frac{1}{N} \sum_{k=1}^{N} \sum_{i=1}^{c} u_{ik}^2 \tag{10}$$

FPC changes between [0, 1] range and the maximum value indicates best clustering quality. Another popular index to measure fuzzy clustering quality was proposed by Xie and Beni (XB) in 1991 [10]. It focuses on the two properties: cluster compactness and separation:

$$XB = \sum_{i=1}^{c} \frac{\sum_{k=1}^{N} (\mu_{ik})^m ||x_k - \vee_i||^2}{N min_{ik} ||x_k - \vee_i||^2} \tag{11}$$

The numerator part shows the strength of the compactness of fuzzy clustering, and the denominator shows the strength of separation between those fuzzy clusters. If a range of clusters $\{k_1, k_2 \ldots k_i\}$ are taken, the k_i minimizing this index will be the optimal number of clusters for the dataset.

4 GloVe Vectors

One of the most known unsupervised learning algorithms to produce vector representations of words is *GloVe*. It is based on a word-word co-occurrence in text corpora. The term stands for Global Vectors as the representation is able to capture global corpus statistics.

4.1 Overview of GloVe

Let us start with defining a word-word co-occurrence counts matrix as X, where X_{ij} is the number of times the word j exists in the context of the word i. Let use denote $X_i = \sum_k X_{ik}$ the number of times any word appears in the context of word i. Lastly, let $P_{ij} = P(j|i) = X_{ij}/X_i$ become the probability in which word j exists in the context of word i. Using a simple example, we demonstrate how aspects of meaning can be extracted from word-word co-occurrences statistics. Pennington et al. shows this with good examples. Assuming we have text corpora related to thermodynamics and we may take words $i = ice$ and $j = steam$. We investigate the relationship of these words by learning from the co-occurrence probabilities with other sample words. For instance, if take word ice and word $t = solid$, we can expect that P_{it}/P_{jt} will be large. Likewise, if we select words t that are related to *steam* but not to *ice* such that *t=gas,* then we expect that the value of ratio should be small. Global vectors try to leverage a series of functions called F that represents those ratios [1] [11]. These F functions for the ratio of P_{it}/P_{jt} depend on words i, j, t to reflect the vector space models with linear structures:

$$F\left(w_i, w_j, w_{t}^{\sim}\right) = \frac{P_{it}}{P_{jt}} \tag{12}$$

where $w \in R$ are real word vectors and $w_t^{\sim} \in R$are context word vectors. In order to attain the symmetry, we require F to be a homomorphism and eventually express Eq. 12 as:

$$\frac{F\left(w_i^T w_t^{\sim}\right)}{F\left(w_j^T w_t^{\sim}\right)} = \frac{P_{it}}{P_{jt}} \tag{13}$$

Adding bias terms for the b_i and $b_t^{\sim}$ for the vectors w_i and $w_t^{\sim}$ and expressing F = *exp*,

$$w_i^T w_t^{\sim} + b_i + b_t^{\sim} = \log (X_{it}) \tag{14}$$

One disadvantage of the Eq. 14 is that the logarithm diverges when its argument becomes 0. Optimal solution to deal with this problem is to represent the right side as $\log(1 + X_{it})$ where it preserves the sparsity of X and avoid the divergence. Based on the above method, the objective function for Glove which combines a least squares regression model with the weight function $f(X_{ij})$ is defined in the following way:

$$J = \sum_{i,j=1}^{V} f(X_{ij}) (w_i^T w_j^{\sim} + b_i + b_j^{\sim} - log\, X_{ij})^2 \tag{15}$$

Here V is the size of the vocabulary and X_{ij} shows the number of times the word j exists in the context of the word i.

4.2 Training GloVe and Data Corpora

The objective to train *GloVe* model is to find appropriate vectors that minimize the objective function Eq. 15. As standard gradient descent algorithm heavily depends on the same learning rate, it does not become helpful to find errors and update them properly. Adaptive gradient algorithm (AdaGrad) has been proposed to solve the problem which adaptively assigns different learning rates to each of parameters [1] [11]. After training, the model produces two sets of vectors: W and $W^{\sim}$. When X is symmetric, the generated word vectors intrinsically perform equally and can become different only owing to random initializations. The authors show the best way to handle with these two vectors is to sum and assign the sum vector as a unique representation of the word:

$$W_{final} = W + W^{\sim} \tag{16}$$

Summing two sets of vectors into one effectively reflects words in the embeddings space. The authors have built the vocabulary of most frequent 400,000 words, and made them publicly available with 50,100, 200 and 300 dimensions, under Public Domain Dedication and License.[1] The source of the training data can be seen at [1].

[1] Pre-trained 400,000 GloVe vectors available in: https://nlp.stanford.edu/projects/glove/.

5 Methodology

Before we provide a description of the obtained results and their analysis, we briefly describe a set of words that have been clustered, as well as a procedure used for determining values of some clustering parameters.

5.1 Gold Standard for Similarity Evaluation

A set of words represented by word embeddings that we cluster has been constructed using words of WordSim-353 [12] dataset. This dataset contains semantic similarity scores of 353 word pairs and contains 437 different words. These pairs have been merged from 153 pairs scored by 13 humans and 200 pairs scored by 16 humans. The semantic similarity scores for the pairs vary in the range of [0–10]. For example, the similarity measures for the words *journey* and *voyage*, *computer* and *internet*, and *media* and *gain* are 9.29, 7.58 and 2.88, respectively. Many researchers have referred to WordSim-353 as a *gold standard* for different word similarity tasks. We have extracted vector representations of those "gold" words from 50, 100, 200 and 300-dimensional versions of *GloVe*, and used them for clustering and further analysis.

5.2 Clustering Parameters

One of the important parameters of a clustering process is a number of clusters. The nature of unsupervised learning means that this number needs to be set a priori. In our experiments, we use t-SNE [13] visualization of the gold standard WordSim-353 words (Fig. 1) to define a minimal number of clusters. The range of number of clusters is determined in the following way.

Lower Boundary. We use a simple visualization of words based on t-SNE, Fig. 1. Based on a visual inspection, we have identified the most obvious groups of words. As you can see, there are ten locations characterized by a higher concentration of words. Therefore, we use ten as our lower boundary for the number of clusters.

Upper Boundary. There are 437 words in the dataset we use in clustering experiments. We have anticipated that a larger number of clusters would provide better performance in the sense of clustering performance measures. However, we would like to avoid creating too small clusters—smaller cluster would be counterintuitive to our need for observing pairs of words in clusters (Sect. 6 B). Therefore, we have established the acceptable smaller size, on average, of a cluster to around 10. That would lead to a maximum of 50 cluster—and this becomes our upper boundary for the number of clusters.

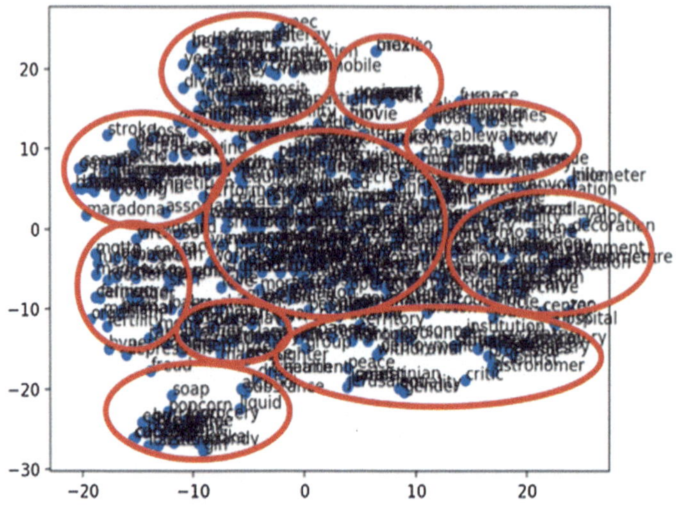

Fig. 1 t-SNE visualization of word vectors

6 Empirical Results

The results of fuzzy clustering of WordSim-353 words represented with *GloVe* embeddings are shown in Tables 1, 2, and 3.

6.1 Quantitative Analysis

It is well-known that fuzzy clustering shows some problems in the case of clustering high-dimensional data [14]. Winkler et al. have shown that performance of fuzzy clustering dramatically changes with the fuzzifier parameter: when using $m = 2$, the majority of prototypes go into the center of the gravity of the whole dataset. Therefore,

Table 1 Clustering results with 50-dimensional *GloVe* embeddings

# of clusters	10	15	20	25	30	40	50
Fuzzy C-Means							
Xie-Beni index	0.0054	0.0053	0.0053	0.0056	0.0053	0.0047	**0.0041**
FPC	0.7246	0.7242	0.7476	0.8021	0.8227	0.8683	**0.8878**
Fuzzy Gustafson-Kessel							
Xie-Beni index	16.48	13.43	10.12	13.43	11.58	9.53	**8.00**
FPC	0.9999	0.9878	0.9863	0.9875	0.9874	0.9879	**0.9892**

Table 2 Clustering results with 100-dimensional *GloVe* embeddings

# of clusters	10	15	20	25	30	40	50
Fuzzy C-Means							
Xie-Beni index	0.0096	0.0122	0.0092	0.0081	0.0108	0.0080	**0.0076**
FPC	0.5917	0.5987	0.6471	0.6980	0.7210	0.7817	**0.8322**
Fuzzy Gustafson-Kessel							
Xie-Beni index	30.03	25.40	21.56	20.58	15.11	12.65	10.87
FPC	**0.9817**	0.9783	0.9708	0.9747	0.9713	0.9648	0.9701

numbers in gray represent unacceptable values

Table 3 Clustering results with 200-dimensional *GloVe* embeddings

# of clusters	10	15	20	25	30	40	50
Fuzzy C-Means							
Xie-Beni index	30997.7	8685.2	10542791	156571.2	641690.1	572974.6	30602.8
FPC	0.2612	0.2896	0.3470	0.3638	0.4935	0.5714	**0.5907**
Fuzzy Gustafson-Kessel							
Xie-Beni index	696.62	731.98	706.72	728.19	782.20	769.02	749.63
FPC	**0.4154**	0.2978	0.2276	0.1842	0.1568	0.1194	0.0951

numbers in gray represent unacceptable values

we neither acquire the expected number of clusters nor sensible clustering results. Adjusting the *fuzzifier* around 1 such as 1.1 substantially improves the performance of the clustering. As the result, we obtain high-quality groupings of data points until some dimensions.

Based on that, we have set the value of *fuzzifier* parameter to $m = 1.1$ for our experiments. As we can see in the tables, clustering of words with 200-dim embeddings (Table 3) results with the performance measures values that start to look quite unreasonable (Xie-Beni index), while values of the index FPC become quite small. Due to such a situation, we consider for further analysis (next subsection) clusters obtained using FCM and FGK with 50-dim *GloVe* embeddings, and only clusters obtained using FCM for 100-dim embedding.

When the fuzzifier parameter m equals 1, the results become crisp. So, one question can naturally be raised: can clustering with $m = 1.1$ be still considered as fuzzy clustering? For this purpose, we have analyzed fuzzy membership matrix to see the distribution of the memberships. Among all words we have clustered, there are 25 words that have at the maximum membership value of 0.75 to a single cluster for

Table 4 GloVe embeddings: number of word pairs found in clusters

GloVe dimensionality	50	100
Fuzzy C-Means	42.40 +/−3.03	47.50 +/−2.42
Fuzzy Gustafson-Kessel	9	5

50 dim embeddings, and 52 such words for 100 dim embedding. We analyze the obtained clusters in Sect. 6.C.

6.2 *Qualitative Analysis*

The results presented in Sect. 6, A describe clusters from the point of view of their quality as measured by the performance indexes. However, these indexes do not show how well the clusters and clustering techniques group semantically similar words. For this purpose, we propose another way of determining the quality of clusters. The proposed approach is done from the point of view of grouping similar—according to humans—words.

The first step in the proposed approach has been to identify a set of pairs of words that are *highly* similar. Here, we use *the gold standard*, i.e., the set WordSim-353. We have assumed that the similarity value of 0.75 could be considered as a reasonable and practical level of treating words as *highly* similar. As a result, we have obtained 93 pairs of words.

The second step of the approach is to determine the number of pairs that are present in the same cluster. Moreover, we look at the distribution of pairs among clusters, i.e., we have identified clusters with zero, one, two and so on a number of pairs. The results of ten experiments are presented in Table 4. It shows an average number of word pairs found in the same clusters. As we can see, FCM is the best performing clustering technique. Also, an increase in the dimensionality of word representation leads to better results. This observation is statistically significant with the value of $p < 0.01$. In the case of FGK, the obtained numbers of pairs in clusters are very low. Also, ten experiments have led to the same numbers: 9 for 50-dim and 5 for 100-dim word representations.

6.3 *Analysis of Fuzzy Clusters*

The usage of fuzzy clustering means that we obtain fuzzy clusters with data point—words in our case—that belong to a cluster to a degree. Therefore, let us analyze some examples of allocating—to a degree—words to different clusters. We show two cases: 1) one that illustrates how two words are 'shared' between four clusters; and 2) other one that demonstrates how two words belong to different degrees to two clusters.

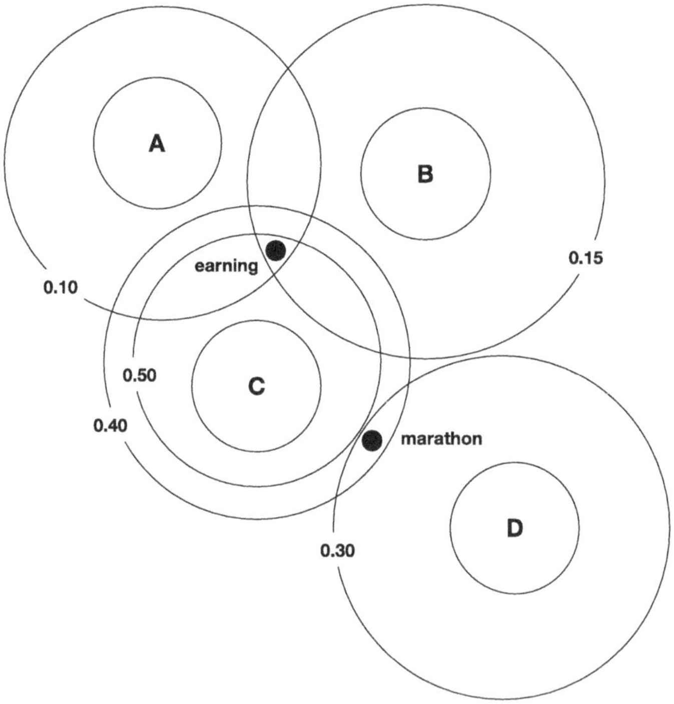

Fig. 2 Visualization of words: **earning** and **marathon** that belong to multiple clusters

The first case is presented in Fig. 2. It shows two words **earning** and **marathon** that belong to different degrees to four different clusters. The content of these clusters is:

A: {'card', 'listing', 'category', 'viewer', 'ticket', 'cd'};

B: {'wealth', 'entity', 'image', 'recognition', 'confidence', 'gain', 'importance', 'prominence', 'credibility'};

C: {'string', 'record', 'number', 'hundred', 'row', 'place', 'five', **'earning'**, **'marathon'**, 'series', 'start', 'year', 'day', 'summer', 'performance', 'seven'}; and

D: {'football', 'soccer', 'basketball', 'tennis', 'star', 'cup', 'medal', 'competition', 'baseball', 'season', 'game', 'team', 'boxing', 'championship', 'tournament', 'world'}.

The word **earning** is a member of three different clusters: **A** to a degree of 0.10, **B** to a degree of 0.15, and **C** to a degree of 0.50. It seems that its main cluster is *C*. While the word **marathon** belongs to **C** to a degree of 0.40, and to **D** to a degree of 0.30. If we look at the words from each cluster and the two considered words, we can easily see that their different degrees of membership to clusters are fully justified. **Earning** 'makes' sense to have some relationship with clusters A and B, while **marathon** could 'easily' be a member of cluster D.

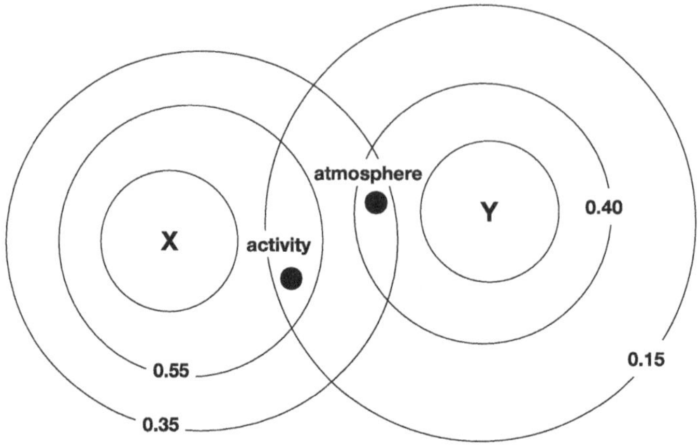

Fig. 3 Visualization of words: **activity** and **atmosphere** that belong to two clusters

The second example shows two words that belong to the same two clusters to different degrees, Fig. 3. Both clusters consist of:

X: {'space', 'example', 'object', 'weapon', 'surface', **'activity'**, 'type', 'combination', 'proximity', 'cell', 'size', 'observation'}, and

Y: {'fear', 'mind', **'atmosphere'**, 'reason', 'problem', 'kind', 'situation', 'direction', 'lesson', 'focus', 'change', 'attitude', 'approach', 'practice', 'experience'}

The word **activity** belongs to **X** to a degree of 0.55 while to **Y** to a degree of 0.15, while **atmosphere** has a membership value of 0.35 to **X**, and 0.40 to **Y**. Once again, we see that different degrees of membership to the clusters are fully justified.

7 Conclusions

In this study, we have examined the fuzzy clustering analysis of word embedding. We analyzed the performance of fuzzy C-means and fuzzy Gustafson-Kessel algorithms on the set of words from WordSim-353 represented using *GloVe* embeddings. Based on the obtained results, fuzzy clustering algorithms were proved to be very sensitive to high-dimensional data. Fuzzy C-means with fuzzifier parameter $m = 1.1$ has provided sensible word clustering results for up to 100-dimensional word embeddings. However, for higher dimensions, it fails to produce both expected number of clusters and plausible word clustering results. The experimental results proved that fuzzy Gustafson-Kessel clustering technique, on the other hand, should be avoided in high-dimensional spaces. Even for the case of 50-dimensional data, very poor performance has been observed. Additionally, we have shown that using fuzzy clustering with a small value of fuzzifier parameter ($m = 1.1$) still provides an interesting and fully justified variation in degrees of membership of words to different clusters.

Acknowledgements The authors express their gratitude to the Ministry of Education of the Republic of Azerbaijan for funding this research under the "State Program on Education of Azerbaijani Youth Abroad in the Years of 2007-2015" program.

References

1. J. Pennington, R. Socher, C. Manning, Glove: global vectors for word representation, in *Proceedings of the 2014 Conference on Empirical Methods in Natural Language Processing (EMNLP)*, (2014), pp. 1532–1543
2. A.K. Jain, M.N. Murty, P.J. Flynn, Data clustering: a review. ACM Comput. Surv. (CSUR) **31**(3), 264–323 (1999)
3. R.O. Duda, P.E. Hart, D.G. Stork, *Pattern Classification*. (Wiley, 2012)
4. N. Dhanachandra, K. Manglem, Y.J. Chanu, Image segmentation using K-means clustering algorithm and subtractive clustering algorithm. Procedia Comput. Sci. **54**, 764–771 (2015)
5. L.V. Bijuraj, Clustering and its application, in *Proceedings of National Conference on New Horizons in IT-NCNHIT*, (2013) p. 169
6. L. Zadeh, Fuzzy sets. Inf. Control **8**(3), 338–353 (1965)
7. J.C. Bezdek, *Pattern recognition with Fuzzy objective Function, Algorithms*. (1981)
8. J.V. De Oliveira, W. Pedrycz (eds.), *Advances in Fuzzy Clustering and its Applications*, (Wiley, 2007)
9. D.E. Gustafson, W.C. Kessel, Fuzzy clustering with a fuzzy covariance matrix, in *1978 IEEE Conference on Decision and Control Including the 17*th *Symposium on Adaptive Processes*, (January 1979.), pp. 761–766, IEEE
10. X.L. Xie, G. Beni, A validity measure for fuzzy clustering. IEEE Trans. Pattern Anal. Mach. Intell. **13**(8), 841–847 (1991)
11. Z. Chen, Y. Huang, Y. Liang, Y. Wang, X. Fu, K. Fu, RGloVe: an improved approach of global vectors for distributional entity relation representation. Algorithms **10**(2), 42 (2017)
12. L. Finkelstein et al., Placing search in context: The concept revisited, in *Proceedings of the 10th International Conference on World Wide Web*, (ACM, April 2001), pp. 406–414
13. L.V.D. Maaten, G. Hinton, Visualizing data using t-SNE. J. Mach. Learn. Res. 2579–2605 (2008)
14. R. Winkler, F. Klawonn, R. Kruse, Fuzzy c-means in high dimensional spaces. Int. J. Fuzzy Syst. Appl. **1**, 1–16 (2013)

Project Management on Base Analysis of Fuzzy Time Series of Project Activity Metrics

I. A. Timina, S. K. Kiselev, and E. N. Egov

Abstract In this article considers fuzzy time series (FTS), which are obtained from the indicators of the project activity. An anomaly prediction method is applied to the FTS. Anomalies are revealed when analyzing the FTS of project activity indicators. Anomalies are used to adjust the forecast of the project. Analyzed the set of project performance indicators. A prediction algorithm based on the fuzzy trends of the TS metric was developed and implemented. In this article considers fuzzy time series (FTS), which are obtained from the indicators of the project activity. An anomaly prediction method is applied to the FTS. Anomalies are revealed when analyzing the FTS of project activity indicators. Anomalies are used to adjust the forecast of the project. Analyzed the set of project performance indicators. A prediction algorithm based on the fuzzy trends of the TS metric was developed and implemented. A method for searching TS anomalies based on entropy is proposed. A formula is constructed for the entropy measure for FTS. The algorithm uses the dependence of the forecast on the measures of entropy.

1 Introduction

Effective management of a set of projects through forecasting their development is one of the most important tasks for the successful organization of project activities. The problem of monitoring the status of project activities can be solved with the help

The article was supported by the Russian Foundation for Basic Research (Grant No. 19-07-00999, 16-47-732070, 16-47-730715).

I. A. Timina (✉) · S. K. Kiselev · E. N. Egov
Ulyanovsk State Technical University, Ulyanovsk, Russia
e-mail: i.timina@ulstu.ru

S. K. Kiselev
e-mail: ksk@ulstu.ru

E. N. Egov
e-mail: e.egov@ulstu.ru

S. N. Shahbazova et al. (eds.), *Recent Developments and the New Direction in Soft-Computing Foundations and Applications*, Studies in Fuzziness and Soft Computing 393, https://doi.org/10.1007/978-3-030-47124-8_45

of a tool for measuring the characteristics of project activities. The tool allows you to monitor the dynamics of a set of projects, identify problems and timely adjust regulatory actions [1]. Also, it is need to reallocate resources between projects. Therefore, the creation of a tool for project management based on forecasting the values of project indicators appears to be an urgent and important scientific and technical task. Often project metrics can be expressed in verbal form. Also, metrics are accumulated in the development process. As a result, project metrics can be represented as FTS.

The objectives of the FTS analysis include the identification of the general trend of the series, the forecast of the next fuzzy label and the evaluation of the series levels at some point in time. There are situations when it is required to assess the trend of FTS change over a certain period of time. The length of the window is set linguistically ("long", "short", etc.). In this case, the series will have an indistinctness not only on the scale of the value, but also on the time scale. These kinds of problems arise when it is necessary to apply expert knowledge about trends in the values of indicators when analyzing FTS in expert systems or to simulate the operation of the system under certain conditions by applying "if-then" rules.

In addition to the classical TS analysis methods, there is a direction for analyzing Time-Series DataMining. The main objectives of this direction are modeling and analysis of processes. These processes have a greater degree of uncertainty (including "non-stochastic" process). The methods are aimed at revealing the hidden regularities and extracting new knowledge from the TS. Based on this methodology, an expanded set of TS analysis tasks is solved.

The article is devoted to solving the problem of project monitoring by means of analyzing the state of metrics of the software project. Metrics are loaded into the version control system with subsequent forecasting and their adjustment.

The project management tool consists of the following steps:

- select metrics, by mathematical nature;
- consider the dependencies;
- construct a formal model;
- build a project management tool.

Metrics are properties of entities. The authors propose to use follow metrics extracted from Control Versions System:

- commit (developer, date);
- release (set of developers, master, date);
- branch;
- bug;
- improvement;
- new feature.

The international standards (ISO 10006, ISO 21500) and national (Russia: GOST R 54869-2011, GOST R 54870-2011, GOST R 54871-2011) management standards are used to make decisions in project management. A significant contribution to the analysis of project activities from the point of view of management was made by Klein G., Charette R., SeungHeonHan, Aucoin B. M., Gold M., Norenkov I. P., Tarasenko F.P., Dulzon A. A. and others.

Foreign scientists who made a significant contribution to the analysis and prediction of TS are Box D., Jenkins G., Chem S. M., Granger C., Herbst G. and etc.

Foreign and Russian scientists who have made a significant contribution to the field of fuzzy modeling, analysis and prediction of TS are Novak V., Perfilieva I. G., Batyrshin I. Z., Kovalev S. M., Yarushkina N. G., Afanasyeva T. V. and etc. [2–4].

2 Analysis of Project Activity Metrics

The model of analysis and management of a set of projects in the process of project activity is developed.

$$\left\{C_t, R_t, B_t, I_t, F_t, R^{BI}, R^{IF}\right\}$$

where C_t—TS commits,

R_t—TS of releases,

B_t—TS of bugs,

I_t—TS of improvement,

F_t—TS of new feature,

R^{BI}—dependence of the number of bugs on improvements,

R^{IF}—the dependence of new functional properties on the number of improvements (new features from improvements).

DiscreteTime Series represent project data.

Model of TS for solving this problem is constructed in two stages:

- A discrete TS $Y = \{t_i, x_i\}, i \in [1, n]$, where t_i—some point in time, x_i—the level of TS, is transformed into an fuzzy TS $\tilde{Y} = \left\{t_i, \tilde{X}_i\right\}, i \in [1, n], X = \{x_i\}, \tilde{x}_i \in \tilde{X}$, where $\tilde{x}_i$—fuzzy label [5].
- Fuzzy TS is transformed into a TS of fuzzy elementary trends. It is indicated by $\tau_i = \mathrm{ETend}(\tilde{x}_i, \tilde{x}_{i+1})$, where ETend—functional, which is implemented on the basis of operations: Ttend—determination of the type of trend and Rtend—determination of the intensity of the trend on a special linguistic scale constructed for the initial TS [6].

Several TS, clear numerical and fuzzy linguistic ones are obtained as a result of the described transformations (for example, a TS of parameters of fuzzy trends).

A generalized model of fuzzy TS of a fuzzy trend is presented [7]:

$$\tau_i = f(\tau_{i-1}, \ldots, \tau_{i-d})$$

where d—fixed number, model parameter;

$\tau_i, \tau_{i-1}, \ldots, \tau_{i-d}$—sequence of fuzzy trends;
f some fuzzy dependency.

Every fuzzy tendency $\tau \in \Im$ of a fuzzy TS is represented by a structural model [8]:

$$\tau = \tilde{v}, \tilde{a}, \Delta t, \mu$$

where τ name of a fuzzy tendency from the set $\Im$, $\tau \in \Im$;
$\tilde{v}$—type of fuzzy trend, $\tilde{v} \in \tilde{V}$, {fall, growth, stability};
$\tilde{a}$—fuzzy intensity, $\tilde{a} \in \tilde{A}$, {strong, medium, weak};
Δt—duration of fuzzy trend, $\Delta t \in \Delta T$;
μ the membership function of an fuzzy TS section bounded by an interval Δt, a fuzzy trend τ.

The fuzzy tendency of the fuzzy TS is classified as a function of Δt:

- elementary fuzzy tendency $T \in \Im$ $(\Delta t = 1)$,
- local fuzzy tendency $NT \in N\Im$ $(1 < \Delta t < n - 1)$
- basic (general) fuzzy tendency $GT \in G\Im$ $(\Delta t = n - 1)$.

The TS of the fuzzy elementary trends is represented in the form (TTend, RTend—procedures for obtaining fuzzy sets $\tilde{v}_t$ and $\tilde{a}_t$):

$$\tilde{v}_t = \mathrm{TTend}(\tilde{x}_t, \tilde{x}_{t+1}), \tilde{a}_t = \mathrm{RTend}(\tilde{x}_t, \tilde{x}_{t+1}), \mu_t = \min(\mu(\tilde{x}_t), \mu(\tilde{x}_{t+1})).$$

The levels of the TS are described by the components of the fuzzy trend—the type and intensity: $\tau = \{t_i, \tau_i\}$.

The clusters $S = \{s_1, s_2, \ldots s_k\}$ of similar fuzzy labels $\tilde{x}_i$ to identify the dominant trend are defined. For each cluster $s \in S$, N is the cardinality.

The FCM-algorithm is used for fuzzy trend clustering. It is based on the minimization of the objective function [9]:

$$J = \sum_{j=1}^{C} \sum_{j=i}^{N} \mu_{ij}^{m} x_i - c_j$$

where N—the number of indicators;
C number of clusters;
μ_{ij}—membership function of the object i to the cluster j;
m real number, >1;
x_i—ith trend of many trends TS;
c_j—jth cluster;
$\|x_i - c_j\|$—the norm characterizes the distance from the center of cluster j to the tendency i.

The fuzzy elementary trend modeling method is used to predict numerical values and fuzzy trends in the state of an organization project in a given product. The forecast uses hypothesis testing.

Hypothesis 1 The hypothesis of conservation of trend. The Forecast is constructed on base the previous period. The formula for the predicted value

$$\tau_{t+1} = \tau_t + \tau_p$$

where τ_{t+1}—forecast for the next period of time;
τ_t—the real value at time t;
τ_p—the value of the trend over the previous period of time.

Hypothesis 2 The hypothesis of stability of the trend. The moving average is used to predict

$$\tau_{t+1} = \tau_t + G\tau_p$$

where $G\tau_p$—the importance of a dominant fuzzy trend.

Consider the trend of the previous selected period. We select the predominant cluster of trends. The forecast for the above formula is calculated. The trend is built. Optimistic forecast for the some number of occurrences of trends used. The highest average trend is selected.

Hypothesis 3 Forecasting for a given period on the basis of fuzzy elementary trends. Stages of the prediction algorithm for the period based on trends.

The expert sets the number of considered trends for the previous period. For example, for half a year—a set of trends A. Either he sets the pattern set of trends. The presumed trend following this set is known.

$$\left\{\tau_{t_{n-m}}, \ldots, \tau_{t_{n-1}}, \tau_{t_n}\right\}$$

1. Search for a set of trends A in all other previous periods.

$$\left\{\tau'_{t_{n-l-k}}, \ldots, \tau'_{t_{n-l-(k-1)}}, \tau'_{t_{n-l}}\right\}$$

2. If such a set of B is found in which the C trend is located after this found set B then trend C is considered into account. The forecast equal to the trend C is constructed.

$$\tau_{t+1} = \tau_t + \tau'_{t_{n-l+1}}$$

3. If the set B, which would coincide with the set A, was not found then the search for the set is repeated, but it is already not looking for its complete coincidence. We select new pattern A is shorter into one trend. This is repeated until a suitable set of trends B [10].

To select the best hypothesis, we proposed to use special Time Series—TS of entropy [11].

3 Forecasting TS Using the Adaptation Algorithm

To solve the prediction problem for the time series $Y = t_i, x_i$ $Y = \{t_i, x_i\}, (i = 1, 2, \ldots, n)$ with the help of fuzzy similarity, provided that the hypothesis of the expert hypothesis exists that the fuzzy tendency of the time series $Z = \{t_i, z_i\}, (i = 1, 2, \ldots, k)$ is a predictor of the time series Y, the forecast hypothesis correction algorithm [12] is used. Algorithm 1 includes three phases.

In the first phase, fuzzy elementary trends in the time series Y are predicted:

$$\tau_{t+1}^{Y} = f(\tau_t^{Y})$$

here τ_{t+1}^{Y} is the predictive fuzzy elementary trend of the time series Y;

τ_{t+1}^{Y}—the current fuzzy elementary trend of the time series Y;

F—dependence the fuzzy elementary trends of the time series Y on previous point Y.

In the second phase, the forecast fuzzy elementary trend of the time series Y is corrected taking into account the components of the main trends of the time series $G\tau_Y$ and the time series of the predictor $G\tau_Z$, respectively:

$$\hat{\tau}_{t+1}^{Y} = r\left(\tau_{t+1}^{Y}, G\tau_Y, G\tau_Z\right)$$

where

τ_{t+1}^{Y}—predictive fuzzy elementary trend of time series Y;

$\hat{\tau}_{t+1}^{Y}$—predictive fuzzy elementary trend of the time series Y after the adjustment;

$G\tau_Y$—basic fuzzy trend of time series Y;

$G\tau_Z$—the main fuzzy trend of time series Z;

r correction rules.

The third phase is used to obtain the estimate of the predicted value of the numerical time series Y.

On this basis, the following algorithm for predicting short-term fuzzy trends in time series is proposed.

Algorithm 2 Step 1. Conversion of the numerical TS $Y = \{t_i, x_i\}, (i = 1, 2, \ldots, n)$ to fuzzy TS $\tilde{Y} = \tilde{x}_t, \tilde{x} \in \tilde{X}, t = 1, 2, \ldots, n$:

$$\tilde{x}_t = \text{Fuzzy}(x_i), x_i \in X, \tilde{x}_t \in \tilde{X}$$

In this case, the intervals at which fuzzy sets are defined, their form and name are specified by the user.

Step 2. Conversion of fuzzy TS $\tilde{Y} = \tilde{x}_t, \tilde{x} \in \tilde{X}, t = 1, 2, \ldots, n$ into an fuzzy time series of fuzzy elementary trends:

$$\tau_t^Y = \tilde{v}_t, \tilde{a}_t, \mu_t$$

$$\tilde{v}_t = \mathrm{TTend}(\tilde{x}_t, \tilde{x}_{t+1})$$

$$\tilde{a}_t = \mathrm{RTend}(\tilde{x}_t, \tilde{x}_{t+1})$$

$$\mu_t = \min(\mu(\tilde{x}_t), \mu(\tilde{x}_{t+1}))$$

We first define the set of types of FT types $\tilde{V} = \{\text{Fall, Growth, Stability}\}$, and a set of names of intensities of FTs $\tilde{A} = \{\text{Strong, Medium, Weak}\}$.

Step 3. Construction of a model for changing the components of fuzzy elementary trends of time series Y and their prediction for one period:

$$\tilde{v}_{t+1} = \tilde{v}_t \times \tilde{v}_{t-1} \times \cdots \times \tilde{v}_{t-p} R_{\tilde{v}}(t, t-p)$$

$$\tilde{a}_{t+1} = \tilde{a}_t \times \tilde{a}_{t-1} \times \cdots \times \tilde{a}_{t-p} R_{\tilde{a}}(t, t-q)$$

Step 4. Forecasting the numerical time series Y with the preliminary defuzzification the component of the fuzzy trend $\tau_{t+1}^Y = \tilde{v}_{t+1}, \tilde{a}_{t+1}, \mu_{t+1}$:

$$x_{t+1} = x_t + v_{t+1} \Delta a_{t+1}$$

Step 5. Application of the algorithm for identifying the main trend (see phase 3. Algorithm 1) for the time series Y and calculating its components $G\tau_Y = \tilde{v}_{G\tau}^Y, \tilde{a}_{G\tau}^Y, \mu_{G\tau}^Y$.

Defuzzification of the components of the main fuzzy trend of the time series Y.

Step 6. Application of the algorithm for identifying the main trend (see phase 3. Algorithm 1) for the time series Z and calculating its components $G\tau_Z = \tilde{v}_{G\tau}^Z, \tilde{a}_{G\tau}^Z, \mu_{G\tau}^Z$.

Defuzzification of the components of the main fuzzy trend of the time series Z.

Step 7. Correction of the predictive fuzzy elementary trend of the time series $Y\tilde{\tau}_{t+1}^Y = r\left(\tau_{t+1}^Y, G\tau_Y, G\tau_Z\right)$:

$$\tilde{\tau}_{t+1}^Y = v_{t+1} \Delta a_{t+1} + v_{G\tau}^Y \Delta a_{G\tau}^Y + v_{G\tau}^Z \Delta a_{G\tau}^Z$$

Step 8. Calculation of the corrected forecast value of the numerical time series Y for one period $x'_{t+1} = x_t + \tilde{\tau}_{t+1}^Y$.

4 Search for TS Anomalies Based on Entropy

Entropy TS can be represented by the following model:

$$\{\tilde{x}_i, H_i^{\mu}, H_i^{T}\},$$

where $\tilde{x}_i$—fuzzy label value at the ith point;

H_i^{μ}—value of the measure of entropy by the membership function at the ith point;

H_i^{T}—value of the measure of entropy by a fuzzy trend.

Entropy TS allows more sensitive response to changes in TS.

5 The Measure of Entropy by the Membership Function

The measure of entropy by the membership function is obtained from [12]:

$$H_i^{\mu} = H\left(\tilde{A}\right) + H\left(\tilde{A}^{C}\right)$$

where

$$H\left(\tilde{A}\right) = -K\sum_{j=1}^{n} \mu_{\tilde{A}}(x_j)\ln(\mu_{\tilde{A}}(x_j))$$

where K—positive constant;

$\tilde{A}^{C}$—addition $\tilde{A}$;

$\mu_{\tilde{A}}(x_j)$—the value of the membership function of the point x to the jth cluster.

The membership function is calculated differently. This depends on the option of converting TS to fuzzy TS. For example, with fuzzification, the membership function is given by formulas. If we take the triangular membership function, then it has the form:

$$\mu(x) = \begin{cases} \frac{x-a}{b-a}, a \le x \le b, \\ \frac{c-x}{c-b}, b < x \le c, \\ 0, x < a, x > c, \end{cases}$$

where x—value of a point in a series;

[a, c]—range of variable change;

B—the most possible value of a variable.

If fuzzy TS is obtained by the methods of clustering, then it can be used the matrix of membership of points to clusters.

If we take FCM-clustering as an example, then the membership matrix $\mu(x)$ is calculated by the following formula:

$$\mu_j = \frac{1}{\sum_{I=1}^{C}\left(\frac{x-c_j}{x-c_I}\right)^{\frac{2}{m-1}}}$$

μ_j—membership function to the jth cluster;
c_j—vector of the center of the jth cluster;
c_I—vector of the center of the ith cluster.
Moreover, if for some cluster j $x - c_j = 0$, then $\mu_j = 1$, and for all other clusters, 0.

The measure of entropy by the membership function is defined in 2 steps:

1. The value of entropy by the membership function is calculated [13].
2. The linguistic interpretation of the measure of entropy is determined on the basis of the value obtained. The value of the measure of entropy close to 0 corresponds to the state "Determined". The value of the measure of entropy close to the maximum corresponds to the state "Uncertain" the value of the entropy measure corresponds to the state "Probably" in the remaining cases.

The measure of entropy is not able to clearly record the change of fuzzy timestamp marks, if it is obtained on the basis of the membership function. It shows how likely the point will be close to the label. If the entropy is close to the maximum value, then the point is in the "boundary" position and it can be related to two different fuzzy marks with equal probability.

$$\Delta\tau_{i-1}^{fact} = \tau_{i-2}^{fact} - \tau_{i-1}^{fact} \tag{1}$$

The code of the point (τ_{i-1}^{fact}, $\Delta\tau_{i-1}^{fact}$) is determined from its position on the coordinate plane of the phase plane.

$$p = \text{CalcCodePoint}(\tau_{i-1}^{fact}, \Delta\tau_{i-1}^{fact}) \tag{2}$$

On the phase plane, there are patterns of behavior of the trend. The most likely trend is chosen. The value of the elementary trend is determined at the next point after the point (τ_{i-1}^{fact}, $\Delta\tau_{i-1}^{fact}$). This value of the trend will be the predicted value of the elementary trend.

The measure of entropy for a fuzzy trend is determined by the following algorithm:

1. The dynamics of the trend is determined at the previous point on the basis of formula (1).
2. The position of the fuzzy trend is calculated on the phase plane by the formula (2) based on the weight of the fuzzy trend and the value of the dynamics of the fuzzy trend at the previous point:

$$p_{i-1} = \text{CalcCodePoint}(\tau_{i-1}^{fact}, \Delta\tau_{i-1}^{fact})$$

3. Three sets of points of the phase plane are determined:

- the most likely points are points (usually one point), in which most often follow after point p_{i-1}:

$$\omega_{mostexpect} = Max(Probability(p_{i-1}))$$

- probable points are points which also follow after the point p_{i-1}, but they did not enter the first set:

$$\omega_{probability} = Probability(p_{i-1})\omega_{mostexpect}$$

- anomalous points are all points not included in the first two sets (transition to them is not expected in normal operating conditions):

$$\omega_{anomaly} = All\,Point\,Probability(p_{i-1})$$

4. The dynamics of the trend is determined at the current point:

$$\Delta\tau_i^{fact} = \tau_{i-1}^{fact} - \tau_i^{fact}$$

5. The point of the phase plane for the fuzzy trend and dynamics is calculated at the current point of the series.

$$p_i = CalcCodePoint(\tau_i^{fact}, \Delta\tau_i^{fact})$$

6. The resulting point p_i is determined to which set of three $\omega_{mostexpect}$, $\omega_{probability}$ or $\omega_{anomaly}$ it refers.

- If the point belongs to the set $\omega_{mostexpect}$, then the value of the entropy measure is set to 0, since the point just obtained was expected. That is, we did not learn anything new.

$$if\left(p_i \in \omega_{mostexpect}\right) then\ H_i^{\tau} = 0$$

- If the point refers to the set of $\omega_{probability}$, then the value of the entropy measure is set to 0.5, since the resulting point, although not expected, was not entirely new.

$$else\ if\ \left(p_i \in \omega_{probability}\right) then\ H_i^{\tau} = 0,5$$

- If the point refers to the set $\omega_{anomaly}$, then the value of the entropy measure is set to 1, since the resulting point was not expected to be seen. That is, at the moment the system being analyzed is in an unknown state:

$$\text{else if } (p_i \in \omega_{anomaly}) \text{then } H_i^{\tau} = 1$$

7. The linguistic interpretation of the obtained numerical value of the entropy measure for the fuzzy trend H_i^{τ} is defined as follows:

$$\text{if}(H_i^{\tau} = 0)\text{then } \tilde{H}_i^{\tau} = \text{Stability}$$

$$\text{else if}(H_i^{\tau} = 0, 5)\text{then } \tilde{H}_i^{\tau} = \text{Change}$$

$$\text{else } \tilde{H}_i^{\tau} = \text{Anomaly}$$

The resulting TS of entropy will allow us to determine the best hypothesis for use. The algorithm for choosing a hypothesis is defined as follows:

- If the "Change" state is rare and the value of the entropy measure by the membership function at the last point corresponds to "Reliably", then the hypothesis of maintaining the current trend is selected:
- If the "Change" state is frequent, but there is no periodicity, then the hypothesis of the stability of the dominant trend is chosen.
- If there is a periodicity, then the hypothesis for a given period is selected.
- If it is not possible to determine the hypothesis according to the degree of entropy by the fuzzy trend, then the choice is made based on the values at the last point.
- If the value of the entropy measure on the membership function is "True" and the fuzzy trend is "Stability," the hypothesis of maintaining the trend is chosen.
- If the value of the entropy measure on the membership function corresponds to "Probably" or "Uncertainly" or the value of the entropy measure for the fuzzy trend is "Change" or "Anomaly", then the hypothesis of the dominant trend is chosen.

The obtained TS of entropy allows to reveal *b* by this algorithm and to correct the most correct hypothesis for the forecast.

6 Measure of Entropy by Fuzzy Trend

This method is based on the assumption that chaos in the system occurs when deviations in the indicator trend begin. The situation is uncertain, when it becomes difficult to predict the further behavior of the system. This situation can be determined by the deviation of the forecast value of the trend. It is obtained based on the difference in the trend values at the two previous points from the actual value of the trend at this point. If the forecast and actual coincide, then we can talk about the stability of the system, otherwise there is a deviation. It can be insignificant or lead the system into a state of chaos and uncertainty [14].

$$H_i^T = \tilde{t}_i^{fact} - \tilde{t}_i^{forecast}$$

where $\tilde{t}_i^{forecast} = \text{Get Forecast Trend}\left(\tilde{t}_{i-1}^{fact}, \tilde{t}_{i-2}^{fact}\right)$

The function Get Forecast Trend$\left(\tilde{t}_{i-1}^{fact}, \tilde{t}_{i-2}^{fact}\right)$, on the basis of the values of elementary trends at the two previous points, calculates the dynamics of the trend.

7 Experiments

For the research, the project MongoDB event metrics data was taken for testing. In particular, its main program part Core Server. Data on errors, improvements and new features were taken from the resource Jira.mongodb.org.

MongoDB is a document-oriented database management system with open source code. It does not require a description of the schema of the tables. The MongoDB system is written in C++.

Table 1 shows the results of the analysis.

We obtain the TS of entropy according to the metrics of the MongoDB project (Table 2).

Based on the data obtained, we can conclude. The hypothesis of preserving the trend is not relevant, since the measure of entropy by the fuzzy trend is often in the "Change" state. This indicates a frequent change of trend in TS. The state of the measure of entropy by the membership function about the expected change of trend says. The high frequency of the of the "Change" state of the entropy measure along a fuzzy trend and the absence of periodicity in these shifts indicate the use of the hypothesis of stability of the dominant trend (Table 3).

Figure 1 shows a variant of prediction using hypotheses, where the x—axis is the TS value, the t—axis is the time; "—◆—" Stands for TS Y; "—■—" Stands for the prediction of hypothesis 1; "—×—" Stands for the hypothesis 2; "—●—" Stands for the hypothesis 3.

Table 1 Analysis of project metrics

TS	The general trend	Dominant tendency
Improvement	Growth	Growth weak
Task	Growth	Stability
Bug	A fall	Growth weak

Table 2 Entropy bp according to the metrics of the mongodb project

FTS point	The measure of entropy by the membership function	Measure of entropy by fuzzy trend
1	Probably	Change
2	Probably	Stability
3	Reliably	Stability
4	Probably	Change
5	Probably	Stability
6	Probably	Change
7	Probably	Stability
8	Reliably	Change
9	Reliably	Stability
10	Reliably	Stability
11	Probably	Change
12	Reliably	Change
13	Probably	Change
14	Probably	Stability

Table 3 Results of prediction of event metrics taking into account entropy measures

TS	Hypothesis 1	Hypothesis 2	Hypothesis 3
Improvement	The fall is strong	Growth strong	The fall is strong
Task	Stability	Growth weak	Stability
Bug	Stability	Growth average	The fall is strong

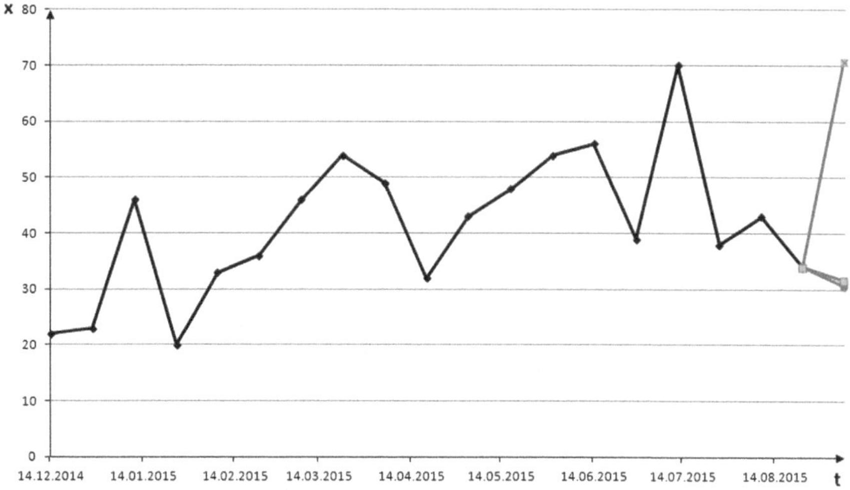

Fig. 1 Forecasting improvement in the MongoDB project

8 Results

The research of a set of metrics series of various projects was conducted with the aim of predicting subsequent values. Based on the obtained values, the average value of MAPE for the developed method was determined. It is no more than 0.5–1.5% of the average value of the forecast metrics.

The proposed approach is a modified method for predicting fuzzy elementary trends. The obtained results show that this approach can be used for short-term forecasting of FTS, if there are anomalies in the FTS considered. The approach uses the capabilities of the FTS. This allows you to predict the values for series with non-numeric values.

References

1. V.S. Moshkin, A.N. Pirogov, I.A. Timina, V.V. Shishkin, N.G. Yarushkina, Intellectual data analysis and the ontological approach in project management. Automat. Control Process. **4**(46), 84–92 (2016)
2. I. Perfilieva, N. Yarushkina, T. Afanasieva, A. Romanov, Time series analysis using soft computing methods. Int. J. Gen. Syst. **42**(6), 687–705 (2013)
3. T. Afanasieva, N. Yarushkina et al, Time series forecasting using fuzzy techniques, in *International Joint Conference IFSA-EUSFLAT*, pp. 1068–1075 (2015)
4. V. Novák, I. Perfilieva, N. Jarushkina, A general methodology for managerial decision making using intelligent techniques. Chapter Recent Advances in Decision Making. Series Studies in Computational Intelligence, vol. 222, pp. 103–120 (2009)
5. G. Herbst, S.F. Bocklish, Online recognition of fuzzy time series patterns, in *2009 International Fuzzy Systems Association World Congress and 2009 European Society for Fuzzy*
6. J. Kacprzyk, A. Wilbik, Using fuzzy linguistic summaries for the comparison of time series, in *2009 International Fuzzy Systems Association World Congress and 2009 European Society for Fuzzy Logic*
7. W. Pedrycz, S.M. Chen, time series analysis, modeling and applications: a computational intelligence perspective (e-book Google). Intell. Syst. Ref. Lib. **47**, 404 (2013)
8. T.V. Afanasieva, Forecasting local time series trends in large data analysis problems. Autom. Control Proces. **4**(38), 42–47 (2014)
9. T.Y. Krol, M.A. Charin, Methods for solving the problem of clustering and forecasting in an electronic archive. Young Sci. **1**(6), 135–137 (2011)
10. I.A. Timina, Correction of forecast hypotheses to extract knowledge of time series, in *XIV National Conference on Artificial Intelligence with International Participation*, pp. 68–76 (2014)
11. E.N. Egov, N.G. Yarushkina, D.V. Yashin, Fuzzy modeling and genetic optimization of time series in the intellectual system of technical diagnostics. Radiotekhnika **9**, 64–71 (2016)
12. N.G. Yarushkina, I.A. Timina, Model and controls for designing an automated system based on the dynamics of program code metrics. Autom. Control Process. **3**(41), 73–81 (2015)
13. N.G. Yarushkina, V.V. Voronina, E.N. Egov, Application of the entropy measure in the diagnosis of technical time series. Autom. Control Process. **2**, 55–63 (2015)
14. I.A. Timina, E.N. Egov, YuP Egorov, D.V. Yashin, S.K. Kiselev, Identification of anomalies of time series of metrics of project activity based on entropic measures. Radio Eng. **6**, 128–135 (2017)

Fuzzy-Interpolative Control for Intelligent Roof-Top Greenhouse Buildings

M. M. Balas, Valentina Emilia Balas, R. Lile, and S. V. Balas

Abstract The Integrated Roof-Top Greenhouse Building (IRTG) is stemming from the conventional roof-top greenhouse. Thanks to the carbon offset due to the hosted plants, IRTGs offer us an effective tool able to oppose the global warm-ing, if applied at a large scale. IRTGs are able to harvest local renewable energy resources (geo-thermal, solar, wind, etc.), to store them and to manage them in an integrated way with the water resources. Besides the CO_2 management, already described in the literature we consider and the O_2 flow. A two ways ventilation system is conveying O_2 enriched air from RTG to the building and CO_2 enriched air from the building to RTG. A tight symbiosis humans-plants is built such way. This multiple-input-multiple-output system is highly nonlinear, temperatures and gas concentrations being interdependent, which is hardening its automated control. A generic model of this system, taking into account the air exchanges between greenhouse, building and environment is presented, along with a self-adaptive fuzzy-interpolative controller. Given the necessity of the Intelligent Control, we adopt a new term: the Intelligent RTG (iRTG).

Keywords Urban agriculture · Roof-top-green-house · Carbon offset · Building metabolism · Fuzzy-interpolative self-adaptive control

1 Introduction

Fuzzy logic is perfectly suited to cope with perceptions. Air-conditioning applications (temperature, humidity, air composition) are among the first and most successful applications of fuzzy sets and logic, because the *thermal comfort* and in general the *comfortable atmosphere* are subjective perception based notions, accepting large and contextual tolerances for temperature, humidity or the oxygen and carbon dioxide concentrations. Our experience with fuzzy control is mostly acquired

M. M. Balas (✉) · V. E. Balas · R. Lile
Aurel Vlaicu University of Arad, Arad, Romania
e-mail: marius@drbalas.ro

S. V. Balas
Coindu Romania SRL, Curtici, Romania

S. N. Shahbazova et al. (eds.), *Recent Developments and the New Direction in Soft-Computing Foundations and Applications*, Studies in Fuzziness and Soft Computing 393, https://doi.org/10.1007/978-3-030-47124-8_46

in air-conditioning installations of railway coaches and passive greenhouses [1–3], etc. Lately our interest shifted from the greenhouse conventional agriculture towards the roof-top greenhouse urban agriculture. This trend is of great actuality [4–7], with promising perspectives.

The concept we are now following is the *Green Skyline City*, a city having all buildings covered by passive greenhouses. The key item here, which is the subject of this paper, is *the Intelligent Roof-Top Greenhouse* (iRTG).

There are two principal reasons that created a recent trend towards the promising concept of *Urban Agri-culture*:

(a) Especially during the last three centuries, wide scale deforestations and carbon dioxide pollution by burning fossil fuels have replaced large amounts of oxygen from the atmosphere, with carbon dioxide. This is causing a global warming by greenhouse effect, with incalculable side sequels [8], etc. After a series of international conferences, a glob-al reaction against this phenomenon was agreed at the United Nations Climate Change Conference UNCCC'15, held in Paris, France (30 Nov.–12 Dec. 2015) [9]. Obviously the strategy to follow is to modify the balance between the fuels burning that is creating now our carbon footprint and the carbon offset that is realized by the plants metabolism—the photosynthesis, in the sense of reducing the carbon footprint and increasing the carbon offset. Since the restoration of the ancient erased forests is impossible because of our demographic growth, the only reasonable approach we have to enhance the insufficient carbon offset existing right now is to increase by all means the number of plants.
(b) Besides the immediate production of vegetables, fruits or flowers, urban agriculture has a strong impact on our quality of life. The presence of plants around us is a major comfort factor, a sure way towards a healthier and more enjoyable life style of the majority of human population, which live in cities.

Urban agriculture may be applied in different ways, starting from small outdoor community gardens, via vertical greenhouses up to roof-top greenhouses (RTG).

If integrated into the building, a classical RTG becomes an *Integrated RTG* (IRTG), bringing in this context two important contributions:

- While the carbon offset provided by any conventional outdoor agricultural technology is low, only greenhouses can match forests in terms of carbon offset, due to the optimal growing conditions plants are benefiting;
- RTGs are improving of our buildings' metabolism by introducing energy, water and CO_2 flows [10, 11].

Our approach is to add renewable energy (sun, geo-thermal, wind, etc.) to this balance and to tighten the symbiosis between the plants of the greenhouse, exposed to the sun, and the people living underneath, in the building. A particular attention will be payed to the O_2 flow, which will close a continuous air circulation between RTG and building.

The common carbon-dioxide/oxygen management will be ensured by means of a two ways ventilation system:

- A RTG to building flow of O_2 enriched by plants air;
- A building to RTG flow of CO_2 enriched by people air, which acts as carbon fertilizer for the plants;

The energy/water management will be realized according to the Watery concept and technology [12].

As it will be shown below, the MIMO (multi-input-multi output) system of the IRTG is highly nonlinear and difficult to be copped by conventional control algorithms. That is why intelligent algorithms, namely fuzzy, should be applied, and the newborn IRTG concept may be re-baptized as *Intelligent RTG* (iRTG).

2 An Integrated Roof-Top Greenhouse Model

The Fig. 1 IRTG system was conceived as a development of the Passive Greenhouse [1, 2], and we firstly discussed it in reference [13].

The roof is occupied by the greenhouse, which is provided with air-to-air heat pumps, rain water collectors, photovoltaic or solar panels or any other renewable energy device that is appropriate for the climate conditions and for the budget.

The water tank (warm) is part of a two tank water system, the main tool for the storage and the management of the solar energy. The second water tank, the cold one is located into the cellar of the building, near the main heating/cooling device, the water-to-water heat pump. The building is provided with heating/cooling radiators and with the characteristic two flows IRGT ventilation system.

The operation of the two tank water systems is briefly described in Ref. [14].

The underground heat exchanger feeds the cellar water-to-water heat pump.

The building and the RTG are provided with ventilation fans, in order to refresh the air.

The passive character of the RTG is extremely important, making possible to extend IRTG at the scale of a whole city.

3 A Mathematical Model for Integrated Roof-Top Greenhouse

The IRTG mathematical model consists of a system of six non-linear equations of first order [15]. We reproduce here the model, in order to ease the reader's effort:

$$V_S \cdot \rho \cdot c_a \frac{dT_{IS}(t)}{dt} = \{[1 - u(t)] \cdot D_S(t) \cdot \rho \cdot c_a + \alpha_S \cdot S_S\} \cdot [T_E(t) - T_{IS}(t)]$$

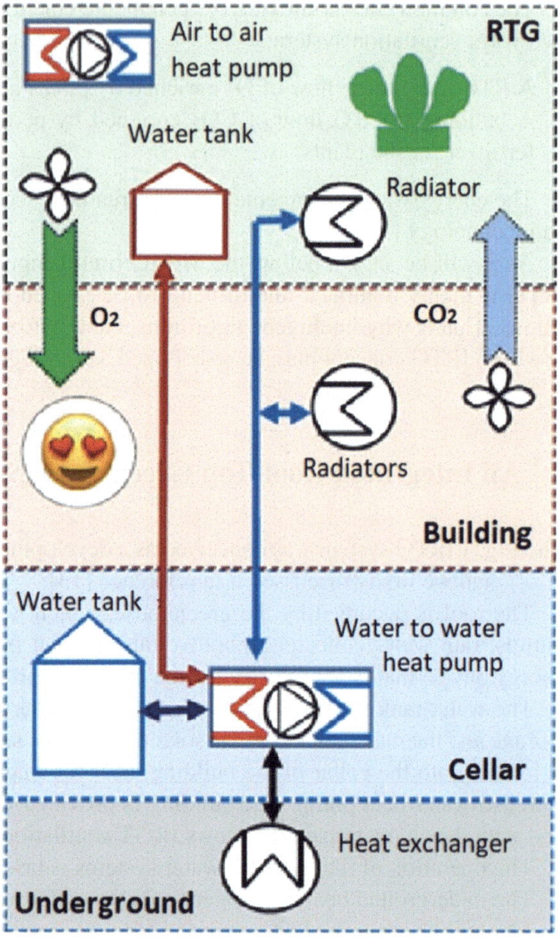

Fig. 1 The integrated roof-top passive greenhouse

$$+ N_S(t) \cdot P_o + P_{ES}(t) \pm P_S(t - \tau S) + D_{RTG}(t - \tau_{RTG}) \cdot \rho \cdot c_a \cdot [\theta_{IC}(t) - \theta_{IS}(t)] \quad (1)$$

$$V_S \cdot \frac{dC_{O2S}(t)}{dt} = [1 - u(t)] \cdot D_S(t) \cdot [C_{O2E}(t) - C_{O2S}(t)] + Q_{O2S} + D_{RTG}(t - \tau_{RTG}) \cdot [C_{O2C}(t) - C_{OZS}(t)] \quad (2)$$

$$V_S \cdot \frac{dC_{CO2S}(t)}{dt} = [1 - u(t)] \cdot D_S(t) \cdot [C_{CO2E}(t) - C_{CO2S}(t)] + N_S(t) \cdot q_{CO2} + Q_{CO2S} + D_{RTG(t-\tau_{RTG})} \cdot [C_{CO2C}(t) - C_{CO2S}(t)] \quad (3)$$

$$V_C \cdot \rho \cdot c_a \cdot \frac{dT_{IC}(t)}{dt} = \{[1 - u(t)] \cdot D_C(t) \cdot \rho \cdot c_a + \alpha_C \cdot S_C\} \cdot [T_E(t) - T_{IC}(t)]$$

$$+ N_C(t) \cdot P_o + P_C(t - \tau_C) + D_{RTG}(t - \tau_{RTG})$$
$$\cdot \rho \cdot c_a \cdot [T_{IS}(t) - T_{IC}(t)] \quad (4)$$

$$V_C \cdot \frac{dC_{O2C}(t)}{dt} = [1 - u_C(t)] \cdot D_C(t) \cdot [C_{O2E}(t) - C_{O2C}(t)]$$
$$+ D_{RTG}(t - \tau_{RTG}) \cdot [C_{O2S}(t) - C_{O2C}(t)] \quad (5)$$

$$V_S \cdot \frac{dC_{CO2}(t)}{dt} = [1 - u(t)] \cdot D_C(t) \cdot [C_{CO2E}(t) - C_{CO2C}(t)] + N_S(t) \cdot q_{CO2}$$
$$+ Q_{CO2C} + D_{RTG}(t - \tau_{RTG}) \cdot [C_{CO2S}(t) - C_{CO2C}(t)] \quad (6)$$

with the following parameters: V [m^3] volumes, ρ [kg/m^3] air density, c_a [J/kg·°K] specific heat of the air, T [°C] temperatures, u recirculation factor, D [m^3/s] air flows, α [$W/m^2 \cdot °K$] mean heat transfer coefficient through the walls, S [m^2] radiant surface, N number of persons, P_o [W] mean power emitted by a person, P_{ES} [W] power of the greenhouse effect, P [W] heating/cooling power, τ [s] delay times, C_{CO2} [kg/m^3] carbon dioxide concentrations, C_{O2} [kg/m^3] oxygen concentrations, q_{CO2} and Q_{CO2} [$kg/m^3 \cdot s$] carbon dioxide emission flows, Q_{O2} [$kg/m^3 \cdot s$] oxygen emission flows, C_{CO2E} [kg/m^3] carbon dioxide concentration outside building and C_{O2E} [kg/m^3] oxygen concentration outside building.

Index S refers the greenhouse, index C the building, index RTG the ventilation system between greenhouse and building and index E the environment.

4 A Simulink Implementation

A Simulink implementation of the (1)–(6) model is presented in Fig. 2 (index CIAS is referring to RTG).

The input parameters (orange colored), may be constants, look-up-tables, functions, sub-systems, etc. The current values of the simulation parameters may be set and observed in Fig. 2. It is to remark the time delays $\tau_S = 30$ s, $\tau_C = 80$ s and $\tau_{CIAS} = 60$ s, which are likely to produce oscillations and overdriving.

A temperature controller is presented in Fig. 3. Timp is the imposed temperature while E is the deployed energy. Ties is the temperature feedback. The same controller may be used for the greenhouse.

In Ref. [15] a linear PID and a nonlinear self-adaptive fuzzy-interpolative PID controller [16] were compared. As expected, the linear PID exhibited serious limitations, manifested by important and energy consuming overdrives after transient regimes. That is why from now on we will use just the fuzzy-interpolative one, shown in Fig. 4.

The fuzzy-interpolative PID controller is realized by a 3D look-up-table (LUT):

$$P: [-2 \; -1 \; 0 \; 1 \; 2]$$

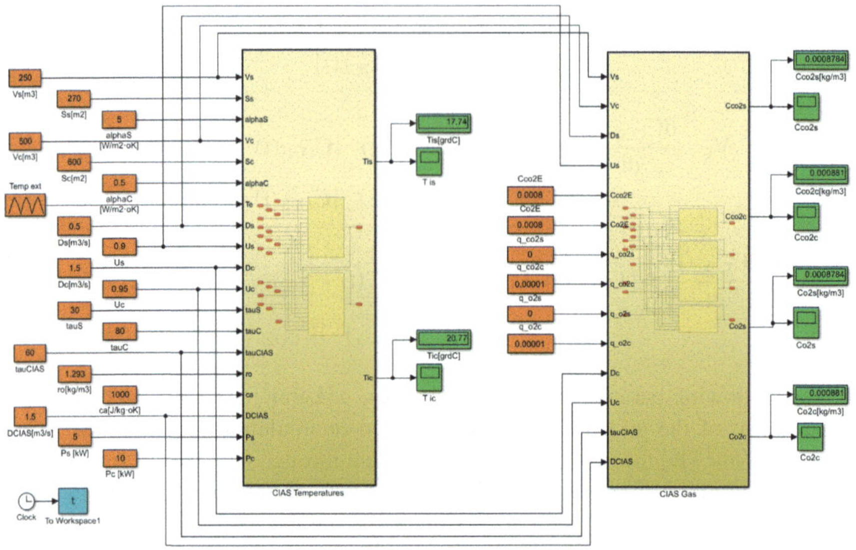

Fig. 2 The Simulink implementation [15]

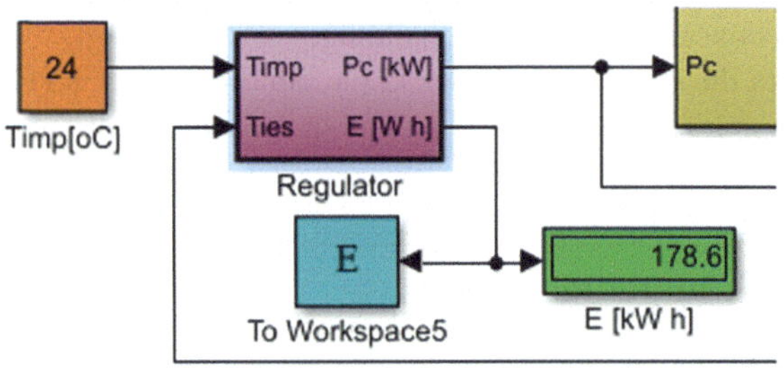

Fig. 3 A temperature controller connected to the building, with Pc as control output

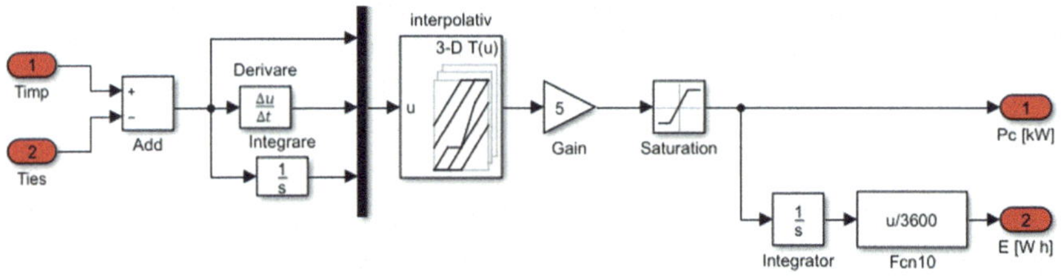

Fig. 4 The nonlinear fuzzy-interpolative PID controller [15]

$$
\begin{aligned}
&\mathrm{I}: [-0.2 \ -0.1\ 0\ 0.1\ 0.2] \\
&\mathrm{D}: [-100\ 0\ 100]
\end{aligned} \tag{7}
$$

$$
\begin{aligned}
\mathrm{PID: cat}(3, &[-2\ -2\ -2\ -1\ 0;\ -2\ -2\ -1\ 0\ 1;\ -2\ -1\ -1\ 1\ 2;\ -1\ 0\ 1\ 2\ 2;\ 0\ 1\ 2\ 2\ 2], \\
&[-2\ -2\ -2\ -1\ 0;\ -2\ -2\ -1\ 0\ 1;\ -2\ -1\ 0\ 1\ 2; -1\ 0\ 1\ 2\ 2;\ 0\ 1\ 2\ 2\ 2], \\
&[-2\ -2\ -2\ -1\ 0;\ -2\ -2\ -1\ 0\ 1;\ -2\ -1\ 1\ 1\ 2; -1\ 0\ 1\ 2\ 2; 0\ 1\ 2\ 2\ 2])
\end{aligned} \tag{8}
$$

The LUT design is driven towards the following objectives:

- Overdriving rejection by decreasing the control action around *error* = 0;
- Accelerating the transitions by strong control actions when error is great;
- Increasing accuracy in steady regimes by strong control action for *error* = 0;

Although this controller was introduced in the early 2000s, the best way to synthetically describe its design was formulated by Jerry M. Mendel in 2018: *Sculpting the State Space* [17].

The IRTG system has the following I/O structure:

- Inputs: P_C, P_S, D_C, D_S, D_{RTG}, u_C and u_S;
- Outputs: T_{IC}, T_{IS}, C_{CO2C}, C_{CO2S}, C_{O2C} and C_{O2S}.

A proper mathematical description of the IRTG system is not yet done so a corresponding MIMO 7 input–6 output controller has to be designed and tested in the future. However, a first approach may be already conceived as a simple collection of 6 independent self-adaptive fuzzy-interpolative controllers, supervised by an expert controller. The knowledge supporting the supervision algorithm will be gradually extended and detailed, with the acquired experience.

This simple approach may be used as the first choice to test a future prototype. This minimal approach is relying on the self-adaptive feature of each controller, expected to react in a balanced way facing multiple possible perturbations: variations of the weather conditions, of the number of people in the building and in the greenhouse, of the plants activity, etc.

The following simulations' scenario is set as in ref. [15] by a simple repeating sequence block, which is defining a daily variation of the external temperature T_E:

$$
\begin{aligned}
&\mathrm{Time[s]} : [0\ 21600\ 43200\ 64800\ 86400] \\
&T_E[^{\circ}C] : [5\ 11\ 15\ 13\ 5]
\end{aligned} \tag{9}
$$

The imposed temperatures will be $T_{ICimp} = 20\ ^{\circ}C$ and $T_{ISimp} = 19\ ^{\circ}C$.

5 Simulation Results

The next simulations cover 24 h, with fuzzy-inter-polative nonlinear PID control. In simulation (a) we have only one controller, for the building temperature T_C while in simulation (b) and the greenhouse temperature T_S is controlled (Fig. 5).

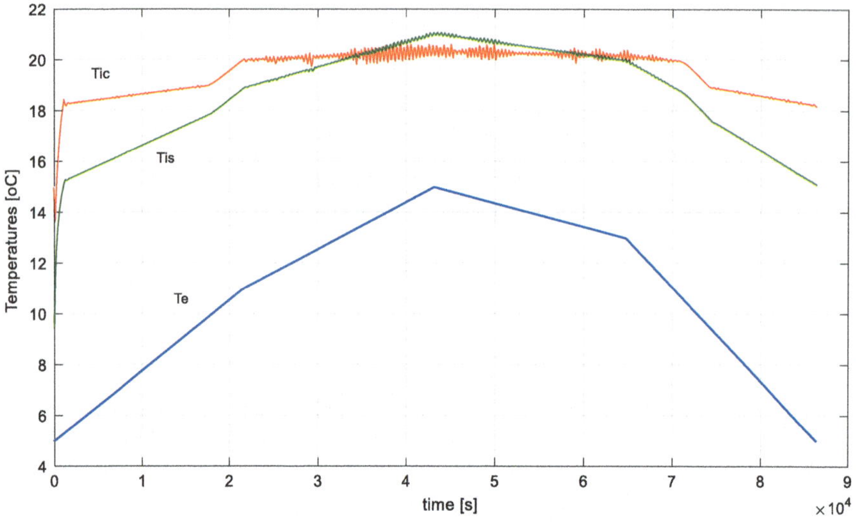

a) Tc temperature controller

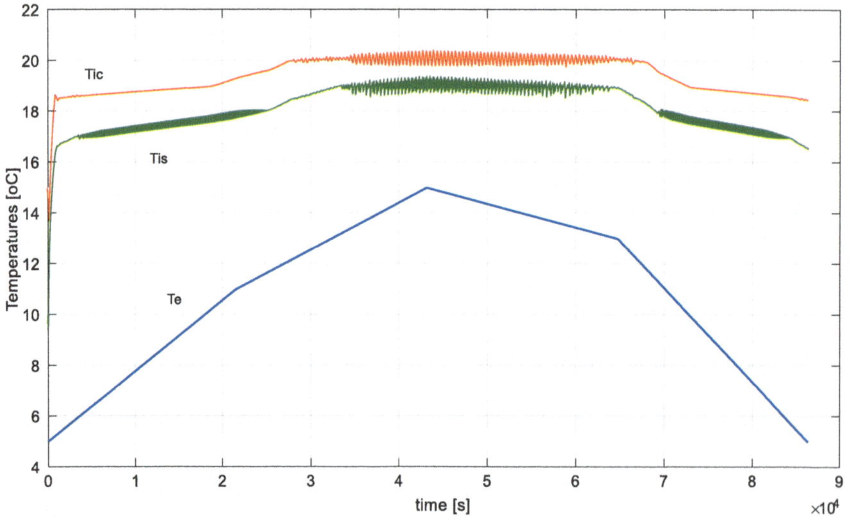

b) Tc and Ts temperature controllers

Fig. 5 iRTG temperature control by fuzzy-interpolative PID controllers

(a) T_C temperature controller
(b) T_C and Ts temperature controllers.

The simulations are showing robust performances in any conditions tested so far, with no malicious interferences between the independent controllers.

Chattering is present in responses but this is not unusual for fuzzy controllers. Chattering as well as the other parameters of the control system may be easily tuned by the usual fuzzy-interpolative methodology (that is by setting the LUT values).

6 Conclusions

The paper introduces a promising concept for a sustainable future: the Intelligent Roof-Top Greenhouse Building, ensuring an integrated management of renewable energies, water resources and atmospheric gas composition. A nonlinear controller able to cope with this high nonlinear and complicated plant is the self-adaptive fuzzy-interpolative one.

References

1. M.M. Balas, S. Musca, D. Toader, C. Mnerie, C. Musca, O. Falcan, On a promising sustainable energy system and its control—the passive greenhouse, in *Proceedings of SOFA'09 the 3rd International Workshop on Soft Computing Applications*, 29 July–1 August 2009, Szeged-Arad, pp. 233–238 (2009)
2. M.M. Balas,C. Musca, S. Musca, The passive greenhouses, in *Chapter 5 in Paths to Sustainable Energy*, ed. by J. Nathwani, A. Ng (InTech Open, 2010), pp. 75–92. http://www.intech-open.com/books/paths-to-sus-tain-able-energy/the-passive-greenhouses
3. M.M. Balas, Seven passive greenhouse synergies. Acta Politehnica Hungarica, Budapest **11**(4), 199–210 (2014)
4. T.K. Rudel et al., Forest transitions: towards a global understanding of land use change. Glob. Environ. Change **15**, 23–31 (2005)
5. S. Taylor Lovell, *Designing a Sustainable Urban Agriculture*. University of Illinois (2014). https://www.multifunctionallandscape.com/uploads/2014_ESA_DesigningUrbanAgriculture.pdf
6. C. Eigenbrod, N. Gruda, Urban vegetable for food security in cities. A review. Agron. Sustain. Dev. **35** (2), 483–498 (2015)
7. B.P. Goldstein, M.Z. Hauschild, J. Fernandez, M. Birkved, Urban versus conventional agriculture, taxonomy of resource profiles: a review. Agron. Sustain. Dev. **36**(9), (2016)
8. NASA: Climate Change and Global Warming. https://climate.nasa.gov/
9. United Nations. Framework Convention on Climate Change. Adoption of the Paris Agreement. https://unfccc.int/resource/docs/2015/cop21/eng/l09r01.pdf
10. O. Pons et al., Roofs of the future: rooftop greenhouses to improve buildings metabolism. Procedia Eng. **123**, 441–448 (2015)
11. J.I. Montero, E. Baeza, P. Muñoz, E. Sanyé-Mengual, C. Stanghellini, Technology for rooftop greenhouses, in *Rooftop Urban Agriculture. Urban Agriculture*, ed. by F. Orsini, M. Dubbeling, H. de Zeeuw, G. Gianquinto (Springer, Cham, 2017), pp. 83–101
12. Watergy International Group. http://www.watergyinternational.com/. Retrieved 7 July (2018)

13. V.E. Balas, fuzzy control for roof-top greenhouses, in *Symposium of Fuzzy Logic and Fuzzy Sets: A Tribute to Lotfi Zadeh*, 5 February 2018, Berkeley, USA, pp. 52–53. https://eecs.berkeley.edu/events/zadeh-symposium-2018?
14. M.M. Balas, M. Buchholz, S. Balas, Expert control for the coupled tanks greenhouse, in *Proceedings of the 6th International Workshop on Soft Computing and Applications SOFA'14*, vol. 2 (2014), Timisoara, July, pp. 939–948
15. M.M. Balas, M. Popa, E.V. Muller, D. Alexuta, L. Muresan, Integrated roof-top greenhouse buildings, in *SOFA'18, the 8th International Workshop on Soft Computing Applications*, Arad, (2018) (in Press)
16. M.M. Balas, The fuzzy interpolative methodology, in *Studies in Computa-tional Intelligence, Soft Computing Based Modeling in Intelligent Systems*, eds. V.E. Balas, A.M. Var-konyi-Koczy, J. Fodor (Springer, 2009), pp. 145–167
17. J.M. Mendel, sculpting the state space, in *Key Note Speech at the 7th World Conference on Soft Computing*, Baku, May 30 (2018)

The manufacturer's authorised representative in the EU is Springer Nature Customer Service Centre GmbH, Europaplatz 3, 69115 Heidelberg, Germany. If you have any concerns regarding our products, please contact ProductSafety@springernature.com

Printed and bound by CPI Group (UK) Ltd, Croydon, CR0 4YY
05/10/2026
02228126-0016